AIP CONFERENCE PROCEEDINGS 323

ATOMIC PHYSICS 14

FOURTEENTH INTERNATIONAL
CONFERENCE ON ATOMIC PHYSICS

BOULDER, CO 1994

EDITORS:
D. J. WINELAND
NATIONAL INSTITUTE OF
STANDARDS AND TECHNOLOGY

C. E. WIEMAN
S. J. SMITH
JOINT INSTITUTE FOR
LABORATORY ASTROPHYSICS
AND UNIVERSITY OF COLORADO

American Institute of Physics   New York

Authorization to photocopy items for internal or personal use, beyond the free copying permitted under the 1978 U.S. Copyright Law (see statement below), is granted by the American Institute of Physics for users registered with the Copyright Clearance Center (CCC) Transactional Reporting Service, provided that the base fee of $2.00 per copy is paid directly to CCC, 27 Congress St., Salem, MA 01970. For those organizations that have been granted a photocopy license by CCC, a separate system of payment has been arranged. The fee code for users of the Transactional Reporting Service is: 0094-243X/87 $2.00.

© 1995 American Institute of Physics.

Individual readers of this volume and nonprofit libraries, acting for them, are permitted to make fair use of the material in it, such as copying an article for use in teaching or research. Permission is granted to quote from this volume in scientific work with the customary acknowledgment of the source. To reprint a figure, table, or other excerpt requires the consent of one of the original authors and notification to AIP. Republication or systematic or multiple reproduction of any material in this volume is permitted only under license from AIP. Address inquiries to Series Editor, AIP Conference Proceedings, AIP Press, American Institute of Physics, 500 Sunnyside Boulevard, Woodbury, NY 11797-2999.

L.C. Catalog Card No. 94-73219
ISBN 1-56396-348-5
DOE CONF-940707

Printed in the United States of America.

CONTENTS*

Preface ix
Memorial to Polykarp Kusch (1911–1993) xiii
Memorial to Laird Schearer (1931–1993) xv
Memorial to Julian Schwinger (1918–1994) xvii
Memorial to Donald R. Yennie (1924–1993) xix

ATOMIC TESTS OF FUNDAMENTAL THEORIES

Electric Dipole Tests of Time Reversal Symmetry 3
    N. F. Ramsey
Atomic Physics in QED and QCD 18
    G. P. Lepage
New Tests of Special Relativity and QED by Laser Spectroscopy in Heavy Ion Storage Rings 30
    T. Kühl, St. Becker, S. Borneis, T. Engel, B. Fricke, M. Grieser, R. Grieser, D. Habs, G. Huber, I. Klaft, D. Marx, P. Merz, R. Neumann, D. Schwalm, and P. Seelig

ATOMIC STRUCTURE AND HIGH RESOLUTION SPECTROSCOPY

Advances in the Theory of Atomic Structure 45
    J. Sapirstein
Precision Spectroscopy of Atomic Hydrogen 63
    T. W. Hänsch
Recent Developments in Laser Spectroscopy of Helium 81
    M. Inguscio, F. S. Cataliotti, P. De Natale, G. Giusfredi, F. Marin, and F. S. Pavone
Light Exotic Atoms—Some Recent Developments 102
    K. P. Jungmann
Experiments with Highly Charged Ions up to Bare U^{92+} on the Electron Beam Ion Trap 116
    P. Beiersdorfer
Precision Measurements of Atomic Lifetimes 130
    C. E. Tanner

*NOTE: Underlining indicates the author who presented the paper.

PREFACE

The Fourteenth International Conference on Atomic Physics (ICAP-14) was held at the University of Colorado, in Boulder, Colorado, USA, July 31 through August 5, 1994. This biennial meeting has become the preeminent forum for presenting new concepts and critical measurements in fundamental atomic physics, a subject that deals with the basic structure of atoms and how they interact with external electromagnetic fields. This is a particularly exciting time to hold such a conference because explosive progress is currently being made in fundamental atomic physics, fueled in part by remarkable advances in optical technology and in part by equally remarkable gains in computational capabilities. Some of the excitement of this meeting arises from the unfolding of revolutionary new measurement capabilities and the appearance of imaginative theoretical perspectives from which to exploit the new computational power and flexibility. One finds in these proceedings powerful confirmation that atomic physics is a full and very effective partner in the ongoing process of testing the foundations of physics. On the other hand, these measurement capabilities, the computation methods, and, indeed, the interests and enthusiasm of many of the participating scientists, diffuse rapidly into the sister fields of science and ultimately into a very broad scientific and technological landscape. In these proceedings we see some of this process at its source point.

The program for this conference consisted primarily of 28 invited papers, all presented in plenary session, each authored by a scientist at the forefront of this development. In this volume we present the written texts submitted by these plenary speakers. These proceedings provide a comprehensive and authoritative overview of the state of the field. Implicit in these presentations are the foundations that will determine some of the principal directions the field of atomic physics will take in the next several years.

This 1994 conference was organized by conference co-chairs Carl Wieman of the University of Colorado and the Joint Institute for Laboratory Astrophysics and David Wineland of the National Institute of Standards and Technology in Boulder, with the able and dedicated assistance of Diana Moreland, the conference secretary. The conference program, which included 340 posters in two sessions in addition to the seven plenary sessions, was determined by the Program Committee: P. Bucksbaum, S. Datz, G. Drake, T. Gallagher, T. Hänsch, D. Kleppner, J. Raimond, and J. Walraven, with conference co-chairs Wieman and Wineland. Policy guidance and numerous suggestions relating to the scientific program were provided by the International Organizing Committee: E. Arimondo, V. Balykin, S. Chu, G. Drake, N. Fortson, M. Gavrila, G. Grynberg, T. Hänsch, S. Haroche, V. Hughes, D. Kleppner, P. Klinkenberg, R. Lewis, I. Lindgren, H. Narumi, P. Sandars, S. Svanberg, H. Takuma, J. Walraven, H. Walther, and J. Woerdman.

The more than 500 registered participants came from 27 countries. The University of Colorado Conference Services handled registration, housing, and meals for these participants, and other conference activities were arranged by members of the Local Committee: J. Bergquist, J. Cooper, E. Cornell, G. Dunn, S. Gilbert, C. Greene, J. Hall, L. Hollberg, W. Itano, F. Kowalski, S. Lee, S. Lundeen, D. Norcross, S. Smith, and P. Zoller.

We are grateful for the sponsorship of the International Union of Pure and Applied Physics, the International Science Foundation, the University of Colorado, the National Institute of Standards and Technology, and agencies of the United States Government: the

National Science Foundation, the Army Research Office, and the Office of Naval Research. The conference planners are also indebted to the University of Colorado and to the National Institute of Standards and Technology for the use of facilities and for numerous services needed to ensure its success. Special thanks are due the authors for their prompt submission of the manuscripts and to the JILA Scientific Reports Office for assistance in preparing this volume for publication.

Carl Wieman
David Wineland
Steve Smith
September 1994

MEMORIALS

Polykarp Kusch
1911 - 1993

Polykarp Kusch (1911–1993)

Polykarp Kusch, Nobel laureate in Physics for 1955, and an outstanding physicist, teacher, and administrator at Columbia University and at the University of Texas at Dallas, died on March 10, 1993 at the age of 82.

Kusch is remembered and admired in the community of physicists and in the history of physics for his important discovery of the anomalous magnetic moment or g-value of the electron and for his central role in the discovery, development, and use of the radiofrequency spectroscopic technique of molecular beam magnetic resonance. Kusch was a great experimenter and could really make an apparatus work. He did research in the days of one or few-men experimentation, and he excelled. His work was always characterized by the highest integrity, thoroughness, uncompromising attention to detail, and enormous effort and enthusiasm.

After receiving his doctorate from the University of Illinois for work in optical molecular spectroscopy with F. Wheeler Loomis and spending a postdoctoral year at the University of Minnesota with J. T. Tate, Kusch came to Columbia University in 1937 where he played a central role in the discovery of the molecular beam magnetic resonance method. The technique was first reported in 1938 and 1939 in papers by I. I. Rabi, Sidney Millman, Kusch, and Jerrold R. Zacharias. Over the following 20-year period the members of the Columbia group and their students developed and brilliantly applied this difficult and demanding technique. These experiments included measurements of nuclear magnetic dipole and electric quadrupole moments, molecular radiofrequency spectra, atomic hyperfine structure, and atomic g-values.

It was in the exciting period at Columbia immediately after World War II, when many fundamental experimental discoveries were made in atomic physics that led to modern quantum electrodynamics, that Kusch and Henry Foley discovered that the spin g-value of the electron as determined from atomic g-values was about 1 part in 10^3 greater than the Dirac value of 2. A theoretical explanation was first provided by Julian Schwinger based on virtual radiative contributions evaluated using the idea of renormalization in quantum electrodynamics.

Kusch had undertaken important administrative jobs at Columbia by 1968 when the first ICAP meeting was held, and later he devoted his energy to teaching at the University of Texas at Dallas. Hence he never attended an ICAP meeting, but he had provided a major heritage of modern scientific results and experimental technique for younger atomic physicists, and he had trained many excellent students.

Vernon W. Hughes

Laird Schearer
1931 - 1993

Laird Schearer (1931–1993)

Laird Schearer died on March 7, 1993 after a heart attack during one of his frequent visits to Ecole Normale Superieure in Paris. He was born in 1931, was raised in eastern Pennsylvania, graduated from Muhlenberg College in 1954, earned an M.S. from Lehigh University in 1958, and a Ph.D. in the laboratory of King Walters at Rice University in 1966. His work in developing magnetometers played a major role in making Texas Instruments a profitable enterprise. He became the chairman of the physics department at the University of Missouri at Rolla in 1971 and successfully fulfilled the mandate to improve its research activities, both by his own work and by inspiring others. He held visiting appointments at universities worldwide during the succeeding 20 years, served as program officer at the NSF in 1979–81, and was honored by being appointed Curator's Professor in 1984, the position he held when he died.

His memory was honored by a special symposium in Paris on June 2–3, 1994, when scientists from many countries and various fields of physics spoke of his accomplishments and his impact on physics. Each of these talks was strongly flavored by fond memories of Laird's unique way of approaching physics problems and his very special personal warmth, generosity, and character. His thesis advisor remembered him as an extremely close friend for over 35 years. His former student David Fahey recalled the close support and encouragement provided by both Laird and his wife of over 40 years, Dotty. His Chancellor, John Park, recalled that Laird and Dotty's house was a frequent meeting place for students and faculty and a continuous guest house for visitors. He was remembered by his colleagues as a superb teacher to students at all levels.

My own personal recollection spans more than 20 years of enormously helpful discussions about development of lasers for optical pumping of helium. I was welcomed into Laird's lab and home, as were my students when they also visited Rolla. Laird was an important presence at Gordon conferences and other opportunities for open discussion of new ideas and experimental techniques. He had the special ability to make things work with a minimum of hardware by depending on his superb intuition for clear and simple design.

Our community has suffered a great loss and will miss him severely, but will continue to benefit from the lives of those he touched both as beloved teacher and as admired colleague.

Harold Metcalf

Julian Schwinger
1918 - 1994

JULIAN SCHWINGER (1918–1994)

With the death of Julian Schwinger on July 16, 1994, at age 76, science lost an intellectual giant. Schwinger made great contributions to quantum mechanics and to atomic, nuclear, particle, and condensed matter physics. Stimulated by the disagreement between the theoretical and experimental values of the atomic hydrogen hyperfine separation, he devised the first successful relativistic theory of quantum electrodynamics (QED), including mass and charge renormalization. This theory and its extensions to higher orders have been the basis of many atomic physics experiments that have confirmed the theory to unprecedented accuracy.

At the age of 21 he received his Ph.D. from Columbia University, and at age 29 became a full professor at Harvard University where he taught and did research for 27 years before going to UCLA in 1972. He directed more than 70 doctoral theses.

Schwinger received many honors including the National Medal of Science and the Nobel Prize. He is survived by his wife Clarice.

Norman Ramsey

DONALD R. YENNIE (1924–1993)

Donald R. Yennie, professor of physics at Cornell University, died on April 14, 1993 at the age of 69. He was an internationally recognized authority on quantum electrodynamics, the fundamental theory of electromagnetic interactions. He will be remembered not only as a leader in the most difficult and precise aspects of this field, but also as a beloved teacher and friend to his many students and colleagues.

Don was born in Paterson, New Jersey, in 1924. He was an undergraduate at the Stevens Institute of Technology, graduating in 1945 with an M.A. degree, and was a physics instructor there during 1946–47. In 1951 he earned his Ph.D. from Columbia University as Yukawa's student. He then joined the Institute for Advanced Study in Princeton. From 1952 until 1957, he was first an instructor and then an assistant professor at Stanford University. He then moved to the University of Minnesota, and in 1964 joined the physics faculty at Cornell University.

Don's work covered a wide range of topics in theoretical physics. He had a strong interest both in fundamental questions about quantum field theory and in applications of field theory to experimental data. Don's incomparable technical skills, together with his profound understanding of quantum field theory, are well illustrated by his high-precision work on simple atoms in quantum electrodynamics. His analyses, particularly of the Lamb-shift and hyperfine splittings in hydrogen and muonium, remain among the most challenging ever attempted in quantum field theory. They provided a starting point for a generation of researchers in this field. Don was actively involved in such work at the time of his death.

Don Yennie will be greatly missed by all who knew him. His influence as a physicist and as a human being will remain with everyone.

Peter Lepage
Jonathan Sapirstein

ATOMIC TESTS OF FUNDAMENTAL THEORIES

Electric Dipole Tests of Time Reversal Symmetry

Norman F. Ramsey

Lyman Physics Laboratory
Harvard University
Cambridge, MA 02138

PACS: 11.30.Er 35.10.Di

Abstract. If there is either T or P symmetry, there can be no electric dipole moment for particles whose orientations are fully specified by the orientations of their spin angular momenta. As a result, there have been extensive searches for electric dipole moments as tests of these symmetries. CP non-conservation has been observed, which implies non-conservation of T if CPT is conserved, but it has only been seen in the decay of the artificially produced K_L^0. The history of the searches for neutron and atomic electric dipole moments is summarized. The earliest experiments utilized a neutron beam magnetic resonance apparatus, but the present neutron experiments study ultra cold neutrons trapped in bottles whose walls totally reflect the neutrons. Recent atomic and molecular searches are sensitive both to particle electric dipole moments and to other related T non-conserving quantities. The various neutron, atomic and molecular experiments are described. The results of these searches are compared with the predictions of theories that account for the known CP non-conservation in the decay of the K_L^0. The experiments directly set limits to T-odd, P-odd interaction terms, but through eleectroweak radiative corrections they also set limits to T-odd, P-even interactions.

INTRODUCTION

The electric dipole moment of a particle is defined as follows. Let the coordinate z be measured from the center of mass of a particle and let ρ_{JJ} be the electric charge density inside the particle, whose angular momentum is **J** with quantum number J and whose orientation state is given by m = J relative to the z axis. The scalar magnitude d_n of the vector dipole moment **d** is then defined by

$$d = \int \rho_{JJ} z \, d\tau \qquad (1)$$

where dt is a differential volume element. If the particle is charged, this definition implies that the center of charge of the particle is displaced from the center of mass if $d \neq 0$. If, on the other hand, the particle has no net charge, the definition implies a greater positive charge in one hemisphere and a correspondingly greater negative charge in the other. The electric dipole moment is ordinarily expressed in units of e cm where e is the charge of the proton. An electric dipole moment **d** in an electric field E will have an interaction energy W = - **d**·**E**.

It is easy to see(1-4) from the following argument that the electric dipole moment must vanish if there is symmetry under either the parity transformation (P) for which **r** → -**r** or the time reversal transformation (T) for which t → -t, provided there is no near degeneracy. Since the orientation of the particle can only be specified by the orientation of its angular momentum, the dipole moment **d** and the angular momentum **J** must transform their signs the same way under P and T if **d** is to have a non-zero value and if there is to be P and T symmetry; if they did not, either of those transformations would produce a neutron of oppositely signed electric dipole moment contrary to the assumption of those symmetries and no degeneracy. But **d** changes sign under P whereas **J** does not, so **d** must vanish if there is P symmetry. Likewise **d** does not change sign under T but **J** does, so **d** must vanish if there is T symmetry. More rigorous proofs of the above can be given(1-4). If one makes the usual assumption of CPT symmetry (where C is the charge conjugation transformation, Ze → -Ze), the existence of an electric-dipole moment would also imply a failure of CP symmetry. In molecules, electric dipole moments do exist and are attributed to degenerate states rather than to a failure of T symmetry; but degeneracy for the neutron would contradict the well-established fact that neutrons obey the Pauli exclusion principle. Since the existence of a particle electric dipole moment would contradict the assumptions of both parity and time reversal invariance, increasingly sensitive searches for an electric dipole moment have been carried out for a number of years to test these assumptions.

Although the electric dipole measurements in first order imply limits to T-odd, P-odd interactions, in higher order they also limit T-odd T-even (TOPE) interactions as first pointed out by Khriplovich (57, 58) and Conti. Electroweak radiative corrections include terms that are products of the usual T-even, P-odd weak interaction and the discussed (57,58) T-odd, T-even interactions which can thereby contribute to the T-odd, P odd electric dipole moment.

The earliest electric dipole searches, and still some of the most sensitive, have been those with neutrons, initially with neutron beams and later with ultra-cold neutrons stored in chambers with reflecting walls. With the increased interest in the theories of T non-conservation, there have been a number of recent experiments to measure electric dipole moments and related T non-conserving properties of various atoms and molecules.

NEUTRON BEAM EXPERIMENTS

In 1950, at a time when parity conservation was almost universally believed, Purcell and Ramsey (3) pointed out that such a belief must be based on experiment and that there was then little experimental evidence for it. They concluded that a search for a particle electric dipole moment would provide such a test since such a moment is forbidden under parity symmetry. However they also noted that an electrically charged particle even with an electric dipole moment would ordinarily have no electric dipole interaction with an electric field since <**E**>=0 if the particle is not accelerated or acted upon by other forces. As a result past experiments were not sensitive in detecting electric dipole moments. On the other hand tests with a neutron are sensitive since the neutron has no electric charge and is consequently not accelerated out of the observation region in a strong electric field. A neutron electric dipole moment can be sensitively detected in a neutron beam magnetic resonance experiment for the following reasons. If the neutron has a magnetic dipole moment μ_M and an electric dipole moment **d**$_n$, the dipole moments will contribute an additional energy -μ_M·**B**-**d**$_n$·**E** if the neutron is in a magnetic field **B** and an electric field **E**. Therefore, if the neutron makes a transition from an orientation state with m = -1/2 to m = +1/2, the Bohr frequency ν_0 for the transition, which in this case is the neutron Larmor precession frequency, is given by

$$h\nu_0 = 2\mu_M \cdot \mathbf{B} + 2\mathbf{d}_n \cdot \mathbf{E} = 2\mu_M |B| \pm 2 d_n |E| \qquad (2)$$

where the upper sign is for **B** and **E** parallel. Therefore, the neutron electric dipole moment is observed by the change in the Larmor precession frequency of the neutron when the relative directions of the electric and magnetic fields are reversed.

Smith, Purcell and Ramsey (4) constructed a neutron beam magnetic resonance experiment with a strong electric field between the two separated oscillatory fields (5) and measured the change in resonance frequencies when the relative orientations of the electric and magnetic fields were reversed. When they started this experiment, the lowest limit that could be set by any existing observations was $d_n = 3 \times 10^{-18}$ e cm and in that first experiment they lowered the limit to $d_n = 5 \times 10^{-20}$ e cm but found no electric dipole moment.

In 1956 Lee and Yang (6) proposed parity non-conservation in the weak interaction as an explanation of an anomaly in the K meson decay. The proposed parity non-conservation in the weak interaction was confirmed in 1957 by Wu et al.(7) in the angular distribution of the electrons in the decay of polarized ^{60}Co. Although the discovery of the failure of parity symmetry eliminated that symmetry as a fundamental argument against the possibility of a particle electric dipole moment, Landau (8) and others (9) argued that even with the failure of P symmetry there should still be conservation of CP and T (where C is the charge exchange operator) and that T symmetry alone would eliminate the possibility of a particle electric dipole moment. However, at that time Ramsey wrote a letter (10) pointing out that the T symmetry assumption must be based on experiment and that a search for a neutron electric dipole moment was a particularly sensitive test of T conservation; Jackson, et al. (11) also emphasized the need for an experimental basis for the assumptions of T and CP symmetry. In 1964, Christenson, Cronin, Fitch and Turlay (12) showed experimentally there was a failure of CP conservation in the decay of the K_L^0 and hence a failure of T if TCP were conserved. The subsequent theories that accounted for the observed CP violation in the K_L^0 predicted a non-zero value for the neutron electric dipole moment, with the initial predictions being usually larger than 10^{-23} e cm. As a result there was a series of experiments (9), mostly by the neutron beam magnetic resonance method, which successively lowered the experimental limit on the neutron electric dipole moment. The most sensitive of the neutron beam experiments (6) was that of Dress, Miller, Pendlebury, Perrin and Ramsey (13). Neutrons from the Grenoble reactor at 180 m/s were polarized and analyzed by reflection from magnetized iron mirrors between which there was a 180 cm long region with a 17 G magnetic field and a 100 kV/cm electric field. Neutron reorientation transitions were induced by the method of separated oscillatory fields (14,15). No electric dipole moment was found and the result was equivalent to an experimental electric dipole moment of $(40 \pm 150) \times 10^{-26}$ e cm.

TRAPPED NEUTRON EXPERIMENTS

In the neutron beam experiments so far described, the neutrons had velocities of about 80 m/s and the neutron resonance widths were typically 80 Hz. It was early realized by Ramsey,(9,16) Zel'dovich (17) and Shapiro (18) that neutron electric dipole moment measurements could be made on neutrons slower than 7 m/s stored in suitable bottles by total reflection of the neutrons at the bottle surface. Such measurements are more sensitive because of the narrower neutron magnetic resonance and are free from uncertainty due to the effective magnetic field E x v/c due to the motion of the neutrons through the applied electric field. The realization of measurements with bottled neutrons was delayed for many years due to the unavailability of neutrons with such low velocity. However, experiments with stored neutrons are now being successfully carried out at both the the Insitut Laue-Langevin in Grenoble, France (20,27) and the Petersburg Nuclear Physics Laboratory in Gatchina, Russia.(19,28).

In the Harvard-Sussex-Rutherford-Washington-ILL experiments at Grenoble (20,27), the ultracold neutrons from the liquid deuterium moderator at the reactor core are further slowed by total reflection from the blades of a rotating

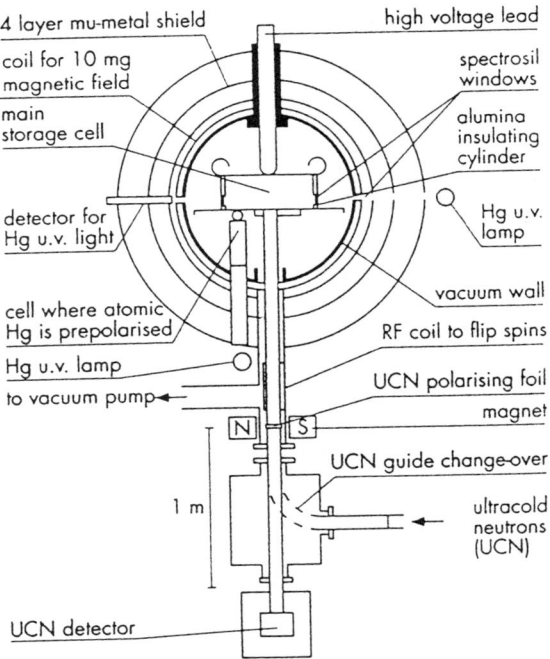

FIGURE 1. Schematic diagram of new Grenoble apparatus. The apparatus used in the most recently completed experiment is similar to that in the diagram except that no Hg magnetometer was provided.

turbine (21) and then transmitted to the resonance apparatus through a guide tube. The resonance apparatus is similar to that in Figure 1. Since the indicated magnetized foil totally reflects the neutrons of one polarization while transmitting the other, it serves as a polarizer for the incident beam and an analyzer for the returning beam. The neutrons are stored by total reflection in a cylinder approximately 25 cm in diameter and 10 cm high with the end plates being of Be and the sides of the cylinder being BeO. After filling for about 10 seconds, a shutter is closed, storing the neutrons for approximately 80s. The neutron density is about 10 cm^{-3}. The oscillatory field is applied in two short initial and final coherent pulses. After the shutter is opened the neutrons again pass through the foil, which this time serves as an analyzing foil which transmits only neutrons whose spin orientation has not changed. Consequently the numbers of neutrons reaching the UCN detector provide a measure of the neutron magnetic resonance. The neutrons are counted at the steepest part of the resonance curve as the directions of the electric and magnetic fields are successively reversed. From the correlation of the detected neutron intensity with the relative directions of the electric and magnetic fields, a limit can be set to the neutron electric dipole moment from Equation (2). The magnetic and electric fields were about 1 μT and 1.6 MV m^{-1}.

The principal uncertainty in the results comes from possible magnetic field changes that might occur when the electric field is reversed. For this reason the current from the high voltage generator is carefully monitored and kept at a low value. In addition, three rubidium magnetometers located as close as possible to the neutron storage volume are continuously monitored for possible magnetic field changes. Initially, the only neutron beam available for this experiment at Grenoble had a moderator at room temperature or above. As a result only a very small fraction of the neutrons were below the 7 m/s velocity required for total reflection storage so the initial neutron counting rate was very low, but with the subsequent upgrading of the reactor including the use of a liquid deuterium moderator counting rate increased to about 16,000 per cycle, with each cycle taking 124 s..

The use of stored ultra-cold neutrons offers two particularly important advantages over the original neutron beam experiments. The resonance curve with stored neutrons is 7000 times narrower than in the best neutron beam experiments. Furthermore, in the neutron beam experiments a large fraction of the running time had to be devoted to eliminating the E x v/c effect, which effectively vanishes in the neutron bottle experiments since the average value of v is very small compared to c when the neutrons enter and exit at the same port with a storage time of 80s.

The Gatchina group (19,28) uses a similar stored neutron method with the principal difference being that their neutrons are stored in a double chamber with the high voltage electrode in the middle so the electric fields in the two chambers are in opposite directions.. This helps to cancel the effects of drifting external magnetic fields but does not compensate as well for local magnetic effects due to sparks and leakage currents from the applied electric field.

The latest published result of the Grenoble group (27) is
$$d_n = (-3\pm2\pm4) \times 10^{-26} \text{ e cm} \qquad (3)$$
where the estimated statistical error precedes the systematic error. This null result translates to an upper limit of $|d_n| < 12 \times 10^{-26}$ e cm to a 95% confidence limit.

The latest published result for the neutron electric dipole moment as measured by the Gatchina group (28) is .
$$d_n = (+2.6\pm4.2\pm1.6) \times 10^{-26} \text{ e cm} \qquad (4)$$

where the first error is from statistics and the second from systematics and the authors conclude $|d_n| < 11 \times 10^{-26}$ e cm.

The combination of the Grenoble and Gatchina experiments together corresponds to an upper limit of

$$|d_n| < 9 \times 10^{-26} \text{ e cm to a 95\% C.L.} \quad (5)$$

As mentioned in the introduction the electric dipole measurements also provide limitations to the T-odd, P-even interactions. The experimental limit on d_n in Equation (5) provides the following limits to the T-odd, P-even parameters defined by Conti and Khriplovich (57,58).

$$\beta_{qe} < 10^{-6} \quad \text{and} \quad \beta_{qq} < 0.3 \times 10^{-6}. \quad (6)$$

Both the Grenoble and the Gatchina groups are planning to increase the sensitivity of their experiments. During repairs on the reactor, the Grenoble experiment has been temporarily discontinued and the apparatus is being rebuilt to incorporate improvements, especially in the monitoring of the magnetic field. A schematic diagram of the new apparatus is shown in Figure 1. In past experiments, the magnetic field monitoring has been done with three rubidium magnetometers no closer than 40 cm to the axis of the bottle, which has lead to uncertainties in their use to correct for magnetic field fluctuations in the neutron storage vessel. The Grenoble group is developing a magnetometer whose working substance is a vapor in the same volume as that in which the neutrons are stored. The original plans (29) were to use ^{3}He, but it now appears (27,38,55,56) that ^{199}Hg should be better because of more suitable energy levels for laser optical pumping. The nuclear spin polarization of the ^{199}Hg gas precesses at about 8 Hz and is monitored directly as a sinusoidal oscillation in the intensity of a horizontal beam of circularly polarized ^{204}Hg resonance radiation that has passed through the storage cell, as shown in Figure 1. Two other changes will be an increase in the neutron storage volume from 5 to 22 liters, and an increase in the electrode voltage. The planned storage time is about 150 seconds. These changes should reduce the systematic errors and increase the sensitivity of the experiment. Similar improvements are planned at Gatchina. More than a year of measurements will be needed for these experiments to lower the experimental errors significantly. Ringo and his associates (54) have discussed a quite different experiment utilizing an interferometer with spatially separated paths.

As a longer term project, Golub, Lamoreaux and others (55) are planning a neutron electric dipole moment experiment with the neutrons in a totally reflecting bottle filled with superfluid helium. A dilute solution of polarized ^{3}He will be used as a polarizer and detector since ^{3}He absorbs neutrons only when the total spin is zero (neutron and ^{3}He spins antiparallel).

ATOMIC AND MOLECULAR ELECTRIC DIPOLE MOMENT EXPERIMENTS

The previously stated theorem of Purcell and Ramsey (3,4) that electric dipole moments of electrically charged particles are ordinarily unobservable does not preclude measurements of atomic electric dipole moments since atoms have no net electric charge. However the Purcell-Ramsey theorem has been generalized by Schiff to what is sometimes called the Schiff Theorem, which states that an assemblage of particles that are nonrelativistic, interact only electrostatically and are such that the electric dipole moment distribution within each particle is the same as its charge distribution will have no permanent electric dipole moment even if the constituent particles do so. The important part of Schiff's paper, however, is not the theorem but the recognition that in real atoms the theorem does not apply due to relativistic effects, presence of non-electrostatic forces, spin-orbit interactions and finite size effects. As first pointed out by Sandars and others (30,42,53), in some paramagnetic atoms an electron electric dipole moment can induce an atomic electric dipole moment 600 or more times bigger than that of the atom. As a result, electric dipole moment measurements with atoms and molecules provide sensitive tests for an electron electric dipole moment and for other T-nonconserving interactions to be discussed later.

For atomic electric dipole moments a relation analogous to Equation (2) applies except that the 2's in Equation (2) should be replaced by 1/F where F is the quantum number for the total atomic spin angular momentum $\mathbf{F} = \mathbf{I} + \mathbf{J}$ where $\mathbf{I}$ and $\mathbf{J}$ are the the nuclear and resultant electronic angular momenta. Consequently, just as with neutrons, the atomic electric dipole moments can be measured by the shift in the magnetic resonance frequency. The resonance pattern for atoms has been observed in different ways, sometimes with a single atomic beam (30-33,35), sometimes with two atomic beams moving in opposite directions (37) and sometimes with circularly polarized lasers beams used for the optical pumping and detecting of atoms stored in bottles with suitable wall coatings (34,36,38,39,40,46,).

The earliest atomic experiments were the 1960's cesium atomic beam experiments of Sandars, Lipworth and their associates (30-31) at Brandeis and Oxford. Sandars (30,31) showed that due to relativity and the admixture of a nearby p state, an electron electric dipole moment in some paramagnetic atoms can induce an atomic moment several hundreds times larger. In this fashion they set an experimental upper limit (30,31) on the electron electric dipole moment of 300×10^{-26} e cm.

There was no work on atom or molecule electric dipole moments in the 1970's, but since the mid 1980's a number of such experiments have recently been completed. With a molecule instead of an atom, Hinds and Sandars at Oxford (32) have set limits on T violating parameters in TlF with multiple resonance transitions. Wilkening, Ramsey and Larson at Harvard (33) improved the sensitivity of these results. Even greater sensitivity has been achieved in later experiments at Yale by Hinds and his associates (35) using a jet beam of TlF.

Fortson, Heckel, Lamoreaux and their collaborators (34,38,39,47) at Seattle have studied ^{199}Hg with the apparatus shown schematically in Fig.2. Each of the two vapor filled cells functions as an optically pumped oscillator based on the Larmor precession of the nuclear spins about B_z. The phase of the precession is detected by the modulation of the transmitted light intensity. The phase shift of this modulation with reversal of the applied electric field measures the electric dipole moment of the atom.

Hunter and his collaborators at Amherst (36), have set a low limit on the electric dipole moment of the electron using optical pumping of a ^{133}Cs cell.

Commins (37) and his associates at Berkeley have used the method of separated oscillatory fields and two counter propagating atomic beams as in Figure 3 to measure the electric dipole moments of the ^{205}Tl atom and of the

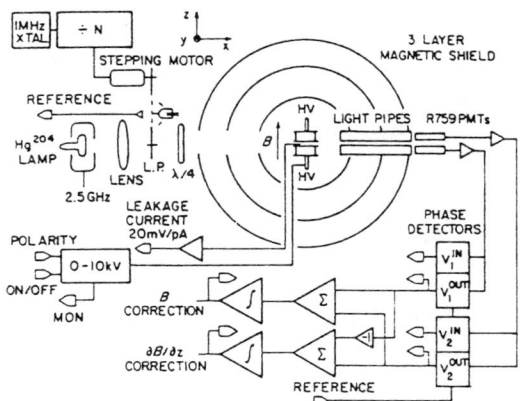

FIGURE 2. Schematic Diagram of the ^{199}Hg Apparatus (39).

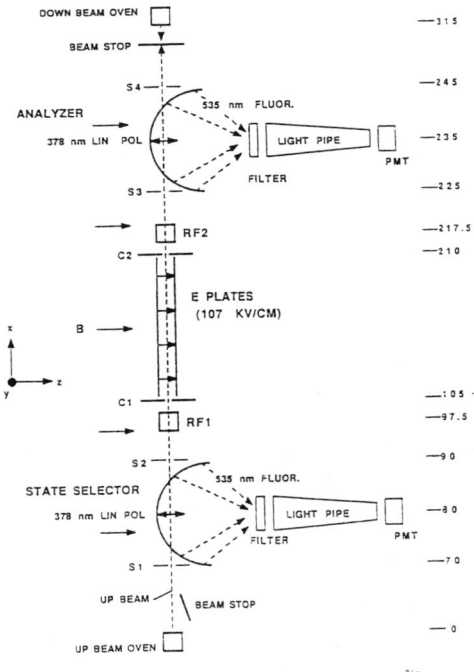

FIGURE 3. Schematic Diagram of the ^{205}Tl Apparatus (37).

electron. The counter-propagating beams are used to minimize the **E x v** effect. Polarized laser beams were used as state selector and analyzer.

Chupp (40) and his students at Harvard and Michigan used ^{129}Xe polarized by spin exchange collisions to determine the atomic electric dipole moment.

No T non-conserving interactions have been seen so far in any of these experiments, but low experimental limits have been set on electric dipole and other T non-conserving parameters as shown in Table I. For Cs and Tl there have also been improved calculations (42) of the sensitivity enhancement factors used in determining the limit to the electron electric dipole moment from that of the atom.

Although electric dipole moments of the constituent particles can produce resultant electric dipole moments of atoms, other T non-conserving interactions between elementary particles also can give rise to atomic or molecular electric dipole moments, especially in heavy paramagnetic atoms where the effects are enhanced (30,41). A resultant electric dipole moment can be due to a a nuclear dipole moment distribution from hadronic interactions between quarks, to a semi-leptonic interaction from a T violating force between electrons and nucleons or to a purely leptonic interaction from an intrinsic electric dipole moment of the electron There are now extensive theoretical calculations (41-44) of the these different contributions to a net electric dipole. There are also several recent reviews of both theories and experiments on atomic and molecular electric dipole moments (45-48). The primary experimental results from the atomic experiments are electric dipole moments of the atoms, but these can be interpreted in terms of T-non conserving interactions between the constituents of the atom and its nucleus.

Khriplovich and others (41-44) have shown that the experimental limits on the various T non-conserving interactions can be parametrized in terms of the atom electric dipole moment d, the electron dipole moment d_e, the proton electric dipole moment d_p, parameters Q_s (Schiff moment), η and η_q of the hadronic interaction terms $4\pi e Q_s \sigma \Sigma \nabla \rho_e$, $i\eta(\bar{n}\gamma_5 n)(\bar{n} n)G_F/\sqrt{2}$ and $i\eta_q(\bar{q}\gamma_5 q)(\bar{q} q)G_F/\sqrt{2}$ and the parameters C_T and C_S of the semi leptonic interaction terms $iC_T(\bar{e}\gamma_5 \sigma_{\mu\nu} e)(\bar{n}\sigma^{\mu\nu}n)G_F/\sqrt{2}$ and $iC_S(\bar{e}\gamma_5 e)(\bar{n}n)G_F/\sqrt{2}$. The neutron and atomic experimental limits on these T non-conserving parameters are given in Table 1. In each of the atomic experiments the directly measured quantity is the limit to the atomic electric dipole moment and all of the other limits are based on calculations using the atomic limit.

All of the groups doing atomic and molecular tests of T non-conservation hope to improve their sensitivities by factors of 10 to 100 in the next few years. Chupp's group for example is planning to operate both a ^{129}Xe and a ^{3}He maser simultaneously in the same cavity so the ^{3}He can serve as a magnetometer while they look for parity non-conserving effects in ^{129}Xe. Hinds has identified YbF as a particularly sensitive molecule for molecular searches for the electron electric dipole moment.

In addition, several groups with no results to report as yet have started different and potentially sensitive electric dipole searches for T non-conserving interactions. Ezhov and his associates (61) at Gatchina plan to use jet molecular beams of TlF and other molecules observing the interference at zero magnetic field of coherent states in of two degenerate levels (Hanle effect) rather than magnetic

TABLE 1. Upper Limits (95% Confidence Levels) on T Non-conserving Interaction Parameters (47).. The Parameters Are Defined in this Section Except For θ_{QCD} and $\varepsilon_{q/e,susy}$ Which are Defined by Barr (53) and the T-odd,P-even Parameters Which Are Defined by Khriplovich and Conti (57,58). This Table Is Based in Part on Tables Prepared by Fortson (39) and Barr (53).

System Units	n	^{199}Hg	TlF	^{205}Tl	^{133}Cs	^{129}Xe
Reference	27,28	39	35,33	37,59	36	40
d ($\times 10^{-26}$ e cm)	<9	<0.13	{d_p < 16,000}	<300	<2000	<21
Hadronic Parameters:						
Q_S ($\times 10^{-11}$ e fm^3)		<3	<100	<30,000		<8,000
η ($\times 10^{-6}$)		<2,000	<20,000			
η_q ($\times 10^{-6}$)	<40	<5				
θ_{QCD} ($\times 10^{-10}$)	<4	<20	<60	<4,000	<4,000	<300
$\varepsilon_{q,susy}$	<0.01	<0.01	<0.08			<1.3
$\varepsilon_{e,susy}$				<0.016	<0.5	
Semileptonic Parameters:						
C_T ($\times 10^{-6}$)		<0.02	<0.5			
C_S ($\times 10^{-6}$)		<1	<20	<0.4	<20	
Leptonic Parameter:						
d_e ($\times 10^{-26}$ e cm)		<9	<40	<0.4	<9	<20,000
T-Odd, P-even Parameters:						
β_e ($\times 10^{-6}$)		<0.7	<3	<0.03	<0.7	
β_{qe} ($\times 10^{-6}$)	<1					
β_{eq} ($\times 10^{-6}$)		<0.2	<1	<0.01	<70	
β_{qq} ($\times 10^{-6}$)	<0.3					
Im($C_T + C_{T'}$)		<0.005	<0.02	<0.0002	<0.005	
Im($C_P + C_{P'}$)		<0.23	<1	<0.01	<0.23	

resonance. Hansch, Weis and their associates (62) in Garching have observed Ramsey fringes in nonlinear Faraday rotation spectroscopy with a thermal Rb atomic beam using separated pumping and probing laser beam regions and they now plan to use such a spectrometer to search for permanent electric dipole moments of atoms. In quite different experiments Hansch, Weis and their collaborators (63) propose to search with magnetic resonance for permanent electric dipole moments of paramagnetic atoms implanted in solid ^{4}He.

THEORETICAL IMPLICATIONS

No electric dipole moments have been observed in any neutron, atom or molecule so far, but the experiments have placed low experimental limits on the moments and on possible internal T non-conserving interactions. The present limits on electric dipole moments are extremely low. An electric dipole moment of 10^{-26} e cm corresponds to a displacement of the proton of only 10^{-26} cm or 10^{-13} of the neutron diameter; if the neutron were scaled up to the dimensions of the earth such a displacement even on the scale of the earth would be only 0.001 mm.

For comparing the sensitivity of neutron electric dipole measurements with limitations on time reversal symmetry from nuclear reaction studies, Boehm (22) uses a simple dimensional argument to obtain the corresponding value of the T violating coupling constant g_T and finds that an electric dipole moment 10^{-26} e cm corresponds to a g_T of 10^{-6} which is far below the limit set by nuclear reaction studies.

The earliest predictions for the neutron electric dipole moment were exactly zero based on assumed parity symmetry. After parity conservation was shown to fail in the weak interaction (6,7) a prediction of zero was continued on the basis of assumed time reversal symmetry. Subsequent to the observed failure (12) of CP conservation in the decay of the K_L^0 most theories predict non-zero values for the neutron electric dipole moment. The predictions for d_n range from 10^{-19} e cm to less than 10^{-33} e cm. Most of the early theories, which attributed the known CP non-conservations to electromagnetic theory or to new milliweak or millistrong forces, predicted values for the neutron electric dipole moment larger than 10^{-23} e cm. These theories have now been mostly abandoned due in part to the disagreements between the predictions and the above experimental limits.

In the most popular current theories (23-26, 48-53) for predicting electric dipole moments, the T non-conservation comes from non-trivial physical complex phases in the theories of the particle fields. Such theories include the following. (a) Theories (23) that attribute CP violations to the exchange of Higgs bosons. Most models with three Higgs doublets and maximal CP violation predict values greater than 10^{-25} e cm. Weinberg (23) points out that, with the present neutron electric dipole moment value, neutral Higgs exchange can not be maximally CP violating. (b) Theories (24,25) that attribute CP non-conservation to the standard model of quantum chromodynamics with a complex phase factor δ in the Kobayashi-Maskawa matrix (24,48) and with the strong phase θ_{QCD} taken to be zero usually predict values between 10^{-31} and 10^{-33} e cm, well below the next experiments. Most theories that go beyond the standard model, however, have more parameters and room for more complex CP non-conserving phases; they are usually consistent with larger electric dipole moment values that should be within the reach of future experiments. (c) Strong quantum chromodynamic theories predict d_n to be approximately equal to $4 \times 10^{-16} \theta_{QCD}$ e cm where θ_{QCD} is the strong CP violating parameter, so the neutron measurements place the limit $\theta_{QCD} < 10^{-9}$ and the limit from ^{199}Hg experiments is only five times higher. The present experimental limits on θ_{QCD} from both atomic and neutron experiments are given in Table 1. (d) In a class of left-right symmetric models (48,51), the predicted moments are $(1 \text{ to } 10) \times 10^{-26}$ e cm. The present experimental limits on (e) Flavor mixing models predict d_n to be of the order of 1×10^{-26} e cm. (f) Supersymmetry (SUSY) permits the transformation of fermions into bosons and vice versa and adds many more possible particles which in turn provides the opportunity for more CP non-conserving phases. As a result T-violating phases occur naturally in supersymmetric theories. In some SUSY theories that account for the observed CP violation in the $K^o - \bar{K}^o$ system, the parameter $\varepsilon_{q,susy}$ should be of order 1 if supersymmetry is broken near the electroweak energy scale (53), whereas by both the neutron and ^{199}Hg experiments in Table I $\varepsilon_{q,susy} < 0.01$. This may indicate that $\varepsilon_{q,susy}$ is not the only source of CP violation (48). (g) cosmological theories (26) that account for the baryon-antibaryon asymmetry and relate the neutron electric dipole moment to the entropy of the universe generated since bariosyntheses are generally consistent with a value smaller than 10^{-26} e cm.

The atomic and molecular experiments on electric dipole moments and related T non-invariant parameters have been extensively analyzed and reviewed by theorists (43-53). Although the standard model theoretical limit for the neutron electric dipole limit is beyond the reach of any proposed experiment, most of the theories that go beyond the standard model can account for values that could be observable in future experiments. Since no one yet knows the way in which CP is not conserved, it is impossible to predict whether atomic or neutron electric dipole moment experiments will be the first to show a failure T symmetry. For some theories the neutron experiments provide the most sensitive tests while other theories are more sensitively probed by atomic experiments. For example, from Table 1 it is apparent that the neutron experiments are now the most sensitive for limiting θ_{QCD} whereas the atomic experiments are needed for C_S and C_T. The neutron and atomic experiments are complementary and both need to be vigorously pursued.

REFERENCES

1. Ramsey, N. F., *Molecular Beams*. Oxford Univ. Press (1956,1985).
2. Golub, R. and Pendlebury, J. M., *Contemp. Phys.* **13**, 519 (1972).
3. Purcell, E. M. and Ramsey, N. F., *Phys. Rev.* **78**, 807 (1950); Ramsey, N. F., *AIP*, **270**, 179 (1993).
4. Smith, J. H., Purcell, E. M. Ramsey, N. F., *Phys. Rev.* **108**, 120 (1957).
5. Ramsey, N. F., *Physics Today* **33**, 25 (1980).
6. Lee, T. D. and Yang, C. N., *Phys. Rev.* **104**, 254 (1956).

7. Wu, C. S., Ambler, E. Hayward, B. W., Hoppes, D. D. and Hudson, R. R., *Phys. Rev.* **105**, 1413 (1957).
8. Landau, L., *Nucl. Phys.* **3**, 127 (1957).
9. Ramsey, N. F., *Ann. Rev. Nucl. Part. Science* **32**, 211 (1982) and **40**, 1 (1990). These articles contain full lists of references.
10. Ramsey, N. F., *Phys. Rev.* **109**, 225 (1958).
11. Jackson, J. D., Treiman, S. B. and Wyld, H. W. Jr., *Phys. Rev.* **106**, 517 (1957).
12. Christenson, J. H. Cronin, J. W. Fitch, V. L. and Turlay, R., *Phys. Rev. Lett.* **13**, 138 (1964).
13. Dress, W. B., Miller, P. D., Pendlebury, J. M., Perrin, P. and Ramsey, N. F., *Phys. Rev. D* **15**, 9 (1977).
14. Ramsey, N. F., *Phys. Rev.* **78**, 695 (1950).
15. Ramsey, N. F., *Phys. Today* **33**, 25 (1980).
16. Ramsey, N. F., *Rev. Sci. Inst.* **28**, 57 (1957).
17. Zel'dovich, Y. A. B., *Zh. Eksp. Teor. Fiz.* **36**, (1959); *JETP* **9**, 1389 (1959).
18. Shapiro, F. L., *Usp. Fiz. Nauk.* **95**, 145 (1968); *Sov. Phys. Usp.* **11**, 345 (1968).
19. Altarev, I. S. Barisov, Uy. V., Brandin, A. B., Egerov, A. I., Ezhov, V. F., Ivanov, S. N., Loboshov, V. M., Nazarenko, V. A., Porsev, G. D., Ryabov, V. L., Serebov, A. P., Taldaev, R. R. and Shulgina *Nucl. Phys.* **A341**, 269 (1980)..
20. Pendlebury, J. M., Smith, K. F., Golub, R., Byrne, J., McComb, T. J. L., Sumner, T. J., Burnett, S. M., Taylor, A. R., Heckel, B., Ramsey, N. F., Green, K., Morse, J., Kilvington, A. I., Baker, C. A. and Clark, S. A., *Phys. Lett.* **136B**, 327 (1984).
21. Steyerl, A., Nagel, H., Schreiber, F. X., Steinhauser,, K. A., Gläser, W., Ageron, P., Astruc, J. M., Drexel, W., Gervais, R. and Mampe, W., *Condensed Matter* **50**, 281 (1983) and ILL Report ICSU/AB:07.
22. Boehm, F., *Comments Nucl. Part Phys.* **11**, 251 (1983).
23. Weinberg, S., *Phys. Rev. Lett.* **37**, 367 (1976) and **63**, 2333 (1989).
24. Kobayashi, M. and Maskawa, K., *Prog. Theor. Phys.* **49**, 652 (1973).
25. Nanopoulos, D. V., Yildiz, A. and Cox, P., *Ann. Phys. (NY)* **127**, 126 (1980).
26. Hinds, E. A. and Sandars,P. G. H., *Phys. Rev.* **A21**, 471 and 480 (1980); Wilkening, D. A., Ramsey, N. F., and Larson, D. J., Phys. Rev. A29, 425 (1984).
27. Smith, K. F., Crampin, N., Pendlebury, J. M., Richardson, D. J., Shiers, D.,, Green, K., Kilvington, A. I., Moir, J., Prosper, H. B., Thompson, D., Ramsey, N. F., Heckel, B. R., Lamoreaux, S. K., Ageron, P., Mampe, W. and Steyrl, A., *Phys. Lett.* **B234**, 191 (1990).
28. Altarev, I. S. Barisov, Uy. V., Borovikova, N. V., Kolomensky, E. A., Lasakov, M. S., Ivanov, S. N., Loboshov, V. M., Nazarenko, V. A., Pirozhkova, A. N., Serebov, A. P., Sobolev, Yu. V., Shulgina, E. V., and Yegerov, A. I., *Phys. Lett.*, **B276**, 242 (1992).
29. Ramsey, N. F., *Acta. Phys.Acad. Sci. Hungerica*, **55**, 117 (1984).
30. Sandars, P. G, H,. Lipworth, E. and Player, M. A., *Phys. Rev. Lett.* **13**, 718 (1964); *Phys. Lett,* **14**, 194 (1965), **19**, 1396 (1967) and **22**, 290 (1966); and *J. Phys.* **B3**, 1620 (1970).

31. Weisskopf, M. C., Carrico, J. P., Gould. H., Lipworth, E. and Stein, T. S., *Phys. Rev. Lett.* **21**, 1645 (1968).and **24**, 1091 (1970).
32. Hinds, E. A. and Sandars, P. G. H., *Phys. Rev.* **A21**, 471 and 480 (1980).
33. Wilkening, D. A. Ramsey, N. F. and Larson, D. J., *Phys. Rev.* **A23**, 425 (1984).
34. Vold, T. G., Raab, F., Heckel, B., and Fortson, E. N., *Phys. Rev. Lett.* **52**, 2229 (1984), *Nucl. Instrum. Methods*, **A284**, 43 (1989).
35. Cho, D., Sangster, D., and Hinds, E. A., *Phys. Rev. Lett.* **63**, 2559 (1989); Hinds, E. A., *Rencontre de Moriond.* **26**, 389 (1991).
36. Murthy, S. A., Krause, D., Jr., Li, Z. L., and Hunter, L. R., *Phys. Rev. Lett.* **63**, 965 (1989); Hunter, L. R., *Rencontre de Moriond.* **26**, 395 (1991)
37. Abdullah, K., Carlberg, C., Commins, E. D., Gould, H., and Ross, S. B., *Phys. Rev. Lett.* **65**, 2347 (1990); *ICAP* **12**, 442 (1991); *AIP*, **270**, 34 (1993); *Physica Scripta*, T46, 92 (1993).
38. Lamoreaux, S. K., Jacobs, J. P., Heckel, B. R., Raab, F. J., and Fortson, N., *Phys. Rev. Lett.* **59**, 2275 (1987).
39. Jacobs, J. P., Klipstein, W. M., Lamoreaux, S. K., Heckel, B., and Fortson, E. N., *Phys. Rev. Lett..* **71**, 3782 (1993)..
40. Hoare, R. J., Oteiga, E. R. and Chupp, T. E.; *ICAP* **13**, 73 (1993) and *AIP* **270**, 84 (1993). Oteiga, E. R., Ph. D. Thesis Harvard University (1993).
41. Bouchiat, C., *Phys. Lett.* **57B**, 284 (1975); Hinds, E. A., Loving, C. E., and Sandars, P. G. H., *Phys. Rev. Lett.* **62B**, 97 (1976).
42. Martensson-Pendrill, A. M., *Phys. Rev. Lett.* **54**, 1153 (1985); Lindroth, E. and Martensson-Pendrill, A. M., *Journ Phys.,* **B23**, 3417 (1990).
43. Flambaum, V. P., Khriplovich, I. B. and Sushkov, O. P., *JETP* **60**, 873 (1984) and *Nucl, Phys.* **A449**, 750 (1986); Khriplovich, I. B., in *Atomic Physics* **11**,113 (1989) [edited by Haroche, Gay, J. C., and Grynberg, G.,World Scientific, Singapore, 1989]; Khatsymovsky, V. M., Khriplovich, I. B., and Yelkhovsky, A. S., *Ann. Phys.* **186**, 1 (1988); Khatsymovsky, V. M. and Khriplovich, I. B., Phys. Lett. **B296**, 219 (1992).
44. Johnson, W. R., Guo, D. S., Idrees, M., and Sapirstein, J., *Phys. Rev.* **A32**, 2093 (1985); **34**, 1043 (1986); Martensson-Pendrill, A. M., Ynnerman, A., and Oster, *P.,J. Phys.* **B22**, 2447 1989).
45. Barr, S. M. and Marciano, W. J., in *CP Violation*, edited by C. Jarlskrog (World Scientific, Singapore, 1989); Barr, S. M. and Zee, A., *Phys. Rev. Lett.* **65**, 21 (1990).
46. Hunter, L. R. Science, **252**, 73 (1991).
47. Fortson,N., private communication.
48. He, X. G. McKellar, B. H. G. and Pakvasa, S., *Int.J. Mod. Phys.*,**A4**, 5011 (1989) and private communication.
49. Wolfenstein, , L., *Ann. Rev. Nucl. Part.Phys.* **35**, 137 (1986).
50. Ellis, J., *Nucl. Inst. & Methods in Phys. Research* **A284**, 33 (1989); and Nature, **344**, 197 (1990).
51. Mohapatra, R. N. and Pati, J. C. *Phys. Rev.* **D11**, 566 (1974).
52. Ellis, J., Gaillard, M. K., Nanopulos, D. V. and Rudaz, S., *Nature* **293**, 41 (1981).
53. Barr, S. M., *Int. J. Mod. Phys.* **A8**, 209 (1993) and private communivcation.
54. Friedman, M. S., Peshkin, P., Ringo, G. R. and Dombeck, T. W., *Proceedings of Third Symposium on Fundamentals of Quant. Mechs.*, **QM3**, 43 (1989).

55. Golub, R. and Lamoreaux, S. K, *J. Physique* **44**, L321 (1983) and *Phys. Reports* **237**, 2 (1994).
56. Pendlebury, J. M., *Nucl. Phys.* **A546**, 359 (1992).
57. Khriplovich, I. B., *Nuclear Physics*, **B352**, 385 (1991).
58. Conti, R. S., *Phys. Rev. Lett..*, **68**, 3262 (1992).
59. Commins, E. D., Ross, S. B., DeMille, D. and Regan, B. C., Private communication to be published in *Phys. Rev. A*.
60. Schiff, L. I., *Phys. Rev.* **132**, 2194 (1963).
61. Ashkinadzi, B. N., Ezhov, V. F., et al., *Petersburg Nuclear Physics Institute Preprint* (1992).
62. Weis, A., Schuh, B. Kanorsky, S. I. and Hansch, T. W., *Technical Digest, EQEC 93*, Florence, Italy (1993) and *Optics Comm.* **100**, 451 (1993)
63. Arndt, M., Kanorsky, S. I., Weis, A. and Hansch, T. W., *Phys. Lett.* **A174**, 298 (1993).

Atomic Physics in QED and QCD

G. Peter Lepage
Newman Laboratory of Nuclear Studies
Cornell University, Ithaca, NY 14853
gpl@hepth.cornell.edu

Introduction

In this talk I describe a new approach to the analysis of QED effects in simple atoms. The general techniques have been successfully applied in studies of muonium, positronium and other hydrogenic systems. They are most likely also valuable in helium and larger multielectron atoms.

The earliest studies of simple atomic systems were based on the quantum hamiltonian developed by Schrödinger, Pauli and others:

$$H = \frac{\mathbf{p}^2}{2m} + e\,\phi - e\,\frac{\mathbf{p}\cdot\mathbf{A}}{m} - e\,\frac{\boldsymbol{\sigma}\cdot\mathbf{B}}{2m} + \cdots \tag{1}$$

where one usually treats the $\mathbf{p}\cdot\mathbf{A}$, $\boldsymbol{\sigma}\cdot\mathbf{B}$, etc terms as perturbations. This theory is very successful: working to first order in the perturbations, it easily accounts for the first six digits in any binding energy of hydrogen. This theory is also very simple to analyze and forms the basis for most people's intuition about atomic structure. However, beyond first-order perturbation theory, this theory leads to ultraviolet infinities; the perturbations are too singular.

By 1950 we had learned how to remove the ultraviolet infinities in Green's functions by using relativistically covariant perturbation theory. This led Bethe and Salpeter, and Schwinger to reformulate the bound-state problem in terms of Green's functions rather than the hamiltonian. The result, the Bethe-Salpeter equation, allows one to use standard QED renormalization methods to remove the infinities in bound-state calculations. This equation, while elegant, is quite obscure to most people and very complicated to use.

To illustrate the problems that arise, consider a traditional Bethe-Salpeter calculation of the energy levels of positronium. The potential in the Bethe-Salpeter equation is given by a sum of two-particle irreducible Feynman amplitudes. One generally solves the problem for some approximate potential and then incorporates corrections using time-independent perturbation theory. Unfortunately, perturbation theory for a bound state is far more complicated than perturbation theory for, say, the electron's g-factor. In the latter case a diagram with three photons contributes only in order α^3. In positronium a kernel involving the exchange of three photons can also contribute to order α^3, but the same kernel will contribute to all higher orders as well:

$$\langle V_3 \rangle = \alpha^3 m \left\{ a_0 + a_1 \alpha + a_2 \alpha^2 + \ldots \right\}. \tag{2}$$

So in the bound-state calculation there is no simple correlation between the importance of an amplitude and the number of photons in it.

Such behavior is at the root of the complexities in high-precision analyses of positronium or other QED bound states. It is a direct consequence of the multiple scales in the problem. Typically a nonrelativistic system has three important scales: the mass m of the consistuents, their three momentum $p \approx mv$, and their kinetic energy $K \approx mv^2$. Any expectation value like that in Eq. (2) will be some complicated function of ratios of these three scales:

$$\langle V_3 \rangle = \alpha^3 m \, F(\langle p \rangle / m, \langle K \rangle / m). \tag{3}$$

Since $\langle p \rangle / m \sim \alpha$ and $\langle K \rangle / m \sim \alpha^2$, a Taylor expansion of F in powers of these ratios generates an infinite series of contributions just as in Eq. (2). A similar series does not occur in the g-factor calculation because there is but one scale in that problem, the mass of the electron.[1]

Traditional methods for analyzing these bound states fail to take advantage of the nonrelativistic character of these systems; and atoms like positronium are very nonrelativistic: the probability for finding relative momenta of $\mathcal{O}(m)$ or larger is roughly $\alpha^5 \approx 10^{-11}$. Our new strategy relies upon renormalization methods to separate relativistic from nonrelativistic scales, allowing us to analyze them separately. The net effect is to reinstate the Pauli hamiltonian as the central object in boundstate analyses by developing a renormalization procedure for removing the infinities. In effect we

---

[1]The bound-state calculation is actually even worse than indicated because QED is a gauge theory. The contribution from a particular kernel is highly dependent upon gauge. In many three-photon kernels, for example, the coefficients a_0 and a_1 in Eq. (2) vanish in Coulomb gauge but not in Feynman gauge. The Feynman gauge result is roughly 10^4 times larger, and it is largely spurious: the bulk of the contribution comes from unphysical retardation effects in the Coulomb interaction that cancel when an infinite number of other diagrams are included. The Coulomb interaction is instantaneous in Coulomb gauge and so this gauge generally does a better job describing the fields created by slowly moving charged particles. On the other hand contributions coming from relativistic momenta are more naturally handled in a covariant gauge like Feynman gauge; in particular renormalization is far simpler in Feynman gauge than it is in Coulomb gauge. Unfortunately most Bethe-Salpeter kernels have contributions coming from both nonrelativistic and relativistic momenta, and so there is no optimal choice of gauge. This is again a problem due to the multiple scales in the system.

are adapting the renormalization procedure to facilitate boundstate analyses (the Bethe-Salpeter formalism does the reverse).

In the following sections we first review recent insights into the nature of quantum field theories and the role of renormalization [1]. Then we adapt these ideas to develop a new field-theoretic approach to nonrelativistic systems that combines the simplicity of a nonrelativistic analysis with the precision of a relativistic analysis. We outline the steps needed in a boundstate analysis using this new formalism [2]. The renormalization techniques that we use here are very general and so are broadly applicable. To demonstrate this point we end by showing how these techniques, first developed for positronium and muonium studies, have led to a breakthrough in the *nonperturbative* analysis of hadron structure in QCD. They have been applied in high-precision studies of the Υ and ψ families of mesons that have important implications for QCD phenomenology [3].

Renormalization Revisited

Despite the complexity of most textbook accounts, renormalization is based upon a very familiar and simple idea: a probe of wavelength λ is insensitive to details of structure at distances much smaller than λ. This means that we can mimic the *real* short-distance structure of the target and probe by *simple* short-distance structure. For example, a complicated current source $\mathbf{J}(\mathbf{r},t)$ of size d that generates radiation with wavelengths $\lambda \gg d$ is accurately mimicked by a sum of point-like multipole currents ($E1$, $M1$, etc). In thinking about the long-wavelength radiation it is generally much easier to treat the source as a sum of multipoles than to deal with the true current directly. This is particularly true since usually only one or two multipoles are needed for sufficient accuracy. The multipole expansion is a simple example of a renormalization analysis.

In QED, the quantum fluctuations probe arbitrarily short distances. This is evident when one computes radiative corrections in perturbation theory. Ultraviolet divergences, coming from loop momenta $k \to \infty$ (or wavelengths $\lambda \to 0$), result in infinite contributions — radiative corrections seem infinitely sensitive to short distance behavior. Even ignoring the infinities, this poses a serious conceptual problem since we don't really know what happens as $k \to \infty$. For example, there might be new supersymmetric interactions, or superstring properties might become important, or electrons and muons might have internal structure. The situation is saved by renormalization theory which tells us that we don't really need to know what happens at very large momenta in order to understand low-momentum experiments. As in the multipole expansion, we can mimic the complex high-momentum, short-distance structure of the real theory, whatever it is, by a generic set of simple point-like interactions.

The transformation from the real theory to a simpler effective theory, valid for low-momentum processes, is achieved in two steps. First we introduce a momentum cutoff Λ that is of order the momentum at which new as yet unknown physics becomes important. Only momenta $k < \Lambda$ are retained when calculating radiative corrections.[2] This means that our radiative corrections include only physics that we understand, and that there are no longer infinities. Of course we don't really know the scale Λ at which new physics will be discovered, but, as we shall see, results are almost independent of Λ provided it is much larger than the momenta in the range being probed experimentally.

The second step is to add local interactions to the lagrangian (or hamiltonian). These mimic the effects of the true short-distance physics. Any radiative correction that involves momenta above the cutoff is necessarily highly virtual, and, by the uncertainty principle, most occur over distances of order $1/\Lambda$ or less. Such corrections will appear to be local to low-momentum probes whose wavelengths $\lambda \approx 1/p$ are large compared with $1/\Lambda$. Thus the correct lagrangian for cutoff QED consists of the normal lagrangian together with a series of correction terms:

$$\begin{aligned}\mathcal{L}^{(\Lambda)} &= \overline{\Psi}\left(i\partial\cdot\gamma - e(\Lambda)A\cdot\gamma - m(\Lambda)\right)\Psi - \tfrac{1}{4}F_{\mu\nu}F^{\mu\nu} \\ &+ \frac{e(\Lambda)c_1(\Lambda)}{\Lambda}\overline{\Psi}\sigma_{\mu\nu}F^{\mu\nu}\Psi \\ &+ \frac{e(\Lambda)c_2(\Lambda)}{2\Lambda^2}\overline{\Psi}i\partial_\mu F^{\mu\nu}\gamma_\nu\Psi + \frac{d_2(\Lambda)}{\Lambda^2}\left(\overline{\Psi}\gamma_\mu\Psi\right)^2 + \cdots,\end{aligned} \quad (4)$$

where couplings $e(\Lambda)$, $c_1(\Lambda)$, $c_2(\Lambda)$ and $d_2(\Lambda)$ are dimensionless. The correction terms are nonrenormalizable, but that does not lead to problems because we keep the cutoff finite. These new interactions are far simpler to work with than the supersymmetric/superstring/... interactions that they simulate.

The correction terms in cutoff QED modify the predictions of the theory. However the modifications are small if Λ is large. Contributions from the $\sigma_{\mu\nu}F^{\mu\nu}$ term, for example, are suppressed by p/Λ, where p is the typical momentum in the process under study. The next two terms are suppressed by $(p/\Lambda)^2$, and so on. In principle there are infinitely many correction terms in $\mathcal{L}^{(\Lambda)}$, forming a series in $1/\Lambda$; but, when working to a given precision (ie, to a given order in p/Λ), only a finite number of these terms is important. Indeed none of these correction terms seems important in any high-precision test of QED. This indicates that the scale for new physics, Λ, is quite large— probably of order a few TeV or larger.

---

[2] For clarity's sake we adopt a simple cutoff as our regulator here. In actual calculations one generally tailors the regulator to optimize the calculation.

One of the goals of high-precision studies of QED is the discovery of new physics. The expansion (4) provides a useful parameterization for the effects of new physics on low-momentum processes. The form of the cutoff lagrangian is independent of the new physics. It is only the numerical values of the couplings $c(\Lambda)$, $d(\Lambda)$... that contain information about the new physics. (The couplings are analogous to the multipole moments of a current in our example above.) Thus the implications of a high-precision test of QED can be expressed in a model-independent way as limits on or values for these couplings.

Nonrelativistic QED (NRQED)

In the previous section we introduced a cutoff into QED in order to exclude unknown physics from radiative corrections. There the value of the cutoff was determined by the energy threshold for new physics. In general, however, we are free to use smaller cutoffs, in which case the extra terms in the cutoff lagrangian compensate for known physics excluded by the cutoff, as well as unknown physics. This means the cutoff can be used as a computational tool to isolate different momentum scales. For nonrelativistic systems, the most important dynamical scales are nonrelativistic. This suggests that we use a cutoff of order the constituent mass. Then the $k < \Lambda$ physics is nonrelativistic and explicit in our effective theory, while $k > \Lambda$ physics is relativistic and implicit (ie, in the coupling constants). In this way we use the cutoff to separate relativistic from nonrelativistic scales.

The cut-off version of QED developed in the previous section is not yet particularly useful in the study of nonrelativistic systems. Even when restricted by the cutoff to nonrelativistic electrons, the theory is still basically a Dirac theory and therefore substantially more complicated than the nonrelativistic theory. The utility of the theory is greatly enhanced if one transforms the Dirac field so as to decouple its upper components, representing the electron, from its lower components, representing the positron. This can be achieved using the Foldy-Wouthuysen-Tani transformation[4]. It transforms the Dirac lagrangian into a nonrelativistic lagrangian. In QED it leads to

$$\overline{\Psi}(iD\cdot\gamma - m)\Psi \;\rightarrow\; \psi^\dagger iD_0\psi + \psi^\dagger \frac{\mathbf{D}^2}{2m}\psi + \psi^\dagger \frac{\mathbf{D}^4}{8m^3}\psi$$
$$+\frac{e}{2m}\psi^\dagger\boldsymbol{\sigma}\cdot\mathbf{B}\psi + \frac{e}{8m^2}\psi^\dagger\nabla\mathbf{E}\psi + \cdots \qquad (5)$$

where $D_\mu = \partial_\mu + ieA_\mu$ is the gauge-covariant derivative, $\mathbf{E}$ and $\mathbf{B}$ are the electric and magnetic fields, and ψ is a two-component Pauli spinor representing the electron part of the original Dirac field. The lower components of the Dirac field lead to analogous terms that specify the electromagnetic interactions of positrons. The Foldy-Wouthuysen-Tani transformation generates an infinite expansion of the action in powers of $1/m$. In an ordinary $\Lambda \to \infty$ field theory this expansion is a disaster: the renormalizability of the theory is completely disguised since it relies upon a delicate conspiracy involving terms of all orders in $1/m$. However, setting $\Lambda \sim m$ implies that the Foldy-Wouthuysen-Tani expansion is an expansion in $1/\Lambda$, and from our general analysis of cut-off theories we know that only a finite number of terms need be retained in the expansion if we want to work to some finite order in $p/\Lambda \sim p/m \sim v/c$. Thus, to work through order $(v/c)^2$, we can replace relativistic QED by a nonrelativistic theory (NRQED) with the lagrangian [5]

$$\begin{aligned}\mathcal{L}_{\text{NRQED}} =\ & -\tfrac{1}{4}(F^{\mu\nu})^2 + \psi^\dagger \Big\{ i\partial_t - eA_0 + \frac{\mathbf{D}^2}{2m} + \frac{\mathbf{D}^4}{8m^3} \\ & + c_1 \frac{e}{2m}\boldsymbol{\sigma}\cdot\mathbf{B} + c_2 \frac{e}{8m^2}\nabla\mathbf{E} \\ & + c_3 \frac{ie}{8m^2}\boldsymbol{\sigma}\cdot(\mathbf{D}\times\mathbf{E} - \mathbf{E}\times\mathbf{D}) \Big\}\psi \\ & + \frac{d_1}{m^2}(\psi^\dagger\psi)^2 + \frac{d_2}{m^2}(\psi^\dagger\boldsymbol{\sigma}\psi)^2 \\ & + \text{positron and positron-electron terms}. \end{aligned} \quad (6)$$

The coupling constants e, m, c_1... are all specified for a particular cutoff with $\Lambda \sim m$. Renormalization theory tells us that there exists a choice for the coupling constants in this theory such that NRQED reproduces all of the results of QED up to corrections of order $(p/m)^3$.

The values of the coupling constants depend upon the choice of cutoff or regulator, since the interactions must compensate for the finiteness of the cutoff. Generally in field theory, one has tremendous freedom in designing regulators, and that freedom should be used to tailor the cutoff to the particular problem of interest. For perturbative calculations in NRQED it is usually most convenient if the regulator respects such symmetries of the theory as gauge invariance, rotation invariance, and parity conservation; breaking one of these symmetries with the cutoff results in a more complicated lagrangian. For example, if the regulator breaks the gauge symmetry of NRQED then extra gauge noninvariant interactions must be added to $\mathcal{L}_{\text{NRQED}}$ to cancel gauge noninvariant effects induced by the cutoff. While it is possible to work with such a regulator, it is usually simpler to maintain gauge invariance, thereby avoiding additional interaction vertices. Insofar as $\mathcal{L}_{\text{NRQED}}$ is not Lorentz invariant, it is not critical that the regulator be Lorentz invariant, although a Lorentz invariant cutoff does simplify the calculation of some of the coupling constants: for example, the $\psi^\dagger(\mathbf{D}^4/8m^3)\psi$ term in $\mathcal{L}_{\text{NRQED}}$ has no further coefficient if the cutoff is Lorentz invariant.

Obvious choices for the regulator include dimensional regularization and the Pauli-Villars regulator. One regulator that has been used successfully with NRQED is based upon a subtraction scheme: the divergences in any diagram (or subdiagram) are removed by subtracting counterterms that are identical in structure to the diagram, or derivatives of the diagram, but evaluated on-shell and at threshold.[5]. This regulator is particularly convenient for numerical work because the subtractions can be made in the integrands of Feynman integrals, before integrating.

Once a regulator has been specified for NRQED, the coupling constants are readily calculated. One simple procedure is to compute scattering amplitudes both in QED and in NRQED, and to adjust the NRQED coupling constants so that the two calculations agree (to a given order in α and v/c). For example, the coefficient c_1 of the $\psi^\dagger \boldsymbol{\sigma} \cdot \mathbf{B} \psi$ interaction can be determined by comparing calculations in the two theories of the spin-flip amplitude for an electron to scatter off an external magnetic field. Once c_1 is tuned for this process, it is correct for all other processes. One finds from such calculations that each of the coupling constants has a perturbative expansion in α; at tree level, $c_1 = c_2 = c_3 = 1$ and $d_1 = d_2 = 0$. The electric charge e is not renormalized in NRQED since vacuum polarization corrections to the low-energy photon propagator involve highly virtual electron-positron states that are omitted from the nonrelativistic theory; there is no vacuum polarization in NRQED. The effects of vacuum polarization on the photon's dynamics are of order $\alpha(v/c)^2$ and smaller, and can be incorporated into NRQED by adding new local interactions to the photon's lagrangian.[3]

NRQED for Simple Atoms

NRQED is far simpler to use than the original relativistic theory when studying nonrelativistic atoms like positronium. The analysis falls into two parts. First one must determine the coupling constants in the NRQED lagrangian. This is easily done by comparing simple scattering amplitudes computed in both QED and NRQED, as discussed in the previous section. Since the couplings contain the relativistic physics, this part of the calculation involves only scales of order m; it is similar in character to a calculation of the g-factor. Furthermore there is no need to deal with complicated bound states at this stage. Having solved the high-energy part of QED by computing the coupling constants in $\mathcal{L}_{\text{NRQED}}$, one goes on to solve NRQED for any nonrelativistic process or system. To study positronium one uses the Bethe-Salpeter equation for this theory, which is just the Schrödinger equation, and ordinary time-independent perturbation theory. One of the main virtues of this

---

[3]The leading correction involves an operator $(\partial_\mu F^{\mu\nu})^2$. The equations of motion imply that such a term can be replaced by four-fermion interactions such as those already present in $\mathcal{L}_{\text{NRQED}}$. In higher orders there are four-photon vertices of the sort computed (long ago) by Euler and Heisenberg.

approach is that it builds directly on the simple results of nonrelativistic quantum mechanics, leaving our intuition intact. Even more important for high-precision calculations is that only two dynamical scales remain in the problem, the momentum and the kinetic energy, and these are easily separated on a diagram-by-diagram basis. As a result infinite series in α can be avoided in calculating the contributions due to individual diagrams, and thus it is trivial to separate, say, $\mathcal{O}(\alpha^6)$ contributions from $\mathcal{O}(\alpha^5)$ contributions.[4]

Positronium is composed entirely of leptons; ordinary atoms like hydrogen can also be analyzed using NRQED. The major difference is that the nucleus in such atoms has structure. NRQED allows us to incorporate the effects of this structure in a systematic and rigorous fashion. The proton in the hydrogen atom, for example, is treated as a point-like particle described by a lagrangian that is identical in form to the electron's lagrangian (Eq. (6)). The proton's structure, being very short-ranged, enters only through the coupling constants $c_1, c_2 \ldots$. Lacking a complete theory of proton structure, we are unable to compute these coupling constants. However, we can measure them: for example, c_1, c_2, and c_3 are completely determined by the proton's magnetic moment and charge radius, and by relativistic invariance. Thus proton structure is for our purposes completely specified by a small number of coupling constants, the actual number depending upon the accuracy needed. Given these couplings, one can forget that the proton has structure and treat it as we treat electrons.

Υ's on the Lattice

The renormalization techniques discussed in the previous section are very general and can be used in nonperturbative as well as perturbative problems. One illustration is a recent analysis of the Υ family of mesons. The (heavy) quarks in these mesons are so strongly bound that the traditional perturbative techniques used to study QED atoms like positronium are useless. A nonperturbative treatment is called for. Currently, the most effective approach to nonperturbative QCD relies upon numerical lattice simulations in which continuous space-time is replaced by a discrete grid with a small lattice spacing a. The lattice is nothing more than a type of cutoff; lattice field theories are cut-off field theories. In a lattice simulation, physics at momenta $k < \pi/a$ is explicit and is analyzed using Monte Carlo evaluations of

---

[4]Furthermore the choice of gauge is now obvious. In computing the coupling constants for NRQED, one uses a covariant gauge like Feynman gauge since this calculation involves only relativistic momenta. These couplings are gauge invariant provided the cutoff used in defining NRQED is gauge invariant. Thus when we use NRQED to compute the properties of positronium, for example, we are free to choose Coulomb gauge, the optimal choice for nonrelativistic systems.

the path integrals that define the theory. Contributions from $k > \pi/a$ enter only implicitly, through the couplings in the lattice lagrangian. In QCD, these couplings are computed using perturbation theory.[5]

The spectrum indicates that the b quarks are nonrelativistic in Υ's, with velocities v such that $v^2 \approx 0.1$. This means the most effective formalism for studying their static properties is the lattice version of NRQCD. The NRQCD lagrangian is almost identical to the NRQED lagrangian, the main difference being that the vector potential is a (nonabelian) matrix operator in NRQCD. On the lattice, derivatives in the lagrangian are replaced by differences, and new correction terms, for cutoff-related errors of order a, $a^2 \ldots$, are added to the lagrangian. The details are described in [6]. Here we describe results from a simulation with a lattice spacing of about $1/12$ fm ($a^{-1} \approx 2.4$ GeV) [7]. This simulation employed the leading terms in the NRQCD lagrangian as well as all of the $\mathcal{O}(v^2)$ corrections.

The Υ spectrum is very well described by the simulation. Simulation results for the low-lying $^3S_1$ and $^1P_1$ energies are shown in Figure 1. These compare well with experimental results (the horizontal lines), as they should since systematic errors are estimated to be less than 20–40 MeV. It is important to realize that these are calculations from first principles. The only inputs are the lagrangians describing gluon and quark dynamics, and the only parameters are the bare coupling constant and the bare quark mass. In particular, these results are not based on a phenomenological quark-potential model. These are among the most accurate lattice results to date.

The relativistic corrections in the NRQCD lagrangian have only a small effect on the overall spectrum, but the spin structure is strongly affected. The lowest-order lagrangian is spin independent, and gives no spin splittings at all. Simulation results for the spin structure of the lowest lying P state are shown in Figure 2. Again these compare very well with the data, giving strong evidence that correction terms in the lagrangian work. Systematic errors here are estimated to be of order 5 MeV.

The correction terms also affect the simulation mass of the Υ. The nonrelativistic action gives only part of the total energy or mass of the meson. The full mass is obtained by adding the masses of the quarks to the meson's nonrelativistic energy E_{NR}:

$$M_\Upsilon = 2\left(Z_m M_b^0 - E_0\right) + E_{\text{NR}} \tag{7}$$

where Z_m and E_0 are ultraviolet renormalizations that are computed using perturbation theory, M_b^0 is a tunable parameter of the theory, and E_{NR} is computed in the simulation. The bare quark mass in this simulation

---

[5]Perturbation theory is useful in QCD for computing high-momentum, short-distance quantities, like couplings, but not for low-momentum, long-distance quantities, like meson structure.

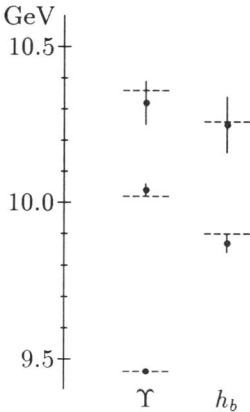

Figure 1: NRQCD simulation results for the spectrum of the $\Upsilon(^3S_1)$ and $h_b(^1P_1)$ and their radial excitations. Experimental values (dashed lines) are indicated for the S-states, and for the spin-average of the P-states. The energy zero for the simulation results is adjusted to give the correct mass to the Υ.

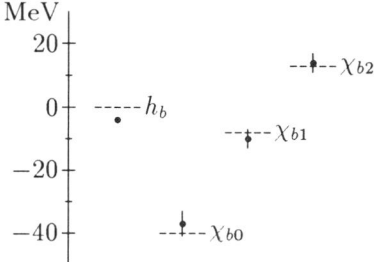

Figure 2: Simulation results for the spin structure of the lowest lying P-state in the Υ family. The dashed lines are the experimental values for the triplet states, and the experimental spin average of all states for the singlet (h_b).

was tuned so that this formula gave $M_\Upsilon = 9.5(1)$ GeV (the correct value is 9.46 GeV). The Υ mass was also determined a second way in the simulation by computing the nonrelativistic energy of an Υ as a function of its momentum p, and fitting its low-momentum behavior to the form

$$E_\Upsilon(\mathbf{p}) = E_{\mathrm{NR}} + \frac{\mathbf{p}^2}{2 M_{\mathrm{kin}}} + \cdots . \tag{8}$$

This determined the kinetic mass M_{kin} of the meson. In a purely nonrelativistic theory the kinetic mass equals the sum of the quark masses. Only when relativistic corrections are included is this mass shifted to include the binding energy, giving M_Υ. The simulation without corrections gave $M_{\mathrm{kin}} = 8.2(1)$ GeV, which disagrees with the upsilon mass determined using Eq. (7). When the relativistic corrections were included, the simulation gave $M_{\mathrm{kin}} = 9.6(1)$ GeV, which is in excellent agreement. All of the spin-independent corrections in the NRQCD lagrangian contribute to the shift in M_{kin}; once again we have striking evidence that cut-off theories work.

This simulation has only two parameters: the bare coupling constant and the bare quark mass. These were tuned to fit experimental data. From the bare parameters we can compute the renormalized coupling and mass. This simulation implies that the renormalized or "pole mass" of the b-quark is

$$M_b = 5.0(2) \text{ GeV}. \tag{9}$$

The renormalized coupling that is obtained corresponds to

$$\alpha_{\overline{\mathrm{MS}}}^{(5)}(M_Z) = 0.115(2), \tag{10}$$

which agrees with results from high-energy phenomenology and is about as accurate. This last result is striking: it shows that the QCD of hadronic structure and the QCD of high-energy quark and gluon jets are really the same theory.

Conclusions

I have described how cut-off field theories can be used as a powerful computational tool, allowing us to isolate and solve separately the short-distance and long-distance parts of a field theory. I have outlined how one adapts these techniques to take advantage of nonrelativistic expansions in perturbative and nonperturbative analysis. These ideas and techniques are broadly applicable in atomic, nuclear and particle physics. In particular they have allowed us to analyze the structure of Υ mesons from first principles — something regarded as completely impossible before these methods were developed. The results are precise and agree well with experiment.

The lattice analysis is particularly significant because it demonstrates that accurate results are possible even on a relatively coarse grid — our grid spacing a was about half the radius of an Υ. This suggests that light hadrons, which are much larger, can be analyzed on still coarser grids, perhaps with lattice spacings as large as 0.4 fm. Such lattices are much coarser than what is commonly used in simulations today, and consequently require much less computing power (by factors of 10^3 or 10^6). If coarse lattices really do work for light hadrons, the shift to small lattices and improved lagrangians will have a revolutionary impact on numerical QCD.

References

[1] For a review of the early history of these developments by one the key figures in that history see K.G. Wislon, Rev. Mod. Phys. **55**, 583(1983). A pedagogical review of the key concepts is given in G.P. Lepage, "What is Renormalization?" in *From Actions to Answers*, edited by T. DeGrand and D. Toussaint (World Scientific, Singapore, 1989).

[2] Sections 3 and 4 in the present paper are adapted from T. Kinoshita and G.P. Lepage, "Quantum Electrodynamics for Nonrelativistic Systems and High Precision Determinations of α" in *Quantum Electrodynamics*, edited by T. Kinoshita (World Scientific, Singapore, 1990).

[3] Section 5 is adapted from G.P. Lepage, "Lattice QCD for Small Computers" to be published in the proceedings of TASI 93, edited by S. Raby (World Scientific, Singapore,1994).

[4] M. H. L. Pryce, Proc. Roy. Soc. **195**, 62 (1948); S. Tani, Soryushiron Kenkyu, **1**, 15 (1949) (in Japanese); Prog. Theor. Phys **6**, 267 (1951); L. L. Foldy and S. A. Wouthuysen, Phys. Rev. **78**, 29 (1950).

[5] W.E. Caswell and G.P. Lepage, Phys. Lett. **167B**, 437 (1986). P. Labelle, G.P. Lepage and U. Magnea, Phys. Rev. Lett. **72**, 2006(1994).

[6] G.P. Lepage and B.A. Thacker, Nucl. Phys. **B** (Proc. Suppl.) 4 199, (1988); B.A. Thacker and G.P. Lepage, Phys. Rev. D **43**, 196 (1991); G.P. Lepage, L. Magnea, C. Nakhleh, U. Magnea and K. Hornbostel, Phys. Rev. D **46**, 4052 (1992).

[7] C. Davies, K. Hornbostel, A. Langnau, P. Lepage, A. Lidsey, C. Morningstar, J. Shigemitsu, J. Sloan, "A New Determination of M_b Using Lattice QCD" submitted to Physical Review Letters (1994); "Precision Υ Spectroscopy from Nonrelativistic Lattice QCD" submitted to Physical Review D (1994); and "A Precise Determination of α_s Using Lattice QCD" submitted to Physical Review Letters (1994).

New Tests of Special Relativity and QED by Laser Spectroscopy in Heavy Ion Storage Rings

T. Kühl[1], St. Becker[1], S. Borneis[1], T. Engel[2], B. Fricke[3],
M. Grieser[4], R. Grieser[5], D. Habs[4], G. Huber[5], I. Klaft[5],
D. Marx[1], P. Merz[5], R. Neumann[1], D. Schwalm[4], P. Seelig[1]

[1] Gesellschaft für Schwerionenforschung GSI,
D-64220 Darmstadt, Germany
[2] Institut für Angewandte Physik, TH Darmstadt,
D-64289 Darmstadt, Germany
[3] Institut für Physik, Gesamthochschule Kassel,
D-34132 Kassel, Germany
[4] Max-Planck-Institut für Kernphysik
D-69029 Heidelberg Germany
[5] Institut für Physik, Universität Mainz,
D-55099 Mainz, Germany

Abstract. The advent of heavy-ion cooler rings has created a completely new scope for the application of laser spectroscopy on accelerator beams. The unique potential opened for laser spectroscopic applications by the exotic properties of the stored ions can be exploited to address fundamental problems of physics.
The high, but very well controlled velocity of the ions has been used as a test ground for the theory of special relativity. The availability of highly-charged ions beyond all previous possibilities has been exploited for precision laser spectroscopy of the ground state hyperfine splitting of heavy hydrogen-like ions. This represents a novel opportunity to study QED corrections in the previously unexplored combination of strong magnetic and electric fields.

Introduction

The use of exotic atoms, i.e. atoms deviating strongly from the varieties found in nature, has, in many cases, allowed the choice of an experimental situation which is interpreted more easily than its standard counterpart. Often the production of such species requires the use of particle accelerators. Typical examples are muonic and radioactive atoms, but also highly charged ions and ions moving at a relativistic speed, as discussed here. The main attraction to performing precision laser spectroscopy in a storage ring is the availability of candidates which cannot be found otherwise. One of the unique properties that can be provided in a storage ring is a beam of particles of relativistic speed but with superb beam quality. These ions can be used as moving clocks with optical frequencies, moving in a well-controlled trajectory, thus providing a test of special relativity. For the first time the requirements for high resolution laser spectroscopy are fullfilled for a test candidate at a speed large compared to the motion of the earth. The extension to higher charged ions was finally successfully demonstrated by performing precision laser spectroscopy on hydrogen-like bismuth ions.

Laser Experiments at Ion Cooler Rings

Limitations in beam quality and interaction time restricted the application of lasers at accelerator beams in the past to exotic single events. The potential offered by an improvement of the experimental conditions was realized very early in the planning of the modern heavy-ion storage rings.

The key experimental feature of heavy-ion cooler rings in comparison to standard accelerators besides the increase in interaction time is the dramatic improvement of the beam quality achieved by electron cooling. This technique was demonstrated for the first time in 1969 at the proton storage ring in Novosibirsk [1] and was then applied to a number of other proton storage rings. The history of the young field of laser spectroscopy at heavy-ion storage rings exhibits a rapid growth which is still going on. Following the inauguration of the TSR in Heidelberg at Christmas in 1987, first results of spectroscopy on lithium ions were already presented at the Workshop of Crystalline Ion Beams in fall 1988 in Wertheim [2]. In the following year, first successes in laser cooling were achieved, both at the TSR and at the ASTRID storage ring in Aarhus.

The transfer of the new experimental techniques onto heavier and more energetic ions at the ESR in Darmstadt proved to be more complicated. Here, at ion energies between 100 and 500 MeV/nucleon, fully stripped ions up to U^{92+} can be produced and stored.

Test of Special Relativity

The concept of space-time in the special theory of relativity is related to the symmetries of the Maxwell equations. The theory had to provide a correct description of electro-magnetic processes in differently moving frames of reference, once the idea of an ether was ruled out by the Michelson-Morley experiment. The transformation between the inertial frames is described by the Lorentz-transformation. According to this, the Doppler-shifted frequency of a fast moving clock is given by

$$\nu = \nu_{1,2}(\gamma(1 + \beta\cos\theta))^{-1}$$

where $\nu_{1,2}$ is the clock frequency at rest, β is the relative velocity, $\beta = v/c$, and θ the angle measured in the laboratory frame. Within the framework of special relativity $\gamma = (1 - \beta^2)^{-1/2}$ is valid. Using a well known transition in a fast moving ion, any deviation given as $\gamma = \gamma_{SRT}(1 + \delta\alpha\beta^2 + \delta\alpha_2\beta^4 + ...)$ can be tested to second order by an exact measurement of $\nu_{1,2}$ and the resonance frequencies observed in parallel and antiparallel geometry ν_a and ν_p by

$$\nu_a = \nu_1 \cdot \nu_2/\nu_p \cdot (1 + 2\delta\alpha(\beta^2 + 2\beta\beta' \cos\Omega))$$

The term β^2 is sensitive on absolute deviations, the term $\beta\beta' \cos(\Omega)$ on sidereal modulations of deviations from special relativity. β' denotes the relative velocity of the earth's reference frame relative to a universal reference frame, usually derived from the anisotropy of the 3K background radiation. A first test of the quadratic Doppler shift was published by Ives and Stilwell in 1938 [3]. In the last several years, a number of high resolution laser experiments have been reported and analyzed for limits on $\delta\alpha$. Three of them [4–6] have been performed as two-photon spectroscopy on a "fast" neon atomic beam at $\beta = 0.0036$. As the best result, an absolute limit for $\delta\alpha \leq 1.6 \cdot 10^{-6}$ [6] and a limit on sidereal modulation $\delta\alpha \leq 1.4 \cdot 10^{-6}$ [5] was reported. Since the anisotropy of the cosmic background radiation implies a velocity of the earth's frame of reference of 350 km/s, experiments at higher velocities like the experiment at LAMPF [7], performed on an accelerated beam of H$^-$ at $\beta = 0.84$, are favourable. With the spectral resolution of $2.7 \cdot 10^{-4}$ achieved, the sensitivity on α_2 gives an upper limit of $1.3 \cdot 10^{-4}$ for this parameter due to the large Doppler shift.

Experimental Set-Up at the TSR

The experiments at the TSR storage ring were performed on singly charged $^7Li^+$ ions accelerated to 13.3 MeV, corresponding to a velocity of $\beta = 0.064$. About 10 - 20 % of these ions are formed in the metastable ^3S-state in the stripper foil of the TANDEM accelerator. The current in the storage ring is increased by multiturn injection. For the experiments 10^6 $^7Li^+$ ions were stored in the ring. The triplet system of helium-like $^7Li^+$ shows a well resolved and precisely known fine and hyperfine structure multiplet. For the stringent test of relativity high-resolution saturation spectroscopy was performed in the Λ system given by the transitions $^3S_1$ F=3/2 $\to$ $^3P_2$ F'=5/2 and $^3S_1$ F=5/2 $\to$ $^3P_2$ F'=5/2. The experimental set-up shown in Fig. 1 is described in detail in [8].

A single-mode Ar$^+$ laser stabilized on a $^{127}I_2$ saturation resonance which is a recommended frequency standard [10] and a tunable dye laser modified

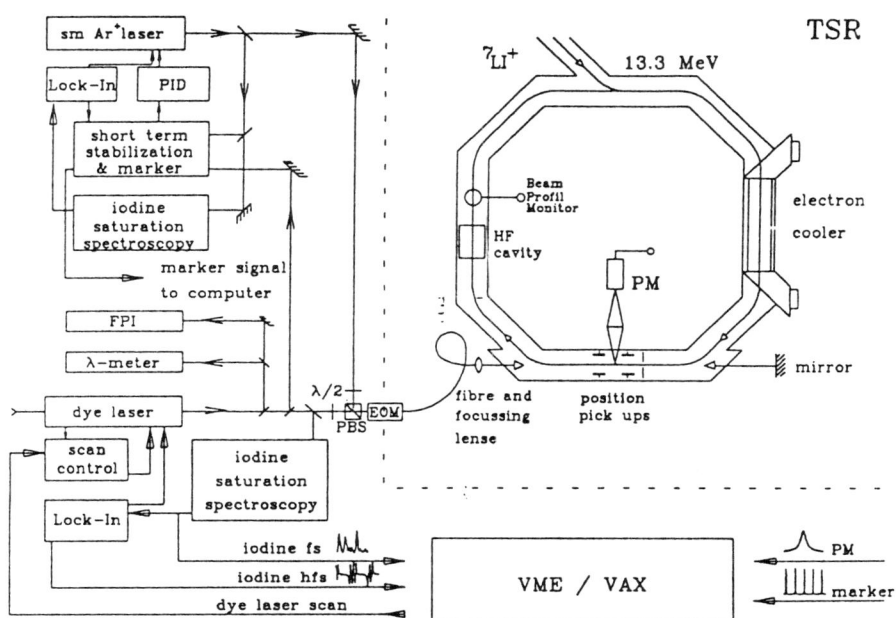

Figure 1: Experimental set-up of the relativity experiment at the TSR

towards a better frequency stability were used. The frequency of the scanning laser was determined by simultaneous recording of $^{127}I_2$ HFS resonances of the R(99) 15-1 transition. The frequency of the i-component of this transition has been calibrated with high accuracy at the Michelson interferometer at the PTB-Braunschweig [9]. In order to assure a well-defined collinearity of the two laser beams, both beams were sent through the same polarization preserving single-mode fiber over a distance of 20 m before entering the experimental section of the TSR. This bichromatic beam was focused on a plane mirror behind the experimental section at 15 m distance. An angular accuracy of $\Delta\phi \leq \pm 40$ µrad was achieved by adjusting the retro-reflected laser beams for maximum transmission back into the fibre. The optical resonances were detected in fluorescence by two photomultipliers, equipped with filters to suppress stray light. The alignment of the laser beams and the ion beam was controlled by four position pick-ups mounted at a distance of 2.4 m in the experimental section of the TSR [11]. The alignment accuracy of the pick-ups corresponds to an angular uncertainty $\delta\theta \leq \pm 370$ µrad.

To perform laser spectroscopy, the injected ion beam was first electron-cooled for 7 sec. During this time, the laser beams were blocked mechanically in order to avoid optical pumping during the longitudinal velocity change associated with the electron cooling. To produce the spectrum displayed in trace

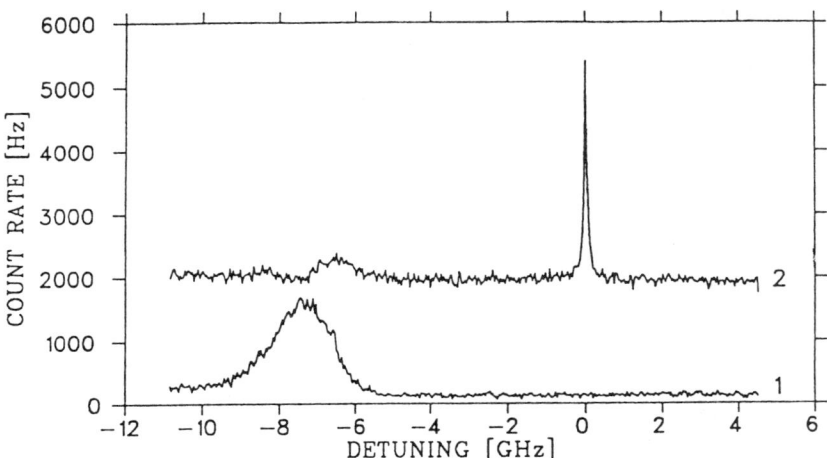

Figure 2: Laser-induced fluorescence spectra of electron-cooled ^{7}Li$^+$ at the TSR. The lower trace shows the Doppler broadened line of the F =5/2 → F'= 7/2 transition excited by the dye laser alone. The upper trace is recorded while additionally the fixed-frequency the Ar$^+$ laser is exciting this same transition, producing a narrow Λ-resonance.

1 of Fig. 2, only the dye laser is sent to the experiment and the Doppler-broadened fluorescence of the F=5/2 → F'=7/2 transition is visible. The FWHM corresponds to $\delta v/v = 3 \cdot 10^{-5}$. To obtain the saturation spectroscopy signal, the Ar$^+$ laser is kept at resonance with the F=5/2 → F'=5/2 transition. Optical pumping depletes the F=5/2 level as probed in the dye laser scan. Instead of the F=5/2 → F'=7/2 transition, the resonance F=3/2 → F'=5/2 now yields a strong fluorescence from the simultaneous excitation of the Λ system. Figure 3 shows the final high-resolution spectrum of this Λ resonance. The frequency scale is relative to the position of the R(99) 15-1 i-component of the iodine spectrum serving as a frequency reference for the antiparallel dye laser beam.

Results

Once a high-resolution scheme as represented by the spectroscopy of the Λ resonance is established the main concern is to exclude sources of systematical uncertainties. Some of these can be traced by a detailed change of the experimental parameters. The effect of AC Stark shift of the Λ resonance is investigated by measurements at different ion velocities, where the fixed frequency Ar$^+$ laser is not resonant at the center of the Doppler profile of the velocity distribution. An influence smaller than $\Delta \nu \leq 1$ MHz was deduced.

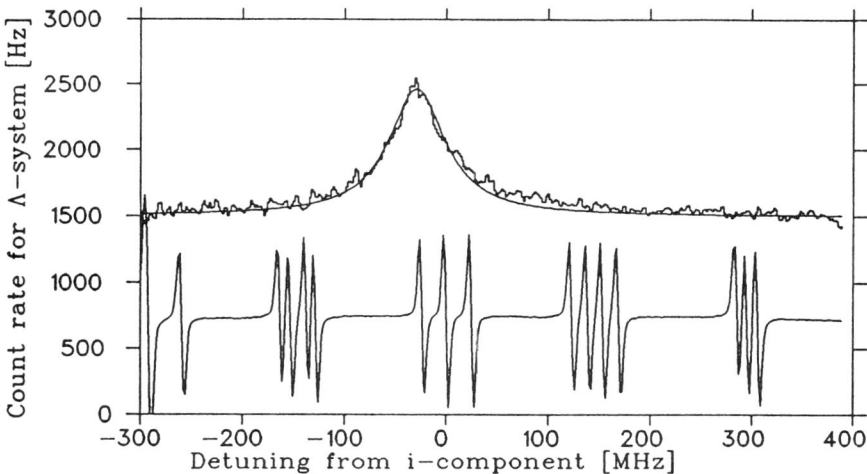

Figure 3: Laser-induced fluorescence spectrum of the Λ-resonance. The lower part shows the simultaneously recorded saturated absorption spectrum of $^{127}J_2$. The frequency scale is relative to the R(99) 15-1 i-component.

The shift due to photon recoil can be experimentally established by changing the role of the two lasers. A maximal shift of $\Delta\nu \leq 0.7$ MHz has been observed. The linewidth of the resonance of $\Gamma = 60$ MHz is twice the expected linewidth. By changing the ion beam intensity an influence of the properties of the storage ring on this broadening could be excluded. An explanation of the linewidth is given by irregularities in the phase front of the laser beams. Assuming that the additional linewidth is caused by such wavefront distortions a possible systematic shift of $\Delta\nu \leq 2.7$ MHz has to be accepted, which constitutes the largest contribution to the experimental uncertainties, including the statistical quality of the data. The measured frequency is finally quoted as $\nu_{exp} = 512667592.4(3.1)$ MHz. The prediction of special relativity can be calculated by taking the product of the frequencies of the F=3/2 → F'=5/2 and the F=5/2 → F'=5/2 transition in $^7Li^+$ at rest [12,13] divided by the Doppler-shifted value given by the frequency of the Ar^+ laser. This leads to $\nu_{SRT} = 512667588.2(7)$. The frequency difference between these values of $\Delta\nu = 4.2(3.2)$ MHz is compatible with zero and confirms the validity of special relativity. For the deviation from the expectation of special relativity a value of $\delta\alpha \leq 8 \cdot 10^{-7}$ is deduced [8].

Considering a precise velocity control by simultaneous laser cooling of the $^7Li^+$ beam, and an increased quality of the laser wavefronts, the final accuracy should be determined by the angular setting of laser and ion beam which has been quoted as $\Delta\nu \leq 0.29$ MHz. Compared to the quoted uncertainty in the present experiment of 3.2 MHz the ultimate limit caused by imperfections of the laser beam, beam alignment and signal distortions by time-of-flight effects in the interaction region should allow an increase in precision by at least one order of magnitude. A further improvement seems possible by an increase in the ion velocity, for instance to $\beta = 0.4$ as in reach at the ESR. Due to the much higher velocity, an increase in sensitivity of 10^2 can be hoped for.

Laser Fluorescence Spectroscopy of the Ground State Hyperfine Splitting

The measurement of the hyperfine splitting (HFS) of hydrogen-like $^{209}\text{Bi}^{82+}$ ions for the first time provides a test of QED corrections to the magnetic dipole interaction between the K-shell electrons and the nucleus in the strong field of the highly charged ion [14]. Since the HFS splitting is proportional to Z^3, the wavelength of the M1-transition is in the optical regime for high Z. For a number of nuclei, including bismuth, where data on the nuclear parameters are available from electron scattering experiments and measurements on muonic atoms, the transition energy without QED corrections can be evaluated with good precision in the frame of Dirac's theory [15]. For $^{209}\text{Bi}^{82+}$ with $Z = 83$, nuclear spin $I = 9/2$ and a magnetic nuclear dipole moment of $\mu_I = 4.1\ \mu_N$, the calculated splitting is approximately 243 nm [16,17], in contrast to 21 cm for the same transition in atomic hydrogen. The calculated lifetime of the upper level is about 0.4 ms [15].

The Experiment

The experiment [19] (Fig. 4) was performed at the ESR in Darmstadt [20]. Up to $1.8 \cdot 10^8$ $^{209}\text{Bi}^{82+}$ ions (4mA) were stored in the ESR with 219 MeV/u ($\beta = 0.59$, $\Delta p/p = 5 \cdot 10^{-5}$). In the experimental section the ion beam was overlapped antiparallel with high-power light pulses (energy 10 mJ at the site of the ions, duration 30 ns FWHM, repetition rate 200 Hz) from an excimer laser pumped tunable dye laser. The laser wavelength in the laboratory frame was ~ 480 nm. The position of laser beam and ion beam in the interaction region could be determined with movable scraper blades in the ESR at two points (distance ~ 6m in the experimental section) in horizontal and vertical direction. The laser beam position was controlled by several removable position monitors.

Due to the long lifetime of the upper HFS level, the decay takes place along the entire beam path of the ESR (108 m). Fluorescence was collected by a 60 cm long elliptical mirror system and detected by three photomultipliers. The mirrror covered a fraction of $5 \cdot 10^{-4}$ of all decay events over the total length. Since the detector system accepts large angles, the Doppler effect leads to a spectral width ranging from about 200 nm to 450 nm. The detection by the

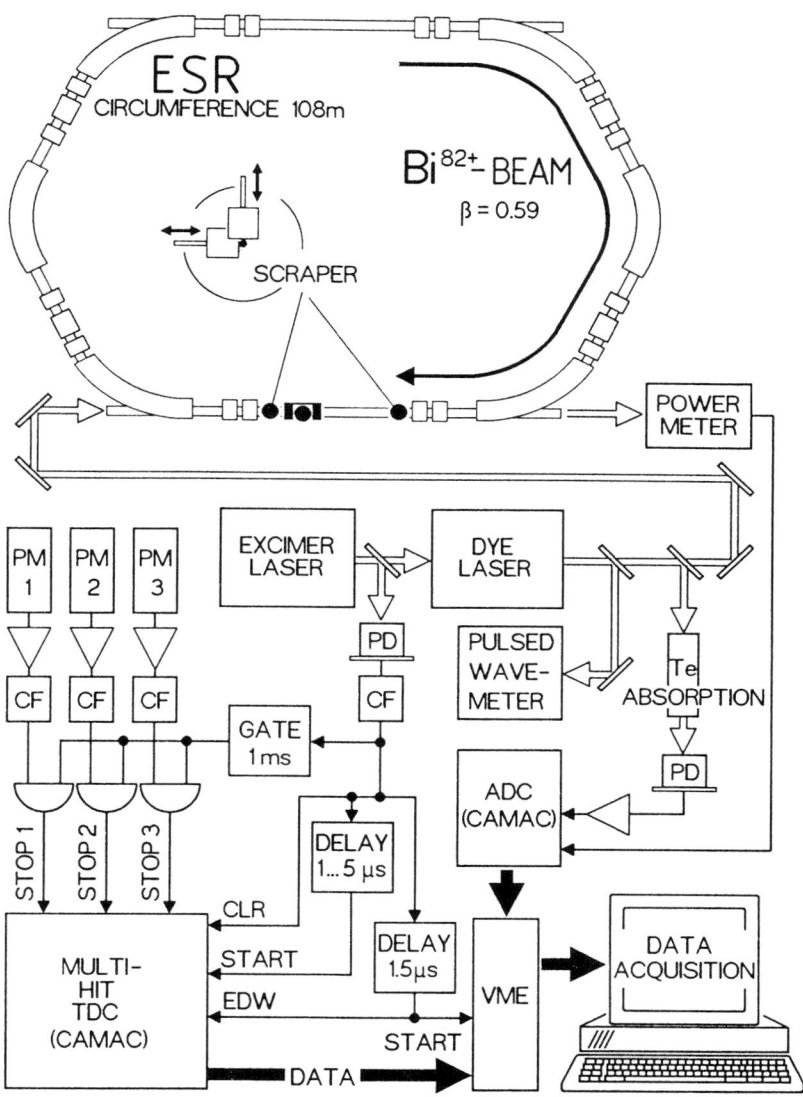

Figure 4: Experimental arrangement at the ESR. a)The pulses of an excimer pumped dye laser are overlapped counterpropagating with the Bi^{82+} ion beam circulating in the ESR. The overlap is adjusted by beam scrapers and observed by laser beam monitors. The fluosrescence signal is detected by 3 photomultipliers. The time delay of every photon event against the laser pulse is recorded via a multi-hit TDC for on-line analysis. b) Elliptical mirror system for efficient fluorescence detection. Fluorescence light is focused on the cooled photomultiplier (PM) cathode.

photomultipliers was limited on the low-wavelength side by the transmission cut-off of the sapphire vacuum window and on the other one by color glass filters, necessary to suppress scattered laser light ($\lambda = 480$ nm).

The time structure of the expected signal requires time resolving data acquisition. The pulsed laser beam (30 ns) overlaps the circulating DC ion beam only over approximately 10 m out of a total length of 108 m. With the revolution frequency of 1.626 MHz the signal structure results in a ~ 100 ns long excited beam section, passing the detectors every 615 ns. The dark count rate of the multiplier can be reduced by a factor of six by time gates synchronized with the circulating laser excited ions.
This was realized via a multi-hit TDC with a resolution of 1ns and a total range of 8 ms. The light pulse of the excimer laser detected by a photodiode was used as a trigger for the measurement window. Every single photon event whithin the gate of 1 ms was recorded as an individual stop of the TDC and read out via CAMAC/VME. This also allowed a precise determination of the lifetime of the excited state.

Results and Conclusions

The resonance in the M1-transition of hydrogen-like ^{209}Bi^{82+} is shown in Fig. 5a. The transition wavelength was measured to be $\lambda = 243.87(4)$ nm, which is close to the value calculated for the extended nuclear charge distribution and taking the distribution of the nuclear magnetization (Bohr-Weisskopf effect) into account, but not correcting for QED effects. In addition the lifetime of the excited state (Fig. 5b) could be determined by keeping the laser in resonance for 90 minutes. The exponential fit yields $\tau = 0.351(16)$ ns, about 15 % shorter than calculated.

A complete theoretical evaluation of the hyperfine splitting is not presently available. Recent calculations [15,16] take into account the distributions of the extended nuclear charge and nuclear magnetization (Bohr-Weisskopf effect), but do not include contributions from QED. A transition wavelength between 242.3 nm and 244.8 nm is obtained, depending mostly on the particular treatment of the Bohr-Weisskopf shift, which has a total magnitude of about 6 nm.

Up to now, only the vacuum polarization (VP) term of the QED corrections has been calculated, resulting in a shift of -1.6 nm [16,17]. Calculations of the self-energy (SE) contributions in a combination of strong magnetic and electric fields are under way. Together with the above VP correction and the splitting energy calculated without QED, the experimental transition wavelength of

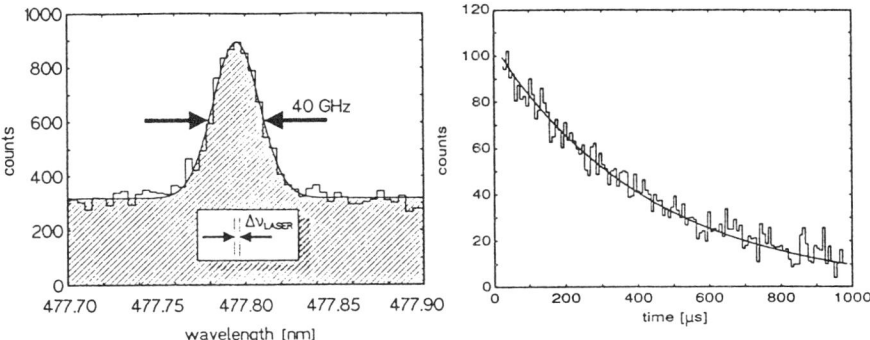

Figure 5: a) Laser-induced Fluorescence spectrum of the (F = 5 → F = 4) - transition of the 1s-state of ^{209}Bi^{82+}. The Doppler width of the line is determined by the ion velocity. The uncertainty of the ion velocity results in a wavelength uncertainty of 0,04 nm indicated by the error bar below the signal. The spectrum was measured with 600 laser pulses per channel. b) Decay of the (F = 5)- level in the ground state of ^{209}Bi^{82+}. The fluorescence is recorded as a function of the time delay after the laser pulse.

243.87(4) nm can be used to extract a range of values for the relative size of the SE and VP contributions. Ratios between SE/VP = -0.4 and -2 are obtained, differing substantially from SE/VP ≈ - 4 found in the computations of the Lamb shift in heavy hydrogen-like ions [18].

In order to disentangle the contributions due to the nuclear size, the QED corrections and the Bohr-Weisskopf effect, the hyperfine structure splitting of the ground states of other hydrogen-like systems should be investigated. Figure 6 shows the size of the hyperfine splitting in the rest frame of a number of candidates for optical spectroscopy at the ESR. Varying the proton number by investigations of several elements and varying the distribution of nuclear charge and magnetism by measurements on different isotopes of the same element, one can hope to get a key to a quantitative understanding of the hyperfine splitting of hydrogen-like heavy ions.

A straightforward application will be ^{207}Pb^{81+}, where the main single-particle contribution is the unpaired neutron rather than the proton in ^{208}Bi^{82+}. The different transition lifetimes of the upper hfs states of lead and bismuth will also address the problem of deexcitation in the ESR. Furthermore, measurements on lead can eliminate most of the uncertainty caused by the ion velocity. The Doppler-shifted laser wavelengths remain in accessible spectral regions for both parallel and antiparallel excitation, allowing a determination of the Doppler shift.

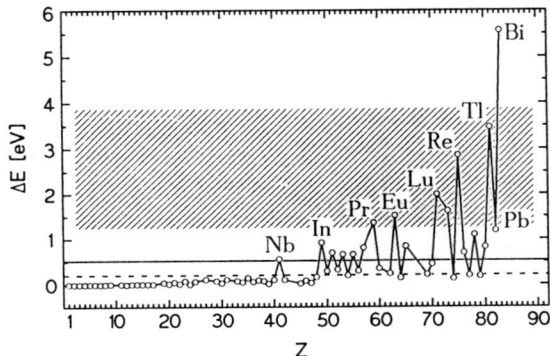

Figure 6: Candidates for laser spectroscopy of the ground state hyperfine structure of hydrogen-like ions. The transition wavelenght λ between the two levels in the rest frame as a function of the nuclear charge.

Outlook

The combination of uncompromising production schemes, high intensities and high beam quality achieved in heavy ion cooler rings has created a completely new field for laser experiments. Besides the high charge state and large speed of the ions discussed here, it has been demonstrated that high-lying hydrogenic Rydberg states can be selectively produced and studied [21]. At the high energy facilities, radioactive species can be produced and stored. At the low energy storage rings, novel experiments with molecules are started. The application to fundamental tests is only at it's beginning, and is likely to produce more stringent results in the near future.

References

[1] G. I. Budker, A. N. Skrinsky, Us. Fiz. Nauk 124 (1978) 561; Sov. Phys. 21 (1978) 277

[2] Proceedings of the Workshop on Crystalline Ion Beams Wertheim, Germany, October 4 - 7, 1988; GSI-Report 89-10 (1989)

[3] H. E. Ives, G. R. Stilwell, J. Opt. Soc. Am. 28 (1938) 215

[4] M. Kaivaola, O. Poulsen, E. Riis, S. A. Lee, Phys. Rev. 54 (1985) 255

[5] E. Riis, L. U. A. Andersen, N. Bjerre, O. Poulsen, S. A. Lee, J. L. Hall, Phys. Rev. Lett. 60 (1988) 81

[6] R. W. McGowan. D. M. Giltner, S. J. Sternberg, S. A. Lee, Phys. Rev. Lett. 70 (1993) 251

[7] D.W. MacArthur, K.B.Butterfield, D.A.Clark, J.B.Donahue, P.A.M.Gram, H.C.Bryant, C.J.Harvey, W.W.Smith, G.Comtet, Phys. Rev. Lett. 56 (1986) 282

[8] R.Grieser, R.Klein, G.Huber, S.Dickopf, I.Klaft, P.Knobloch, P.Merz, F.Albrecht, M.Grieser, D.Habs, D.Schwalm, T.Kühl, Appl. Phys. B 59 (1994) 127

[9] R.Grieser, S.Dickopf, G.Huber, R.Klein, P.Merz, G.Bönsch, A.Niclolaus, H.Schnatz, Z. Phys. A 348 (1994) 147

[10] Documents concerning the New Definition of the Metre, Metrologia 19:163, 1984 Comite International des Poids et Mesures (CIPM), 81^e session, 1992,

[11] F. Albrecht, Diplomarbeit, Heidelberg 1993

[12] J.Kowalski, R.Neumann, S.Noethe, K.Scheffzek, H.S uhr, G. zu Putlitz, Hyperf. Interact. 15/16:159, 1983

[13] E. Riis et al., Phys.Rev. A49: 207, 1994

[14] T. Kühl, Physica Scripta T22, 144 (1988)

[15] M. Finkbeiner, B. Fricke ,T. Kühl, Phys. Lett. 176 A, 113 (1993)

[16] S.M. Schneider, J. Schaffner, W. Greiner, G. Soff, J. Phys. B: At. Mol. Opt. Phys. 26,L581 (1993)

[17] S.M. Schneider, W. Greiner, G. Soff, Phys. Rev. A, in print

[18] W. Johnson, G. Soff, At. Data Nucl. Data Tables 42, 189 (1985)

[19] I. Klaft et al., submitted to Phys. Rev. Lett. (1993)

[20] B. Franzke, Nucl. Instr. Meth. Phys. Res. B24, 18 (1987)

[21] S. Borneis et al., Phys. Rev. Lett. 72, 207(1994)

ATOMIC STRUCTURE AND HIGH RESOLUTION SPECTROSCOPY

Advances in the Theory of Atomic Structure

J. Sapirstein

Department of Physics, University of Notre Dame, Notre Dame, IN 46556

Abstract. Atomic structure is discussed in the context of expansion parameters. For simple atoms, an expansion in the fine structure constant allows the most precise tests of Quantum Electrodynamics. For more complicated atoms such an expansion is not generally available. In the special case of highly charged ions, however, an expansion in $1/Z$ again allows precision tests of QED. It is argued that neutral systems require a different kind of expansion, and applications to the properties of cesium are discussed. Throughout, the role of nuclear structure as a limit of precision QED tests is stressed.

I. INTRODUCTION

One of the distinguishing characteristics of atomic physics is precision, both experimentally and, in certain cases, because of the well understood nature of electromagnetic interactions, also theoretically. It is of interest to question the value of such precision. To illustrate, consider the extremely accurate measurement of ground state hyperfine splitting in hydrogen [1]:

$$\Delta\nu_{\text{exp}} = 1420.405\ 751\ 766\ 7(9)\text{MHz}. \tag{1}$$

This should be contrasted with the present theoretical result [2]

$$\Delta\nu_{\text{th}} = 1420.405\ 1(8)\text{MHz}. \tag{2}$$

The theoretical error here is dominated by strong interaction uncertainties. The principal source of uncertainty comes from the nonrelativistic size correction,

$$\delta_p(\text{Zemach}) = \frac{2\alpha m_e}{\pi^2} \int \frac{d^3 p}{p^4} [\frac{G_E(-p^2)G_M(-p^2)}{1+\kappa} - 1]. \tag{3}$$

where G_E and G_M are elastic form factors of the proton and κ its anomalous magnetic moment. At the present stage of QCD, these form factors cannot be calculated from first principles, and are available only from experiment. They are actually known with sufficient accuracy so that the correction is known to 1.4%, but in order to reach the level of experimental error, this strong interaction effect would have to calculated to 16 ppb! Thus while it may be useful for metrological purposes to increase the experimental accuracy, no useful new information about the physics of hydrogen hyperfine splitting would result. Nevertheless, the value of the extensive theoretical and experimental efforts expended in reaching this level of precision is the following. Firstly, at the 1 ppm level, the anomalous magnetic moment of the electron is being tested to about four digits. Further, higher order QED corrections of order $\alpha^2 E_F$ and $\alpha m_e/m_p E_F$ are being tested at the few percent level. Thus if we had only this system to test QED, our understanding of radiative corrections would have been advanced to a fairly high level, although we will see below that muonium provides an even more refined test of these corrections. Finally, a strong interaction effect of about 30 ppm has been measured to the percent level. However, the uncertainty in the latter masks any new physics entering at under the ppm level: this could be radiative corrections of higher order, weak interaction effects, or perhaps even completely new, unanticipated physics. At this point the field of atomic physics merges into strong interaction physics. In fact, I would like to argue that this is the ultimate fate of atomic physics. We are going to get better and better theoretical control over calculations of atomic structure, but at some level of precision the fact that atoms contain strongly interacting nuclei will require advances in QCD to proceed.

In the above example, our ability to get to the 1 ppm level is largely dependent on the simplicity of the hydrogen atom. In this case, the atomic structure is completely understood, so that one can cleanly study radiative corrections and strong interaction effects. This is of course not always the case: in general one deals with many-electron atoms, and cannot solve the Schrödinger equation analytically. However, the basic interactions of many-electron atoms are still understood at a very fundamental level, so that if the difficulties of solving the Schrödinger equation can be overcome, the possibility of testing QED and perhaps detecting new physics in more complicated atoms exists. There are two situations that I wish to discuss here. The first is the recent progress in understanding of the spectroscopy of highly charged ions. Experiments of increasing accuracy using beam-foil techniques and EBITs [3] have made available a body of spectroscopic data that has been complemented by considerable theoretical progress in calculating structure and QED effects, which progress exploits the existence of a $1/Z$ expansion. The second is the observation of parity nonconserving (PNC) transitions in heavy neutral atoms. In this case interpretation of the experimental results provides a major challenge to atomic theory, because in this case there is no $1/Z$ expansion. It will be argued that a different kind of expansion, in terms of the number of electrons excited, can lead to high accuracy calculations. In both cases strong interaction uncertainties play a significant role.

The plan of this talk is as follows. In section II, an example of QED in which strong interaction uncertainties are very small and both theory and experiment are quite advanced, muonium hyperfine splitting, is discussed. In section III the QED of highly charged uranium ions is treated. Finally, in section IV we discuss neutral cesium. The status of the most sophisticated calculations is reviewed along with the implications for particle physics arising from the measurement of PNC in this atom.

II. MUONIUM HYPERFINE SPLITTING

One way of reducing strong interaction uncertainties is to consider exotic atoms in which the nucleus is either a positive muon or a positron. Note however that these uncertainties can never be entirely eliminated because they always enter vacuum polarization loops. However, such loops are highly suppressed in atoms because of the small mass scale of atoms compared to the strong interaction scale. Muonium hyperfine splitting provides one of the most refined tests of QED available. It has been measured [4] to be

$$\Delta\nu_{\rm exp} = 4\ 463\ 302.88(16){\rm kHz}. \tag{4}$$

This is to be compared with the Fermi splitting

$$E_F = \frac{16}{3}(Z\alpha)^2 c R_\infty \frac{m_e}{m_\mu}(\frac{m_r}{m_e})^3 \tag{5}$$

where m_r is the reduced mass and we follow the convention of setting the charge of the muon to be $Z|e|$. Using $\alpha^{-1} = 137.035\ 989\ 5(61)$, $m_\mu/m_e = 206.768\ 262(62)$, and $R_\infty = 10\ 973\ 731.568\ 30(31){\rm m}^{-1}$ this splitting is

$$E_F = 4\ 453\ 839.60(1.33)(0.40){\rm kHz}. \tag{6}$$

The errors quoted here are firstly from the uncertainty in the muon mass and secondly from the uncertainty in the fine structure constant. The expansion parameters for this system are the fine structure constant and the electron-muon mass ratio, and because of their smallness only a few orders of perturbation theory should account for the 9463 kHz discrepancy between experiment and the Fermi splitting. While the bulk of this (10 345 kHz) arises from the Schwinger correction, which augments the g factors of the electron and muon by a factor of $\alpha/2\pi$, 882 kHz remain to be accounted for by higher order corrections. QED corrections to muonium hfs are conventionally split into two parts. The first, called the non-recoil part, accounts for radiative corrections that would be present for an infinite mass muon. The second part is any effect proportional to the electron-muon mass ratio, with such effects

Figure 1: Ladder and crossed ladder contributions to muonium hfs

called recoil corrections. There has been recent progress in the evaluation of the former, which is given by

$$\Delta\nu_{\text{non-recoil}} = (1 + a_\mu)[1 + a_e + \frac{3}{2}(Z\alpha)^2 + \alpha(Z\alpha)(\ln 2 - \frac{5}{2})$$
$$- \frac{8\alpha(Z\alpha)^2}{3\pi}\ln(Z\alpha)(\ln(Z\alpha) - \ln 4 + \frac{281}{480})$$
$$+ \frac{\alpha(Z\alpha)^2}{\pi}(15.38\,(0.29)) + \frac{\alpha^2(Z\alpha)}{\pi}D_1]E_F. \tag{7}$$

That progress is the complete evaluation of D_1, which is a binding correction to the two-loop g-2 contribution, by Kinoshita and Nio [5], who found $D_1 = 0.82(4)$. Subtracting $\delta\nu_{\text{non-recoil}}$ from experiment leaves 795 kHz to be accounted for by recoil. This is accounted for predominantly by the first term in the following expression for the recoil corrections:

$$\Delta\nu_{\text{recoil}} = [-\frac{3Z\alpha}{\pi}\frac{m_e m_\mu}{m_\mu^2 - m_e^2}\ln\frac{m_\mu}{m_e} + \frac{(Z\alpha)^2 m_r^2}{m_e m_\mu}(-2\ln(Z\alpha) - 8\ln 2 + \frac{65}{18})$$
$$+ \frac{\alpha(Z\alpha)}{\pi^2}\frac{m_e}{m_\mu}(-2\ln^2\frac{m_\mu}{m_e} + \frac{13}{12}\ln\frac{m_\mu}{m_e} + \frac{21}{2}\zeta(3) + \frac{\pi^2}{6} + \frac{35}{9}$$
$$+ 2.15(14) + \frac{\alpha}{\pi}(-\frac{4}{3}\ln^3\frac{m_\mu}{m_e} + \frac{4}{3}\ln^2\frac{m_\mu}{m_e}))]E_F. \tag{8}$$

That term, which arises from the graphs of Figure 1, gives -800.3 kHz of the total -794.9 kHz. The excellent agreement with experiment thus confirms this lower order term to very high accuracy. The reason we emphasize this point is that the graphs of Fig. 1 play an important role in many different QED tests. In particular similar graphs applied to Rydberg states of helium are associated with Casimir forces, which have received considerable attention [6]. The extra terms associated with retardation are automatically included in these graphs, but in this short distance effect play a much larger role. Alternatively, accepting the validity of the one-loop diagram, the higher order recoil terms, which amount to 5 kHz, are being tested at the 20 % level. Further progress in muonium hyperfine splitting as a test of QED will require a more accurate determination of the muon mass.

III. HIGHLY CHARGED IONS

There has been considerable interest in the study of the Lamb shift in high-Z hydrogenic ions. Part of the motivation for this interest is that one is testing QED in an extreme environment, in the sense that the energy scale of the highest-Z ions becomes a sizable fraction of the electron-positron production threshold. A second is that the calculation becomes highly nonperturbative at high Z. At small Z, the Lamb shift is usually expressed as a power series in $Z\alpha$, and calculations up to order $m\alpha(Z\alpha)^6$ have been carried out. However, at high Z this expansion breaks down and totally numeric methods are required, as will be discussed further below. However, experiments on hydrogenic ions are quite difficult to perform with high precision. A recent development that bears on these questions is the realization, despite the additional complexity of having more than one electron present, that experiments on many-electron highly charged ions can yield precision tests of high-Z QED. On the experimental front, we note first that notable progress has been made at the Berkeley HILAC in the determination of the spectrum of heliumlike [7] and lithiumlike [8] uranium. More recently a set of accurate measurements of $2s_{1/2} - 2p_{3/2}$ transitions in lithiumlike through neonlike uranium was carried out at the Livermore EBIT [9]. At the same time considerable progress has been made on lower Z heliumlike ions [10]. We choose the EBIT measurements to illustrate the present status of QED in many-electron ions.

Before QED can be studied in these ions, the Schrödinger equation must be solved to high accuracy. However, the demands on accuracy in high-Z ions are not as stringent as in neutral systems. This is because the energy scales as Z^2 and the Lamb shift as Z^4. At uranium this effect makes the Lamb shift enter at the percent level as opposed to the ppm level characteristic of hydrogen. The experimental accuracy of the EBIT measurements is on the order of 0.1 eV, and unless atomic structure calculations can reach this level, QED consequences can not be drawn. Now, while it is possible to start directly from a QED approach, it is useful to start from the Schrödinger equation. In the relativistic case, the Hamiltonian for an N-electron system is

$$H = \sum_{i=1}^{N}(\vec{\alpha}_i \cdot \vec{p}_i + \beta_i m - \frac{Z\alpha}{r_i}) + \frac{\alpha}{2}\sum_{ij}\frac{1}{|\vec{r}_i - \vec{r}_j|}. \qquad (9)$$

The interpretation of this Hamiltonian requires care because of the role of negative energy states. We illustrate this point in the context of many-body perturbation theory (MBPT). In this approach to the many-body problem the Hamiltonian is split up as $H = H_0 + V_C$, where

$$H_0 = \sum_{i=1}^{N}(\vec{\alpha}_i \cdot \vec{p}_i + \beta_i m - \frac{Z\alpha}{r_i} + U(r_i)) \qquad (10)$$

and
$$V_C = -\sum_{i=1}^{N} U(r_i) + \frac{\alpha}{2}\sum_{ij}\frac{1}{|\vec{r}_i - \vec{r}_j|}. \tag{11}$$

Here $U(r)$ is a central potential that is frequently chosen to be the Hartree-Fock potential. Because we want to subsequently put MBPT in the framework of QED, however, we will use instead local potentials similar to Hartree-Fock, as the nonlocality of the latter causes difficulties. The exact choice of $U(r)$ should not matter if one can go to sufficiently high order, and we will illustrate this point below. MBPT consists of Rayleigh-Schrödinger perturbation theory in V_C. This perturbation theory has been extensively applied to alkali isoelectronic sequences [11], and more recently to *particle-hole* states [12]. These latter are appropriate for closed shell ions in which one electron, the hole, is excited to a valence state, the particle. In lowest order they are described in second quantization as

$$|JM> = F_{av}a_a a_v^\dagger |0_C> \tag{12}$$

where F_{av} is a Clebsch-Gordon coefficient coupling the states to angular momentum JM and $|0_C>$ represents a closed shell. The energy of this state compared to the core is

$$E^{(0)} = \epsilon_v - \epsilon_a \tag{13}$$

and the first order correction is

$$E^{(1)} = \sum_c \tilde{g}_{cvcv} - \sum_c \tilde{g}_{acac} + F_{av}F_{bw}\tilde{g}_{awvb}. \tag{14}$$

The operator F_{bw} differs from F_{av} only in that the sum over the magnetic quantum numbers $m_a m_v$ is replaced by a sum over $m_b m_w$. We have introduced matrix elements of the Coulomb interaction,

$$g_{ijkl} \equiv \alpha \int \frac{d^3r d^3r'}{|\vec{r} - \vec{r}'|}\psi_i^\dagger(r)\psi_k(r)\psi_j^\dagger(r')\psi_l(r') \tag{15}$$

and the commonly occuring combination $\tilde{g}_{ijkl} \equiv g_{ijkl} - g_{ijlk}$. Finally, the second-order energy is given by the following set of terms:

$$E_A^{(2)} = F_{av}F_{bw}\delta_{ab}\sum_{mcd}\frac{g_{cdvm}\tilde{g}_{wmcd}}{\epsilon_m + \epsilon_v - \epsilon_c - \epsilon_d} \tag{16}$$

$$E_B^{(2)} = F_{av}F_{bw}\sum_{mn}\frac{g_{awmn}\tilde{g}_{mnbv}}{\epsilon_m + \epsilon_n - \epsilon_a - \epsilon_v} \tag{17}$$

$$E_C^{(2)} = -F_{av}F_{bw}\delta_{ab}\sum_{mnc}\frac{g_{wcmn}\tilde{g}_{mnvc}}{\epsilon_m + \epsilon_n - \epsilon_v - \epsilon_c} \tag{18}$$

$$E_D^{(2)} = -F_{av}F_{bw}\sum_{mc}\frac{\tilde{g}_{macv}\tilde{g}_{wcbm}}{\epsilon_a + \epsilon_m - \epsilon_c - \epsilon_v} \tag{19}$$

$$E_E^{(2)} = F_{av}F_{bw}\sum_{cd}\frac{g_{cdvb}\tilde{g}_{wacd}}{\epsilon_a + \epsilon_v - \epsilon_c - \epsilon_d} \tag{20}$$

$$E_F^{(2)} = \left[F_{av}F_{bw}\sum_{mc}\frac{\tilde{g}_{cwmv}\tilde{g}_{macb}}{\epsilon_m - \epsilon_c} + c.c.\right] \tag{21}$$

$$E_G^{(2)} = F_{av}F_{bw}\sum_{mc}\frac{\tilde{g}_{cavm}\tilde{g}_{wmbc}}{\epsilon_m + \epsilon_v - \epsilon_a - \epsilon_c} \tag{22}$$

$$E_H^{(2)} = F_{av}F_{bw}\delta_{vw}\sum_{mcd}\frac{g_{cdbm}\tilde{g}_{macd}}{\epsilon_m + \epsilon_a - \epsilon_c - \epsilon_d} \tag{23}$$

$$E_I^{(2)} = F_{av}F_{bw}\delta_{vw}\sum_{mnc}\frac{g_{camn}\tilde{g}_{mncb}}{\epsilon_m + \epsilon_n - \epsilon_a - \epsilon_c} \tag{24}$$

$$E_{\Delta 1}^{(2)} = F_{av}F_{bw}\left[\sum_i\frac{\Delta_{ai}\tilde{g}_{wivb}}{\epsilon_i - \epsilon_a} + \sum_i\frac{\Delta_{wi}\tilde{g}_{aibv}}{\epsilon_i - \epsilon_v} + c.c.\right] \tag{25}$$

$$E_{\Delta 2}^{(2)} = F_{av}F_{bw}\left\{\delta_{ab}\sum_i\frac{\Delta_{wi}\Delta_{iv}}{\epsilon_v - \epsilon_i} + \delta_{vw}\sum_i\frac{\Delta_{ai}\Delta_{ib}}{\epsilon_i - \epsilon_a}\right.$$
$$\left. + \left[\delta_{vw}\sum_{mc}\frac{\Delta_{mc}\tilde{g}_{camb}}{\epsilon_m - \epsilon_c} + \delta_{ab}\sum_{mc}\frac{\Delta_{mc}\tilde{g}_{cwmv}}{\epsilon_c - \epsilon_m} + c.c.\right]\right\}. \tag{26}$$

In the above sums over c,d range over occupied core states, m,n over excited states, and i over all states. The terms proportional to δ_{ab} reproduce the energy of a valence electron, appropriate for alkalis, and those proportional to δ_{vw} the energy of a hole, appropriate for halogens. Returning to the point that this Hamiltonian has to be carefully treated, consider any of the second order energies in which there is a sum over two excited states. If the sum is allowed to range only over positive energy states, the second order energies are well defined: this restriction is equivalent to using a *no-pair* Hamiltonian, as emphasized by Sucher [13]. However, without this restriction the denominators can vanish when one state has positive energy and the other negative. In order to resolve this problem a QED approach must be taken, as will be discussed below. When this is done these terms vanish, though terms involving two negative energy states do not, and contribute interesting QED effects. Before considering QED, we show how MBPT works.

At high Z, one is close to the hydrogenic approximation and it is straightforward to estimate the size of the above energy contributions. We do this by noting that the bound state energies are proportional to $m(Z\alpha)^2$, and the Coulomb matrix elements to $\alpha/r = mZ\alpha^2$. Then it is simple to show that, in atomic units (1 a.u. = $m\alpha^2$), $E^0 \sim Z^2$ a.u., $E^{(1)} \sim Z$ a.u., $E^{(2)} \sim 1$ a.u., and $E^{(3)} \sim 1/Z$ a.u.. This is an example of the well-known $1/Z$ expansion [14]. An important point to make at this point is that regardless of the order of MBPT, the overall order of an energy shift is always an atomic unit: thus for neutral systems, where the factor $1/Z$ need not be small, it is not at all clear how valuable MBPT will be. However, for highly charged ions, a calculation through second order will miss only terms of order $1/Z$ a.u., and can be expected to be quite accurate.

We now wish to concentrate on one of the transitions observed at the Livermore EBIT, the transition in carbonlike uranium. Because the $2p_{3/2}$ state is strongly split from $2p_{1/2}$, the ground state can be treated as a closed shell $(1s^2_{1/2}2s^2_{1/2}2p^2_{1/2})$. The measured transition of 4548.32±0.16 eV (167.148±0.006 a.u.) is between this ground state and a state in which a $2s_{1/2}$ electron is excited to the $2p_{3/2}$ state with total angular momentum 1. As discussed above, the exact nature of the starting potential is not in principle important. In a recent calculation [15] two potentials denoted as U_A and U_B were used, where

$$U_A(r) = 2v_0(1s,r) + 2v_0(2s,r) + 2v_0(2p_{1/2},r) \tag{27}$$

and

$$U_B(r) = 2v_0(1s,r) + 2v_0(2s,r) + v_0(2p_{1/2},r). \tag{28}$$

Here

$$v_0(a,r) = \frac{1}{r}\int_0^r (G_a^2(r') + F_a^2(r'))dr' + \int_r^\infty (G_a^2(r') + F_a^2(r'))\frac{dr'}{r'} \tag{29}$$

and G_a and F_a are the upper and lower radial Dirac wavefunctions determined self-consistently. At asymptotic distances, because $v_0(a,r) \to 1/r$, these potentials have an effective charge of $Z-6$ for U_A and $Z-5$ for U_B. We present in Table 1 the behavior of MBPT for the two potentials. We have however replaced $E^{(1)}$ with the quantity $F^{(1)}$ in which the Coulomb matrix element is replaced with

$$g_{ijkl} \equiv \alpha \int \frac{d^3r d^3r'}{|\vec{r}-\vec{r}'|}\bar{\psi}_i(r)\gamma_\mu\psi_k(r)\bar{\psi}_j(r')\gamma^\mu\psi_l(r')e^{ik|\vec{r}-\vec{r}'|} \tag{30}$$

and $k = |\epsilon_i - \epsilon_k| = |\epsilon_j - \epsilon_l|$. This is equivalent to working in Feynman gauge, and because when working with local potentials we have gauge invariance, this corresponds to a Coulomb gauge calculation, in which $E^{(1)}$ is included along with the Breit interaction including frequency dependence. In addition, smaller effects to do with one Breit interaction together with one Coulomb interaction ($B^{(2)}$) and the finite mass of the nucleus have been included. Note that while the lowest order energies differ by 10.2 eV, after all the corrections have been added in, the $1/Z$ expansion has led to extremely close agreement. Thus, because of the existence of a small expansion parameter, a relatively simple calculation allows the accurate solution of the Schrödinger equation. In this case, we see a relatively very large discrepancy with experiment of about one percent, attributable to radiative corrections.

Before we turn to their calculation, we return to the theme of strong interaction uncertainties. The finite size of the nucleus is built into the term $-Z\alpha/r$ term in the MBPT Hamiltonian, and changes the pointlike nucleus

Table 1: Structure and QED contributions to the $2s_{1/2}-2p_{3/2}$ energy difference in carbonlike uranium: units a.u.

Term	$2s_{1/2} - 2p_{3/2}(U_A)$	$2s_{1/2} - 2p_{3/2}(U_B)$
$E^{(0)}$	164.7607	164.3856
$F^{(1)}$	3.8121	4.1958
$E^{(2)}$	-0.0277	-0.0395
$B^{(2)}$	0.0096	0.0092
Reduced mass	-0.0019	-0.0019
Structure	168.5528	168.5492
self-energy	-1.9358	-1.9547
Uehling potential	0.5537	0.5595
Higher order	-0.0270	-0.0273
QED	-1.4091	-1.4224
Total	167.1437	167.1268

energy of the 2s state by 36 eV. Fortunately, very accurate muonic X-ray experiments [16] have been carried out, so that the associated uncertainty is about 0.01 eV. However, another strong interaction uncertainty comes from nuclear polarization, which is relatively large for the uranium nucleus. This has been calculated to be [17] to be 0.20 eV with an estimated uncertainty of 0.05 eV. Fortunately that uncertainty is substantially smaller than the QED corrections of interest.

We have seen in the above discussion that the $1/Z$ expansion allows the consideration of a relatively small number of MBPT diagrams to adequately describe carbonlike uranium. This same expansion applies to a QED treatment, though in this case the diagrams are Feynman diagrams. This QED treatment is known as the S-matrix approach. It is based on a generalization due to Sucher [18] of the Gell-Mann-Low formula,

$$\Delta E = \frac{i\epsilon}{2} \lim_{\epsilon \to 0} \lim_{\lambda \to 1} \frac{\partial}{\partial \lambda} \ln < S_{\epsilon,\lambda} > \quad (31)$$

where

$$S_{\epsilon,\lambda} = T(e^{-i\lambda \int dx_0 H(x_0) e^{-\epsilon|x_0|}}). \quad (32)$$

Now, it is useful in making a connection with the previous MBPT calculation to work with the same potential. This can be done by using the QED Hamiltonian $H = H_0 + H_I$:

$$H_0 = \int d^3x \psi^\dagger(x)[\vec{\alpha} \cdot \vec{p} + \beta m - \frac{Z\alpha}{|\vec{x}|} + U(|\vec{x}|)]\psi(x) \quad (33)$$

and

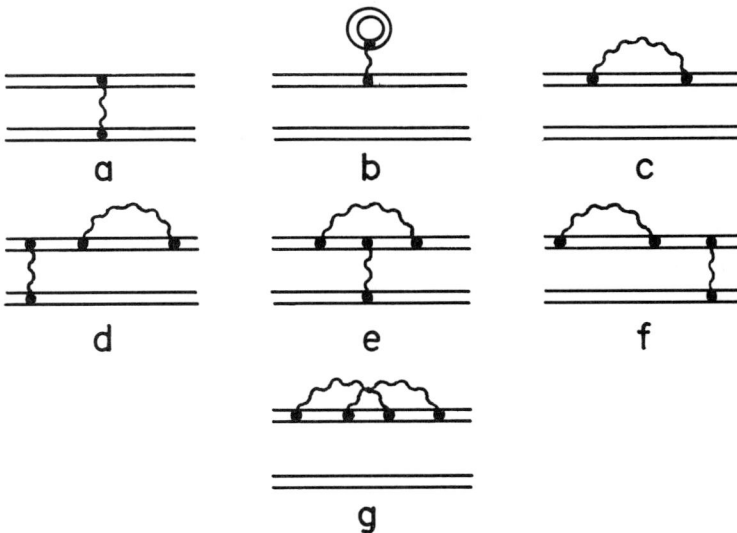

Figure 2: First and second order Feynman diagrams contributing to energy shifts

$$H_I = -\int d^3x \psi^\dagger(x) U(|\vec{x}|) \psi(x) - e \int d^3x \psi^\dagger(x) \vec{\alpha} \cdot \vec{A}(x) \psi(x)$$
$$+ \frac{\alpha}{2} \int \frac{d^3x d^3x'}{|\vec{x} - \vec{x}'|} \psi^\dagger(x) \psi(x) \psi^\dagger(x') \psi(x'). \quad (34)$$

At this point one can start discussing Feynman diagrams. When one photon is involved, it can either be exchanged between two different electrons (Fig. 2a), which can easily be shown to lead exactly to $F^{(1)}$, or else be in a radiative correction. Now, until recently a standard approach to evaluating the large QED corrections in highly charged ions was to do some kind of interpolation of the hydrogenic QED corrections calculated by Mohr [19]. However, in the S-matrix approach one is dealing with the electron self-energy and vacuum-polarization in realistic potentials, which incorporate automatically the screening of the nuclear potential by the other electrons. While the vacuum-polarization diagram, Fig. 2b, is relatively easy to evaluate, the numerical evaluation of the self-energy in non-Coulomb potentials, Fig. 2c, was until recently quite difficult. Over the last few years, however, these difficulties have been overcome [20], and it is now straightforward to evaluate both diagrams. We show the result for the potentials U_A and U_B in Table 1. While it is seen that they correctly account for the bulk of the QED effect, note that they differ by 0.36 eV. This is not surprising, since we have not carried out corrections to QED at the same level as done in MBPT. These corrections are described by the remaining diagrams of Fig. 2, which we now discuss.

We expect the bulk of the difference to come from Figs. 2d and 2f. These can be interpreted as modifying the wave function. They have been treated by Blundell [21] in his very complete treatment of alkalis. A particularly interesting effect comes from the ladder and crossed ladder diagrams of Fig. 1 introduced in the discussion of muonium hfs. This is because the second order energy of MBPT, which had to be treated in an ad hoc manner above, is contained within these diagrams. When the exchanged photons are Coulomb, these diagrams can be evaluated with contour methods. The undefined terms discussed above in which one electron is positive energy and the other negative can be shown to vanish, but when both are negative energy states there results a term of order $1/Z$ of the order of the hydrogenic Lamb shift, i.e. $Z^3\alpha^3$ a.u.. The full calculation has recently been completed for ground state helium [22]. In that calculation the MBPT results are subtracted from the full calculation so that QED can be added on in a clean manner to MBPT. However, extension of these calculations to more complicated systems has not yet been made. The vertex correction Fig. 2e was not treated by Blundell, but has been evaluated for lithiumlike uranium by Indelicato and Mohr [23]. The final diagram of interest is Fig. 2g. This is of order $(Z\alpha)^4$ a.u., and again has not yet been evaluated at high Z. While considerable effort will be required to complete the evaluation of these diagrams, the task is a well-defined one involving a finite number of Feynman diagrams. Thus, even when many electrons are present, this field is on the same footing as more traditional QED tests. It will particularly valuable for measurements of even higher accuracy to be carried out: such experiments will lead to an extension of QED tests to a very interesting region, in which one simultaneously is forcing the advancement of the theory of many-electron structure and higher order QED in intense Coulomb fields.

IV. NEUTRAL CESIUM

We turn finally to neutral cesium, which is of particular interest because of the possibility of indirectly detecting new particle physics effects through the measurement of PNC transitions. The demands of accuracy are less stringent here because of the relatively large size of radiative corrections in electroweak theories. In particular, there are radiative corrections associated with possible new physics that enter at the percent level. These have been parameterized by the quantities S and T [24], which parameterize the effect of new physics on the self-energy diagrams of W and Z bosons. They modify the fundamental quantity of atomic PNC, the weak charge Q_W [25], as follows:

$$Q_W = -73.20(13) - 0.8S - 0.005T. \tag{35}$$

The weak charge enters the parity-violating Hamiltonian

$$H_W = \frac{G_F}{\sqrt{8}} Q_W \rho_{\text{nuc}}(\mathbf{r}) \gamma_5 \qquad (36)$$

that dominates cesium PNC. The measurement of PNC then allows a determination of Q_W if the atomic structure calculations can be carried out with sufficient accuracy. As an example of the accuracy needed, we note that values of S of around 2 arise in certain models of the dynamical origin of electroweak symmetry breaking, which would enter Q_W at the 2 percent level. The challenge to atomic theory is then a prediction of the PNC amplitude resulting from H_W at under this level. Calculations at Notre Dame [26] and Novosibirsk [27] claim 1 % accuracy, and are in agreement at this level. Other calculations [28] with larger error estimates are all consistent with these results. Combining the prediction [26]

$$E_{\text{PNC}} = -0.905(9) \times 10^{-11} i |e| a_0 (-Q_W/N) \qquad (37)$$

with the experimental determination of E_{PNC} [29] leads to a determination of the weak charge

$$Q_W = -71.04(1.58)[0.88], \qquad (38)$$

where the first error is experimental and the second from atomic theory. While this is consistent with the theoretical prediction (though a large positive S is disfavored), it is clearly important to decrease both errors. Before we turn to a description of the calculational method used and the prospects for better theoretical accuracy, we note that again nuclear physics provides a barrier. The quantity $\rho_{\text{nuc}}(\mathbf{r})$ in the definition of H_W depends dominantly on the distribution of neutrons in the cesium nucleus. While this distribution is not directly available from experiment, the charge distribution is. This allows a test of nuclear shell model calculations, which can predict both. Fortunately E_{PNC} is relatively insensitive to this distribution, shifting only by 0.1 % if the theoretical neutron distribution is replaced with the proton distribution. A potentially more serious nuclear physics issue arises in the proposal to measure PNC in different isotopes of cesium, which would allow elimination of wave function uncertainties by taking ratios. The problem is that variations in neutron distributions in different isotopes must be understood. This issue has recently been addressed [30], with the conclusion that such experiments can still be interpreted, because the nuclear physics uncertainties enter at only the 0.1 percent level.

We have described MBPT in the high-Z section, and presented explicit formulas for energies through second order. As mentioned there, there is no guarantee for neutral systems that this will give high accuracies, and indeed the result for the 6s ionization energy, experimentally known to be 0.14310 a.u., is 0.14519 a.u. at this order when starting from the Hartree-Fock potential. At this point a very dangerous situation is encountered. The 1.46 % discrepancy is of the size of various third order energies (all 84 of which have been evaluated) and certain fourth order diagrams. It is then possible to simply evaluate fourth order diagrams until good agreement is found. If all of these several hundred terms were known the validity of this could be gauged, but no such calculation has been carried out. Even if that were to work, one can always imagine diagrams of fifth or higher order playing a significant role. Thus one is on much less firm ground than in the previous discussion. However, a large fraction of higher order diagrams can be picked up with *all-orders* methods, to which we now turn.

In many-body perturbation theory, the first-order correction to the ground state wave function, $\Psi_0 \equiv a_v^\dagger |0_C>$, is

$$|1_C> = (\sum_{amn} \frac{g_{nmva}}{\epsilon_{mn} - \epsilon_{av}} a_m^\dagger a_n^\dagger a_a a_v + \frac{1}{2} \sum_{abmn} \frac{g_{mnba}}{\epsilon_{mn} - \epsilon_{ab}} a_m^\dagger a_n^\dagger a_b a_a) \Psi_0 \qquad (39)$$

where the shorthand $\epsilon_{ij} \equiv \epsilon_i + \epsilon_j$ has been used. When $|2_C>$ is calculated, similar structures arise, such as

$$|2_C>_a = \sum_{amnrs} \frac{g_{nmrs} g_{rsav}}{(\epsilon_{mn} - \epsilon_{av})(\epsilon_{rs} - \epsilon_{av})} a_m^\dagger a_n^\dagger a_a a_v \Psi_0. \qquad (40)$$

Such terms are referred to as *doubles*, because two electrons (one core electron a and the valence electron v) have been destroyed and replaced with excited states. They enter in each order of perturbation theory, and it would of course be impractical to explicitly evaluate them order by order. However, it is possible to do this in another way, in which the coefficient of $a_m^\dagger a_n^\dagger a_a a_v \Phi_0$ is treated as an unknown coefficient ρ_{abmv}. Explicitly, we assume the following form for the wavefunction:

$$\Psi = N_v \left(1 + \sum_{am} \rho_{ma} a_m^\dagger a_a + \sum_m{}' \rho_{mv} a_m^\dagger a_v + \right.$$
$$\frac{1}{2} \sum_{abmn} \rho_{mnab} a_m^\dagger a_n^\dagger a_b a_a + \sum_{amn} \rho_{mnva} a_m^\dagger a_n^\dagger a_a a_v$$
$$\left. + \sum_{abcmnr} \rho_{mnrabc} a_m^\dagger a_n^\dagger a_r^\dagger a_c a_b a_a + \sum_{abmnr} \rho_{mnrvab} a_m^\dagger a_n^\dagger a_r^\dagger a_b a_a a_v \right) \Psi_0, \qquad (41)$$

where N_v is a normalization factor. The terms on the first line describe single excitations, on the second double excitations, and those on the third triple excitations. Substituting this form for the wave function into the Schrödinger equation one obtains a set of coupled equations for the expansion coefficients. With the neglect of the ρ_{mnrabc} term a rather complex set of equations results:

$$(\epsilon_m - \epsilon_a - \delta E_v)\rho_{ma} = \sum_{bn} \rho_{nb} \tilde{g}_{bman} + \sum_{bcn} \tilde{g}_{bcan} \rho_{nmcb} + \sum_{bnr} \tilde{g}_{bmnr} \rho_{rnba} \qquad (42)$$

$$(\epsilon_m - \epsilon_v - \delta E_v)\rho_{mv} = \sum_{an} \rho_{na}\tilde{g}_{amvn} + \sum_{abn} \tilde{g}_{abnv}\rho_{nmab} + \sum_{anr} \tilde{g}_{amnr}\rho_{nrva}$$
$$+ \sum_{abnr} \tilde{g}_{abnr}(\rho_{rnmvab} - \rho_{rmnvab} + \rho_{mrnvab}) \quad (43)$$

$$(\epsilon_{mn} - \epsilon_{av} - \delta E_v)\rho_{mnva} = -g_{mnva} + \sum_{rs} g_{mnrs}\rho_{srva} + \sum_{bc} g_{bcva}\rho_{nmbc} +$$
$$\sum_r \rho_{ra}g_{mnrv} + \sum_r \rho_{rv}g_{nmra} + \sum_b \rho_{nb}g_{bmav} + \sum_b \rho_{mb}g_{bnva} +$$
$$\sum_{br} \tilde{g}_{bmrv}\tilde{\rho}_{rnab} + \sum_{br} \tilde{g}_{bnra}\tilde{\rho}_{rmvb} + \sum_{brs} \tilde{g}_{bmrs}(\rho_{snrvab} + \rho_{rnsvab} - \rho_{nrsvab})$$
$$+ \sum_{bcr} \tilde{g}_{cbra}(-\rho_{rnmvbc} + \rho_{nrmvbc} - \rho_{nmrvbc}) \quad (44)$$

$$(\epsilon_{mn} - \epsilon_{ab} - \delta E_v)\rho_{mnab} = -g_{mnab} + \sum_{rs} g_{mnsr}\rho_{rsab} - \sum_{cd} g_{cdab}\rho_{mncd} +$$
$$\sum_c (\rho_{nc}g_{cmba} + \rho_{mc}g_{cnab}) + \sum_r (\rho_{rb}g_{mnra} + \rho_{ra}g_{nmrb}) +$$
$$\sum_{cr} (\tilde{g}_{cmbr}\tilde{\rho}_{rnac} + \tilde{g}_{cnar}\tilde{\rho}_{rmbc}) \quad (45)$$

$$(\epsilon_{mnr} - \epsilon_{abv} - \delta E_v)\rho_{mnrvab} = -\rho_{rb}g_{mnva} + \frac{1}{2}\rho_{rv}g_{mnba} + \sum_c \tilde{g}_{crbv}\rho_{mnac} +$$
$$\sum_s g_{mnsv}\rho_{srab} - \sum_c g_{cmba}\rho_{nrvc} + \sum_s g_{mnsa}\tilde{\rho}_{rsvb} + \ldots \quad (46)$$

where

$$\delta E_v = \sum_{amn} \tilde{g}_{vamn}\rho_{mnva} + \sum_{abm} \tilde{g}_{abvm}\rho_{mvab} + \sum_{am} \tilde{g}_{vavm}\rho_{ma} +$$
$$\sum_{abmn} \tilde{g}_{abmn}(\rho_{vmnvab} + \rho_{nvmvab} + \rho_{mnvvab}). \quad (47)$$

We begin the discussion of these equations by neglecting the triple excitation terms, referred to as *triples* in the following. If one approximates the equation for ρ_{mnav} by dropping δE_v and keeping only the first term on the right-hand side, one sees that this reproduces the analogous term in $|1_C>$. Further, by taking this form and adding it in perturbatively into the second term, one can see the origin of $|2_C>_a$. However, by iterating the entire set of equations, an approximation to the wave function results that encompasses large parts of arbitrarily high order contributions to the wave function. In practice, between 5 and 10 iterations lead to convergence to six digits. However, this does not mean that agreement with experiment at this level need be expected. This is because of the neglected triples. It is possible to trace the effect of this approximation on the third order energy. While the second order energy is entirely accounted for, 36 of the 84 third order diagrams are left out. While these can be added in perturbatively, they also can be seen to arise from the inclusion of the triples in the equation for the single coefficient ρ_{mv}. It is this procedure that is followed in the calculations of PNC and oscillator strengths. However, the effect of the triples on the equation on the doubles coefficient ρ_{mnav} is not included, so that terms of fourth order and higher are missed, which will be discussed below.

Of course, even with the exact inclusion of triples one is still not done with cesium. This is because of the neglect of more complicated changes to the wave function involving quadruple and higher excitations. However, when only a few electrons are present, all-orders approaches can become 'exact' (although of course radiative corrections are not accounted for). For example, the equations described above will account for every order of MBPT when applied to lithiumlike ions. This is because quadruple and higher excitations vanish for the simple reason that four and more destruction operators applied to a three electron system give zero. Similarly, the triples do not contribute for heliumlike ions. In this latter case all-orders [31] and closely related CI calculations [32] have been applied to the helium isoelectronic sequence, and interesting QED consequences found.

We have throughout this talk emphasized the role played by expansion parameters. In the case of neutral cesium, we propose that the expansion parameter be the number of electrons excited. By including two, then three, then four excitations, a more and more complete wave function should be formed. For this to work, it is essential that terms involving 5 excitations contribute at a very small level. If this is not the case, some more powerful theoretical method must be introduced. A practical way of testing this idea is simply comparing with experiment. One has in cesium a number of accurately determined experimental properties, the most accurate of which are hyperfine constants and energies. While the former are of course extremely well known, the theme of nuclear uncertainty actually limits this as a test of many-body theory. This is because the analog of the Zemach correction discussed in the Introduction, the shift arising from the finite size of the cesium nucleus, is quite large, being about 0.6 % for the 6s state. At some point, expected to be about 0.1 %, the same uncertainties that cloud the interpretation of hydrogen hfs should enter. Fortunately this level is more than adequate for interpretation of PNC experiments in cesium. Another test, less sensitive to nuclear uncertainties, that has received considerable attention recently [33] is oscillator strengths in cesium. A calculation carried out at Notre Dame [34], using the above equations with the neglect of the effect of triples on doubles, found for the lifetime of the $6p_{1/2}$ state

$$\tau(6p_{1/2}) = 35.22 \text{ns} \tag{48}$$

as compared to the experimental result [33] 34.934 ± 0.094 ns. As a way of crudely estimating the neglected effects, the quantity ρ_{mv} was scaled by a factor λ determined from comparison of the calculated 6s energy to the experimental value. When this is done, we find $\tau(6p_{1/2})_{\text{scaled}} = 34.76$ ns. Because of the phenomenological nature of this procedure, we use the latter lifetime only as a means of estimating the error associated with the terms neglected in our calculation. We note that a similar calculation has been done by the Novosibirsk group [35], that includes a slightly different set of diagrams. In the parts of the calculation that can be compared, good agreement is found.

The next step in the cesium calculation is a more complete inclusion of triples, specifically their effect on the doubles coefficient ρ_{mnva}. This is a much more computationally demanding task, firstly because of the storage requirements for the large basis sets required for high-accuracy calculations, and secondly because of the large number of computations required. Both problems are becoming more tractable as more powerful computers become available. An important test of the technique will be provided by lithium. As mentioned above, the equations presented above become exact in this case. Their evaluation may shed light on the present discrepancy between theory and experiment for lifetimes in this atom, as discussed further by Tanner in this conference [33].

While the original motivation for applying all-orders techniques to cesium was to calculate PNC, the approach is of course of general utility for calculations of atomic structure. While the accuracy of calculations completely including triples remains to be determined, they may turn out to be extremely precise. In that case, the interesting possibility arises that the kind of QED tests for high-Z ions discussed in the previous section will become possible for a wide range of low-Z and neutral systems. While strong interaction uncertainties will eventually limit any atomic physics calculation, the physics of atoms and ions is sufficiently rich that a great deal more remains to be learned about QED and the many-body problem before that stage is reached.

ACKNOWLEDGMENTS

I would like to thank Peter Beiersdorfer, Steve Blundell, K.T. Cheng, Peter Mohr, and Walter Johnson for helpful conversations. Some of the research described here was supported in part by NSF grant PHY-92-04089.

References

[1] L. Essen, R.W. Donaldson, M.J. Bangham, and E.G. Hope, Nature **229**, 110 (1971).

[2] G.T. Bodwin and D.R. Yennie, Phys. Rev. D**37**, 498 (1988).

[3] See the contribution by P. Beiersdorfer in these proceedings.

[4] F.G. Mariam, W. Beer, P.R. Bolton, P.O. Egan, C.J. Gardner, V.W.Hughes, D.C. Lu, P.A. Souder, H. Orth, J. Vetter, U. Moser, and G. zu Putlitz, Phys. Rev. Lett **49**, 993 (1982).

[5] T. Kinoshita and M. Nio, Phys. Rev. Lett. **72**, 3803 (1994).

[6] See the contributions in *Long-Range Casimir Forces: Theory and Recent Experiments in Atomic Systems*, edited by F. S. Levin and D. A. Micha (Plenum, New York, 1993).

[7] C.T. Munger and H. Gould, Phys. Rev. Lett. **57**, 2927 (1986).

[8] J. Schweppe, A. Belkacem, L. Blumenfeld, Nelson Claytor, B. Feinberg, Harvey Gould, V.E. Kostroun, L. Levy, S. Mishawa, J.R. Mowat, and M. Prior, Phys. Rev. Lett. **66**, 1434 (1991).

[9] P. Beiersdorfer, D. Knapp, R.E. Marrs, S.R. Elliot, and M.H. Chen, Phys. Rev. Lett. **71**, 3939 (1993).

[10] H.G. Berry, R.W. Dunford, and A.E. Livingston, Phys. Rev. A**47**, 698 (1993).

[11] W.R. Johnson, S.A. Blundell, and J. Sapirstein, Phys. Rev. A**37**, 2764 (1988); ibid., Phys. Rev. A**38**, 2699 (1988); ibid, Phys. Rev. A**42**, 1087 (1990).

[12] E. Avgoustoglou, W.R. Johnson, D.R. Plante, J. Sapirstein, S. Sheinerman, and S.A. Blundell, Phys. Rev. A **46**, 5478 (1992).

[13] J. Sucher, Phys. Rev. A**22**, 348 (1980).

[14] H.T. Doyle, *Advances in Atomic and Molecular Physics* (Academic Press, New York 1969), vol 5., p. 337.

[15] W.R. Johnson, J. Sapirstein, and K.T. Cheng, submitted to Phys. Rev. A.

[16] J.D. Zumbro, E.B. Shera, Y. Tanaka, C.E. Bemis, Jr. R.A. Naumann, M.V. Hoehn, W. Reuter, and R.M. Steffen, Phys. Rev. Lett. **20**, 1888 (1984).

[17] Günter Plunien, Berndt Müller, Walter Greiner and Gerhard Soff, Phys. Rev. A **43**, 5853 (1991).

[18] J. Sucher, Phys. Rev. **107**, 1448 (1957).

[19] P.J. Mohr, At. Data Nucl. Data Tables **29**, 453 (1985).

[20] See the contributions in *Heavy-Ion Spectroscopy and QED Effects in Atomic Systems* (World Scientific, Singapore, 1993).

[21] S.A. Blundell, Phys. Rev. A**46**, 3762, 1992.

[22] S.A. Blundell, P.J. Mohr, W.R. Johnson, and J. Sapirstein, Phys. Rev. A**48**, 2615 (1993): I. Lindgren, H. Persson, S. Salomonson, and L. Labzowsky, Goteborg preprint (1994).

[23] P. Indelicato and P.J. Mohr, Theor. Chem. Acta **80**, 207 (1991).

[24] Michael E. Peskin and Tatsu Takeuchi, Phys. Rev. Lett. **65**, 964 (1990), and W. Marciano and J. Rosner, Phys. Rev. Lett. **65**, 2963 (1990).

[25] M.A. Bouchiat and C. Bouchiat, J. Phys. (Paris) **35**, 899 (1974).

[26] S.A. Blundell, W.R. Johnson, and J. Sapirstein, Phys. Rev. D**45**, 1602 (1992).

[27] V.A. Dzuba, V.V. Flambaum, P.G. Silvestrov, and O.P. Sushkov, Phys. Lett. A **141**, 147 (1989).

[28] A.C. Hartley, E. Lindroth, and A.-M. Martensson-Pendrill, J. Phys. B**23**, 3417 (1990); C. Bouchiat and C.A. Piketty, Europhys. Lett **2**, 511 (1986); A.C. Hartley and P.G.H. Sandars, J. Phys. B**23**, 1961 (1990).

[29] M.C. Noecker, B.P. Masterson, and C.E. Wieman, Phys. Rev. Lett. **61**, 310 (1988).

[30] B.Q. Chen and P. Vogel, Phys. Rev. C**48**, 1392 (1993).

[31] D.R. Plante, W.R. Johnson, and J. Sapirstein, Phys. Rev. A**49**, 3519 (1994).

[32] M.H. Chen, K.T. Cheng and W.R. Johnson, Phys. Rev. A**47**, 3692 (1993).

[33] See the contribution by C. Tanner in these proceedings.

[34] S.A. Blundell, W.R. Johnson, and J. Sapirstein, Phys. Rev. A**43**, 3407 (1991).

[35] V.A. Dzuba, V.V. Flambaum, and O.P. Sushkov, Phys. Lett. A**142**, 373 (1989).

Precision Spectroscopy of Atomic Hydrogen

Theodor W. Hänsch

Max-Planck-Institut für Quantenoptik, Hans-Kopfermann-Str. 1, 85748 Garching and Sektion Physik, University of Munich, Schellingstr. 4, 80799 Munich, Germany

Abstract. The simple hydrogen atom permits unique confrontations between spectroscopic experiment and fundamental theory. The experimental resolution and measurement accuracy continue to improve exponentially. Recent advances include a new measurement of the Lamb shift of the 1S ground state which provides now the most stringent test of QED for an atom and reveals unexpectedly large two-loop binding corrections. The H-D isotope shift of the extremely narrow 1S-2S two-photon resonance is yielding a new value for the structure radius of the deuteron, in agreement with nuclear theory. The Rydberg constant as determined within 3 parts in 10^{11} by two independent groups has become the most accurately known of any fundamental constant. Advances in the art of absolute optical frequency measurements will permit still more precise experiments in the near future.

INTRODUCTION

Hydrogen is the simplest of the stable atoms and the most common element in the visible universe. Spectroscopy of hydrogen has played a pivotal role in the development of atomic physics and quantum mechanics.[1-3] The simple Balmer spectrum has inspired many pathbreaking discoveries, from Bohr and Sommerfeld to De Broglie, Schrödinger, Dirac, Lamb and on to the development of modern quantum electrodynamics (QED). More than once small discrepancies between observation and theoretical prediction have led to major conceptual breakthroughs. Today, QED permits us to predict the energy levels and resonance frequencies of the hydrogen atom with unmatched precision. QED is generally considered the most successful theory of physics, despite some remaining small dicrepancies between experiment and theory, e.g. for the magnetic moment of the electron or the annihilation life time of hydrogen-like positronium.

Why is spectroscopy of hydrogen still interesting today? Hydrogen is so far the only atom for which we have learned to carry out QED calculations to a level that can do justice to the precision achievable with modern laser spectroscopy. Hydrogen provides a unique testing ground where one might discover conceivable limits or flaws of our fundamental theory. Even if we do not find any further surprises, however, precision spectroscopy of hydrogen can yield accurate values of important physical

© 1995 American Institute of Physics

quantities such as the Rydberg constant, the charge radius of the proton, or the electron to proton mass ratio.

To us experimentalists, spectroscopy of hydrogen has long provided a challenging frontier, motivating us to advance the technical limits of optical spectroscopy. For more than two decades, this quest has been quite successful. Optical spectroscopy of hydrogen has been revolutionized in the early 70s by the advent of tunable dye lasers.[4] Since then, we have improved the resolution and measurement accuracy by almost 5 orders of magnitude. The rate of progress is still growing exponentially, fueled in part by dramatic advances in laser science, nonlinear optics, and opto-electronics. These fields are receiving much worldwide support since they provide key technologies for lucrative ventures such as telecommunications. On the other hand, the pursuit of hydrogen spectroscopy has led to new spectroscopic instruments and techniques which have found wide applications much beyond their original purpose. Examples include novel methods of Doppler-free spectroscopy, such as two-photon spectroscopy,[5] polarization spectroscopy,[6] or even the original proposal for laser cooling of gases.[7]

The experimental advances are calling for more precise QED computations of hydrogen energy levels. Unfortunately, the physics of the hydrogen atom is no longer simple at the achieved level of scrutiny. It becoms apparent that the electron is furiously interacting with a sea of virtual photons and virtual electron-positron pairs so that we are no longer dealing with a simple two-particle-system. The required calculations are notoriously difficult and time consuming. As outlined by P. LEPAGE elsewhere in these Proceedings, new approaches are being developed which promise to make the bound state problem much more manageable. And even to the theorist, technology is coming to the rescue. Symbolic manipulation programs are now available even on small personal computers, which facilitate the tedious bookkeeping during analytical calculations that may involve thousands of terms. In addition, we are gaining access to ever more powerful computers for the numerical evaluation of previously forbidding multidimensional integrals.

My own interest in laser spectroscopy of hydrogen dates back to the early 70s, when I was a young postdoc working in the laboratory of Art Schawlow at Stanford University. At that time, we applied the then new method of Doppler-free saturation spectroscopy to the prominent red Balmer-α line, observed in a Wood-type gas discharge.[4] It was an unforgettable thrill to resolve, for the first time, the long elusive single fine structure components and to observe the n=2 Lamb shift directly in the optical spectrum. With my student M. Nayfeh we proceeded to measure the absolute wavelength and to achieve an order-of-magnitude improvement in the accuracy of the Rydberg constant, the universal scaling factor for any spectroscopic transition and an important cornerstone in the system of fundamental constants.[8]

Soon afterwards, we observed transitions from the 1S ground state to the metastable 2S state by Doppler-free two-photon spectroscopy,[9] a method first proposed by the late Veniamin Chebotaev and his coworkers.[5] This technique which gives access to dipole-forbidden transitions between long-living states has been the basis for almost all recent optical precision experiments with atomic hydrogen. It eliminates Doppler broadening without the need to select slow atoms by exciting an atom with two counterpropagating laser beams whose first order Doppler shifts cancel.

TWO-PHOTON TRANSITION 1S-2S

Much of our work at Garching is centered around the extremely sharp two-photon transition from the 1S ground state of hxdrogen to the metastable 2S level. Fig. 1 shows the lowest two energy levels of hydrogen, with the Dirac fine structure, Lamb shifts, and hyperfine splittings magnified. The 1S-2S two-photon transition is the most intriguing resonance in the hydrogen atom, since the 1/7 s life time of the upper level implies a natural line width of only 1.3 Hz. The Q-factor of this line is thus larger than 10^{15}, and one can hope to find the line center ultimately to a part in 10^{18}.

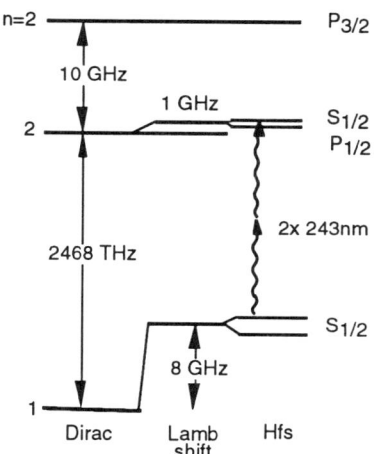

FIGURE 1. Energy levels of atomic hydrogen with 1S-2S two-photon transition.

For a long time, the main obstacle to such spectroscopy has been the lack of an intense monochromatic laser source at the required ultraviolet wavelength near 243 nm, and early experiments had to resort to the frequency-doubled output of rather crude pulsed dye lasers.[9-11] The first cw experiments at Stanford, starting around 1984, relied on sum frequency mixing of a dye laser and an ultraviolet argon ion laser.[12]

When I returned to my native Germany in 1986, we started a new generation of precision experiments at the Max-Planck-Institut für Quantenoptik in Garching, and recent progress of this work will be at the main focus of this report. Around that time, nonlinear crystals of BBO (beta-barium-borate) had just become commercially available. This material permits efficient angle-tuned frequency doubling of a 486 nm cw dye laser, when used inside an external build-up resonator. We are now routinely producing ultraviolet powers up to 20 mW in this way with 300 to 500 mW of fundamental power. A detailed description of this source and our 1S-2S two-photon spectrometer is given elsewhere.[13] Very recently, C. ZIMMERMANN in our laboratory has demonstrated similar power levels starting with a high power semiconductor diode laser at 972 nm followed by two stages of frequency doubling. Such a light source could be made portable and small enough to fit into an attaché case rather than occupying a 5 m long optical table.

The adequate stabilization of the laser frequency presents a rather challenging problem. We have equipped our commercial cw dye laser with an internal electro-optic phase modulator, which compensates for any fluctuations in optical path length. The laser frequency is locked to a passive reference cavity, using the POUND-DREVER-HALL FM sideband method. We employ a 45 cm long massive Zerodur spacer with gyroscope quality mirrors optically contacted to the end faces. The entire cavity is suspended from soft springs inside a temperature-stabilized vacuum tank at 10^{-8} mbar. With a finesse near 60 000 the resonance width is about 5 kHz. With a servo bandwidth of 1 MHz, we easily achieve tracking of the laser frequency to this cavity at the sub-Hz level, but residual vibrations in our rather noisy laboratory cause a remaining frequency jitter of the order of 1 kHz.

With improving laser stability it becomes increasingly important to avoid collision broadening and to ensure a sufficiently long interaction time of the atoms with the light. This is not trivial, because the mean thermal velocity of hydrogen atoms at room temperature is 3000 m/s. So far, the Garching experiments rely on collinear excitation of a cold atomic beam, as illustrated in Fig. 2 The atomic beam apparatus is evacuated with a large cryo-pump, which is turned off during data taking to avoid vibrations. A standing wave build-up-cavity inside the vacuum chamber supports up to 100 mW of circulating UV light. The atoms are produced by dissociation of H_2 in a microwave discharge and slowed by collisions with the walls of a cold nozzle mounted at the bottom of a helium cryostat. After a 15 cm long interaction zone we detect excited 2S atoms by applying an electric quench field and counting Lyman-α photons with a solar blind photomultiplier or a sensitized channeltron detector. A graphite-coated wire net shields the beam from stray electric fields.

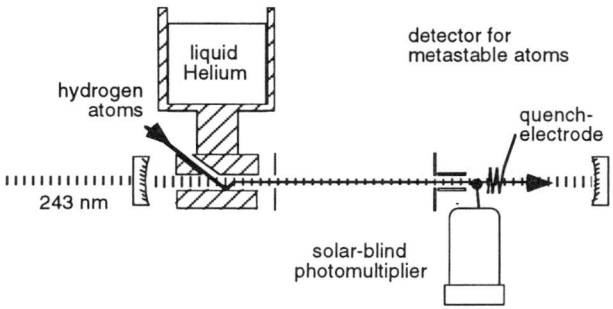

FIGURE 2. Hydrogen atomic beam apparatus for two-photon spectroscopy of 1S-2S.

To find the 1S-2S two-photon resonance, we take advantage of the rich absorption spectrum of the diatomic $^{130}Te_2$ molecule near the blue fundamental wavelength of the dye laser. Even the first hydrogen resonances were much narrower than the best available calibration lines in that molecule as recorded by Doppler-free saturation spectroscopy. Fig. 3 shows spectra of the F=1 hyperfine component of the hydrogen

1S-2S transition recorded at different nozzle temperatures. At room temperature, we observe a line width of 80 kHz (at 243 nm) and a pronounced asymmetry, a red-shifted tail due to the relativistic Doppler effect of second order, which is not cancelled in the excitation with counterpropagating beams. Near liquid helium temperature, the line becomes much more symmetric, and the width is only 12 kHz. This resolution of one part in 10^{11} corresponds to the thickness of a human hair compared to the diameter of the earth.

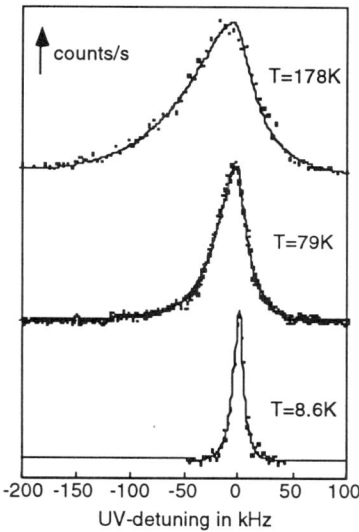

FIGURE 3. Doppler-free two-photon spectra of the hydrogen 1S-2S transition observed in an atomic beam. The F=1 hyperfine component has been recorded at different nozzle temperatures. The resolution at 8.6 K reaches 10^{-11}.

Recently, D. LEIBFRIED in our laboratory has recorded still narrower lines with two rather simple tricks. A small aperture in front of the 2S detector eliminates those atoms which cross the laser field at large angles. More importantly, a chopper turns off the UV light field periodically, and time-delayed detection selects the signal from only the slowest atoms. Even near 80 K, the resonance width could thus be reduced to 3 kHz, corresponding to a resolution of 3 parts in 10^{12}. Unfortunately, cooling of the nozzle to helium temperature or longer delay times did not improve this resolution any further. We have now reached the limits imposed by laser frequency fluctuations, but we hope to observe line widths down to a few hundred Hz with better acoustic and seismic isolation of the reference cavity.

To approach the ultimate resolution, one may have to resort to tricks such as laser cooling or magnetic atom trapping, as discussed by J.T.M. WALRAVEN elsewhere in this volume. We are currently exploring the possibility of generating quasi-continuous Lyman-α radiation for laser cooling of hydrogen with the pulse train of a mode-locked

Ti:Sapphire laser. Some years ago, R.G. BEAUSOLEIL at Stanford showed with model calculations that two-photon Ramsey spectroscopy of a fountain of cold hydrogen atoms could reach a resolution near 1 Hz with trajectory heights of a few inches.[14]

GROUND STATE LAMB SHIFT

Precision measurements of the hydrogen 2S Lamb shift by rf spectroscopy have long provided one of the best tests of QED.[15,16] But further improvements have become difficult, because the short life time of the 2P state implies a 100 MHz natural line width of the 2S-2P resonance.

We report here on recent advances in the measurement of the 1S ground state Lamb shift which do not suffer from this limitation. The 1S Lamb shift is eight times larger than the 2S shift, but it is not accessible to rf spectroscopy, since there is no 1P state. The first experiments at Stanford relied on a comparison of the 1S-2S transition with the blue Balmer-β line.[10,11] At Garching, we have now determined this shift by a direct comparison of the two-photon transitions 1S-2S and 2S-4S.[17,18] In the simple Bohr model the two transition frequencies have the precise ratio 4:1. Any deviation is due to the Lamb shifts of the participating levels, together with well-understood relativistic effects.

We observe the 2S-4S resonance by collinear two-photon excitation of a second atomic beam. Following F. BIRABEN et al.[19], we excite hydrogen atoms to the metastable 2S state with an electron gun. A standing wave of about 50 W circulating power near 972 nm is generated by coupling the output of a Ti:sapphire laser into a linear build-up resonator. When the atoms travelling along this light field are excited to the 4S state, the observed flux of metastable atoms drops slightly, since the 4S atoms

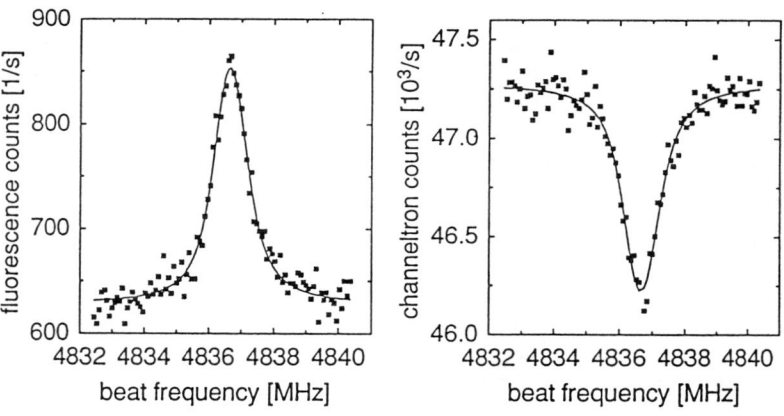

FIGURE 4. Two-photon spectra of the hydrogen 2S-4S (F=1) resonance, observed via fluorescence (left) and via decrease of the metastable flux (right).

tend to decay to the ground state via radiative cascades. In more recent experiments, we have observed instead the blue 4S-2P fluorescence radiation from a 15cm long section of the atomic beam.[18] An elliptical mirror together with a system of lucite light-guides collects this radiation onto the cathode of a photon counting photomultiplier. Despite residual background light from the electron gun this method gives considerably better signal contrast and somewhat narrower lines. Fig. 4 shows typical line profiles of the hydrogen 2S-4S (F=1) resonance as detected via Balmer-β fluorescence or via the decrease of the metastable flux.

Frequency doubling of some of the 972 nm radiation in a crystal of $KNbO_3$ produces about 500 nW of blue light near the frequency of the dye laser of the 1S-2S spectrometer. The two frequencies can be compared precisely via a radiofrequency beat signal detected with a fast photodiode. We keep the dye laser frequency servo-locked to the maximum of the 1S-2S resonance, while the IR laser is slowly and repeatedly scanning across the 2S-4S resonance. As is evident from Fig. 4, the beat frequency at the 2S-4S resonance is not zero but amounts to about 4836 MHz for hydrogen. This beat frequency is a direct measure for the "hyper-Lamb-shift" $1/4\ L_{1S} - 5/4\ L_{2S} + L_{4S}$, from which we can derive the 1S Lamb shift L_{1S}.

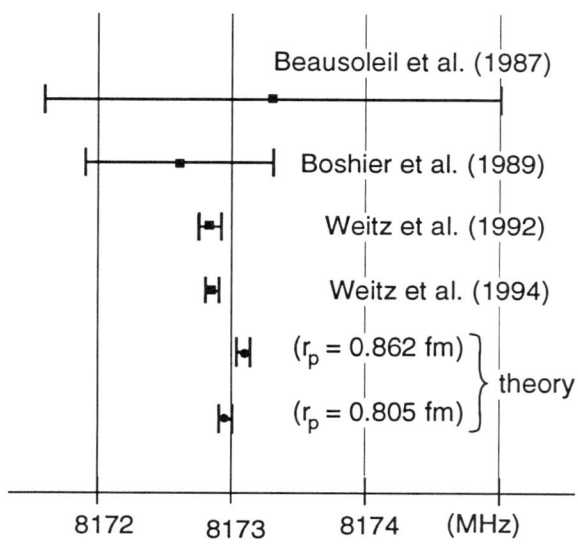

FIGURE 5. Measurements of the hydrogen 1S ground state Lamb shift and two theoretical predictions, based on different proton charge radii r_p.

Unfortunately, the strong IR field saturates the 2S-4S resonance for those atoms which travel very nearly along the axis. The net ac Stark shift becomes thus a nonlinear function of intensity. We are grateful to F. BIRABEN and L. JULIEN for their assistance with the early computer analysis of the required systematic correction.

Fig.5 shows the hydrogen 1S ground state Lamb shift, as derived from the Garching experiments in 1992[17] and 1994,[18] in a comparison with earlier measurements.[20,21] Our most recent value is 8172.86(6) MHz. The accuracy of 6 parts in 10^6 exceeds now that of the best rf measurements of the 2S Lamb shift.[15,16] For deuterium, we have similarly measured a 1S Lamb shift of 8184.00(8) MHz, which exceeds the accuracy of earlier experiments by an order of magnitude.[18]

Also shown in Fig. 5 are two different theoretical predictions of the hydrogen 1S Lamb shift, based on two different values for the rms charge radius r_p of the proton. Since the nuclear size effect scales with n^{-3} it is traditionally included in the definition of the Lamb shift. The proton radius has been determined from electron scattering experiments at high energies. With the more recent and larger Mainz value[22] of 0.862(12) fm, the agreement has been rather poor at the time of publication of our experimental data. An earlier and contradicting value[23] of 0.805(11) fm appeared to give better agreement. However, this old measurement has been performed with such large values of momentum transfer that the proton is deformed by mesonic and quark resonances, and this value is now generally considered as unreliable.

In the past few months, we have seen some remarkable advances in the QED calculation of hydrogenic energy levels. K. PACHUCKI in Garching has used a novel semi-analytical method to compute the long-elusive two-loop binding corrections.[24] Trying to explain their origin to a non-specialist, we may recall that the dominant contribution to the Lamb shift is the "self-energy" of the electron: the law of energy conservation is not violated, according to the uncertainty principle, if an electron emits a virtual photon as long as it reabsorbs it within a sufficiently short time. Or the electron may first absorb such a virtual photon and later re-emit it. It is well known that this interaction with the vacuum fluctuations of the electromagnetic field and the electron´s own radiation field leads to a diverging energy shift if we integrate over all frequencies to infinity, assuming that the electron is a point-like particle. However, an accurately predictable finite correction is obtained, if we are only interested in the difference of this radiative shift between a bound electron and a free electron, so that the contributions at very high frequencies cancel. In particular, for an s electron which comes close to the nucleus, the vacuum fluctuations tend to smear out the Coulomb potential, so that the binding strength is lessened. For the 1S state, this self energy amounts to 8396.461.. MHz. Symbolically, the self-energy for an electron bound in a Coulomb potential is represented by the top Feynman graph in Fig.6.

A smaller counter-acting QED effect arises from vacuum polarization. Both the electron and the proton are partially shielded by a cloud of polarizable virtual electron-positron pairs. In an s-state, the electron penetrates this shield and experiences thus a somewhat stronger Coulomb attraction. Vacuum polarization shifts the 1S state by -215.169... MHz. This effect is symbolized by the second diagram in Fig.6.

Higher order corrections are described by graphs which involve combinations of these two effects. Most of the two-loop corrections depicted in Fig.6. have been calculated during the past year by two independent groups[25,26] and turn out to be rather small. The one term missing has been the computationally most difficult graph shown in the third row of Fig.6., which accounts for the interaction of the electron with two virtual photons and leads to a QED shift of the order $\alpha^2(Z\alpha)^5$. K. PACHUCKI has now completed a first calculation of this term by solving a multidimensional integral with troublesome singularities with a Monte-Carlo method, employing an N-Cube parallel computer with 64 processors which the Max-Planck-Institute for Plasma Physics in Garching made kindly available to us for an entire month. The interaction with two virtual photons gives rise to a surprisingly large energy shift of -328 kHz for the 1S state. With the total two-loop correction of -291(7) kHz, the theoretically predicted 1S Lamb shift becomes 8172.83(6) MHz, which is now in excellent agreement with the measured value of 8172.86(6) MHz, provided we rely on the larger Mainz value of the proton radius.

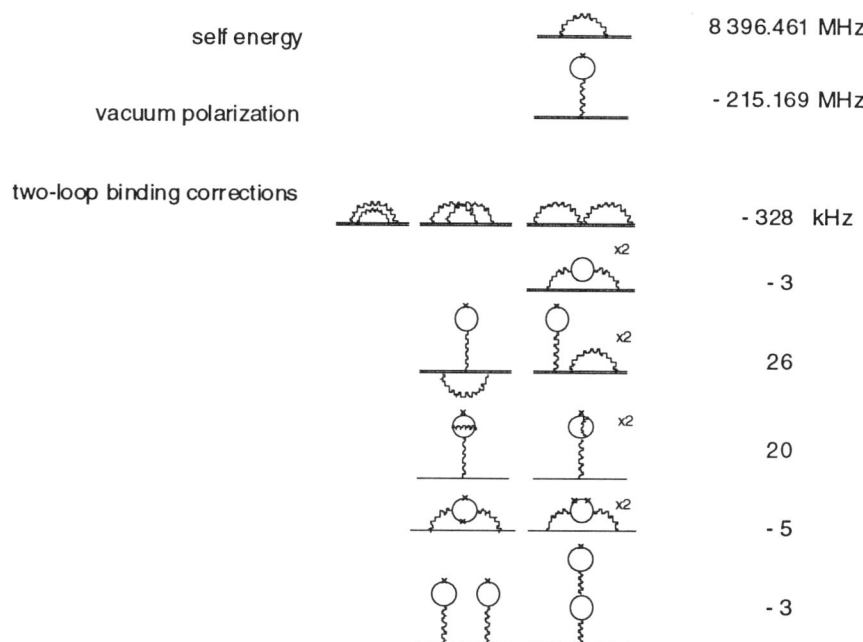

FIGURE 6. One-loop and two-loop contributions to the hydrogen 1S Lamb shift.[22-24]

Troublesome dicsrepancies between theory and experiment have also long persisted for the classical 2S Lamb shift of hydrogen, as recently remeasured by E. HAGLEY and F. PIPKIN at Harvard.[16] With the new two-loop correction of -36.5(9) kHz for the 2S state, the agreement becomes very good. Unfortunately, however, as noted by M. BOSHIER,[27] the application of the same two-loop correction to the 2S Lamb shift of He⁺, measured by a quench asymmetry method,[28] destroys the originally found good agreement between theory and experiment. Although

K. PACHUCKI has meanwhile repeated his computations by a somewhat different method, an independent calculation of the two-loop binding corrections is certainly desirable. The experimental value of the He$^+$ Lamb shift is so far based on one single accurate measurement, and an independent experiment would clearly be welcome. For hydrogen, we have now several different and consistent measurements of both the 2S and 1S Lamb shifts. At this conference, M. BOSHIER and coworkers from Yale are reporting on a new independent measurement of the hydrogen 1S Lamb shift which is in excellent agreement with the most recent Garching results.

Of course one has to ask if there might not be other significant theoretical corrections. Recently, K. PACHUCKI and H. GROTCH[29] have made much progress in the computation of nuclear recoil effects. It is well known that hydrogen is a relativistic two-body system which cannot be fully described by solving the Dirac equation for an electron of reduced mass in a fixed Coulomb potential. Relativistic recoil corrections up to order α^5 have been summarized by ERICKSON.[30] Recoil corrections of α^6 have been analyzed by DONCHESKI et al.[31] who predicted a shift of +3 kHz for the 2S state. Recently, KHRIPLOVICH ET AL.[32,33] have pointed out that the evaluated expression is incomplete. Since then, GROTCH and PACHUCKI have determined a corrected expression which gives a shift of -1 kHz for the hydrogen 2S level and -7.4 kHz for the 1S state.[29]

The most serious obstacle to future more stringent tests of QED is the current inaccurate knowledge of the charge radius of the proton. Electron scattering experiments with large accelerators have proven notoriously difficult, and nuclear physicists have been reluctant to carry out new experiments, even though the possible systematic errors are now much better understood. A rather promising approach for a future independent determination of the proton radius is a measurement of the 2S Lamb shift in muonic hydrogen, as proposed by D. TAQQU and coworkers at the Paul Scherrer Institute (PSI) in Switzerland.[34] A source of intense slow muons now under construction at the PSI will facilitate the production of the muonic atoms. The Lamb shift is here much larger than for hydrogen, so that the 2S - 2P transition becomes accessible to infrared laser radiation near 6 µm. Since the muon is 200 times heavier than the electron it comes much closer to the proton, and the Lamb shift is dominated by vacuum fluctuations, with a nuclear size effect as large as 1.75 percent. or 60 times the natural line width. A measurement of the proton charge radius to within 1 part in 10^3 appears therefore quite feasible.

HYDROGEN-DEUTERIUM 1S-2S ISOTOPE SHIFT

The demonstrated high resolution of the 1S-2S two-photon resonance in hydrogen and deuterium has opened the possibility for very accurate measurements of the H-D isotope shift, which amounts to 678 GHz or 168 GHz at the dye laser frequency near 486 nm. In a first experiment,[35] we have employed an ultrafast electrooptic modulator[36] to measure this shift with radio-frequency accuracy.

An auxiliary second dye laser is locked to a mode of a stable refence resonator halfway in between the hydrogen and the deuterium resonances, and the modulator operating at 84 GHz produces two sidebands which serve as marker frequencies of accurately known spacing close to the two isotopic resonance lines. The dye laser of the 1S-2S spectrometer is then alternatingly tuned to the hydrogen or deuterium line, and a rf beat signal between this dye laser and the nearby marker line is registered with a photodiode. In this way, we have determined an experimental isotope shift of 670.994 337(22) GHz. The accuracy of 3.7 parts in 10^8 exceeds that of earlier experiments[37] by a factor of 25 and is limited only by nonlinear drifts of the reference cavity. Improvements by one or two orders of magnitude should be feasible with the present apparatus.

The H-D isotope shift is mainly a mass effect. The mass ratio of proton to deuteron is very well known, so that this mass shift can be predicted with high precision. As shown in Fig.7, the experimental value falls below this prediction by a small but significant amount. This discrepancy can be accounted for by the difference of the rms charge radii of the two nuclei, with small corrections due to relativistic nuclear recoil effects. From the Garching measurement, we can determine an accurate experimental value of the structure radius of the deuteron.[29] As shown in Fig 8, this value is significantly larger than the structure radius derived from electron scattering experiments.[38] But intriguingly, it agrees well with a recent theoretical prediction by J.L. FRIAR et al., based on Nijmegen potentials as derived from nucleon-nucleon scattering experiments.[39] As theorists learn to predict the charge radii of few-nucleon systems, such as deuterium, tritium, or the two helium isotopes with ab-initio calculations, precise spectroscopic isotope shift measurements become increasingly important, because they can provide more accurate comparisons of these radii than electron scattering experiments with large accelerators.

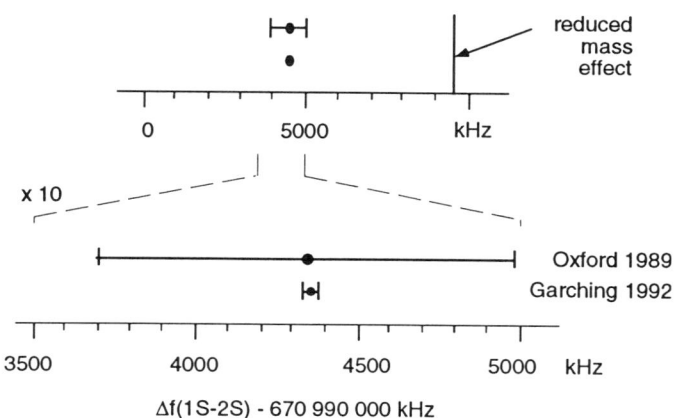

FIGURE 7. Measurements of the H-D isotope shift of the 1S-2S transition.[36,37]

A more detailed theory of the H-D isotope shift[35,40,41] has to include a small correction due to the polarizability of the deuteron. With a rather simple model of the nuclear structure, we have estimated a correction of 19 kHz. More recently, J. MARTORELL et al.[42] have investigated more sophisticated nuclear models which confirm our estimate and predict a deuteron polarizibility correction of the 1S-2S interval of 19.33(21) kHz.

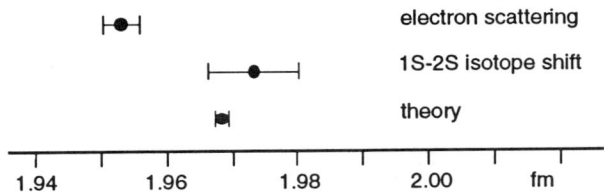

FIGURE 8. Deuteron rms structure radius, as derived from electron scattering experiments,[37] from the 1S-2S H-D isotope shift,[29,36] and as predicted by theory.[39]

RYDBERG CONSTANT

The most challenging experiment at Garching to date has been a precise measurement of the absolute optical frequency of the 1S-2S two-photon resonance.[43] Traditionally, spectroscopic experiments measure optical wavelengths rather than frequencies. Even the most advanced methods of wavelength interferometry suffer from unavoidable systematic phase shift errors due to diffraction, disperion or mirror imperfections. Ideally, we would like to count the number of optical cycles of a light wave during one second, as presently defined by the Cs atomic clock, a microwave oscillator near 9 GHz. However, the state of the art of optical frequency metrology is still rather primitive, comparable to the early stages of mechanical clock making, when people just learned to build and assemble gears so that the rapid motion of a pendulum could be translated into the slow motion of the clock hands.

Accurate frequency measurements are possible with harmonic frequency chains, as they are being maintained in a few national standards laboratories. Such chains rely on the fortuitous coincidence of a frequency to be measured with a high harmonic of a known reference and are typically engineered to measure just one particular frequency. Three years ago, the highest frequency measured witch a phase-coherent frequency chain has been that of a 3.39 μm He-Ne laser, locked to a CH_4 absorption line. Since then, experiments in Garching and in Paris have extended harmonic frequency chains to visible frequencies.[43,44]

Fig. 9 gives an overview over the Garching Rydberg experiment. The right hand side shows the layout of our frequency chain. This chain starts with a portable 3.39 μm He-Ne laser oscillator, developed by V.P. CHEBOTAEV and coworkers at the Institute for Laser Science in Novosibirsk. It has been calibrated in June of 1993 at the PTB in Braunschweig to within 3 parts in 10^{13} in a direct comparison with the Cs atomic clock, thanks to the accommodating help of G. KRAMER and coworkers. Its oscillation frequency was found to be f = 88 376 181 599 991(30) Hz.

We take advantage of a fortuitous coincidence of the 1S-2S interval with the 28th harmonic of the CH_4- standard, i.e. our 486 nm dye laser operates near the 7th harmonic of the He-Ne laser. A blue auxiliary frequency, the eighth harmonic near 424 nm, is produced in three successive steps of second harmonic generation with two transfer laser oscillators to boost the power: a NaCL:OH⁻ color center laser near

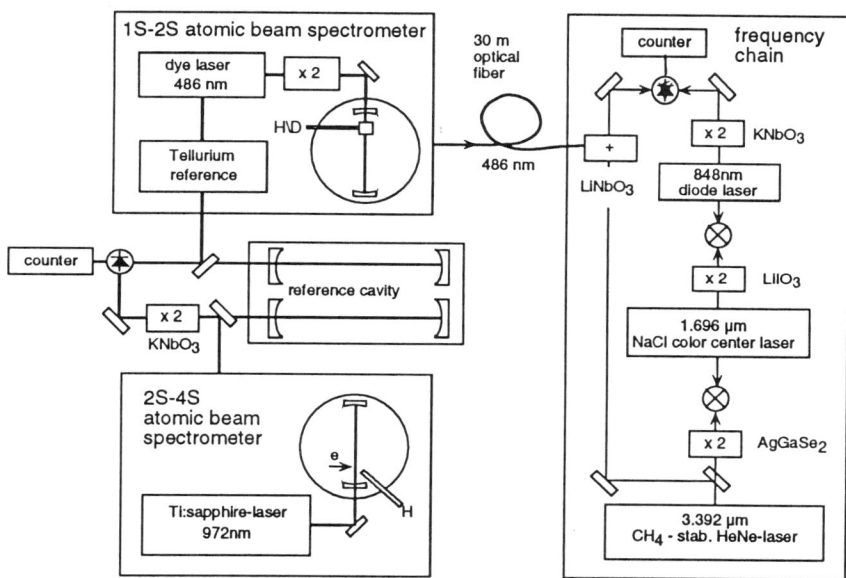

FIGURE 9. Overview over the Garching Rydberg experiment.

1.69 μm, and a grating-tuned diode laser near 848 nm. Electronic servo systems control the frequencies and phases of these slave lasers so that they follow their weak optical control signals without cycle slips.

To measure the frequency of the 1S-2S transition, we send some blue light from the dye laser through a 30 m long optical fiber to the frequency chain. Here, a $LiNbO_3$ crystal produces some 200 nW at the sum frequency of the dye laser and the He-Ne laser, so that a beat note with the directly produced 8th harmonic of the He-Ne standard can be counted. Unfortunately, the hydrogen resonance does not precisely coincide with the 7th harmonic. In the first experiments, we have bridged a frequency mismatch of 2 THz near 486 nm with the help of the precisely calibrated comb of axial modes of our stable reference cavity. This compromise limits the accuracy of the experiments to 1.8 parts in 10^{11}.

In this way, we have found a 1S-2S frequency of 2 466 061 413.182(45) MHz. This result is 18 times more accurate than the best[45] of the previous measurements. A comparison of the 1S-2S frequency with theory yields a new value of the Rydberg constant, $R_\infty = 109\ 737.315\ 684\ 4(31)$ cm^{-1}. Further improvements will require an improved knowledge of the nuclear size effect. To arrive at the quoted uncertainty of 3 parts in 10^{11}, we have taken advantage of the most recent experimental values for the Lamb shifts of the 1S and 2S states.[18]

Even the original Garching result[43] obtained in May of 1992 with an uncertainty near 4 parts in 10^{11} represented almost an order-of-magnitude improvement over the best previous measurements of the Rydberg constant by F. BIRABEN et al. in Paris, based on two-photon spectroscopy of 2S-nS/D transitions in a metastable hydrogen beam.[46] Soon afterwards, A. CLAIRON and coworkers in Paris succeeded in recalibrating the I$_2$-stabilized He-Ne-laser with the help of an independent newly developed frequency chain,[44] so that another more accurate Rydberg value could be derived from the 2S-nS/D measurements at Paris. After an initial brief period of disagreement, the Paris results fully confirmed the Garching measurements.[47] In a more recent experiment, F. BIRABEN et al. have completed a phase-coherent frequency measurement of the 2S-8S/D transitions near 778 nm.[48] They take advantage of a coincidence of the frequency of a 633 nm I$_2$-stabilized He-Ne laser with the sum frequency of the 778 nm hydrogen transition and a CH$_4$-stabilized He-Ne laser at 3.39 μm, and they were able to bridge a remaining frequency mismatch of 80 GHz with the help of a fast Schottky diode. These recent Rydberg measurements are compared in Fig. 10. It is very satisfying that different and independent measurements agree now to within a few parts in 10^{11} so that the Rydberg constant has become, by a wide margin, the most accurately known of the fundamental constants. The new results fall well outside the error limits of the official value recommended in the 1986 least squares adjustment of the fundamental constants (Codata).[49].

In Garching, we are now perfecting a novel frequency divider chain that promises to become a universal instrument for the measurement and synthesis of arbitrary optical frequencies. The basic building block is an optical interval divider:[50,51] A laser is electronically phase-locked to operate at the exact midpoint between two given optical frequencies. The servo signal is obtained from the beat note between the second harmonic frequency of this laser and the sum of the two input frequencies. A true optical to microwave frequency divider can be realized by cascading a number of such divider stages, and using a laser frequency and its second harmonic as the starting frequencies. After n stages, we reach a frequency interval equal to the original frequency divided by 2^n. In contrast to a traditional harmonic frequency chain, all stages can be technically similar, since the frequencies quickly converge to a chosen region. The divider path can be selected so that interesting frequencies are synthesized along the way, as in a proposed "artificial hydrogen atom."[50] In Garching, we have chosen to work at wavelengths in the convenient region around 850 nm, where good diode lasers are available, and KNbO$_3$ provides an almost ideal nonlinear crystal material. If the two input frequency are closer than a few THz, one single temperature-tuned crystal is sufficient for simultaneous 90° phase-matched second harmonic and sum frequency generation.

W. KÖNIG, M. PREVEDELLI, and T. ONAE at Garching have now completed a divider chain of 4 stages which remain reliably phase-locked over hours. In the measurement of the hydrogen 1S-2S transition we will use this chain to divide the frequency mismatch of 1 THz near 850 nm down to 66 GHz. If we can measure this remaining interval with a Schottky diode we will obtain the 1S-2S frequency to within better than one part in 10^{12}, and we can expect an order-of-magnitude improvement in the accuracy of the H-D isotope shift.

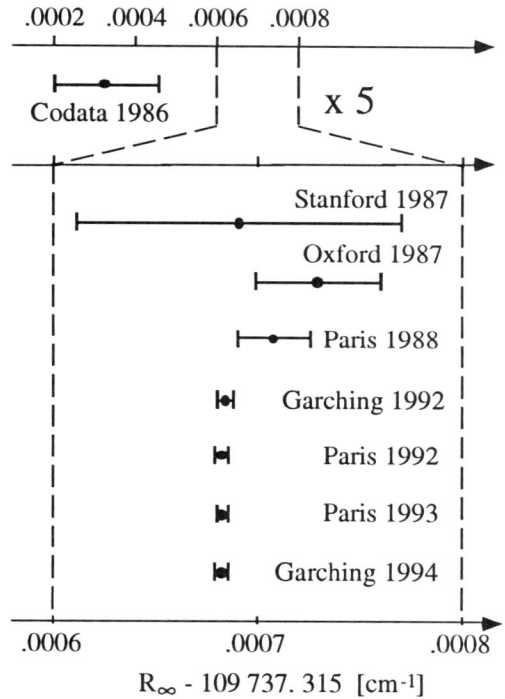

FIGURE 10. Comparison of recent measurements of the Rydberg constant.

OUTLOOK

Many future improvements in precision spectroscopy of atomic hydrogen can be envisioned. We are currently working towards improved Lamb shift measurements by observing 2S-nS two-photon resonances in an optically excited cold metastable hydrogen beam. We have already recorded signals with a doubly resonant cavity which enhances both the 243 nm radiation for excitation of 1S-2S and 972 nm radiation for spectroscopy of 2S-4S.

As pointed out in the past, we hope to determine very accurate values of the Rydberg constant and the electron/proton mass ratio from linear combinations of measured hydrogen frequencies which are chosen so that nuclear size effects scaling with n^{-3} cancel. Other possible goals include hydrogen atom interferometers to measure the photon recoil energy and thus m/h or the fine structure constant α.

Considerable further reductions of the 1S-2S linewidth are expected from improvements of the laser frequency stability. With advanced semiconductor-based laser sources it may become feasible to build a reliable and compact hydrogen optical clock. The frequency of such a clock would be dominated by electromagnetic interactions, while the hyperfine frequency of the Cs atomic clocks depends on the nuclear magnetic monment and is thus sensitive to strong interactions. A long-term comparison of such different clocks might reveal conceivable slow changes of fundamental constants.

Although it will clearly be interesting to carry out similar experiments with other hydrogen-like atoms, notably He+, perhaps the most intriguing goal might be high resolution spectroscopy of anti-hydrogen,[52] (p^-e^+), the antimatter twin of the hydrogen atom which annihilates in contact with ordinary matter, but could be kept and observed over extended periods inside a magnetic trap. We are currently exploring electromagnetic traps which can store not only the paramagnetic atoms but also the charged building blocks, i.e. antiprotons and positrons. Laser-induced recombination may provide a method for producing antihydrogen inside such a trap. A suitable source of slow antiprotons is still available at the LEAR storage ring at CERN, even though there appears to be an acute danger that this unique resource will soon be closed for budgetary reasons.

It would be fascinating if only the slightest diffence between matter and antimatter could be detected spectroscopically. This would be proof of the violation of some fundamental symmetry laws of physics. Such experiments could yield unprecedented tests of CPT symmetry or of the Einstein equivalence principle. It has been speculated that the gravitational attraction between antimatter and matter may differ from that of ordinary matter. This might lead to a detectable differential gravitational red shift of the sharp 1S-2S resonance, some kind of "isotope shift", which could be interpreted as revealing a difference of the inertial masses of the antiproton and proton.

High resolution spectroscopy of atomic hydrogen will continue to hold fascinating challenges for many years to come. Perhaps the biggest surprise in this endeavor would be if we found no surprise.

ACKNOWLEGEMENTS

This work has been supported in part by the Deutsche Forschungsgemeinschaft and within an EC SCIENCE program cooperation, contract SCIT 92-0816.

REFERENCES

1) T.W. Hänsch, A.L.Schawlow, and G.W. Series, Scientific American 240, 94 (1979)
2) "The Spectrum of Atomic Hydrogen, Advances", G.W. Series, Ed., World Scientific, Singapore, 1988
3) "The Hydrogen Atom", G.F. Bassani, M. Inguscio and T.W. Hänsch, Eds., Springer Verlag, Heidelberg, 1989
4) T.W. Hänsch, I.S. Shahin, and T.W. Hänsch, Nature 235, 63 (1972)
5) L.S. Vasilenko, V.P. Chebotaev, and A.V. Shishaev, JETP Lett. 12, 113 (1970)
6) C. Wieman and T.W. Hänsch, Phys. Rev. Lett. 34, 307 (1975)
7) T.W. Hänsch and A.L. Schawlow, Optics Comm. 13, 68 (1975)
8) M.H. Nayfeh, S.A. Lee, S.M. Curry, and I.S. Shahin, Phys. Rev. Lett. 32, 1396 (1974)
9) T.W. Hänsch, S.A. Lee, R. Wallenstein, and C. Wieman, Phys. Rev. Lett. 34, 307 (1975)
10) S.A. Lee, R. Wallenstein, and T.W. Hänsch, Phys. Rev. Lett. 35, 1262 (1975)
11) C. Wieman and T.W. Hänsch, Phys. Rev. A22, 192 (1980)
12) C.J. Foot, B. Couillaud, R.G. Beausoleil, and T.W. Hänsch, Phys. Rev. Lett. 54, 1913 (1985)
13) F. Schmidt-Kaler, D. Leibfried, S. Seel, C. Zimmermann, W. König, M. Weitz, and T.W. Hänsch, Phys. Rev. A, submitted for publication
14) R.G. Beausoleil and T.W. Hänsch, Phys. Rev. A33, 1661 (1986)
15) S.R. Lundeen and F.M. Pipkin, Phys. Rev. Lett. 46, 232 (1981)
16) E.W. Hagley and F.M. Pipkin, Phys. Rev. Lett. 72, 328 (1994)
17) M. Weitz, F. Schmidt-Kaler, and T.W. Hänsch, Phys. Rev. Lett. 68, 1120 (1992)
18) M. Weitz, A. Huber, F. Schmidt-Kaler, D. Leibfried, and T.W. Hänsch, Phys. Rev. Lett. 72, 328 (1994)
19) F. Biraben, J.C. Garreau, L. Julien, and M. Allegrini, Rev. Sci. Instr. 61, 1468 (1990)
20) R.G. Beausoleil, D.H. McIntyre, C.J. Foot, B. Couillaud, E.A. Hildum, and T.W. Hänsch, Phys. Rev. A35, 4878 (1987)
21) M.G. Boshier, P.E.G. Baird, C.J. Foot, E.A. Hinds, M.D. Plimmer, D.N. Stacey, J.B. Swan, D.A. Tate, D.M. Warrington, and G.K. Woodgate, Phys. Rev. A40, 6169 (1989)
22) G.G. Simon, Ch. Schmitt, F. Borkowski, and V.H. Walther, Nucl. Phys. A333, 381 (1980)
23) L.N. Hand, D.J. Miller, and R. Wilson, Rev. Mod. Phys. 35, 335 (1963)
24) K. Pachucki, Phys. Rev. Lett. 72, 3154 (1994)
25) K. Pachucki, Phys. Rev. A48, 2609 (1993)
26) M.I. Eides, S.G. Karshenboim, and V.A. Shelyuto, Phys. Lett. B312, 389 (1993)
27) M. Boshier, private communication
28) A. v. Wijngaarden, L. Kwela, and G.W.F. Drake, Phys. Rev. A43, 3325 (1991)
29) K. Pachucki and H. Grotch, to be published
30) G. Erickson, J. Chem. Phys. Ref. Data 6, 831 (1977)
31) M. Doncheski, H. Grotch, and G.W. Erickson, Phys. Rev. A43, 2125 (1991)
32) I.B. Khriplovich, A.I. Milstein, and A.S. Yelkhovsky, Phys. Scrip. T46, 252 (1993)
33) R.N. Fell, I.B. Khriplovich, A.I. Milstein, and A.S. Yelkhosky, Phys. Lett. A181, 172 (1993)
34) D. Taqqu, private communication
35) F. Schmidt-Kaler, D. Leibfried, M. Weitz, and T.W. Hänsch, Phys. Rev. Lett. 70, 2261 (1993)
36) D. Leibfried, F. Schmidt-Kaler, M. Weitz, and T.W. Hänsch, Appl. Phys. B56, 65 (1993)
37) M.G. Boshier, P.E.G. Baird, C.J. Foot, E.A. Hinds, M.D. Plimmer, D.N. Stacey, J.B. Swan, D.A. Tate, D.M. Warrington, and G.K. Woodgate, Phys. Rev. A40, 6196 (1989)
38) S. Klarsfeld, J. Martorell, J.A. Oteo, M. Nishimura, and D.L.W. Sprung, Nucl. Phys. A456, 373 (1986)

39) J. L. Friar, G.L. Payne, V.G.J. Stoks, and J.J. de Swart, Phys. Lett. B311, 4, (1993); J.L. Friar, *Few-Body Physics - Then and Now*, Summary talk presented at the "XIVth International Conference on Few Body Problems in Physics," Williamsburg, VA, May 1994, F. Gross, Ed., AIP Conference Proceedings (to appear)
40) K. Pachucki, D. Leibfried, and T.W. Hänsch, Pys. Rev. A48, R1 (1993)
41) K. Pachucki, M. Weitz, and T.W. Hänsch, Phys. Rev. A49, 2255 (1994)
42) J. Martorell, D.W. Sprung, and D.C. Zheng, Phys. Rev. C, submitted for publication
43) T. Andreae, W. König, R. Wynands, D. Leibfried, F. Schmidt-Kaler, C. Zimmermann, D. Meschede, and T.W. Hänsch, Phys. Rev. Letters 69, 1923 (1992)
44) O. Acef, I.J. Zondy, M. Abed, D.G. Rovera, A,H, Gerard, A. Clairon, P. Laurent, Y. Millerioux, and P. Juncar, Opt. Comm. 97, 29 (1993)
45) M.G. Boshier et al., Phys. Rev. A39, 4591 (1989)
46) F. Biraben, J.C. Garreau, L. Julien, and M. Allegrini, Phys. Rev. Lett. 62, 621 (1989)
47) F. Nez, M.D. Plimmer, S. Bourzeix, L. Julien, and F. Biraben, Phys. Rev. Lett. 69, 2326 (1992)
48) F. Nez, M.D. Plimmer, S. Bourzeix, L. Julien, F. Biraben, R. Felder, Y. Millerioux, and P. de Natale, Europhys. Lett. 24, 635 (1993)
49) E.R. Cohen and B.N. Taylor, Rev. Mod. Phys. 59, 1121 (1987)
50) D. McIntyre and T.W. Hänsch, in *Digest of the Annual Meeting of the Optical Society of America*, 1988, paper THG3; see also T.W.Hänsch, in ref. 3, p. 93
51) H.R. Telle, D. Meschede, and T.W. Hänsch, Opt. Lett. 26, 858 (1990)
52) "Antihydrogen", *Proceedings of the Antihydrogen Workshop, Munich, Germany, July 30-31, 1992*, Hyperfine Interactions, Vol. 76 (1993)

Recent Developments in Laser Spectroscopy of Helium

M. Inguscio[a], F.S. Cataliotti[a], P. De Natale, G. Giusfredi[b], F. Marin, and F.S. Pavone

European Laboratory for Nonlinear Spectroscopy (LENS)
University of Firenze, Largo E. Fermi 2, Firenze, Italy

Abstract. Recent progress in high resolution laser spectroscopy of Helium is reviewed. The direct determination of the frequency of the $2^3S_1 \rightarrow 3^3P_J$ transition in the near ultraviolet at 389 nm is illustrated. The novel scheme uses the heterodyne with a laser diode locked to 5S→5D two-photon transition in Rubidium which serves as optical frequency reference in the near infrared at 778 nm. A precise value for the 2^3S_1 level Lamb shift is extracted, more than two orders of magnitude more accurate than theoretical calculations. Measurements include ^{3}He-^{4}He isotope shift and ^{3}He hyperfine structure. Results for the $2^3S_1 \rightarrow 3^3P_J$ transition are analyzed in conjunction with those for the $2^3S_1 \rightarrow 2^3P_J$ transition for a precise test of QED corrections. Perspectives for a helium atom determination of the fine structure α are also discussed.

INTRODUCTION

A long tradition of high precision measurements in Helium has provided fundamental tests of atomic systems more complicated than Hydrogen. Progressive improvements in techniques for Helium to the level of 100 KHz bring the precision close to the best measurements in Hydrogen. At the same time, theoretical uncertainties are well below the experimental accuracy for all but the quantum electrodynamics (QED) part of the problem. As a consequence, comparisons between theory and experiment for Helium provide stringent tests of the QED corrections and precise measurements provide an important stimulus for future improvements in the theory. It is worth noting that in Helium specifically two-electron QED effects can be studied, which are not present at all in Hydrogen. Accurate spectroscopic measurements in Helium have also been considered as a possibility for the evaluation of the Rydberg constant. Even if not as precise as the one inferred from the measurements in Hydrogen, such a value would allow a comparison between physical constants involved in different atomic systems. Helium experimental accuracy has indeed followed the improvements in the knowledge of the Rydberg constant, remaining worse by only one order of magnitude or less. The evolution of the R_y and Helium accuracies can be appreciated in Fig. 1.

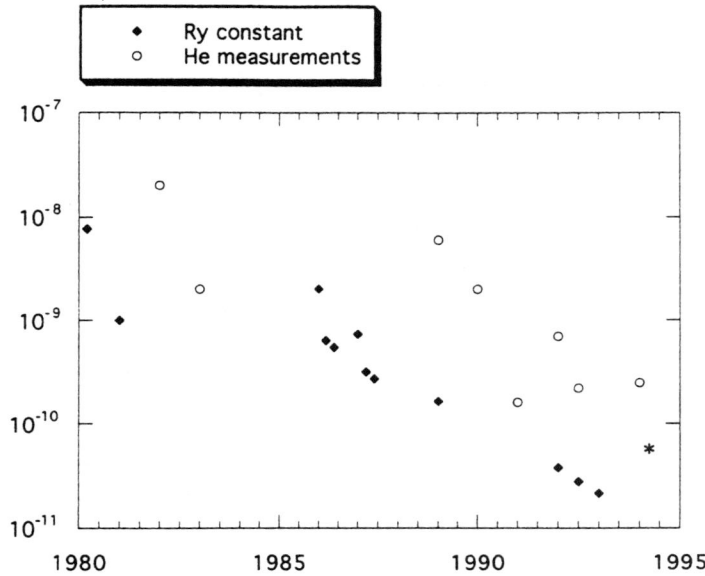

FIGURE 1. Qualitative comparison between the improving accuracy of the Rydberg constant and of ^{4}He spectrum measurements, during the last 15 years. *: this point (present work) reflects the 6×10^{-11} accuracy obtained by direct measurement of the optical frequency.

Full list of references can be found in ref. 1. At present, the Rydberg constant is known (2) with an accuracy of 2.2 parts in 10^{11}, while a preliminary analysis of the pure optical frequency measurements reported in this work for Helium in a metastable atomic beam shows an accuracy of about 6 parts in 10^{11}, a factor of four better than our original measurement in a discharge cell (3).

A simplified energy level scheme, both for singlets and triplets, is shown in Fig. 2 together with references to some of the transitions involved in laser spectroscopy studies. As for the singlet states, one and two-photon investigations (4,5,6) have recently allowed the absolute ionisation energy of the 2^1S state to be inferred. Also, the experimental value is in satisfactory agreement with the theoretical prediction (13), modified with the improved Bethe logarithm evaluated in ref. 14. The 2^1S Lamb shift calculation is now confirmed at the level of 0.05%. Very recently, new perspectives for the investigation of the singlet system have been opened by the direct excitation (7) of the $1^1S \rightarrow 2^1P$ at 58.4 nm, using harmonics of pulsed radiation in the visible.

Subdoppler laser investigations of triplet system were opened by the excitation (8,9) of the two-photon transitions from the 2^3S_1 metastable level to n^3S or n^3D levels. However, also single photon transitions to n^3P levels are suited to observe narrow

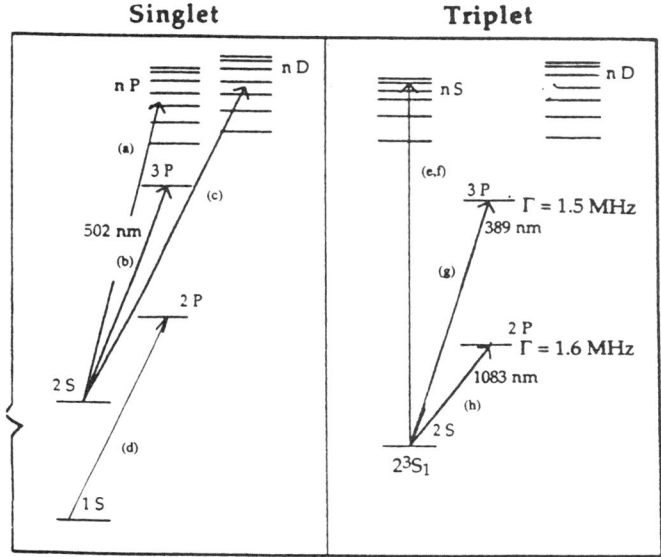

FIGURE 2. Simplified level scheme of singlet and triplet systems in Helium. Some of the single and two-photon transitions involved in high precision experimental investigations have been indicated (a - ref. 4, b - ref. 5, e - ref. 6, d - ref. 7, e - ref. 8, f - ref. 9, g - refs 3,10, h - refs. 11,12).

resonances since, differently from Hydrogen, ^{3}P states are relatively long lived. Indeed, radiative widths are 1.6 and 1.5 MHz for 2^3P and 3^3P states respectively.

In the present work, we shall concentrate mostly on the $2^3S_1 \to 3^3P_J$ transitions in the near ultraviolet at 389 nm. There are several reasons for this choice. First, the uv radiation is obtained by frequency doubling a laser source at 778 nm, in close coincidence with two photon transitions in Rubidium which now constitute a precise optical frequency reference (15). Indeed this allowed the first pure frequency measurement of an optical transition in helium to be performed (3). Also, the energy of the upper level of the transition, n=3, is accurately predicted by theory (16). As a consequence, the precise measurement of the transition results in an accurate determination of the 2^3S_1 metastable level energy and Lamb shift.

As is well known, 2^3S_1 can be considered as a sort of "auxiliary" ground state for helium. Indeed closed transitions in the optical region originate from it, which allow not only high resolution investigations, but also laser cooling or trapping and atomic interferometry experiments. For both $2^3S_1 \to 3^3P_J$ and $2^3S_1 \to 2^3P_J$ transitions highly precise measurements of isotope effect and hyperfine structure can be performed and the comparison with the accurate predictions of the nuclear size contributions to the energy allows a determination of the relative nuclear radii of ^{3}He and ^{4}He. Finally, differently from singlet states, triplet ones display fine

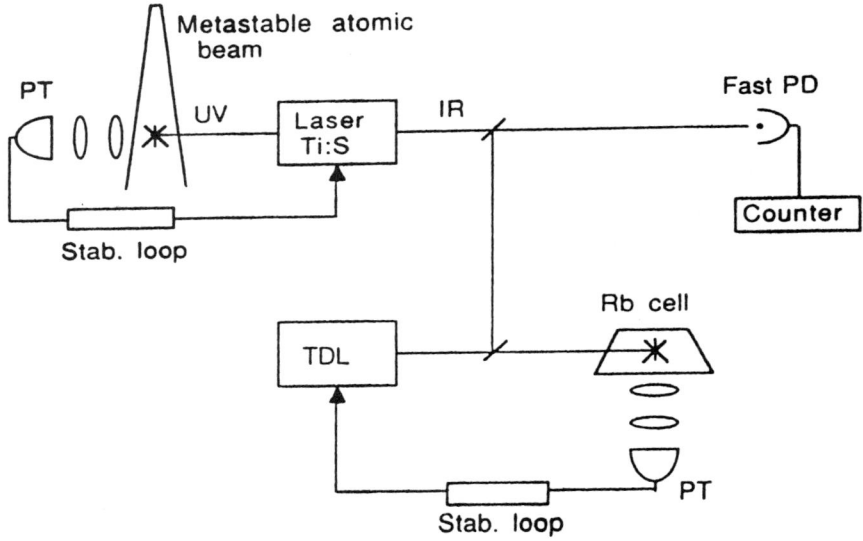

FIGURE 3. General scheme of the experimental apparatus for the investigation of Helium $2^3S_1 \to 3^3P_J$ transitions at 389 nm.

structure splittings. Of course these enter both in the experiments and in the theory and, as we shall discuss, there are good prospects to obtain a value for the fine structure constant α comparable in accuracy to those from the quantum Hall effect or from the electron magnetic anomaly.

PURE FREQUENCY MEASUREMENT OF THE $2^3S_1 \to 3^3P_J$ TRANSITIONS

The general scheme of our experimental apparatus for the high resolution and high precision investigation of the $2^3S_1 \to 3^3P_J$ transitions is shown in Fig.3. Four different sections can be individuated: the 389 nm laser source, the two-photon Rubidium reference diode laser spectrometer, the metastable Helium atomic beam, the optical-microwave frequency mixing apparatus.

As in our recent works (3,17), radiation at 389 nm is provided by a commercial intracavity frequency doubled Ti - doped sapphire laser. Typically we have 20 mW with a frequency jitter of about 1 MHz. At the same time, 15 mW of the fundamental oscillation at 778 nm are available for the frequency measurement section of the experiment.

Two-Photon Rubidium Frequency Standard at 778 nm

As for the 778 nm laser frequency needed to excite, after doubling, the Helium transitions, there are some interesting coincidences which have made possible the direct measurement. Indeed it is only 42 GHz higher than those involved in the Hydrogen 2S-8S/8D two-photon transitions used (18) for a determination of the Rydberg constant. As is well known, the latter absolute frequency measurements were performed using the close coincidence with the difference in frequency between the iodine stabilized HeNe laser and the methane stabilized HeNe laser, at 473 THz (633 nm) and 88 THz (3.39 µm), respectively. By means of the same frequency scheme, the group in Paris (15,20) has measured the frequency of the $5S_{1/2} \rightarrow 5D_{3/2,5/2}$ two-photon transitions in Rubidium. In this case the coincidence is with a residual difference of about 5 and 50 GHz for the $D_{3/2}$ and $D_{5/2}$ components respectively. The absolute optical frequencies of the Rb hyperfine components were measured with an uncertainty of 5 KHz, i.e. 1.3×10^{-11}. Light-shift and the effect of changing the temperature of the Rb cell were also carefully investigated. In general, the reproducibility was found to be very good and the perspectives for a Rubidium two-photon optical frequency standard are now quantitatively investigated (20,21).

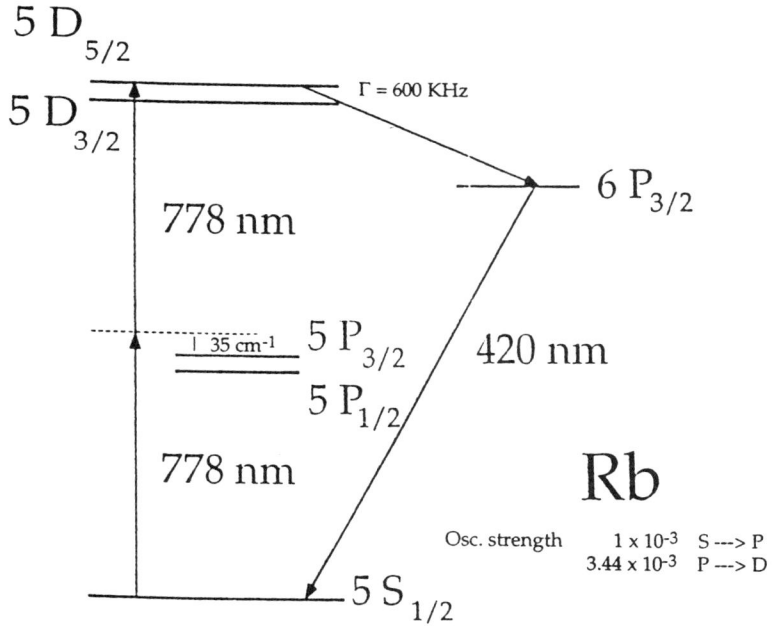

FIGURE 4. Simplified level scheme of Rb, showing the two-photon transitions used as optical frequency references at 778 nm.

This new standard enters now new frequency sum/difference schemes like the one by J. Hall and co-workers for the measurement of the frequency doubled Nd-Yag laser in the green (22). Due to the presence of nearly-resonant intermediate 5P levels, as schematically shown in Fig. 4, the two-photon Rb transitions are fairly strong and easy to be induced also by means of low power diode lasers. While broadband diode laser excitation of two-photon transitions in Rubidium were reported (23), we instead were able (3,24). in collaboration with F. Biraben, to use only 1.6 mW from a AlGaAs laser mounted in a grating feedback scheme (25) for line narrowing. In the present experiment, we used a little more powerful diode laser and the scheme of the apparatus is shown in Fig. 5.

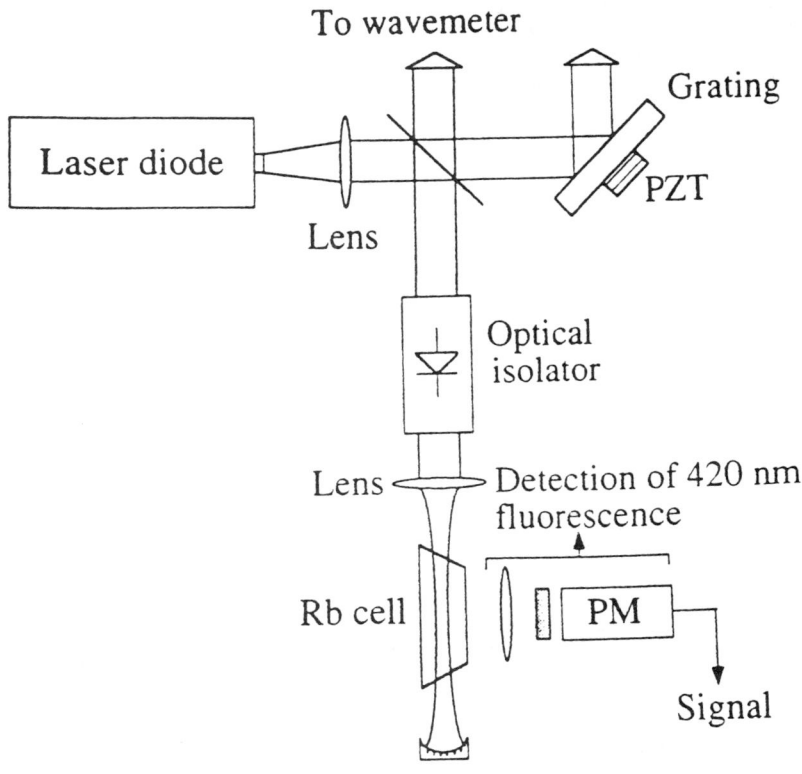

FIGURE 5. Diode laser based Rubidium two-photon spectrometer.

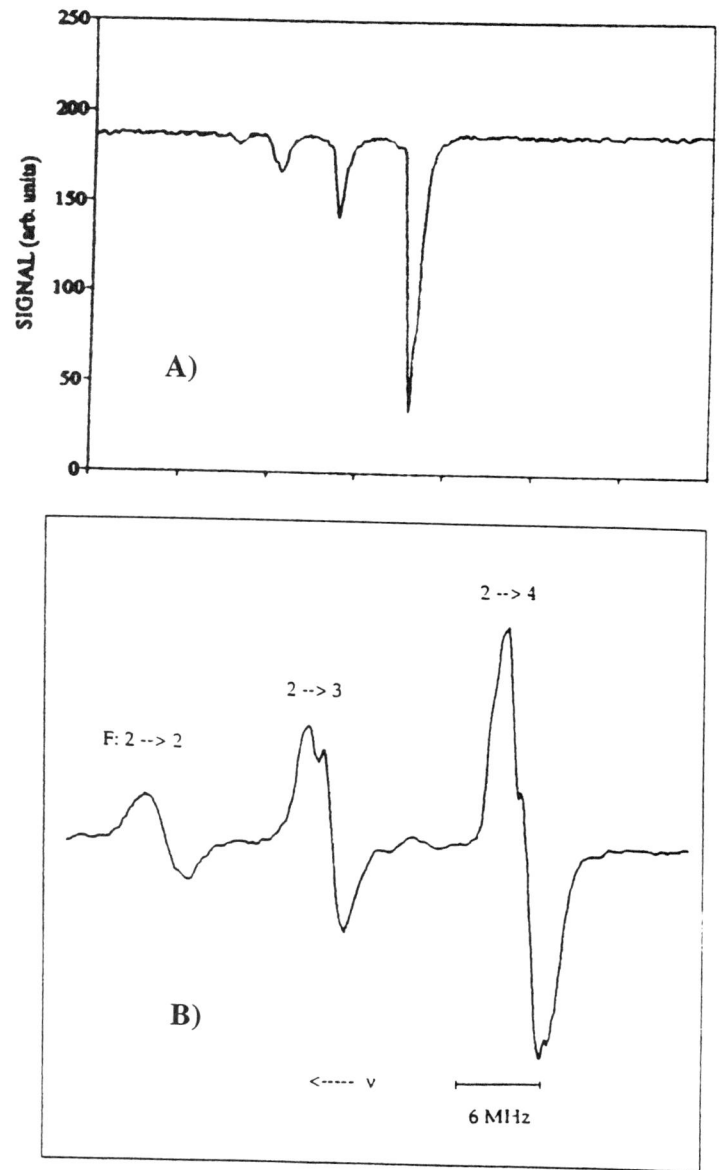

FIGURE 6. a) Two-photon subdoppler recording of the hyperfine components of the ^{87}Rb $5S_{1/2}$ (F=2) -> $5D_{5/2}$ (F=1-4). b) First derivative signal. Time constant 1 msec. The frequency lock of the diode laser on the F= 2->4 component provides a reference at ν = 385284566366±6 KHz, after light shift is subtracted.

The laser radiation for the two-photon excitation comes from the intracavity beam splitter, while two other outputs, from the other side of the beam splitter and from the zeroth order of the grating are used for the wavelength control and for the mixing with the radiation from the Ti:sapphire laser. A 60 dB optical isolator is necessary to avoid the feedback from the two-photon section. The laser beam is focused into the Rubidium cell by means of a lens and then retroreflected by a spherical mirror. The geometry is controlled in order to carefully evaluate the beam waist, which is necessary to estimate the light shift effect. The two-photon transition is detected by monitoring the fluorescence at 420 nm, as indicated in Fig. 4. The Rb cell is heated at about 90 °C and is placed at the center of three pairs of Helmholtz coils for the compensation of the residual magnetic field along the three axis. A typical subdoppler recording of a portion of the hyperfine structure is shown in Fig. 6. The linewidth of each resolved component is about 1 MHz, almost twice the natural linewidth possibly because of a non perfect compensation of the Doppler effect, some residual magnetic field and fast frequency jitter of the laser. The first derivative signal is well suited for the frequency lock of the diode laser. Indeed, to create the reference for the helium experiment, we locked the laser on the strongest F=2→4 component.

Metastable He Atomic Beam and Laser Interaction

The accuracy of our original (3) Helium frequency measurement at 389 nm, as well as that of the previous (10) wavelength determination, was affected by factors related to the environment, in both case a discharge cell, such as collisions or spurious electric and magnetic fields, under control only to a certain extent. As a consequence we decided to implement the apparatus with a metastable Helium atomic beam.

The original mainframe of the beam had been developed by the group in Napoli to study atomic oxygen. In the present work much care has been devoted to the excitation of the helium atoms in the triplet metastable state and to optical configuration for the interaction with the laser beam and its detection.

In principle, a simple way to produce metastable atoms is to induce collisions with electrons which must be rather monoenergetic as a consequence of excitation cross sections showing a narrow peak with the collision energy. However, for Helium the cross sections for this process are quite small, of the order of $0.01 Å^2$ at a peak of about 35 eV. Due to this fact, an elevate electron density current is necessary to reach a high level of atomic excitation, condition which is not easy to get with a standard electron gun. Moreover Helium atoms are light and are sensibly deflected by elastic and inelastic collisions with the electrons.

For all the above reasons we chose to use as excitation method a longitudinal discharge just at the exit of an effusive atomic source (a 1 cm long channel with a diameter of 0.3 mm), with a scheme similar to that demonstrated (27) highly efficient for the production of metastable magnesium atoms. Over a threshold on the atomic flux, the discharge is self-sustained by a relatively large production of ions. These almost neutralise the negative charges allowing electron currents of amperes over cm^2 at a relatively low voltage. The molybdenum toroidal cathode, coated by a mixture of BaO and SrO, is positioned around the channel aperture, where the atomic density has its maximum, while the anode has the shape of a cylinder coaxial with the atomic beam and is made by two stainless steel rings at the ends and a number of tungsten wires as the lateral wall. A collimator (with a slide of 1 mm) separates the excitation chamber from the metastable beam chamber and is positioned at 6 cm from the channel aperture. To strongly enhance the electron density, and therefore the production of metastable atoms, the discharge is inside a permanent magnet that produces a longitudinal field (maximum ≈ 400 G) with strength lines forming "funnels" that have their narrow section at the anode. The motion of the electrons in these non uniform electric and magnetic fields is quite complex, and in particular the Larmor loop changes into a cycloid with a drift motion perpendicular to both the local electric and magnetic fields. Because of the axial symmetry of the system, the resulting motion is then mainly a spiral around the axis. The electrons go forward and backward several times inside the anode until, because of the collisions, they diffuse reaching the anode wall. In this way inside the anode the electric potentials grow radially, with a maximum at the axis.

In conclusion, this discharge configuration allows a high current density and a low potential along the beam axis, hence optimising the production of metastable atoms. The starting atomic flux from the source channel is about 3×10^{19} atoms/sec, the anode current is nearly 500 mA, with an anode voltage which may drop to less than 60 V for high flux (in this case the atomic mean free path becomes too low in the production chamber), and is around 70 V in good working conditions. In the beam we could estimate, also using a metastable detector at the end, a flux of at least a few 10^{14} met. at./(secxsterad). A second collimator (again a slide of 0.3 mm) is placed just before the laser/atoms interaction zone limiting the transverse Doppler broadening to about 15 MHz.

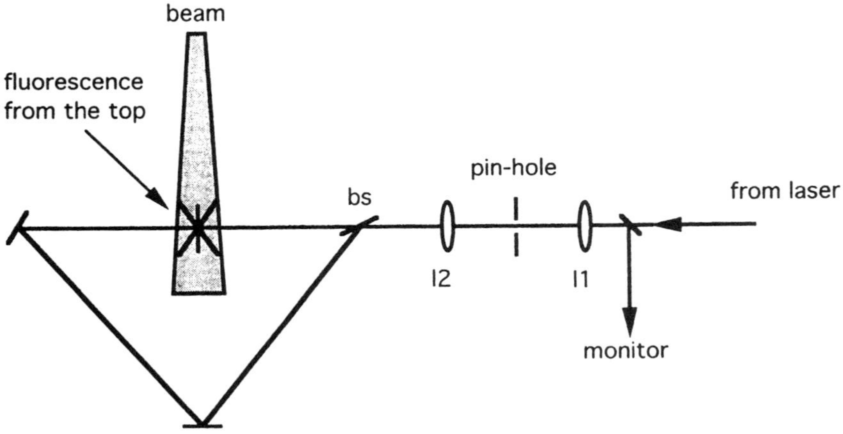

FIGURE 7. Scheme of the geometry of the interaction between laser and atomic beam. Fluorescence at 389 nm is detected from the top.

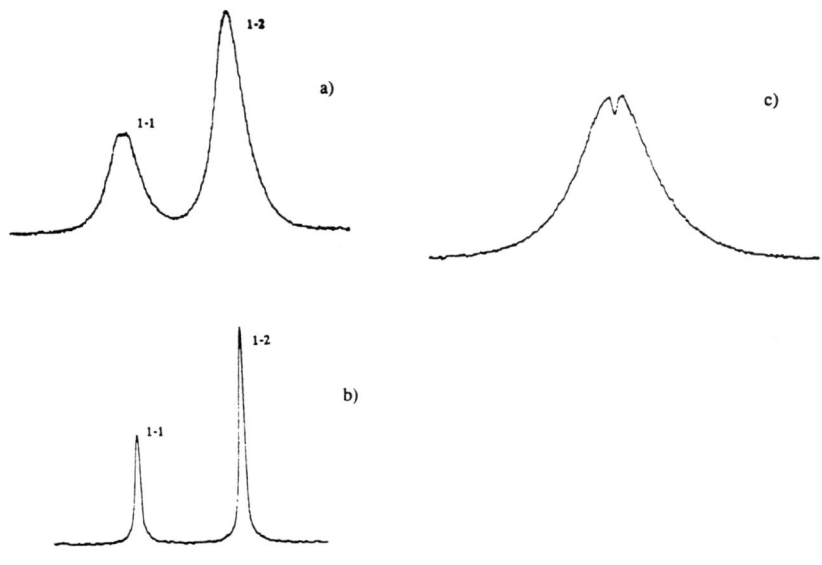

FIGURE 8. Typical recordings of the Helium transitions at 389 nm. In a) and b) signals are obtained using a single laser beam, with an improved atomic beam collimation in b). The Doppler width is reduced by nearly a factor of 500. In c) two counterpropagating beams are used and the saturation dip is detected.

Much care is devoted to the careful control of relative orthogonality and forward-backward alignment between the laser and the atomic beams. This is achieved with the aid of a pin-hole and of a "triangle" configuration, as schematically shown in Fig. 7. Fluorescence at 389 nm is then detected from the top. Typical recordings of the (partial) fine structure of the $2^3S_1 \rightarrow 3^3P_J$ transition are shown in Fig. 8. "Single laser beam" recordings are shown together with a saturated absorption signal obtained in presence of two counterpropagating beams. For the preliminary measurements reported in this work we preferred to use the "single" beam configuration: the linewidth was much reduced by means of a better collimation and the good signal to noise ratio allowed the frequency lock of the laser at the center with a jitter of few hundred kilohertzs. Frequency measurements were then performed alternatively using the laser beam from the right and from the left in order to evaluate, compensate and average possible minor differences caused by possible residual misalignment between the beams. A significant improvement in the linewidth and in the signal intensity could come from a cooled and transversally compressed atomic beam.

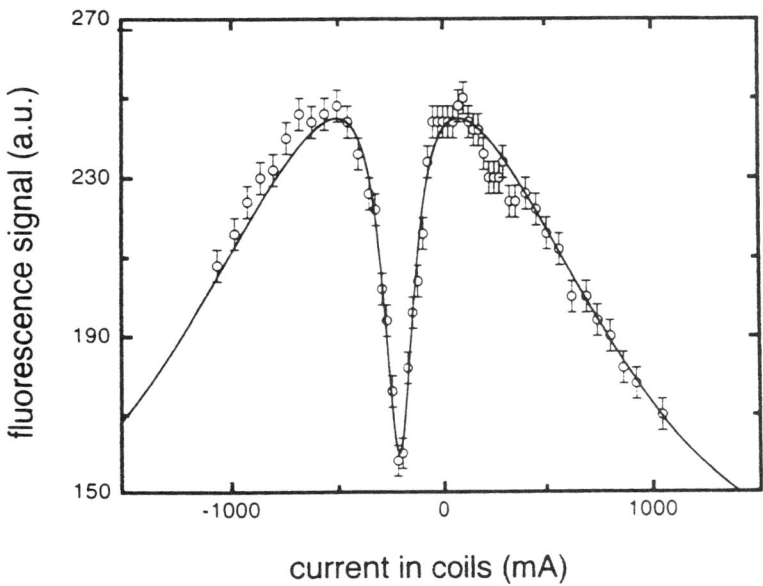

FIGURE 9. Typical recording of the nonlinear Hanle effect used to control the residual magnetic field. This could be compensated at better than 10 mG using three pairs of Helmholtz coils. The $2^3S_1 \rightarrow 3^3P_0$ transition was used.

Three pairs of Helmholtz coils in the interaction region allow to compensate the residual magnetic field. The "zero" value in each direction is determined by sending a laser beam polarised in such way to induce $\sigma^+ + \sigma^-$ transitions and hence detecting the change in fluorescence caused by the nonlinear Hanle effect (28), as typically shown in Fig. 9.

Optical-Microwave Frequency Mixing and Measurements

The measurement of the transition at 389 nm is obtained by locking the frequency doubled Ti:Sapphire on He, the semiconductor diode on Rb, and by heterodyning the two radiations at 778 nm. However, the two frequencies are still much different (82 GHz) and therefore a fast nonlinear detector is needed. For this purpose, we have used a Schottky-barrier diode formed by the contact between an electrolytically sharpened tungsten wire and a GaAs semiconductor base. Discovered as a broadband mixer of visible light ten years ago (29,30), the Shottky diode has been recently introduced by L. Hollberg and co-workers (31) as efficient device for heterodyning diode lasers and high harmonics of microwave sources.

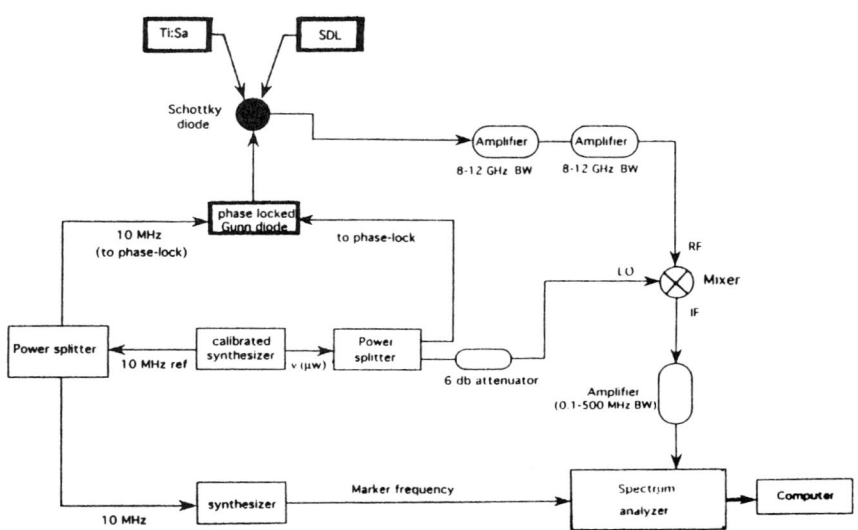

FIGURE 10. Scheme of the mixing experiment of Ti:Sapphire and diode lasers in a Shottky diode. The beat note frequency is lowered by mixing with microwaves at 82 GHz from a Gunn diode phase locked to a low noise quartz oscillator.

In our experiment, schematically shown in Fig. 10, the two laser beams are superimposed and focused on the Shottky diode by means of a microscope objective with magnification 60 and numerical aperture 0.7. The coupling angle is about 30° from the tungsten whisker. The beat note frequency is lowered by mixing with radiation from an intermediate frequency oscillator. In particular, microwave radiation from a Gunn diode oscillating at 82 GHz is coupled through a waveguide, with the electric field parallel to the whisker. The Gunn oscillator was phase locked to a low noise quartz oscillator following the scheme, also shown in Fig. 10, developed in collaboration with A. Godone from IEN "G. Ferraris", Torino. Low frequency beat notes between the Gunn diode frequency and the frequency differences of the lasers were observed with a typical signal to noise ratio of 25 dB (300 KHz bandwidth). For the preliminary experiment here reported, some electronic components (for instance the bandwidth of the microwave amplifiers) were optimised for the measurement of the frequency of the $2^3S_1 \rightarrow 3^3P_0$ fine structure component.

The results for several independent measurements of the low frequency beat note are shown in Fig. 11. The different values are scattered within 200 KHz and from their average a value of 770732840.325 (45) MHz can be inferred for the frequency of the doubled Ti:Sapphire laser locked on Helium. However, in order to obtain the frequency of the $2^3S_1 \rightarrow 3^3P_0$ transition, some corrections must be taken into account.

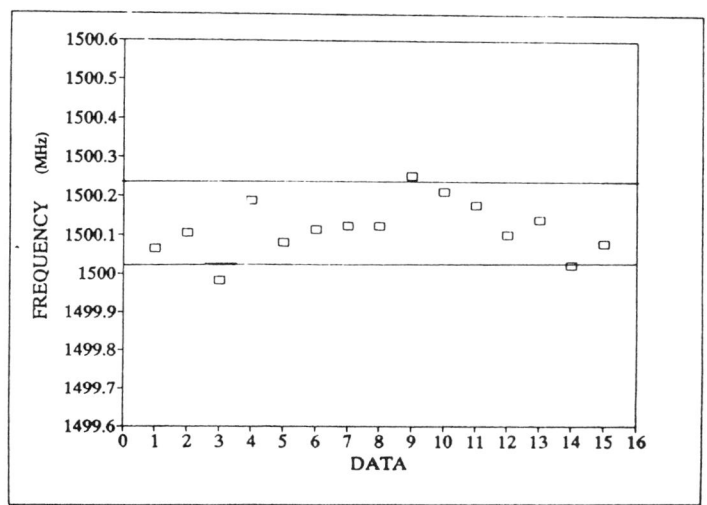

FIGURE 11. Results of independent low frequency beat notes with the frequency doubled Ti:Saph. laser locked to the $2^3S_1 \rightarrow 3^3P_0$ He transition. The total spread is about 200 KHz.

They are the Rb and He light shifts, the photon recoil which for this transition is rather large (-325 KHz ≈ Γ/4), the 2nd order Doppler shift. The latter is at present the largest source of uncertainty before a quantitative analysis of the velocity distribution will be completed. However a preliminary value for He $2^3S_1 \rightarrow 3^3P_0$ transition can be given: 770732840151 (50) KHz which shows an accuracy of 6×10^{-11}. This value can be compared with a wavelength measurement (10) of the crossover resonance among the He $2^3S_1 \rightarrow 3^3P_1$ and $2^3S_1 \rightarrow 3^3P_2$ transitions, using the existing (32) fine structure values for the 3^3P state and taking into account the correction for the HeNe secondary standard (33). Even though our value must be considered as preliminary, the comparison in Fig. 12 clearly indicates the significant improvement obtained by the direct measurement of the frequency, as already discussed also for our previous measurements in a discharge cell (3).

It will be desirable to repeat the pure optical frequency measurement also for the others $2^3S_1 \rightarrow 3^3P_1$ and $2^3S_1 \rightarrow 3^3P_2$ transitions. Indeed, with a confirmed accuracy of a few tens of kilohertz, it will be possible to extract by difference the fine structure separations in the 3^3P state. This would allow a meaningful comparison with the data from a level crossing experiment (32) using high magnetic fields and theoretical values for the giromagnetic factors.

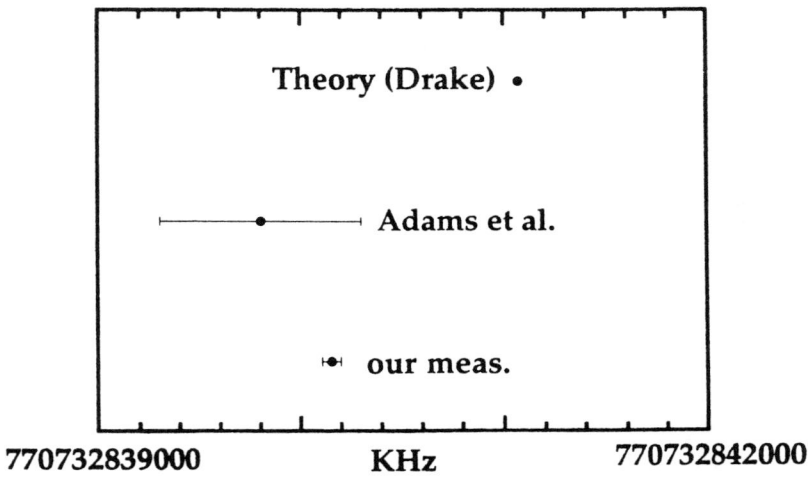

FIGURE 12. Comparison between the $2^3S_1 \rightarrow 3^3P_0$ transition frequency values as directly measured in an atomic beam in this work (preliminary) or inferred from a wavelength measurement (Adams et al.) combined with fine structure data. The theoretical prediction takes into account a "renormalization" factor as explained in ref. 16.

A recent analysis (34) of ^{3}He hyperfine structure and isotope shift indicates that the fine structure data should be possibly corrected, but it is clear that a direct measurement would be important.

2^3S_1 and 2^3P Levels Lamb Shift

As for the theoretical prediction reported in Fig. 12, it is worth noting that the uncertainty of 163 MHz is not shown. This uncertainty is mostly caused by the evaluation of the 2^3S_1 level Lamb shift, in particular of the "one electron" contribution (16). Also, a renormalization for the energy position of the 2^3S_1 level is introduced (16), as inferred from the analysis of all previous experimental results. However, the accuracy of theory improves significantly going to higher excited levels and in particular for n>2, as shown in Table I where we report a selection of the computed (16) uncertainties of the energy levels due to QED terms and Rydberg constant determination. This means that energy, and Lamb-shift, of the metastable 2^3S_1 level can be obtained from accurate measurements of transitions to higher excited levels. Considering a theoretical uncertainty of 600 KHz for the 3^3P level and a most accurate measurement of the $2^3S_1 \rightarrow 3^3P_0$ transition, we were able (3) to extract the 2^3S_1 level Lamb-shift with an accuracy of 1.9 parts in 10^4.

Table I. Theoretical uncertainties (in KHz) in the ^{4}He energy level position due to QED terms and Rydberg constant determination.

Level	Uncertainties	
	Lamb Shift	Rydberg Constant Determination
2^1S_0	180	---
3^1P_1	600	---
2^3S_1	163000	---
2^3P_0	1800	---
2^3P_1	1800	---
2^3P_2	1800	---
3^3P_0	600	---
3^3P_1	600	---
3^3P_2	600	---
3^3D_3	17	8
4^3D_3	10	4
5^3D_3	6	3
6^3D_3	3	2
3^3D_1	17	8
4^3D_1	10	4
5^3D_1	6	3

The present measurements in the beam, even if preliminary, further reduce the uncertainty, which now is essentially determined by the theoretical accuracy in the 3^3P level. The experimental value for the Lamb-shift of the 2^3S_1 level is more than two orders of magnitude more accurate than the theoretical prediction (16). This should stress an improved calculation of the Bethe logarithm for the 2^3S_1 level, as done by Baker et al. (35) for the corresponding singlet level. In Fig. 13 we compare our results with the values which can be deduced from the investigations of other transitions. The upper level of each transition is also indicated since the uncertainty of its theoretical prediction (Tab. I) constitutes the ultimate limitation to the accuracy of the Lamb-shift which can be extracted for the 2^3S_1 level. In this respect, the theory is much more accurate for D states and two-photon transitions to them are the best candidates for further significant improvements. For instance, the uncertainty from the investigation of Hlousek et al. (9) is mostly dominated by the accuracy of the wavelength measurement, while the theory for the upper level would allow an improvement of more than one order of magnitude. Particularly interesting could be to repeat the measurement of the two-photon transition to the 4^3D level, which opened the list of high resolution investigations in Helium (Giacobino and Biraben (8)). In this case the wavelength is 632.8 nm and a direct frequency measurement could now be performed taking advantage of the close coincidence (112 GHz) with the I_2 stabilized HeNe laser (36).

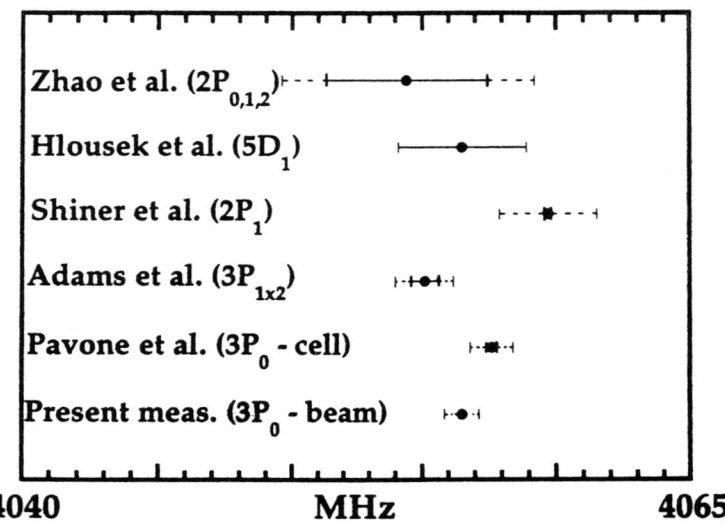

FIGURE 13. Lamb-shift of the He 2^3S_1 level as deduced from measurements on different optical transitions (Zhao et al. ref. 11, Hlousek et al. ref. 9, Shiner et al. ref. 12, Adams et al. ref. 10, Pavone et al. ref. 3). When necessary, the corrections for the new value (33) of the HeNe frequency standard have been done.

On the contrary, the uncertainty which can be achieved from the $2^3S_1 \rightarrow 2^3P$ transition is already mostly limited, in particular for Shiner et al. (11), by the poor knowledge of the Lamb-shift of the 2^3P upper level. Instead, these spectroscopic measurements at 1.08 μm can be used to obtain an accurate value for the energy of this level. Indeed, we can use our determination of the energy of the 2^3S_1 level and the measurements (11,12) on the $2^3S_1 \rightarrow 2^3P$ transition to extract an experimental value of the energy of the 2^3P level. By subtracting the non QED contributions (16), we obtain the Lamb shift. The result is compared in Fig. 14 with the theoretical prediction which in this case seems rather satisfactory. It is evident that the availability of precise measurements on different transitions could allow other "loop" calculation for a further test of the theoretical model and, maybe, an alternative determination of fundamental quantities such as the Rydberg constant or the Casimir forces contribution to the energies.

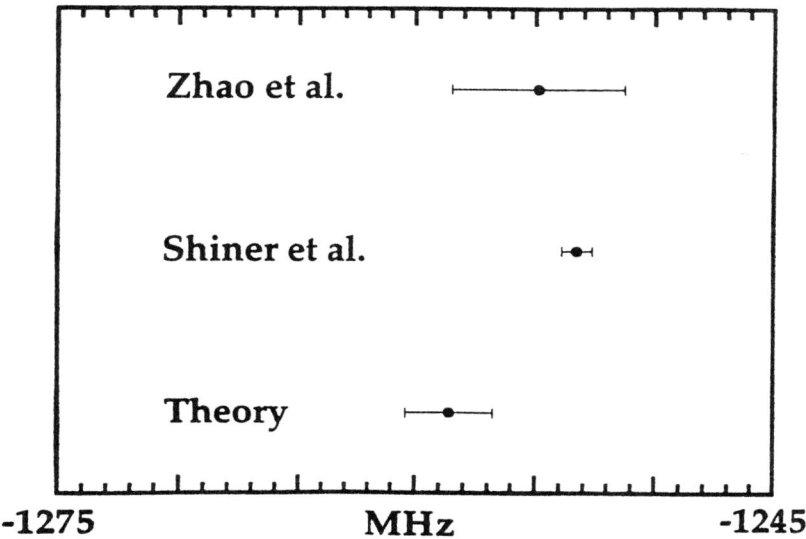

FIGURE 14. Lamb-shift of the 2^3P level as deduced combining spectroscopic measurements at 389 nm (ref. 3 and present work) and at 1.08 μm (two different determinations by Zhao et al. (11) and Shiner et al. (12)). The theoretical prediction (16) is also shown.

ISOTOPE SHIFT AND HYPERFINE STRUCTURE

Precise measurements on the $2^3S_1 \rightarrow 3^3P$ and $2^3S_1 \rightarrow 2^3P$ transitions were also extended (17,37,38,39) to ^{3}He for the determination of isotope shift and hyperfine structure. Most interest comes from the fact that the partial cancellation of the less accurately known terms contributing to the energy allows to give predictions (16) for the isotope shift with an uncertainty of 1 KHz, not counting that originating from the finite nuclear size. This contribution (field shift) can be evaluated in a first-order perturbative scheme, and a laser spectroscopy experiment, if sufficiently accurate, can reach to effects of nuclear structure, and in particular to give values of the nuclear radii. In particular, ^{3}He nuclear charge radius could be determined, using as a reference the precise value available for ^{4}He from muonic Helium (39). In this way, values from table-top atomic physics experiments, in a regime of low energy and momentum transfer, can be compared with the results extrapolated from high-energy scattering measurements. The accuracy achievable from atomic physics shows to be higher even though an unambiguous analysis of the measurements still requires a better understanding of both fine and hyperfine structures.

α – FINE STRUCTURE CONSTANT AND HELIUM PRECISION SPECTROSCOPY

We have so far evidenced the importance of a precise knowledge of the fine structure separations in 2 and 3 ^{3}P levels. Indeed these are involved in the quantitative evaluation of different effects such as QED corrections, nuclear size contributions and hyperfine interactions. Even more important, the comparison with theoretical calculations could be used to determine an "atomic physics" value for the fine structure constant α.

The list of precise measurements of helium fine structure was opened by the early work by Lamb and Maiman (40) in 1957. More recently, in 1981, Hughes and co-workers (41,42) measured the separations between the 2 ^{3}P$_J$ sublevels using an optical-microwave atomic beam magnetic resonance technique. They could extract a value of α with an accuracy of 0.8 ppm. This was possible thanks to a comparison with the theory of Lewis and Serafino (43) taking into account perturbative terms up to α^4 a.u. At present, a measurement of the α constant must compete with the accuracy of 0.008 ppm, 0.024 ppm and 0.056 ppm achieved respectively from the QED analysis of the electron anomaly, from the Quantum Hall effect and from the ac Josephson effect. An exhaustive comparison of the different most precise determinations of α was performed by Kinoshita (44).

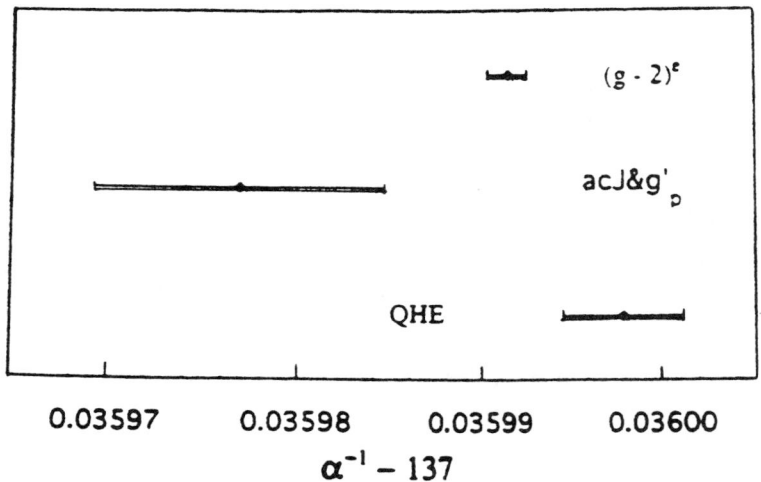

FIGURE 15. Most precise values of the fine structure constant a as extracted from different physical systems. (g-2) = electron magnetic dipole anomaly; (QHE) = quantized Hall effect; (acJ&γ$_p$) = Josephson effect.

The situation is summarised in Fig. 15. As can be noticed the agreement is not satisfactory (values scattered over 0.15 ppm) and could reflect the substantial differences between the physical systems in which measurements are performed. A measurement of the Helium fine structure separations with a precision of 1 KHz would provide an α value with an accuracy of 0.016 ppm (2P separations) or 0.06 ppm (3P separations). It is worth noting that, with respect to Hydrogen, the Helium P states are longer lived and natural linewidths are of the order of 1.5 MHz. As a consequence the 1 KHz accuracy in the separations measurements is accessible at least in principle. Of course a sensitive and precise technique must be implemented and in our opinion the value must be extracted by a direct frequency measurement of the separation. Probably the disagreement between the values reported in refs. 41,42 and those in ref. 12 comes from the fact that these more recent values are extracted only as differences of the 2S→3P optical wavelengths determined separately. However, also the existing measurements using the double resonance scheme (41,42) could have suffered by the presence of a strong magnetic field which was added to tune the atomic structure into resonance with a fixed frequency microwave radiation.

It is worth noting that 1 KHz accuracy measurements are likely to be comparable with a similar accuracy theoretical predictions. In the last years, the dominant α^2 term has been computed (45) with a 0.03 KHz accuracy. The α^4 corrections come from the second order Breit terms and by the Douglas and Kroll terms (46). These corrections have been calculated to better than 100 Hz, together with $\alpha^5 \ln Z\alpha$ terms (47) and hence the theory at the α^4 level is now complete. The contribution of the pure α^5 terms is less than 50 KHz and it is likely to be accurately evaluated (48) also because it is sufficient to take into account only the spin dependent terms.

CONCLUSION

In conclusion, we have shown how pure frequency measurements can bring the uncertainty of optical energy separations in Helium down to the level of few parts in 10^{11}. Combined with the dramatic progress in the theory, these experiments make possible a quantitative analysis of QED and finite nuclear size corrections. Lamb shifts are obtained for 2^3S_1, 2^3P and 3^3P levels, as well as the nuclear charge radius for ^{3}He. Like Hydrogen for two-body systems, Helium can now provide high accuracy tests of the fundaments of the theory of three-body systems and multielectron atoms. Furthermore, the investigation of triplet states opens the fascinating perspective of an "atomic physics" determination of the fine structure constant α with an accuracy of few parts in 10^8.

[a] - Also Department of Physics, University of Firenze, Italy.
[b] - Also Istituto Nazionale di Ottica, Firenze, Italy.

REFERENCES

1. F. Marin *"Precision spectroscopy of Helium at 389 nm"* Ph. D. Thesis, Scuola Normale Superiore, Pisa, Italy, 1994.
2. F. Nez, M.D. Plimmer, S, Bourzeix, L. Julien, F. Biraben, R. Felder, Y. Millerieux, and P. De Natale: *Europhys. Lett.* **24**, 635 (1993).
3. F.S. Pavone, F. Marin, P. De Natale, M. Inguscio, and F. Biraben: *Phys. Rev. Lett*, **73**, 42 (1994).
4. C.J. Sansonetti and J. D. Gillaspy: *Phys. Rev. A* **45**, R1 (1992).
5. C. J. Sansonetti, J. D. Gillaspy, and C. L. Cromer: *Phys. Rev. Lett.* **65**, 2539 (1990).
6. W. Lichten, D. Shiner, and Zhi-Xiang Zhou: *Phys. Rev. A* **43**, 1663 (1991).
7. K. S. E. Eikema, W. Ubachs, W. Vassen, and W. Hogervorst: *Phys. Rev. Lett.* **71**, 1690 (1993).
8. E. Giacobino and F. Biraben: *J. Phys.* **B15**, L385 (1982).
9. L. Hlousek, S. A. Lee, and W. M. Fairbank, Jr.: *Phys. Rev. Lett.* **50**, 328 (1983).
10. C. S. Adams, E. Riis, A. I. Ferguson, and W. R. C. Rowley: *Phys. Rev. A* **45**, R2667 (1992).
11. Ping Zhao, J. R. Lawall, A. W. Kam, M. D. Lindsay, F. M. Pipkin, and W. Lichten: *Phys. Rev. Lett.* **63**, 1593 (1989).
12. D. Shiner, R. Dixon, and P. Zhao: *Phys. Rev. Lett.* **72**, 1802 (1994).
13. G. W. F. Drake, I. B. Khriplovich, A. I. Milstein, and A. S. Yelkhovsky: *Phys. Rev. A* **48**, R15 (1993).
14. J. D. Baker, R. C. Forrey, J. D. Morgan III, and R. N. Hill: *Bull. Am. Phys. Soc.* **38**, 1127(1993).
15. F. Nez, F. Biraben, R. Felder, and Y. Millerioux: *Opt. Commun.* **102**, 432 (1993).
16. G. W. F. Drake; F. S. Levin and D. A. Micha Eds., : in *"Long-Range Casimir Forces: Theory and recent Experiments on Atomic Systems"*, Plenum Press, New York 1993.
17. F. Marin, F. Minardi, F. Pavone, and M. Inguscio: *Phys. Rev. A* **49**, R1523 (1994).
18. F. Nez et al.:*Phys. Rev. Lett.* **69**, 2326 (1992).
19. O. Acef, R. Felder, and Y. Millerioux : private communication.
20. Y. Millerioux, D. Touahri, L. Hilico, A. Clairon, R. Felder, F. Biraben, and B. de Beauvoir: *Opt. Commun.* **108**, 91(1994).
21. Y. Millerioux, R. Felder, D. Touahri, O. Acef, L. Hilico, A. Clairon, F. Biraben, B. de Beauvoir, L. Julien, and F. Nez: Proc. CPEM '94, Boulder (USA) and IEEE Trans. Instrum. Meas., to be published.
22. P. Jungner, S. Swartz, J. Ye, J. L. Hall, and S. Waltman: Proc. CPEM '94, Boulder (USA) and IEEE Trans. Instrum. Meas., to be published.
23. R. E. Ryan, L. A. Westling, and H. J. Metcalf: *J. Opt. Soc. Am. B* **10**, 1643 (1993).
24. F. Biraben, M. Inguscio, F. Marin, and F. Pavone: *Laser Physics*, vol. 2, April 1994, special issue in honour of V. P. Chebotayev.
25. G. M. Tino, L. Hollberg, A. Sasso, M. Inguscio, and M. Barsanti: *Phys. Rev. Lett.* **64**, 25(1990).
26. G. Pesce, A. Sasso, and G. M. Tino: unpublished.
27. G. Giusfredi, A. Godone, E. Bava, and C. Novero: *J. Appl. Phys.* **63**, 1279 (1988).
28. See for instance: G. Moruzzi and F. Strumia Eds. *"The Hanle Effect and Level-Crossing Spectroscopy"*, Plenum Press, New York and London, 1991.
29. H. U. Daniel, B. Maurer, and M. Steiner: *Appl. Phys. B* **30**, 189 (1983).

30. J.Bergquist and H. U. Daniel: *Opt.Commun.* **48**, 327 (1984).
31. See for instance: R. Fox, G. Turk, N. Makie, T. Zibrova, S. Waltman, M. P. Sassi, J. Marquandt, A. S. Zibrov, C. Weimer, and L. Hollberg; M. Inguscio and R. Wallenstein Eds : in *"Solid State lasers: new developments and applications"* , Plenum Press, New York and London, 1993.
32. D. H. Yang, P. Mc Nicholl, and H. Metcalf: *Phys. Rev. A* **33**, 1725 (1986).
33. O. Acef, J. J. Zondy, M. Abed, D. G. Rovera, A. H. Gerard, A. Clairon, Ph. Laurent, Y. Millerioux, and P. Juncar: *Opt. Commun.* **97**, 29 (1993).
34. F. Marin, F. Minardi, F. Pavone, M.Inguscio, and G. F. Drake: *Z. Physik D*, in press.
35. J. D. Baker, R. C. Forrey, J. D. Morgan III, and R. N. Hill: *Bull. AM. Phys. Soc.* **38**, 1127 (1993).
36. F. Biraben, private communication.
37. P. Zhao, J. R. Lawall, and F. M. Pipkin : *Phys. Rev. Lett.* **66**, 592 (1991).
38. D. Shiner, R. Dixon, and V. Vedantham: to be published.
39. G. Carboni, G. Gorini, L. Palffy, F. Palmonari, G. Torelli, and E. Zavattini: *Nucl. Phys. A* **278**, 381 (1977).
40. W. E. Lamb and T.H. Maiman: *Phys. Rev.* **105**, 573 (1957).
41. A. Kponou, V. W. Hughes, C. E. Johnson, S. A. Lewis, and F. M. J. Pichanick: *Phys. Rev. A* **24**, 264 (1981).
42. W. Frieze, E. A. Hinds, V. W. Hughes, and F. M. J. Pichanick: *Phys. Rev. A* **24**, 279 (1981).
43. M. L. Lewis and P. H. Serafino: *Phys. Rev. A* **18**, 867 (1978).
44. T. Kinoshita; G. F. Bassani, M. Inguscio, and T. W. Hänsch Eds.: in *"The Hydrogen Atom"*, , Springer-Verlag, Berlin, Heidelberg, New York, London, Paris, Tokyo, 1989.
45. G. W. Drake and Z. -C. Yan: *Phys. Rev. A* **46**, 2378 (1992).
46. M. Douglas and N. M. Kroll: *Ann. Phys.* (N.Y.) **82**, 89 (1974).
47. Tao Zhang and G. W. F. Drake: *Phys. Rev. Lett.* **72**, 4078 (1994).
48. G. W. F. Drake: private communication.

Light Exotic Atoms - Some Recent Developments

Klaus P. Jungmann

Physikalisches Institut, Universität Heidelberg, D-69120 Heidelberg, Germany

Abstract. Simple exotic atomic systems render the possibility of studying fundamental interactions and testing basic symmetries in physics and of determining precise values for fundamental constants. Of particular interest in this connection are hydrogen-like systems. Examples for systems, where some progress has been made recently, are among others pionic hydrogen, antiprotonic helium, muonic helium, positronium, and muonium.

INTRODUCTION

In atomic systems the dominating interaction is electromagnetic. For studying the contributions of other interactions it is of advantage to use simple two and three body atoms, where the theory of quantum electrodynamics (QED) provides the framework for most accurate calculations of the electromagnetic part of the interaction. Particularly in exotic atoms non-electromagnetic interactions can be strongly pronounced or suppressed depending on the composition of the object under investigation.

HADRONIC SYSTEMS

In the purely hadronic systems, like protonium atom ($p\bar{p}$) [1] and the pionic hydrogen atom ($p\pi^-$) [2], the strong interaction contributes significantly to the energies and the decay width of the atomic states. The strong interaction can be studied in such atoms at zero energy.

Pionic Hydrogen

The 1S level of pionic hydrogen ($p\pi^-$) is shifted due to the strong interaction (Fig. 1a) by

$$\epsilon_{1S} = -\frac{4E_{1S}}{3r_B}(2a_1 + a_3) \tag{1}$$

and broadened by

$$\Gamma_{1S} = \frac{16QE_{1S}}{9r_B}\left(1 + P^{-1}\right)\left(a_1 - a_3\right)^2,\qquad(2)$$

where E_{1S} is the 1S binding energy of the π^-, r_B the Bohr radius, Q the reduced mass factor, P the Panofsky ratio and a_1 and a_3 the pion-nucleon s-wave scattering lengths for isospin 1/2 respectively 3/2 [3]. The scattering length can be determined, in principle, from low energy pion-nucleon scattering data with an extrapolation to zero energy. However, such procedures have not led to consistent results.

An experiment presently under way at the Paul Scherrer Insitut, PSI, in Villigen, Switzerland, aims for a clarification of the situation by determining a_1 and a_3 from spectroscopic data from 3P-1S transitions in pionic hydrogen and pionic deuterium $(d\pi^-)$ [2] using a newly build high resolution reflecting crystal x-ray spectrometer with CCD readout [4]. The resolution is substantially better than 1 eV (FWHM). The 3P-1S transition lines were chosen because of nearby calibration lines in argon (K_α) and beryllium $(\pi^-\text{Be}(4\text{f-3d},4\text{d-3p}))$ as shown in Fig. 1b. From previous experiments the shift is known as $\epsilon_{1S,\pi^-p} = (-7.12 \pm 0.32)eV$ and $\epsilon_{1S,\pi^-d} = (5.5 \pm 1.3)eV$. The new experiment aims for a 1% accuracy in the shift and a 10% accuracy in the width which is expected to be 0.9 eV for pionic hydrogen and could not been measured so far.

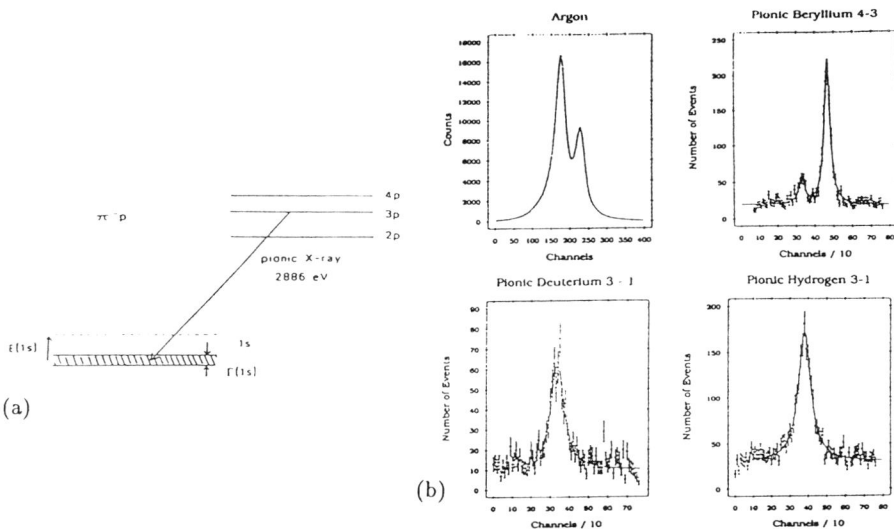

Figure 1: (a) Energy levels of pionic hydrogen. (b) Pionic X-ray signals from pionic Hydrogen and deuterium together with calibration lines in Argon (line splitting 2.13 eV) and Beryllium (line splitting 2.26 eV) [2].

SYSTEMS CONSISTING OF HADRONS AND LEPTONS

The progress in modern spectroscopy allows measurements on atomic hydrogen (pe^-) with highest accuracy [5]. There are various proposals for producing and trapping its antiatom, antihydrogen ($\bar{p}e^+$), at the Low Energy Antiproton Storage Ring (LEAR) facility at CERN [6]. Spectroscopy of similar accuracy may be possible in the future providing a test of the CPT theorem similar in accuracy to the one reached in the mass difference in the $K^0 - \overline{K^0}$ system. Whereas precision experiments on antihydrogen will be reserved for the future, muonic helium and antiprotonic helium already have experimental records.

Muonic Helium

Polarized muonic helium(3) atoms ((μ^{-3}He)$^+e^-$) can be obtained by stopping unpolarized negative muons in a gas volume containing unpolarized helium(3) and a sufficient density of Rb-atoms which have been polarized by laser spectroscopy. Charge exchange and spin exchange reactions result in an average polarization of ^3He of 27% [7]. The experiments were conducted at the Clinton P. Anderson Meson Physics Facility, LAMPF, in Los Alamos, USA.

Polarized muonic helium(3) is of interest for testing fundamental symmetries, for example T-invariance [8], and for investigating the spin dependence of nuclear muon capture in ^{3}He: $\mu^- + ^3$He$\rightarrow ^3$He $+\nu$ which will allow to study the weak interaction between the muon and the nucleon. The measurement will be sensitive to the induced pseudoscalar coupling g_p of the muon to the nucleon and possible second class currents [7].

Antiprotonic Helium

The observation that antiprotons stopped in liquid or gaseous helium exhibit long lifetimes can be explained, if one assumes that the antiprotons are trapped in metastable states of high principal quantum number n and high angular momentum l [9]. A recent laser experiment carried out at the LEAR has strongly confirmed that model [10]. A pulsed tunable dye laser was fired about 1.8 μsec after the formation of the metastable system. A resonant transition was induced between metastable states and short lived states the lifetime of which is mostly determined by Auger processes (Fig. 2a,b). On resonance an immediate increase in the antiproton annihilation rate was observed. The transition at 597.259(2) nm has been assigned to the quantum numbers $(n, l)=$ (39,35) $\rightarrow$ (38,34). By using a second laser, which was pulsed after a (variable) delay with respect to the first one, the re-population from higher states could be demonstrated [11].

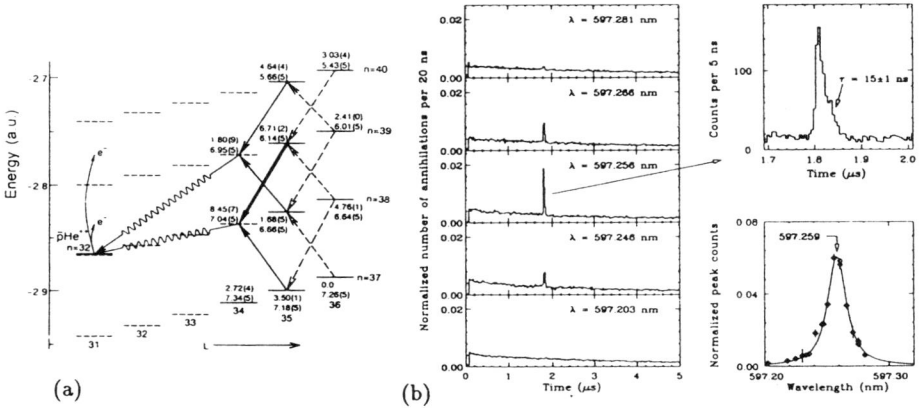

Figure 2: (a) Excited states of antiprotonic helium. (b) Annihilation of antiprotons can be forced by laser radiation which excites the metastable antiproton state to a state with lower angular momentum [10].

With the observation of long lived antiprotonic states it appears not longer impossible to perform precision spectroscopic measurements even in systems consisting of more than one hadronic particle, since immediate annihilation does not appear to be unavoidable.

LEPTONIC SYSTEMS

The energy levels of naturally abundant hydrogen isotopes, hydrogen-like ions and exotic systems with hadronic nuclei are influenced by the finite size and the internal structure of the hadrons and the dynamics of the charge carrying constituents within these particles. The interpretation of highly precise measurements in those systems is limited by the insufficient knowledge of the nuclear structure effects. The positronium atom (e^+e^-) [12] and muonium atom (μ^+e^-) [13] each consist of two different leptons for which no internal structure is known to date [14]. Both systems can be explained up to high accuracy by bound state Quantum electrodynamics. Positronium as a particle-antiparticle system is subject to real and virtual annihilations, which broaden the lines and present an additional challenge for calculations. The low mass of the system requires the Bethe Salpeter formalism for a fully adequate description. In muonium the weak decay of the muon within 2.2 μsec sets a lower bound for all linewidths of $\Delta\nu_{nat,\mu^+e^-}$=145 kHz. The heavy "nucleus" allows the use of the Furry picture in the theoretical description. An important step forward on the experimental side was the observation that muonium atoms in vacuum with thermal energies can be produced from SiO_2 powder targets [13].

Positronium

1S-2S Transition

A measurement of the 1^3S_1-2^3S_1 energy interval in positronium has been performed at the Bell Laboratories in Murray Hill, USA, by Doppler-free two-photon spectroscopy using high power (2.5 kW) cw laser light at 486nm in a high finesse build up cavity [15]. The excited state population was probed using a pulsed laser at 532 nm to photo-ionize the 2S state. The liberated positrons were counted (Fig. 3a,b).

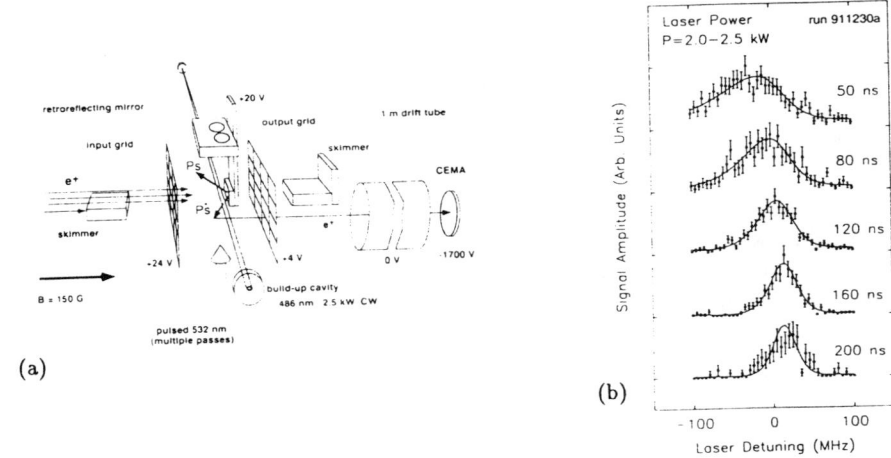

Figure 3: Apparatus (a) and signals (b) in the positronium 1S-2S experiment [15].

The 1S-2S transition frequency was determined to 1 233 607 216.4(3.2) MHz in excellent agreement with QED calculations [16], which yield 1 233 607 221.7 MHz with an estimated uncertainty of order 10 MHz due to uncalculated terms of order $\alpha^2 R_\infty$. The measurement constitutes a test of the $\alpha^2 R_\infty$ QED contributions at the 35 ppm level. The experiment may be interpreted as as test of the the positron-electron mass ratio at the level of 8 ppb.

The cw excitation of the transition circumvents the problems associated with precise frequency measurements (see below: Muonium 1S-2S Transition) and yields higher accuracy. Further improvements are expected, if laser cooled positronium atoms will be used [15].

n=2 State Fine Structure

In an experiment at the slow positron Facility, TEPOS, in Giessen, Germany, microwave transitions between the metastable 2^3S_1 state and the 2^3P_J $J=0,1,2$, states were induced inside of a microwave waveguide (Fig. 4). They were observed by increased Lyman-α-radiation after a transition [17]. The measured frequencies $\nu_0 = 18\,499.65(1.20)(4.00)$ MHz, $\nu_1 = 13\,012.42(0.67)(1.54)$ MHz, and $\nu_2 = 8\,624.38(0.54)(1.40)$ MHz, are in good agreement with the theoretical calculations $\nu_0 = 18\,497.10$ MHz, $\nu_1 = 13\,011.86$ MHz, and $\nu_2 = 8\,626.21$ MHz where the uncertainty is on the MHz level. There is only poor agreement with earlier independent measurements[18].

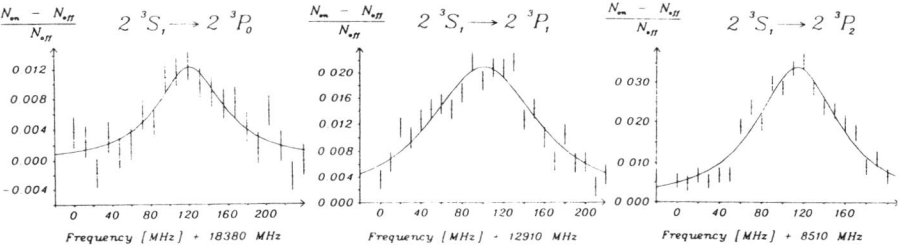

Figure 4: Microwave transition signals from fine structure transitions in the first excited state of positronium [17].

Since positronium is an eigenstate of the charge conjugation operator C the 2^3S_1-2^1P_1 is strictly forbidden. In an external magnetic field, however, it can be induced due to the admixing of other states. The transition frequency was found to be 11 180(5)(4) MHz and from the magnetic field dependence of the transition probability the CP violating matrix element was determined to be less than 76 MHz [19].

Positronium Decays

The large discrepancy between measured Orthopositronium decay rate into three γ's of $\lambda_T = 7.048\,2(16)$ μs^{-1} [20] and theoretical calculations of $99(16) \times 10^{-4}$ [21] constitutes a puzzle now for several years which demands for a satisfactory solution. Recent estimates of terms of order $m\alpha^8$ have given evidence that the problem might be on the theoretical side. The discrepancy gave rise to numerous speculations proposing exotic decays of orthopositronium [22], none of which could be supported by experimental evidence, so far.

A new experimental value is available for the parapositronium decay rate into two γ's, $\lambda_S = 7\,990.9(1.7)\mu s^{-1}$, which is at an accuracy of 215 ppm sufficient to test the $m\alpha^7 \ln \alpha^{-1}$ contributions [23].

Muonium

Ground State Hyperfine Structure Splitting

Important progress has been made in the calculations of the muonium ground state hyperfine structure splitting by numerous theorists calculating terms of order $\alpha^2(Z\alpha)$ in part analytically and in part numerically.. Kinoshita and Nio find $\Delta\nu_{HFS,\mu^+e^-}$=4 463 302.63(1.34)(0.21)(0.17) kHz [24]. This figure includes a 65 Hz contribution from weak interaction due to axial-axial vector coupling. The largest uncertainty arises from the uncertainty in the muon mass, the 210 Hz error reflects the uncertainty in the fine structure constant α measured by the quantum Hall effect, and the 170 Hz uncertainty is due to uncalculated terms. The value compares well with the experiment which yielded $\Delta\nu_{HFS,\mu^+e^-}$=4 463 302.88(16) kHz [25]. If a value is used for the fine structure constant which can be derived from the anomalous magnetic moment of the electron, the improved theoretical calculations allow to extract from the latest measurement the muon mass m_μ with 5.6 times higher accuracy than it has been known so far.

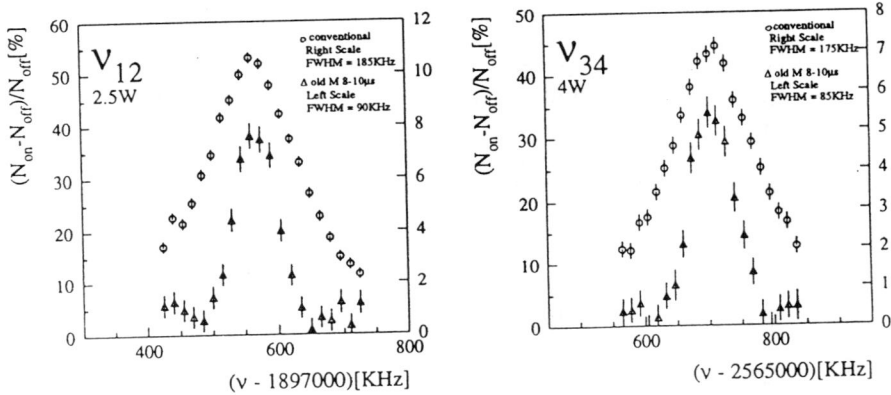

Figure 5: Microwave resonance lines obtained by conventional and by "old muonium" techniques for two Zeeman transitions in the ground state of muonium at 1.7 T magnetic field.

A new experiment is on its way at LAMPF which aims for an improvement over the present value of at least a factor of five [13]. Two Zeeman transitions will be induced by microwaves in an external magnetic field produced by a 1.7 Tesla superconducting MRI magnet, which will allow to obtain in addition to the hyperfine splitting a precise value for the muon magnetic moment, respectively the muon mass. The experiment uses a line narrowing technique by observing atoms only which have been interacting with the microwave field for a time interval large compared to the muon life time. Linewidth of about one half of the natural linewidth of 145 kHz have been already demonstrated (Fig. 5).

1S-2S Transition

In muonium the optical 1S-2S two-photon transition is of particular interest because the transition offers the highest resolution of all possible electromagnetic transitions and hence provides for sensitive experimental tests of the theory. The experiment was performed at the worldwide brightest pulsed surface muon source at the Rutherford Appleton Laboratory, RAL, in Chilton, UK. Doppler-free two-photon absorption was employed for exciting the $1^2S_{1/2}$,F=1–$2^2S_{1/2}$,F=1 transition using a high power pulsed laser system. The population of the 2S state was probed by photo-ionization with a third photon from the same laser field. The slow muon released in the process was observed as the signal (Fig. 6) for a two-photon absorption [26].

The experiment yielded $\Delta\nu_{1S-2S}$ = 2 455 529 002 (33)(46) MHz for the centroid 1S-2S transition frequency in agreement with theory within two standard deviations [27]. This value compares well with an independent measurement at an earlier stage of the experiment [28] and the final result obtained by a Japanese-American collaboration at KEK [29]. The accuracy of the results quoted above as well as the transition rate are limited by systematic effects which are due to the properties of the pulsed laser system (ac-Stark effect and and rapid changes of the refractive index in the dye solutions of the laser amplifier).

The Lamb shift contribution to the 1S-2S splitting has been determined to be $\Delta\nu_{LS}$ = 6 988 (33)(46) MHz. This is the most precise experimental Lamb shift value which could be obtained for muonium so far. The present accuracy is of the size of the 71.6 MHz difference between atomic hydrogen and muonium. The mass of the positive muon has been determined to be m_μ = 105.658 80 (29)(43) MeV/c using the muonium-hydrogen and muonium-deuterium isotope shifts in the resonance signals. It is about one order of magnitude less precise than the best measurements of m_μ [30]. The isotope shifts were obtained by measuring the 1S-2S two-photon transition signals in the same apparatus. The experiment may alternatively be interpreted as a measurement of the ratio of electric charges of electron (q_{e^-}) and positive muon (q_{μ^+}). Relying in the correctness of the QED calculations we can extract q_{μ^+}/q_{e^-} = $-1 - 1.4(1.2)\cdot 10^{-8}$. This test of the ratio of charge quantization in the first two lepton generations is an order of magnitude more precise than

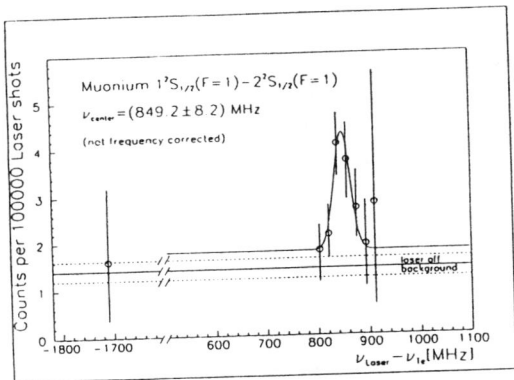

Figure 6: The number of slow muons observed versus the laser frequency offset from the d_4 line in Te$_2$, which serves as frequency reference [26].

the one which has recently been extracted from the muonium ground state hyperfine structure splitting [31].

Independent studies reveal that the influence of systematic line shifts can be understood and controlled to allow a two-photon transition frequency measurement at the 1 MHz level which would allow the extraction of a competitive muon mass value.

Muonium Spectroscopy and the Anomalous Magnetic Moment of the Muon

At the Brookhaven National Laboratory, BNL, in Upton, USA, a precision storage ring is being set up for a precise determination of the muon magnetic anomaly, which may contain contributions from physics beyond the standard model and promises a clean test of the renormalizability of electroweak interaction [32].

The spectroscopic experiments in muonium, the ground state hyperfine structure splitting, the Zeeman effect in the ground state and the 1S-2S energy difference, are closely inter-related with the determination of the muon magnetic anomaly a_μ through the relation $\mu_\mu = (1 + a_\mu)\, e\hbar/(2m_\mu c)$. The results from all experiments establish a self consistency requirement for QED and electroweak theory and the set of fundamental constants involved as indicated in Fig. 7. In this context, the constants α, m_μ, μ_μ are the most stringently tested parameters. The only necessary external input are the hadronic corrections to a_μ which can be obtained from a measurement of the ratio of cross sections $(e^+e^- \to \mu^+\mu^-)/(e^+e^- \to hadrons)$. Although, in principle, the system could provide the relevant electroweak constants, the Fermi coupling

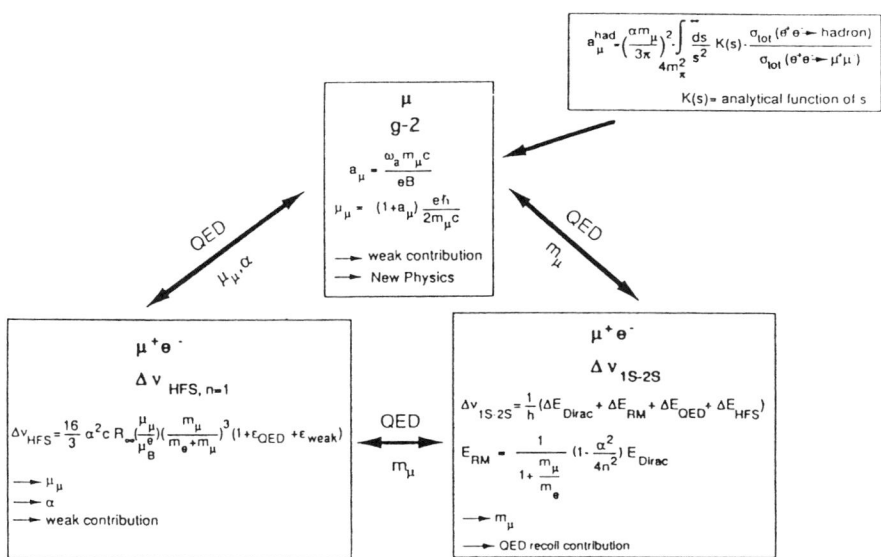

Figure 7: Relation between a measurement of the muon anomalous magnetic moment and spectroscopy of the energy levels in muonium.

constant (G_F) and the weak mixing angle ($\sin^2 \theta_W$), the use of more accurate values from independent measurements may be chosen for higher sensitivity to new physics.

Muonium to Antimuonium Conversion

A spontaneous conversion of muonium atoms into antimuonium atoms ($\overline{M}=\mu^- e^+$) violates separate additive muon and electron number conservation. The process is not provided in the standard model of the electroweak interactions as are other lepton flavour violating processes, in general. However, in many speculative theories, which try to extend the standard model in order to explain some of its features, lepton flavour violation appears to be natural. M–$\overline{M}$ conversion is particularly allowed in some left-right symmetric models [33] and supersymmetric models with broken R-parity [34]. The coupling constant $G_{M\overline{M}}$ in an effective four fermion coupling could be as high as the present experimental limit. Previous experiments found $G_{M\overline{M}} \leq 0.13 \cdot G_F$ (90% C.L.) [35], where G_F is the Fermi coupling constant of the weak interaction. In the framework of minimal left-right symmetric theory a lower bound of $G_{M\overline{M}} \geq 7 \cdot 10^{-5} G_F$ has been estimated which even could become higher, if the present limit on the mass of the muon neutrino could be lowered [33].

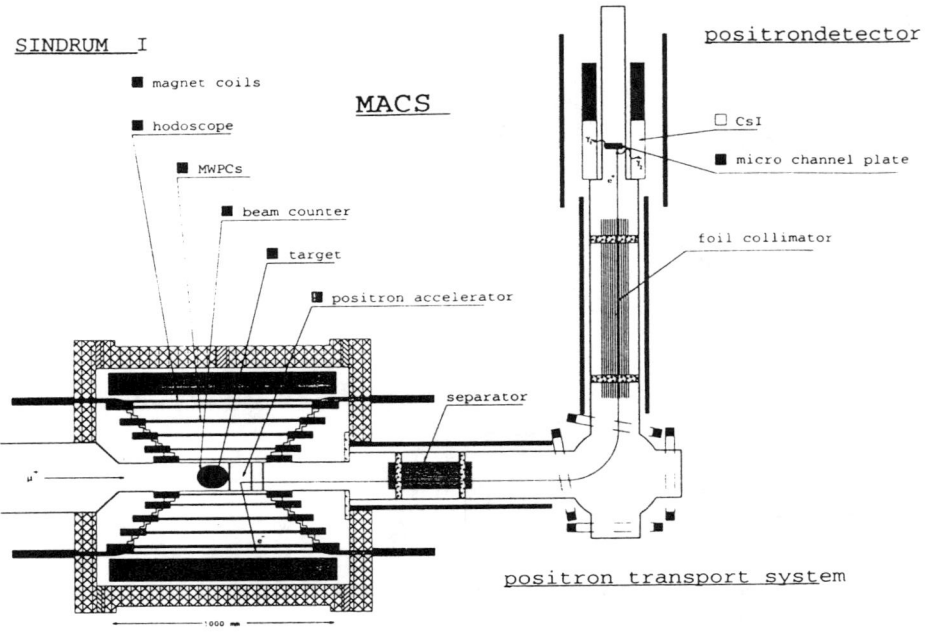

Figure 8: MACS - The spectrometer was built at PSI to search for a spontaneous conversion of muonium into antimuonium [36].

A new experiment is presently carried out at PSI which ultimately aims for an improvement of the experimental sensitivity to the coupling constant by two orders of magnitude over the present value [36]. It utilizes the powerful signature for a conversion event which had been developed at LAMPF [35] and which detects both constituents (μ^-, e^+) of the antiatom in its decay. The experimental setup was tested for the first time in 1992 in the πE3 beam area of PSI. Within 24 hours of searching for antimuonium decays we had $3.9(3)\cdot 10^7$ muonium atoms in the interaction region. No antimuonium event could be observed. Taking into account correction factors representing differences in the relevant muonium respectively antimuonium detection efficiencies (0.64), corrections for the finite observation time (0.85), and the suppression of the M-$\overline{\text{M}}$ conversion in an external magnetic field (0.35) we find at 90% confidence level the M$\overline{\text{M}}$ conversion probability to be $P_{M\overline{M}} \leq 3.3 \cdot 10^{-7}$. The conversion probability is related to the coupling constant through [35,36]

$$G_{M\overline{M}} = G_F \cdot \sqrt{\frac{P_{M\overline{M}}}{2.56 \cdot 10^{-5}}} \quad . \tag{3}$$

In our first test we were able to establish a new experimental limit on the coupling constant of $G_{M\overline{M}} = 0.11$ G_F (90%C.L.). This value is a slight improvement over the results obtained previously [35] and confirms them.

Up to now data has been accumulated for approximately 400 hours while searching for antimuonium. Although the analysis is not yet completed it is known already that the sensitivity to the conversion probability has been improved by two orders of magnitude. Further improvements by at least one order of magnitude are expected within the next two years. This will allow to put a stringent test on the minimal left-right symmetric model predicting a lower limit for the coupling constant $G_{M\overline{M}}$.

CONCLUSIONS

The exotic systems mentioned in this article represent only a selected fraction of all the research work, both experimental and theoretical, that is presently carried out in numerous laboratories worldwide. They proof that atomic physics can make relevant contributions towards a better understanding of fundamental questions in physics. Information can be obtained which is complementary to the results in nuclear and high energy particle physics.

ACKNOWLEDGMENTS

We would like to thank V.W. Hughes and G. zu Putlitz for fruitful discussions and W. Schwarz for carefully reading the manuscript. The support by the organizers of the conference and the BMFT of Germany is gratefully acknowledged.

REFERENCES

1 E. Klempt,"Antiprotonic Hydrogen", in Hydrogen Atom, G.F. Bassani et al (eds.), Springer, Berlin , 211-220 (1989)

2 E.C. Aschenauer et al.,"A Measurement of the strong interaction shift and width of the 1S Levels in Pionic Hydrogen and Deuterium", *PSI Newslett.* **1993** , 57-58 (1994)

3 S. Deser et al.,"Energy Level Displacement in Pi-Mesonic Atoms", *Phys.Rev* **96**, 774-776 (1954)

4 A. Badertscher et al.," A high resolution reflecting crystal spectrometer to measure 3 keV pionic hydrogen and deuterium X-rays", *Nucl.Instr.Meth* A **335**, 470-478 (1993)

5 T.W. Hänsch, "Precision spectroscopy of atomic hydrogen", this volume

6 M. Charlton et al., "Antihydrogen Physics", *Phys.Rep.* **241**, 65-117 (1994)

7 P. Souder ,"Measuring g_p with Laser Polarized ^{3}He", in *Proceedings of the International Workshop on Low Energy Muon Science - LEMS 93*, Los Alamos Conference Report LA-12698-C, M. Leon (ed.), 106-112 (1994)

8   J. Deutsch, "Muon-Capture in Nuclei: The Case of $^3He^+$", in *Proceedings of the workshop on Fundamental Muon Physics: Atoms, Nuclei, and Particles*, C.M. Hoffman et al. (eds.), Los Alamos Conference Report LA-10714-C, 47-55 (1986)

9   S.N. Nakamura, et al. "Delayed annihilation of antiprotons in helium gas", *Phys.Rev.* A**49**, 4457-4465 (1994)

10 N. Morita et al., " First Observation of Laser-Induced Resonant Annihilation in Metastable Antiprotonic Helium Atoms", *Phys.Rev.Lett.* **72**, 1180-1183 (1994)

11 R.S. Hayano et al., " Laser Studies of the decay Chain of Metastable Antiprotonic Helium Atoms", submitted for publication (1994)

12 A.P. Mills and S. Chu, "Precision Measurements in Positronium", in *Quantum Electrodynamics*, T. Kinoshita (ed.), World Scientific, Singapore, 774-821 (1990)

13 V.W. Hughes and G. zu Putlitz, "Muonium", in *Quantum Electrodynamics*, T. Kinoshita (ed.), World Scientific, Singapore, 822-904 (1990); and V.W. Hughes, "Muonium", *Z.Phys.C* **56**, S35-S43 (1992)

14 T. Kinoshita and W.J. Marciano, "Theory of the Muon Anomalous Moment", in *Quantum Electrodynamics*, T. Kinoshita (ed.), World Scientific, Singapore, 774-821 (1990)

15 M.S. Fee et al. "Measurement of the 1^3S_1-2^3S_1 interval by continuous-wave two-photon excitation", *Phys.Rev.* A **48**, 192-219 (1993)

16 R.N. Fell, "Order $\alpha^4 \ln \alpha^{-1} F_{RYD}$ Corrections to the n=1 and n=2 Energy Levels of Positronium", *Phys.Rev.Lett* **68**, 25-28 (1992)

17 D. Hagena, et al.,"Precise Measurement of n=2 Positronium Fine-Structure Intervals", *Phys.Rev.Lett* **71**, 2887-2890 (1993)

18 S. Hatamian, "Measurements of the 2^3S_1-$2^3P_J (J=0,1,2)$ Fine-Structure Splitting in Positronium",*Phys.Rev.Lett.***58**, 1833-1836 (1987)

19 D. Hagena, et al.,"Untersuchung des CP-verletzenden 2^3S_1-2^1P_1 Übergangs im Positronium", Verhandl.DPG(VI) 29, 399 (1994)

20 J.C. Nico et al., "Precision Measurement of the Orthopositronium Decay Rate Using the Vacuum Technique", *Phys.Rev.Lett.* **65**, 1344-1347 (1990)

21 P. Labelle et al., " Order $m\alpha^8$ contributions to the decay rate of Orthopositronium" *Phys.Rev.Lett.* **72**, 2006-2008 (1994)

22 S.N. Gninenko, "Limit on disappearence of orthopositronium in vacuum", *Phys.Lett.* B **326**, 317-319 (1994)

23 A.H. Al-Ramadhan and D.W. Gidley, "New Precision Measurement of the Decay Rate of Singlet Positronium", *Phys.Rev.Lett.* **72**, 1632-1635 (1994)

24 T. Kinoshita and M. Nio,"Improved Theory of the Muonium Hyperfine Structure", *Phys.Rev.Lett.* **72**, 3803-3806 (1994)

25 F.G. Mariam, et al. "Higher Precision Measurement of the hfs Interval of muonium and of the Muon Magnetic Moment", *Phys.Rev.Lett* **49**, 993-996 (1982)

26 F.E. Maas et al., "A Measurement of the 1S-2S Transition Frequency in Muonium", *Phys.Lett.A* **187**, pp. 247-254 (1994)

27 D.R. Yennie, "Two-body QED bound states", *Z.Phys.C* **56**, S13-S23 (1992)

28 K. Jungmann et al., "Two-photon laser spectroscopy of the muonium 1S-2S transition", *Z.Phys.D* **21**, 241-243 (1991)

29 K. Danzmann, M. S. Fee, and Steven Chu, "Doppler-free laser spectroscopy of positronium and muonium: Reanalysis of the 1S-2S measurements", *Phys. Rev. A* **39**, 6072-6073 (1989); and S. Chu et al., "Laser excitation of the muonium 1S-2S transition", *Phys. Rev. Lett.* **60**, 101-104 (1988)

30 Particle Data Group, "Review of Particle Properties", *Phys. Rev. D* **45**, special volume (1992)

31 X. Fei, "Electric charges of positive and negative muons", *Phys. Rev. A* **49**, 1470-1472 (1994)

32 R.M. Carey, "A Report on the New g-2 Experiment at BNL", in *Proceedings of the International Workshop on Low Energy Muon Science - LEMS 93*, Los Alamos Conference Report LA-12698-C, M. Leon (ed.), 180-187 (1994); and V.W. Hughes, "The anomalous magnetic moment of the muon", in *Particles, Strings and Cosmology*, P. Nath and S. Redcross (eds.), World Scientific, Singapore, 868-879 (1992)

33 P. Herczeg and R.N. Mohapatra, "Muonium to Antimuonium Conversion and the Decay $\mu^+ \to e^+ \overline{\nu}_e \nu_\mu$ in Left-Right Symmetric Models", *Phys. Rev. Lett.* **69**, 2475 (1992).

34 A. Halprin and A. Masiero, "Muonium-antimuonium oscillations and the exotic muon decay in broken R-parity SUSY models", *Phys. Rev.* **D48**, 2987 (1993)

35 B.E. Matthias et al., "New Search for the Spontaneous Conversion of Muonium to Antimuonium", *Phys. Rev. Lett.* **66**, 2716 (1991); see also V.A. Gordeev et al., *JETP Lett.* **57**, 270-275 (1993)

36 PSI proposal R-89-06, "Search for Spontaneous Conversion of Muonium to Antimuonium", K. Jungmann and W. Bertl spokespersons (1989)

Experiments with Highly Charged Ions up to Bare U^{92+} on the Electron Beam Ion Trap

Peter Beiersdorfer

Lawrence Livermore National Laboratory, P.O. BOX 808, Livermore, CA 94551

Abstract. An overview is given of the current experimental effort to investigate the level structure of highly charged ions with the Livermore electron beam ion trap (EBIT) facility. The facility allows the production and study of virtually any ionization state of any element up to bare U^{92+}. Precision spectroscopic measurements have been performed for a range of $\Delta n = 0$ and $\Delta n = 1$ transitions. Examples involving 3–4 and 2–3 as well as 3–3 and 2–2 transitions in uranium ions are discussed that illustrated some of the measurement and analysis techniques employed. The measurements have allowed tests of calculations of the the quantum electrodynamical contributions to the transitions energies at the 0.4% level in a regime where $(Z\alpha) \approx 1$.

I. INTRODUCTION

The strong nuclear field of high-Z, highly charged ions provides an ideal environment for measuring the nonperturbative aspects of QED theory where $(Z\alpha) \approx 1$. This is especially true for the contributions from the electron self-energy. The self-energy contribution cannot be tested in muonic atoms, and higher order terms in $(Z\alpha)$ are too small in low-Z multi-electron ions and cannot be measured relative to the size of the uncertainties in solving the Schroedinger equation.

Highly charged, high-Z ions have been produced in a variety of sources. In tokamaks and high-power laser-produced plasmas the highest attainable charge state of uranium, the highest-Z naturally occurring element, approaches that of nickellike U^{64+}. Yet higher charge states, have been produced in heavy-ion accelerators, as attested by two recent measurements of the K-shell transitions in hydrogenlike and heliumlike uranium [1,2]. While accelerators combined with storage-ring facilities can now provide virtually any ion imaginable, the experimental conditions are very challenging. Ions are moving at relativistic speeds, and there typically is a high-radiation environment. As a result, only two other spectroscopic measurements of very highly charged uranium ions have been reported apart from the two measurements already mentioned. These are the measurements of the $2s_{1/2}$–$2p_{1/2}$ energy splittings in lithiumlike U^{89+} [3] and in heliumlike U^{90+} [4].

During the past six years the electron beam ion trap (EBIT) has been operated as a spectroscopic source at the Lawrence Livermore National Laboratory. Like ion-beam accelerator facilities it also is able to produce very highly charged ions. This included ions as highly charged as neonlike U^{82+} [5]. Recently, a high-energy version, dubbed Super-EBIT, started operation. This machine has produced ions as highly charged as bare U^{92+} [6]. Unlike accelerator sources, the ions in EBIT are stationary and trapped in the space-charge potential of the electron beam. Their thermal temperature was shown to be less than a few hundred eV [7]. As spectroscopic sources the EBIT devices thus use the inverse of the accelerator-based beam-foil or beam-gas jet technique. Instead of a stationary electron target interacting with relativistic ions, the ions in EBIT form a stationary target interacting with a relativistic electron beam. Under these conditions, "standard" spectroscopic techniques can be applied to measuring the emission from highly charged ions [8], and systematic measurements over a wide range of ions and transitions are possible.

Experiments on EBIT so far include high-resolution crystal spectrometer measurements of the x-ray emission from M-shell transitions in nickellike U^{64+} and L-shell transitions in neonlike U^{82+}, as well high-resolution crystal spectrometer measurements of $3s_{1/2}$–$3p_{3/2}$ transitions in near sodiumlike and of $2s_{1/2}$–$2p_{3/2}$ transitions in near heliumlike uranium. Moreover, measurements of K-shell transitions have been made with solid-state detectors that prove the production of hydrogenic and bare uranium. The relatively easy access to highly charged ions and the precision achieved in these measurements, which rivals and exceeds that of other sources, indicates that experimental atomic physics has finally entered an era, where all ions of all elements in the periodic table are accessible to scrutiny.

In the following, we present an overview of spectroscopic measurements of highly charged ions using the EBIT facility. We discuss the focus of our current research effort, and we illustrate some of the measurements and analysis techniques employed. In doing so we use experiments involving uranium as examples. The range of elements studied on EBIT, however, spans virtually all elements starting from oxygen.

II. THE EBIT DEVICE

EBIT uses an electron beam to generate, trap, and probe very highly charged ions [9]. The ion trap is kept short (≤ 4 cm) to minimize instabilities, which might affect the electron beam and reduce the maximum achievable charge state. The electron beam passing through the trap is compressed to a radius of about 30 µm by a 3-T magnetic field generated by a pair of superconducting Helmholtz coils. Six radial slots allow direct viewing of the trapped ions in a plane perpendicular to the electron beam.

Neutrals or ions are injected into the trap from a metal vapor arc [10] or through a gas valve. Typically, the trapping region contains about 50,000 ions, i.e., the ion density is about 10^9 cm^{-3}. By contrast, the electron density is around 5×10^{12} cm^{-3}. Hence, the trap operates in a low-collisional regime, and metastable ions do not affect the operation under most circumstances.

Because collisions with the electron beam results in the heating of the ions, light-ion cooling is used to prevent detrapping of heavy ions [11]. For this purpose a low-Z gas, such as nitrogen, is introduced to the trap to spoil the high vacuum. Fully stripped nitrogen (N^{7+}) or neon (Ne^{10+}) can escape the trap much more readily than, for example, neonlike uranium (U^{82+}). As light ions leave the trap, they carry away heat and thus prevent the build-up of ion temperatures sufficient to detrap heavy ions. Trapping times of up to 4 hours were observed for neonlike gold with light-ion cooling [12]. Without cooling the observed trapping time is only a few seconds. A measurement of the ion temperature was performed recently showing that temperature at least as low as 100 eV can be achieved [7].

Super-EBIT functions almost identically to EBIT. However, modifications to the electron gun and collector assembly allow the use of electron beams with energies exceeding 200 kV [12].

III. SPECTROSCOPIC INSTRUMENTATION

Because EBIT has been designed as an x-ray source, instrumentation emphasizes x-ray detection and analysis. Solid-state detectors provide a highly efficient means for x-ray observation because of their near 100% quantum efficiency. Efficiency is especially important, as the number of photons produced in EBIT is low compared to plasma or beam-foil sources. Another advantage is their wide energy coverage. Crystal spectrometers provide much higher-spectral resolution than solid-state detectors. Their counting efficiency, however, is greatly reduced mainly because of low crystal reflectivities.

Because of the 50-μm width of its electron beam EBIT represents a line source. It is thus well suited for the deployment of flat-crystal spectrometers in the von Hámos geometry. The von Hámos geometry provides focusing of x rays in the non-dispersive direction. As a result, its efficiency surpasses that of a flat-crystal spectrometer [13], and, by employing large-radius crystals, very high resolving powers can be attained. Flat-crystal spectrometers, by contrast, are easier to operate and to align to different Bragg angles. A flat-crystal spectrometer that operates in vacuo was installed on EBIT just recently [14], complementing the set of von Hámos-type spectrometers that have been standard for many years [13].

IV. MEASUREMENTS OF NEAR-NICKELLIKE U^{64+}

The $n=4 \rightarrow n=3$ M-shell transitions span a wide energy band from about 2.5 to 4.5 keV in uranium. A survey of the M-shell emission on EBIT was made with a Si(Li) detector and a low-resolution flat-crystal spectrometer by DelGrande et al. [15]. Much higher resolution was achieved in subsequent measurements. In particular, a measurement that achieved a precision of 80 ppm enabled the first identification of magnetic octupole decay in an atomic system [16].

The first excited level in the closed-shell lithiumlike ion, the level $1s2s\ ^3S_1$, decays to the $^1S_0$ ground state via a magnetic dipole transition [17]. The lowest excited level in the closed-shell neonlike ion, the level $(2p^5_{3/2}\ 3s_{1/2})_{J=2}$, decays to the ground state via a magnetic quadruple transition [18]. Similarly, the first excited level in the closed-shell nickellike ion, the level $(2p^5_{3/2}\ 4s_{1/2})_{J=3}$, decays to the $^1S_0$ ground state via a magnetic octupole transition [16]. A high-resolution spectrometer is necessary to resolve this transition from neighboring electric quadruple transition from level $(3d^9_{3/2}\ 4s_{1/2})_{J=2}$ in nickellike U^{64+}, as illustrated in Fig. 1. Its wavelength was measured as 4.6103(3) Å; that of the neighboring electric quadruple transition was measured to be 4.6058(3) Å.

In Fig. 1 we also present a synthetic spectrum for comparison with the measurements. In many of our measurements we rely on such calculations to unambiguously identify observed features. The synthetic spectrum was constructed from line intensities calculated with a detailed collisional-radiative model that incorporates wave functions, energy levels, and radiative transitions rates calculated by using the relativistic, multi-configuration parametric potential method with full configuration interaction [19]. The collisional excitation rates were calculated in the distorted-wave approximation. For this an efficient technique, which performs an angular factorization of both the direct and exchange contributions to excitation cross sections and interpolates the necessary radial integrals as a function of threshold energy [20] was used. The resulting model includes all singly excited levels with a $3s^{-1}$, $3p^{-1}$, or $3d^{-1}$ core and an optical electron in the $n=4$ or 5 shell.

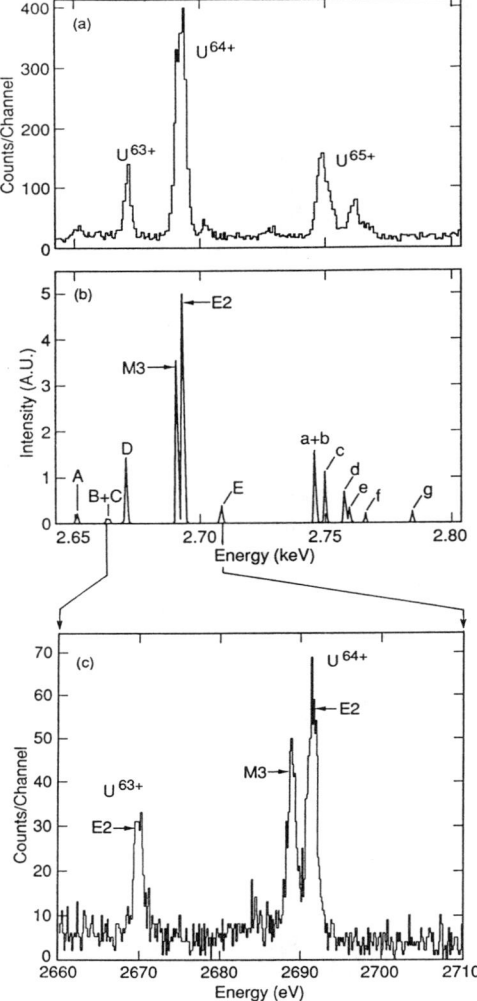

FIGURE 1. (a) Survey spectrum of the 3–4 x-ray transitions in the low-energy region from near-nickellike uranium ions obtained at an electron-beam energy of 9.2 keV. (b) Model spectrum which includes transitions from copperlike U^{63+} (uppercase letters), nickellike U^{64+} (lines M3 and E2), and cobaltlike U^{65+} (lowercase letters). The assumed charge balance is $U^{63+}:U^{64+}:U^{65+} = 1:1:1$. (c) High-resolution spectrum of nickellike U^{64+} in the region 2660–2710 eV recorded at an electron-beam energy of 7.1 keV. The E2 and M3 transitions are clearly resolved. Also seen is an electric quadruple transition in copperlike U^{63+}, labeled D in (b).

Level populations are calculated from a balance of all radiative transitions and electron-impact excitations connecting these levels, of which there are more than 40,000. The model predicts that electron collisions do not affect the population of these excited levels for electron densities below about 10^{14} cm^{-3}. Moreover, the low-lying excited levels are not populated by direct electron-impact excitation from the ground state. Instead, they are populated almost exclusively by radiative cascades form higher levels. Direct electron collisions contribute to the excitation of the $(3d_{5/2}^{-1}4s)_{J=3}$ level, for example, less than 1%. By contrast, radiative cascade feeding from high-lying levels, involving many intermediate levels, is highly effective in populating the $J=3$ level, making the $M3$ line the sixth most intense line in the nickel-like M-shell spectrum.

V. MEASUREMENTS OF NEAR-NEONLIKE U^{82+}

The n=3 → n=2 L-shell transitions in neonlike U^{82+} are situated in the energy band 12-19 keV. The observation of such transitions thus requires spectrometers that can operate efficiently at small Bragg angles.

Recently, we have implemented a von Hámos-type spectrometer that employed a detector with high quantum efficiency for photon energies exceeding 10 keV [21]. This allowed us, for example, to measure the electric dipole and magnetic quadruple transitions from levels $(2p^5{}_{3/2}\,3s_{1/2})_{J=1}$ and $(2p^5{}_{3/2}\,3s_{1/2})_{J=2}$ to ground, respectively, as shown in Fig. 2 (a). These lines were recorded in second order Bragg reflection. The wavelengths of these lines are determined relative to the wavelength of the Lyman-α lines in manganese measured in first order, whose energies are given by Johnson and Soff [22]. Similar use of hydrogenic reference lines is used in virtually all of our crystal-spectrometer measurements. A spectrum of the manganese transitions is shown in Fig. 2(b). We find 12,877.20(80) eV for the electric dipole and 12,866.07(90) eV for the magnetic quadruple transition. As the comparisons with theoretical values in Table I shows, there is a significant difference with relativistic calculations employing a multiconfiguration Dirac-Fock (MCDF) code [23]. The disagreement might be caused by correlation effects as well as by inaccurate estimates of the QED terms from a semi-empirical scaling of hydrogenic values [24]. We discuss these possibilities in more detail in the next section.

The relative contribution of QED to the transition energy is enhanced, if we study 3s–3p transitions instead of 2p–3s transitions. Such transitions can be studied most readily in charge states lower than neonlike, especially in sodiumlike and magnesium-like ions. With the help of many-body perturbation theory, it is now possible to calculate the Dirac energies of one-valence-electron ions, such a sodium like ions, with very high precision [25]. In contrast to neonlike ions,

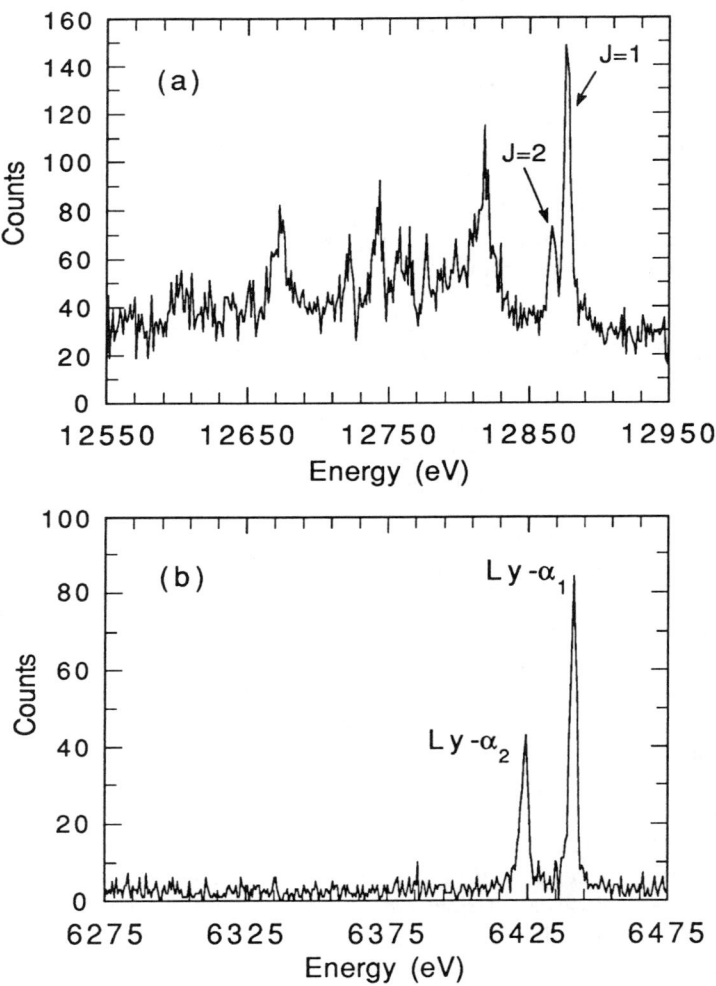

FIGURE 2. High-resolution spectra of (a) the $2p_{3/2}$–$3s_{1/2}$ transitions in neonlike U^{82+}, recorded in second order Bragg reflection, and (b) of the associated calibration lines in hydrogenic Mn^{24+}, recorded in first order. The electric dipole (J=1) and magnetic quadruple (J=2) transitions are measured. Unlabeled features are from lower charge states.

TABLE I. Comparison between theoretical and experimental energies of 3s → 2p transitions in neonlike U^{82+}. The theoretical values were obtained with a multi-configurational Dirac-Fock (MCDF) code [23]. $E_{Coulomb}$ is the relativistic Coulomb energy, E_{Breit} is the transverse Breit correction, E_{VP} is the vacuum polarization energy, E_{SE} is the self-energy, and E_{total} is the sum of the preceding columns. ΔE is defined as the difference between the experimental energies E_{meas} and E_{total}. All values are in eV.

Transitions	$E_{Coulomb}$	E_{Breit}	E_{VP}	E_{SE}	E_{total}	E_{meas}	ΔE
$(3s_{1/2} \to 2p_{3/2})J=1$	12897.8	-31.9	-4.4	10.5	12872.0	12877.2	5.2±0.8
$(3s_{1/2} \to 2p_{3/2})J=2$	12886.9	-32.8	-4.4	10.5	12860.1	12866.1	6.0±0.9

where the calculated Dirac energies are not yet known with high precision, the QED terms in sodiumlike ions can be isolated from the non-QED terms and measurements thus directly compared to QED calculations by subtracting off the Dirac energies.

The $3s_{1/2}$–$3p_{3/2}$ transitions in high-Z ions fall into the energy region from about 800 eV for lead to 1100 eV for uranium. To measure such low-energy transitions vacuum spectrometers are necessary. We have recently implemented such a spectrometer on EBIT [14] and made a preliminary measurement of the $3s_{1/2}$–$3p_{3/2}$ transitions in near sodiumlike lead. The result is shown in Fig. 3. The spectrum was recorded for only one hour of observation time, but clearly showed features from sodiumlike, magnesiumlike, aluminumlike, and siliconlike lead. As in the case of the uranium M-shell spectra, a detailed collisional-radiative model was constructed to aid in the analysis. Results from an energy determination of the observed lines are given in Table II.

TABLE II. Summary of measured energies of $3s_{1/2}$–$3p_{3/2}$ transitions in sodiumlike through siliconlike Pb (in eV). Values in parentheses indicate the uncertainty in the last digit. Theoretical energies are from MCDF calculations [23].

Key	Ion	Transition	$E_{measured}$	E_{MCDF}
Si	Pb^{68+}	$(3s_{1/2}3p^2_{1/2}3p_{3/2})J=1 \to (3s^23p^2)J=0$	806.87(16)	806.93
Al-1	Pb^{69+}	$(3s_{1/2}3p_{1/2}3p_{3/2})J=3/2 \to (3s^23p_{1/2})J=1/2$	803.16(31)	803.37
Al-2	Pb^{69+}	$(3s_{1/2}3p_{1/2}3p_{3/2})J=1/2 \to (3s^23p_{1/2})J=1/2$	814.77(21)	816.65
Mg	Pb^{70+}	$(3s_{1/2}3p_{3/2})J=1 \to (3s^2)J=0$	810.78(11)	812.11
Na	Pb^{71+}	$3p_{3/2} \to 3s_{1/2}$	798.92(26)	799.43

Despite the short observation time our measurement is already comparable in accuracy to a much more elaborate measurement of 3–3 transitions in Na-like Pb performed on the Unilac accelerator facility and reported very recently [26]. The Unilac measurement reported 798.65±0.13 eV for a blend of the Na-like 3s–3p transition and several Ne-like transitions, which is in agreement with our value of 789.92±0.26 eV. The QED contributions to this transition are calculated to be 6.46 eV [26] so our measurement represents a 4%-test of these calculations. An extension of our 3–3 measurement to near-sodiumlike uranium ions is planned for the near future.

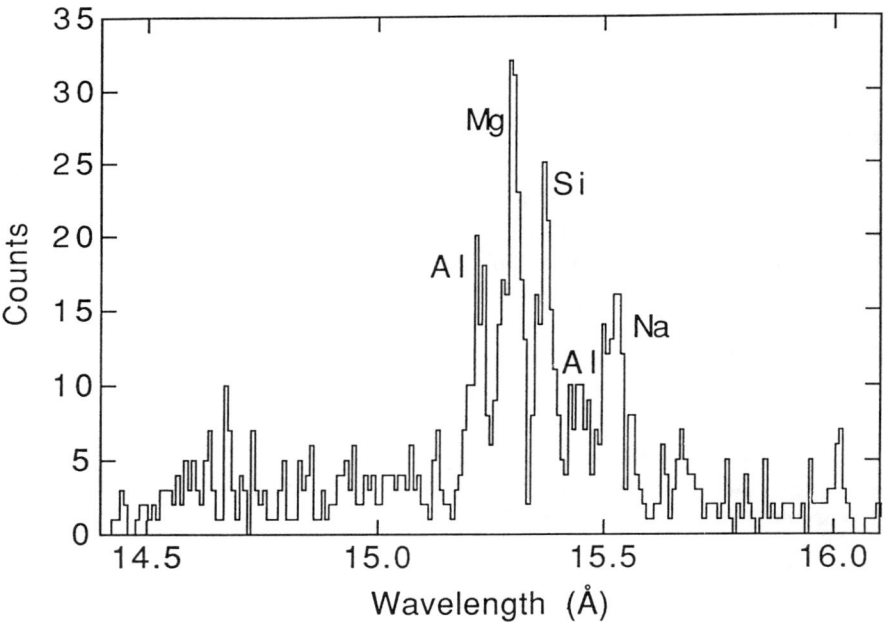

FIGURE 3. Spectrum of the $3s_{1/2}$–$3p_{3/2}$ transitions in sodiumlike, magnesiumlike, aluminumlike, and siliconlike lead, Pb^{68+}–Pb^{71+}. The spectrum was recorded during a 57-minute interval with the beam energy and current set to 29 kV and 140 mA, respectively.

VI. MEASUREMENTS OF NEAR-HELIUMLIKE U^{90+}

QED terms scale approximately as n^{-3}, where n is the principal quantum number. The QED contribution to a transition involving a 4s electron in near-nickellike U^{64+} is about 3 eV, while it is about 6.5 eV for transitions involving a 3s electron, about 42 eV for those involving a 2s electron, and about 460 eV for K-shell transitions. As a percentage of the transition energy QED effects are largest for $\Delta n = 0$ intrashell transitions in the $n = 2$ shell. Although the L-shell QED energies are more than eight times smaller, the Dirac energies are at least 25-fold smaller than in the case of K-shell transitions. For this reason, we have initially concentrated on measuring $n = 2 \rightarrow 2$ transitions [27]. A high-resolution spectrum of the $2s_{1/2}$–$2p_{3/2}$ transitions in lithiumlike U^{89+} through neonlike U^{82+} is shown in Fig. 4. Nevertheless, measurements of the QED contributions to the uranium 1s electron are in progress [28] and promise to provide very high precision results.

In recording the $2s_{1/2}$–$2p_{3/2}$ transitions the ionization balance in the trap was varied by changing the electron energy, the neutral gas density (i.e., the charge-

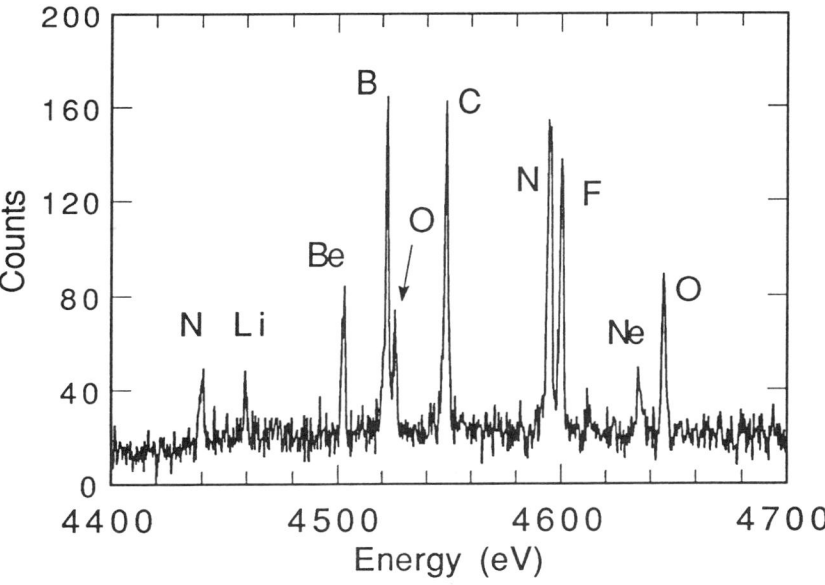

Figure 4. Crystal-spectrometer spectrum of the $2s_{1/2}$–$2p_{3/2}$ electric dipole transitions in U^{82+} through U^{89+}. Lines are labeled by the charge state of the emitting ion.

exchange recombination rate), and the effective current density [27]. This allowed a study of the full range of charge states from U^{82+} to U^{89+}. The beam energies used were well above the 33-keV ionization potential of lithiumlike U^{89+} and below the 130-keV ionization potential of heliumlike U^{90+}. As a result, we observed and unequivocally identified the features produced by all thirteen $2s_{1/2}$-$2p_{3/2}$ electric dipole transitions from the eight ionization stages U^{82+} through U^{89+} that were predicted to fall in our spectral region [27].

From these measurements we determined 4459.37±0.35 eV for the $2s_{1/2}$-$2p_{3/2}$ transitions in lithiumlike U^{89+} with a 90% confidence level [27]. More recently, we measured the $2s_{1/2}$–$2p_{3/2}$ transition in lithiumlike Th^{87+} through neonlike Th^{80+} [29]. Here, the $2s_{1/2}$ –$2p_{3/2}$ transitions in lithiumlike Th^{87+} was determined to be 4025.14±0.14 eV at a 1-σ confidence level.

The measurements of lithiumlike U^{89+} and Th^{87+} are in excellent agreement with recent *ab initio* calculations of the QED energies [30] and represent a check of these calculations at the 0.4% level, i.e., a check with 10-times higher accuracy

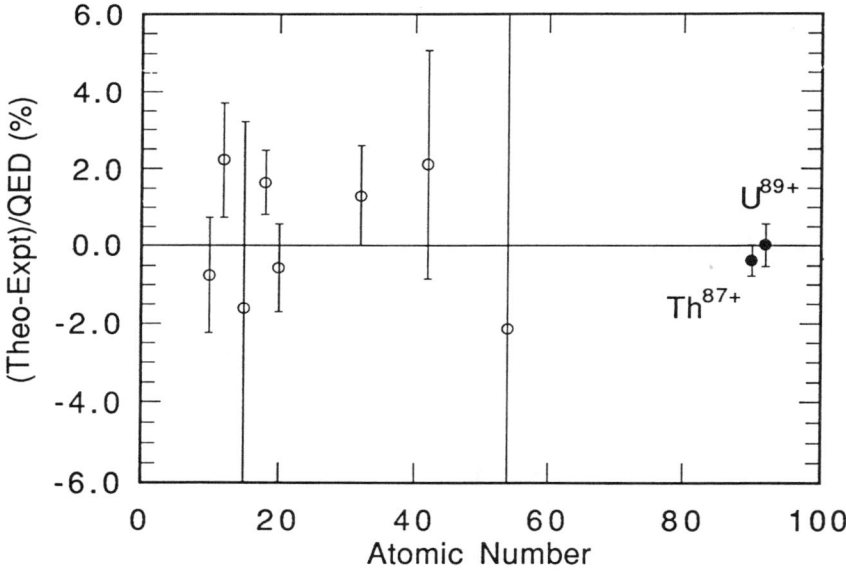

Figure 5. Difference between theory and experiment for the $2s_{1/2}$–$2p_{3/2}$ transition energy in the lithiumlike isoelectronic sequence expressed as a percentage of the theoretical QED energy. The thorium and uranium points are from SuperEBIT and represent 68% confidence limits.

than that achieved in the 3–3 measurements. The agreement with theory is illustrated in Fig. 5, where we have plotted the fractional difference between measured and calculated QED contributions to the $2s_{1/2}$-$2p_{3/2}$ transitions in lithiumlike ions.

The $2s_{1/2}$–$2p_{3/2}$ transition measurements in berylliumlike U^{88+} through neonlike U^{82+} allows us to test relativistic correlation energies and QED calculations in highly charged ions with more than one valence electron. Comparing our measured transition energies to values computed with standard multi-configuration Dirac-Fock (MCDF) calculations, we find that such calculations significantly overestimate the transition energies, as illustrated in Fig. 6. The discrepancy increased with the number of spectator electrons in the $n = 2$ shell and can be attributed to residual correlation energies unaccounted for by the atomic structure calculations. This is true for both measurements thorium and uranium. Recently, efforts have been initiated that attempt to apply the principles of many-body perturbation theory to the calculations of the Dirac energies of such systems with more than one valence electron [31]. If successful, the Dirac energies of ions with more than one valence electron can be calculated with a precision now reserved for ions with a single valence electron. This would enable isolating the QED contributions and the development and testing of *ab initio* calculations of the QED terms for such ions.

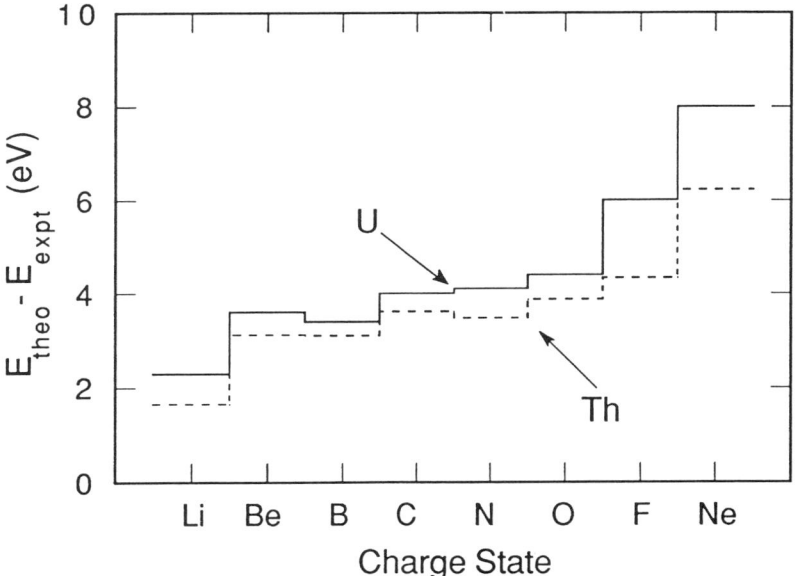

Figure 6. Average difference between MCDF calculations and measured 2–2 transition energies for different charge states of thorium and uranium. The difference increases for the lower charge states due to difficulties in calculating the electron correlation terms.

VII. CONCLUSION

A vigorous, world-wide effort is currently under way to perform precise measurements of the atomic structure of highly charged, high-Z ions. At the Livermore EBIT facility this effort ranges from crystal-spectrometer measurements of near nickellike U^{64+} to measurements of the 1s-binding energy of hydrogenic U^{91+}. The large number of measurements in turn have spurred the development of formalisms to accurately calculate the Dirac and QED energies of any type of highly charged high-Z ion. In many cases, such calculations were impossible and unneeded just a few years ago.

In the case of those ions with one valence electron the measurements have provided precise values of the electron self-energy contribution to the level energies. Tests on the 0.4-% level of the total QED contribution were possible in a regime where $(Z\alpha) \approx 1$.

While our discussion so far has focused on measurements of atomic structure, experiments have also been performed that centered on others aspects of atomic physics. These included the measurement of the 1s-ionization cross sections of heliumlike U^{90+} and hydrogenic U^{91+}, which were found to be twice the size predicted by recent theory [6], and the first observation of quantum interference of radiative and dielectronic recombination in berylliumlike U^{88+} and boronlike U^{87+} [31]. These measurements demonstrate that experiments involving very highly charged ions are merely in their infancy and much work needs to be done.

ACKNOWLEDGMENTS

The continued support of Mark Eckart and Andrew Hazi is greatly appreciated. This work was performed under the auspices of the U.S. Department of Energy by Lawrence Livermore National Laboratory under contract No. W-7405-ENG-48.

REFERENCES

[1] J. P. Briand, P. Chevallier, P. Indelicato, K. P. Ziock, and D. Dietrich, Phys. Rev. Lett. **65**, 2761 (1990).
[2] Th. Stöhlker, P. H. Mokler, K. Beckert, F. Bosch, H. Eickhoff, B. Franzke, M. Jung, T. Kandler, O. Klepper, C. Kozhuharov, R. Moshammer, F. Nolden, H. Reich, P. Rymuza, P. Spädtke, and M. Steck, Phys. Rev. Lett. **71**, 2184 (1993).
[3] J. Schweppe, A. Belkacem, L. Blumenfeld, N. Claytor, B. Feinberg, H. Gould, V. E. Kostroun, L. Levy, S. Misawa, J. R. Mowat, and M. H. Prior, Phys. Rev. Lett. **66**, 1434 (1991).
[4] C. T. Munger and H. Gould, Phys. Rev. Lett. **57**, 2927 (1986).

[5] P. Beiersdorfer, Nucl. Instrum. Methods **B56/57**, 1144 (1991).
[6] R. E. Marrs, S. E. Elliott, and D. A. Knapp, Phys. Rev. Lett. **72**, 4082 (1994).
[7] P. Beiersdorfer, V. Decaux, S. Elliott, K. Widmann, and K. Wong, Rev. Sci. Instrum. **66** (January 1995) (in press).
[8] P. Beiersdorfer et al., in *UV and X-Ray Spectroscopy of Astrophysical and Laboratory Plasmas*, ed. by E. Silver and S. Kahn (Cambridge University Press, Cambridge, 1993), p. 59.
[9] R. E. Marrs and M. A. Levine and D. A. Knapp and J. R. Henderson, Phys. Rev. Lett. **60**, 1715 (1988).
[10] I. G. Brown, J. E. Galvin, R. A. MacGill, and R. T. Wright, Appl. Phys. Lett. **49**, 1019 (1986).
[11] M. B. Schneider, M. A. Levine, C. L. Bennett, J. R. Henderson, D. A. Knapp, and R. E. Marrs, in *International Symposium on Electron Beam Ion Sources and their Applications - Upton, NY 1988*, AIP Conference Proceedings No. 188, edited by A. Hershcovitch (AIP, New York, 1989), p. 158.
[12] D. A. Knapp, R. E. Marrs, S. R. Elliott, E. W. Magee, and R. Zasadzinski, Nucl. Instrum. Methods **A334**, 305 (1993).
[13] P. Beiersdorfer, R. E. Marrs, J. R. Henderson, D. A. Knapp, M. A. Levine, D. B. Platt, M. B. Schneider, D. A. Vogel, and K. L. Wong, Rev. Sci. Instrum. **61**, 2338 (1990).
[14] P. Beiersdorfer and B. J. Wargelin, Rev. Sci. Instrum. **65**, 13 (1994).
[15] N. K. Del Grande, P. Beiersdorfer, J. R. Henderson, A. L. Osterheld, J. H. Scofield, and J. K. Swenson, Nucl. Instrum. Methods **B56/57**, 227 (1991).
[16] P. Beiersdorfer, A. L. Osterheld, J. Scofield, B. Wargelin, and R. E. Marrs, Phys. Rev. Lett. **67**, 2272 (1991).
[17] H. R. Griem, Astrophys. J. **156**, L103 (1969)
[18] M. Klapisch, J. L. Schwob, M. Finkenthal, B. S. Fraenkel, S. Egert, A. Bar Shalom, C. Breton, C. de Michelis, and M. Mattioli, Phys. Rev. Lett. **41**, 403 (1978).
[19] M. Klapisch, Comput. Phys. Commun. 2, 239 (1971); M. Klapisch, J. L. Schwob, B. S. Fraenkel, and J. Oreg, Opt. Soc. Am. **61**, 148 (1977).
[20] A. Bar-Shalom, M. Klapisch, and J. Oreg, Phys. Rev. A **38**, 1773 (1988).
[21] D. Vogel, P. Beiersdorfer, V. Decaux, and K. Widmann, Rev. Sci. Instrum. **66** (January 1995) (in press).
[22] W. R. Johnson and G. Soff, At. Data Nucl. Data Tables **33**, 405 (1985).
[23] I. P. Grant, B. J. McKenzie, P. H. Norrington, D. F. Mayers, and N. C. Pyper, Comput. Phys. Commun. **21**, 207 (1980); M. H. Chen, private communication.
[24] P. Beiersdorfer, M. H. Chen, R. E. Marrs, M. A. Levine, Phys. Rev. A **41**, 3453 (1990).
[25] W. R. Johnson, S. A. Blundell, and J. Sapirstein, Phys. Rev. A **38**, 2699 (1988).
[26] A. Simionovici, D. D. Dietrich, R. Keville, T. Cowan, P. Beiersdorfer, M. H. Chen, S. A. Blundell, Phys. Rev. A **48**, 3056 (1993).
[27] P. Beiersdorfer, D. Knapp, R. E. Marrs, S. R. Elliott, and M. H. Chen, Phys. Rev. Lett. **71**, 3939 (1993).
[28] R. E. Marrs, private communication (1994); contributed presentation this conference.
[29] P. Beiersdorfer et al., in preparation.
[30] S. A. Blundell, Phys. Rev. A **47**, 1790 (1993).
[31] J. Sapirstein, private communication (1994); invited presentation this conference.
[32] D. A. Knapp, P. Beiersdorfer, M. H. Chen, D. Schneider, and J. H. Scofield, Phys. Rev. Lett. (submitted).

Precision Measurements of Atomic Lifetimes

Carol E. Tanner

Department of Physics, University of Notre Dame, Notre Dame, IN 46556

Abstract. Precision measurements of excited-state lifetimes play important roles in many areas of physics. Measurements of atomic lifetimes are used to determine fundamental interaction parameters, test atomic and molecular theory, and place constraints on models of stellar evolution and big-bang nucleosynthesis. A variety of techniques for measuring atomic and molecular lifetimes are presented along with a discussion of associated systematic effects. A comparison is made between experimental and theoretical results for a variety of alkali and alkali-like systems. Lifetime measurements in cesium are used to test atomic-structure calculations necessary for the interpretation of parity nonconservation experiments.

INTRODUCTION

A large body of experimental work has been performed measuring lifetimes of excited states in a variety of atomic and molecular systems. Throughout all fields of physics, lifetime measurements in general provide fundamental information about interactions and/or initial and final state wave functions. A variety of motivations and techniques for measuring atomic and molecular lifetimes is represented by the contributed papers at this conference.

Astrophysical motivations include the need for accurate oscillator strengths (or f-values) for interpreting interstellar and stellar spectroscopic data at the 1% level. Often relative values of oscillator strengths are known precisely, but for interpretation of the data absolute values are needed. One of the most accurate methods for normalizing the relative measurements uses precisely and accurately determined atomic lifetimes. The work presented by Bergeson and Lawler addresses these needs in the UV and VUV spectral regions [1]. The applications of lifetimes to the interpretation of data in plasma research and geophysics are similar. The work of Hutton *et al.* also falls into this category where they have measured lifetimes of highly charged Na-like Nb [2]. Escalante *et al.* calculate transition rates in atomic oxygen and apply their results to the interpretation of nightglow in the earth's atmosphere [3].

Tests of atomic and molecular theory require comparison with experimental data. Because of the sophistication of current calculational techniques [4], the necessary experimental precision for atomic lifetimes is often below the one-percent level. The theoretical work presented by Olsen *et al.* addresses the difficulty of finding agreement for both transition energies and oscillator strengths [5] in the same calculation. Plante *et al.* present calculations of oscillator strengths and transition rates in helium like ions [6] and comparisons are made with experiments. In the contributed work by Biémont *et al.*, calculations of lifetimes in CII, NII, NIII are compared to experimental results and are in good agreement [7]. Liaw performed *ab initio* calculations of 4p ^{2}P state lifetimes in Ca$^+$ [8] and found agreement with recent experimental work of Jin and Church [9] at the 1% level. Molecular theory is also being challenged by the experimental work of Ray and Lafyatis [10] who have measured lifetimes of excited states in H$_2$, the simplest molecule.

My own research interests involve precision measurements [11, 12] necessary for the interpretation of parity nonconservation experiments (PNC). Interpretation of atomic PNC experiments in all elements requires accurate knowledge of associated transition matrix elements. The accuracy of atomic theory used for this purpose is dependent upon comparisons with experimentally measured transition amplitudes and hyperfine structure. Neutral alkali systems have been the atoms of choice for many PNC experiments because of their calculational simplicity. Precise lifetime measurements in alkali and alkali-like systems can provide information important to further theoretical development. Many experimental techniques have been applied to lifetime measurements of the same state in the neutral alkalis lithium and cesium, and some experimental results are compared below. The most precise measurements in lithium and cesium are compared with recent atomic structure calculations. These two atoms represent one of simplest and one of most complex alkali systems for evaluating the calculational capabilities of atomic theory applied to many electron systems.

MANIFESTATIONS OF A FINITE LIFETIME

The experimental techniques used to measure the lifetimes of excited atomic states can be classified by how the lifetime is manifest, examples of which are discussed below in turn. Many of these techniques can be applied to both atoms and molecules. In some cases, as for high ionization states, only beam-foil techniques are used extensively. A number of experimental techniques are described briefly along with some of the systematic effects peculiar to each. More detailed explanations of these techniques can be found in the excellent review article by Imhof and Read [13]. For comparing experimental and theoretical results, the relevant formulas are given below.

Exponential Decay

There is a characteristic exponential decay of population, N, for a single excited state with a finite lifetime. The population as a function of time, t, is given by Equation 1, where τ is the lifetime.

$$N(t) = N_0 \, e^{-t/\tau}. \tag{1}$$

Many of the atomic lifetime experiments performed in recent years have been designed to generate a population in an excited state and then time-resolve the exponential decay of this population. These experiments typically involve allowed decays with lifetimes from fractions of a nanosecond to several microseconds and tend to fall into two classes.

In one class of experiments, the atoms or molecules being studied are "at rest" in the laboratory, and the delayed coincidence between excitation and emission is observed. The excited state can be populated by pulsed excitation or by cascade from another excited state. Pulsed excitation may be accomplished with a variety of techniques including pulsed lasers or pulsed electron beams. A single decay is often observed by detecting the emission of a photon, and the time delay between excitation and emission is resolved electronically. The process is repeated to obtain an exponential distribution of times. DeVoe and Brewer have recently applied this technique to the lifetime of the $6^2P_{1/2}$ state of Ba^+ in an ion trap [14]. Similarly, if the state under study is populated by cascade, then the time delayed coincidence between the emission of the cascade photon and the decay photon is measured. Photomultiplier tubes are typically used as single photon detectors. These time resolved techniques have some sources of uncertainty in common such as calibration and linearity of the time scale. Calibration usually entails detailed evaluation of fast-coincidence electronics including a time-to-amplitude converter and pulse height analyzer. Extracting a decay constant from the observed coincidence spectrum may involve deconvoluting the spectrum with an instrument response function. With pulsed electron excitation, the selectivity of the excited level is limited allowing cascades from unwanted levels to contribute. With laser excitation, the selectivity of the excited state is much improved and cascading is eliminated. However, high peak-power short-pulsed lasers can introduce other problems. The high power introduces scattered light leading to large signals at short times that can produce nonlinearities in the detectors. In cascade delayed coincidence experiments, signal rates are much lower, but scattered light is typically very low and can be filtered away. In either case, often the sample under study is contained in a fluorescence cell or thermal beam at densities where quenching collisions and radiation trapping may become apparent. A thermal velocity distribution can also limit the selectivity of laser excitation.

In a second class of experiments, ions are accelerated into a beam and velocity selected. The ions can be neutralized by charge exchange, if desired. The excitation takes place during passage through a foil, gas target, or laser beam. The exponential decay is observed by detecting the number of photons emitted as a function of position, and the speed is used to convert the distance scale to a time scale. In these fast-beam experiments, the time scale is determined by the precision of the position and velocity measurements. Ion beam experiments have their own set of peculiarities. Variation in ion-beam current is a typical problem. Excitation of residual gas in the vacuum system leads to an observable background. Beam-foil excitation is a widely used and versatile technique and may be the only means available to excite very highly-ionized systems. However, the nonselective excitation from a foil or gas target can induce cascades from unwanted excited states. Curtis *et al.* have reported a technique that can eliminate cascade effects in some cases [15]. In a contributed paper, Pinnington and Berends present a new technique for eliminating cascades which uses laser excitation from metastable levels [16]. Foil excitation can increase the angular divergence of the ion beam possibly changing light collection efficiency as a function of position. The narrow velocity distribution and low density in a fast beam are advantages that minimize unwanted collisional and radiation-trapping effects. Laser excitation of a fast beam offers additional advantages including selectivity of excitation, high signal rates due to large excited state populations, and elimination of cascades. However, lasers tend to increase scattered light.

In both pulsed-laser excitation and fast-beam laser experiments, the finite excitation time transforms into a spread in excitation energy leading to coherent excitation of closely spaced energy levels. The coherence introduces the possibility of quantum beats that can cause systematic shifts in the measurements. Quantum beats between coherently populated hyperfine levels can be suppressed by carefully selecting observation angles and polarizations. Residual magnetic fields cause quantum beats between magnetic sublevels, but these can be minimized by shielding or cancellation of fields and selecting observation angles and polarizations.

Decay Rate

The lifetime may also be determined by observing the decay rate of an excited-state population. For states that have a lifetime much longer than a typical observation time, the rate of decay can be observed by generating a known number of excited states then detecting the number of decays in an interval of time short compared to the lifetime.

$$\frac{dN}{dt} = -\frac{1}{\tau} N(t) . \qquad (2)$$

This approach is often used for measuring half-lives of radioactive nuclei and particles, but is not typically applied to allowed atomic decays where lifetimes are on the order of 10^{-8} s. Long lived or metastable atomic states with time constants on the order of seconds or more can be measured this way. Experiments in this category have their own particular systemic uncertainties. The total number, N, of excited atoms must be determined. When emission is observed, the solid angle of the collection optics and detector efficiencies must be determined. For long lifetimes, focused collection optics can skew the measurement of a time constant when atoms or molecules pass out of the field of view during an observation. In a recent example of this approach, laser cooled and trapped Xenon atoms were used to measure the metastable 6s $[3/2]_2$ state lifetime. Walhout *et al.* [17] measured the 43 s lifetime by observing the rate of photons emitted into a known solid angle from a sample of cold trapped atoms. A discrepancy between experiment and theory was found to be more than a factor of two, thus stimulating additional theoretical work.

Finite Line Width

The finite Lorentzian line width of a resonance transition, Equation 3, is another signature of the finite lifetime of an excited state. The excitation and resonance frequencies in radians are ω_γ and ω_0, respectively.

$$L(\omega_\gamma) = \frac{1}{(\omega_\gamma - \omega_0)^2 + (\Gamma/2)^2}, \quad \Gamma = \frac{1}{\tau}. \tag{3}$$

The level crossing technique is a well known method for determining the width of an atomic resonance. In this technique, two excited state hyperfine levels are forced to cross by the application of an external field, usually a magnetic field but can be an electric field or sometimes both. The atoms are excited by photon absorption from the ground state, then the intensity or polarization of emitted light is observed as a function of field strength. The observed emission as a function of field can be related back to the energy width of the crossing levels. The level crossing technique is generally not used for molecules because of the large number of closely spaced levels. The Hanle effect is often referred to as a zero-field level-crossing technique because similar emission features are observed as an applied magnetic field is scanned through zero. In the Hanle effect, atoms or molecules are polarized in the excited state such that the excited sample has a finite magnetic moment that precesses in the field. By observing the decay of the polarization, one can determine the exponential decay constant relative to the precession rate. The Hanle effect and level crossing techniques are usually applied to the measurement of lifetimes from nanoseconds to microseconds, and are subject to a variety of similar uncertainties. Optical and field alignments are important. Accurate field calibration and homogeneity is necessary. Systematic uncertainties can arise from the sensitivity of detectors to magnetic fields, admixtures from unwanted states, the spectral distribution of the exciting light, and optical pumping. The samples are usually neutral atomic species contained in cells where radiation trapping, collisional de-excitation, and wall collisions can affect the measurement of lifetimes.

A completely new approach for directly observing the Lorentzian lineshape of an atomic transition is made possible by recent developments in capabilities for cooling atomic velocities and stabilizing lasers. Otes *et al.* [18] are presently using a stabilized laser to scan the Na 3s $^2S_{1/2} \to 3p\ ^2P_{3/2}$ resonance line shape of laser cooled Na atoms.

Phase Shift

A lifetime can also be revealed by what is known as the phase shift technique. In the presence of a modulated excitation rate, R, for populating an excited state, there is a phase shift between the population of the excited state, N, and the modulation imposed on the excitation. The origin of the shift is illustrated by the solution to the following differential equation for a sinusoidal modulation.

$$\frac{dN}{dt} = -\frac{N}{\tau} + R\, e^{i\omega_m t}. \tag{4}$$

The steady state solution, Equation 5, exhibits a similar modulation on the excited state population with a phase shifted relative to that of the excitation process. The intensity of emitted light follows the population of the excited state. Thus the shift is usually measured by observing the relative phase between the modulation that appears on the emitted light and that of the excitation.

$$N(t) = \frac{R\, e^{i\omega_m t}}{1/\tau + i\omega_m}. \tag{5}$$

Maximum sensitivity is achieved when the modulation frequency is chosen to give a phase shift of 45^0. Excitation can be achieved optically or by electron impact. With optical excitation the phase shift can be measured in an interferometer with equal optical path lengths. Electron excitation is far less selective, and observations of transitions with known lifetimes must be used for calibration. Early optical experiments were performed with lamps and Kerr cells for modulation. These have been replaced by lasers and a variety of more sophisticated modulation techniques including Pockel cells, and acousto-optic and electro-optic modulators. The phase shift technique has been used to measure lifetimes in neutral atoms and molecules. Typically, thermal beams and vapor cells are used so collisional de-excitation and radiation trapping must be considered. There has not been much recent work using the phase shift technique, but with improved techniques in laser stabilization and modulation there is potential for greater precision than has been achieved in past experiments.

Comparing Experimental Measurements with Theoretical Calculations

The "A" coefficient describing spontaneous emission of a photon from an excited atomic or molecular state can be derived quantum mechanically. First order time dependent perturbation theory gives the following result for the transition probability per unit time.

$$W = \frac{2\pi}{h} |\langle F| H_{INT} |I\rangle|^2 \rho(E). \tag{6}$$

In the case of a single final state, the transition rate is simply related to a single matrix element involving the initial and final state wave functions and an interaction Hamiltonian. In particle physics, one often knows the initial and final state wave functions and a measurement of a lifetime yields information about the interaction. For example, the muon lifetime is known most precisely to two parts in 10^5 [19] and is used to determine the value of the Fermi coupling constant of weak interactions. In atomic physics, the electromagnetic Hamiltonian is known very well [20], and the precision measurement of a lifetime yields information about the initial and final state wave functions. The electromagnetic interaction can be expanded in multipole moments representing decreasing transition rates.

$$H_{INT} = \frac{-q}{c} \mathbf{A} \cdot \mathbf{p} = H_{E1} + H_{M1} + H_{E2} + H_{M2} + \ldots \tag{7}$$

Most of the examples discussed in this paper are of transitions due to the electric dipole (E1) interaction. However, the previously mentioned work of Walhout *et al.* was concerned with an allowed magnetic quadrupole (M2) transition [17].

The A coefficient for photon emission is derived from Equations 6 and 7 after performing the appropriate integrals over the photon density of states. For an E1 transition, the A coefficient is proportional to the square of a single radial matrix element. The results of theoretical work are commonly presented as calculations of reduced matrix elements that can be converted to an A coefficient through the following formula.

$$A_{n'L'J' \to nLJ} = \frac{4}{3} \alpha \frac{\omega_0^3}{c^2} \frac{|\langle nLJ \| r \| n'L'J' \rangle|^2}{2J'+1} \cdot \text{length} \quad (8)$$

$$A_{n'L'J' \to nLJ} = \frac{4}{3} \alpha \frac{\omega_0}{c^2} \frac{|\langle nLJ \| p/m \| n'L'J' \rangle|^2}{2J'+1} \cdot \text{velocity} \quad (9)$$

Also shown is the velocity form of the dipole interaction. As an initial test of the quality of calculated wave functions, two forms of the interaction are usually calculated. The length and velocity forms of the matrix elements are related by a commutation relation and should yield identical results if the theory is exact. Some authors quote their results in terms of a line strength or an oscillator strength (f-value). The following formulae are given so that one can convert easily from one form to another.

$$\text{Line Strength} \quad S_{n'L'J' \to nLJ} = \frac{|\langle nLJ \| r \| n'L'J' \rangle|^2}{a_\infty^2}. \quad (10)$$

$$\text{Oscillator Strength} \quad f_{nLJ \to n'L'J'} = \frac{2}{3} \frac{\hbar \omega_0}{\left(\frac{e^2}{a_\infty}\right)} \frac{1}{2J+1} S_{n'L'J' \to nLJ}. \quad (11)$$

In the case of multiple final states, the inverse lifetime of the excited state is the sum of the A coefficients involved. The cases discussed in the remainder of this paper are for an excited state decaying to a single final state where the reciprocal lifetime is equal to a single A coefficient. The following expressions are used for comparing theory to experiment.

Lifetime in terms of a reduced matrix element or line strength:

$$\frac{1}{\tau_{n'L'J' \to nLJ}} = \frac{\left\{\frac{4}{3} \alpha (2\pi)^3 c\, a_\infty^2\right\}}{2J'+1} \frac{1}{\lambda^3} \frac{|\langle nLJ \| r \| n'L'J' \rangle|^2}{a_\infty^2}. \quad (12)$$

Lifetime in terms of an oscillator strength:

$$\frac{1}{\tau_{n'L'J' \to nLJ}} = \left\{ 2 (2\pi)^2 \alpha^2 c\, a_\infty \right\} \frac{1}{\lambda^2} \frac{2J+1}{2J'+1} f_{nLJ \to n'L'J'} . \tag{13}$$

Throughout this section, a_∞ is the atomic unit of length and e^2/a_∞ is the atomic unit of energy. $\hbar$ is Plank's constant, α is the fine structure constant, and c is the speed of light. λ and ω_0 are the transition wavelength and frequency, respectively.

MEASUREMENTS OF Li 2p $^2P_{1/2, 3/2}$ STATE LIFETIMES

The lithium 2p state is presented here as an example of an atomic lifetime that plays an important role in the interpretation of astrophysical data, as a notable achievement in precision measurements of atomic lifetimes, and as a test of atomic and molecular calculations. In astrophysics, stellar surface abundances of lithium are determined relative to that of hydrogen by observing the emission spectra of stars. The data is interpreted using the oscillator strength for the lithium 2s-2p transition at 670 nm. Since the 2p $^2P_{1/2, 3/2}$ states have only a single decay mode, the determination of lithium abundances depends directly on the accepted value of the 2p lifetime. In recent studies of stars in the Hyades cluster [21], the authors emphasize the importance of lithium abundances in models of stellar evolution, in determining the primordial lithium abundance, and in understanding big-bang nucleosynthesis.

The 2p lifetime of neutral lithium has been studied by a variety of experimental techniques, as presented in Figure 1. Most authors have measured the 2p $^2P_{3/2}$ lifetime whereas Gaupp *et al.* measured the 2p $^2P_{1/2}$ lifetime.

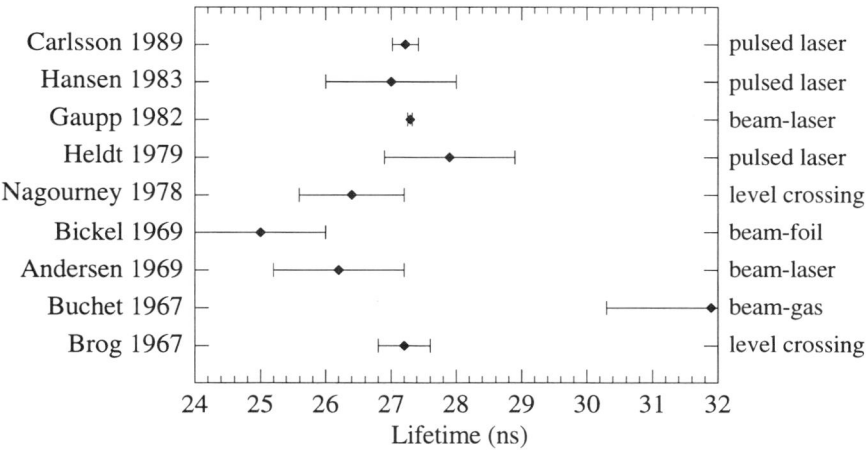

Figure 1. The experimental measurements of the 2p ^2P state lifetime in atomic lithium represent a wide cross section of the established techniques. References [22-30].

Since the transition energies and matrix elements for these two fine structure states are so similar, the difference in their lifetimes is less than three parts in 10^4. The most striking result shown is the measurement by Gaupp et al. [22] with an exceptional precision of 0.15% using laser excitation of a fast atomic beam. A more recent pulsed-laser delayed-coincidence experiment by Carlsson et al. [23] agrees with Reference [22] but only reaches a precision of 0.7%.

An important role of lifetime measurements is to provide a means for testing atomic structure calculations. Alkali atoms with a single electron outside of a closed shell provide some of the simplest open shell systems for comparison between experiment and theory. The allowed electric dipole transitions from the lowest lying $^2P_{1/2,3/2}$ states to the $^2S_{1/2}$ ground state in alkali-like systems each depend on a single radial matrix element that is sensitive to the behavior of the electron wave function far from the nucleus. Core correlation and polarization effects have a strong influence on radial matrix elements in these cases. Precision measurements of lifetimes in these systems can guide further theoretical development by testing the wave functions at large radii. In the lithium atom, core correlation effects can be calculated in the framework of a variety of theoretical approaches. In alkali systems with more electrons, the correlation effects are larger, and approximation techniques must be developed for treating these cases. Figure 2 shows the most recent calculations of the 2s-2p radial matrix elements [31-36] represented in terms of the 2p state lifetime.

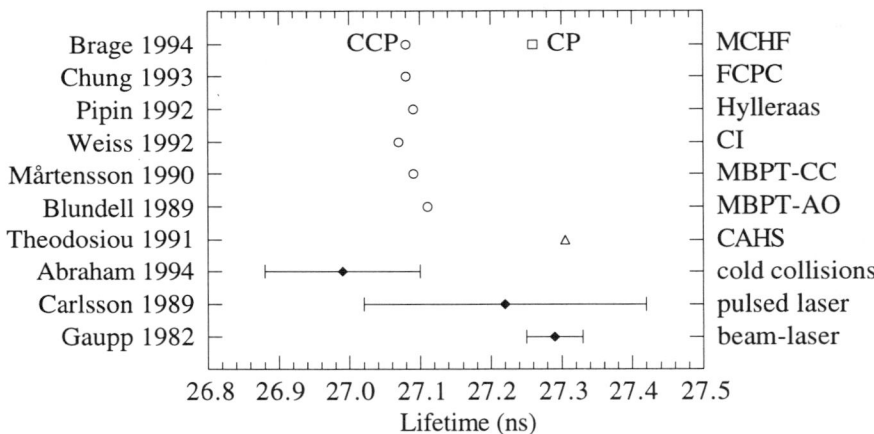

Figure 2. Comparison between recent calculations and the most precise experimental results to date. (O) *ab initio* including core polarization and correlations [31-36], (□) not including core correlations, (Δ) semi-empirical [37], (♦) experimental results [39, 22, 23].

There are a number of older calculations not presented here which can be found in Reference [22]. The *ab initio* calculations in Figure 2 all include core polarization and core correlation effects, and all fall within 0.15% of each other. Theodosiou et al. [37] used a semi-empirical model potential that includes only core polarizations explicitly. Brage et al. calculated the effects of core correlations separately using an approximation that can be transferred to heavier systems [31]. The importance of including correlations is illustrated by comparing the points labeled CCP and CP. Both include core polarizations, but the effects of core correlations are only included in the point labeled CCP. For this simple atom, one would expect good agreement between theory and experiment. All *ab initio* calculations disagree with the experimental results of Gaupp *et al.* by about 0.75%, that is five times the stated experimental precision. The number of recent calculations indicates the intensity of the interest that exists in resolving this discrepancy. Disagreements between experiment and theory in this and other systems have been discussed by Wiese [38].

Also shown in Figure 2 is the lifetime derived from an experimental measurement of the 2s-2p radial matrix element using cold lithium atoms. Abraham *et al.* observe the spectra of ^{6}Li atoms during ultra-cold collisions between atoms in a magneto-optical trap [39]. The molecular potential at large interatomic distances is proportional to the 2s-2p matrix element which is revealed in the vibrational spectra. The precision in this determination of the lifetime is 0.4% where the majority of the uncertainty is in the molecular potential at intermediate distances. One would expect agreement with the results for ^{7}Li, because the difference in the lifetimes for the two isotopes is negligible.

In my own collaboration, we are pursuing lifetime measurements of the 2p $^2P_{1/2,3/2}$ states of lithium [40]. We employ a fast-beam laser technique similar to that of Gaupp *et al.* [22], but with a substantially different approach for measuring the atomic velocity and collecting the decay fluorescence. Neutral lithium atoms are prepared from a 30-50 keV beam of ^{7}Li$^+$ ions by electron capture in a gas target. A dye laser at 670 nm, aligned perpendicular to the beam, is used to excite the atoms. After excitation, the fluorescent signal from the selected 2p $^2P_{1/2,3/2}$ state decays in flight as the atom moves along the beam line. Fluorescence is monitored precisely as a function of position using a translating fiber-optic light-collection system, shown schematically in Figure 3. A second collection system remains stationary, and that signal is used to normalize for ion beam fluctuations.

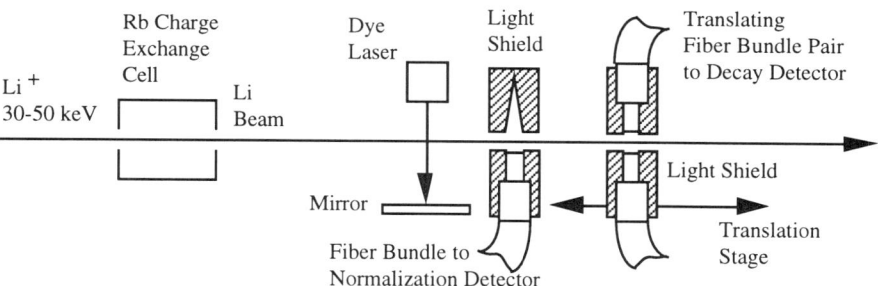

Figure 3. Schematic diagram of the experiment. The fiber optic bundles couple the collected light through the vacuum chamber wall and into the photomultiplier tubes.

The atomic velocity is measured when the laser beam is aligned antiparallel to the atomic beam. The Doppler shift of the atomic resonance determines the velocity. The capabilities of our beam-laser technique have been established by our measurements in cesium [12]. Changes to our apparatus now allow us to measure the lifetime of the $2^2P_{1/2,3/2}$ states in lithium. The changes include interference filters in the detectors at 670 nm, and an ion source that produces Li$^+$ instead of Cs$^+$. In addition, the dye laser replaces two diode lasers used to drive the cesium transitions $6^2S_{1/2} \rightarrow 6^2P_{1/2,3/2}$ at 895 nm and 852 nm respectively. The primary advantages of our fast-beam laser technique are the selectivity of the excitation, the intensity of the fluorescent signal, the stable light collection optics, and the precision time scale obtained by direct measurement of the neutral-atom velocity.

MEASUREMENTS OF ^{133}Cs 6p ^{2}P$_{1/2, 3/2}$ STATE LIFETIMES

We began our series of lifetime measurements with the 6p $^2P_{1/2,3/2}$ states in neutral cesium because of their importance to the interpretation of parity nonconservation (PNC) experiments in the $6^2S_{1/2} \rightarrow 7^2S_{1/2}$ transition [41]. In atomic cesium, PNC experiments are reaching a sensitivity that can test the standard model of weak interactions at the level of radiative corrections making it possible to probe for new physical effects [42]. The interpretation of a new PNC experiment in cesium [43] requires accurate knowledge of selected atomic transition matrix elements below the 0.5% level. The theoretical aspects of atomic PNC are discussed by Sapirstein [4] in these proceedings. Atomic theory plays an important role in this interpretation. The best way to assess the accuracy of atomic theory is by comparing calculated quantities involving radial matrix elements and hyperfine constants to precision measurements. Although the actual calculation of the PNC $6^2S_{1/2} \rightarrow 7^2S_{1/2}$ transition amplitude, E_{PNC}, is quite complex [44], the contributions made by various matrix elements are illustrated in first order perturbation theory by the sum:

$$E_{PNC} = \sum_n \frac{\langle 7S_{1/2} | er | nP_{1/2} \rangle \langle nP_{1/2} | H_{PNC} | 6S_{1/2} \rangle}{E_{6S_{1/2}} - E_{nP_{1/2}}} \quad (14)$$
$$+ \frac{\langle 7S_{1/2} | H_{PNC} | nP_{1/2} \rangle \langle nP_{1/2} | er | 6S_{1/2} \rangle}{E_{7S_{1/2}} - E_{nP_{1/2}}}.$$

The effective weak interaction atomic Hamiltonian, H_{PNC}, is very short range and is sensitive to the wave functions only very close to the nucleus. Hyperfine structure constants are sensitive to the wave functions in the same region, and provide the only means of testing the wave functions at small radii. As shown in Equation 14, radial matrix elements play an equally important role in this sum. At present there is little data for transitions in cesium precise enough to test radial matrix elements below the 1% level. Measurements of the $6^2P_{1/2}$ and $6^2P_{3/2}$ state lifetimes provide a test for only one term in this sum. Other terms in the sum are also highly weighted by their corresponding energy denominators, and we are planning measurements of these quantities.

Our lifetime measurements in Cs have extended the application of the fast-beam laser technique to heavy alkali atoms and include the first application of diodes lasers in fast-beam laser spectroscopy [45]. With a beam energy of 50 keV the 30 ns lifetime of the $6\,^2P_{3/2}$ state has a 1/e decay length of 8 mm, enabling us to monitor the signal over several decay lengths. Detailed attention is given to systematic uncertainties which are comparable to statistical uncertainties, and these are combined in quadrature. Our new result for the $6\,^2P_{3/2}$ state lifetime is 30.499±0.070 ns where the uncertainty is 0.23% [12]. In addition, we have measured the lifetime of the $6\,^2P_{1/2}$ state to be 34.934±0.094 ns where the fractional uncertainty is 0.27% [12]. The only previous measurement of the $6\,^2P_{1/2}$ state lifetime is that of Dodd *et al.* [46] with an uncertainty of 4%. The $6\,^2P_{3/2}$ lifetime has been measured previously by several techniques [47-50]. Figure 4 shows a comparison of our results with previous measurements that have uncertainties of less that 5%. A more complete list of older citations is given in Reference [45]. Our present measurement of this state [12] agrees with our previous result [45] and improves upon the uncertainty by a factor of four as shown in Figure 4. Also included in the Figure 4 is a recent pulsed-laser delayed-coincidence measurement by Young *et al.* [51].

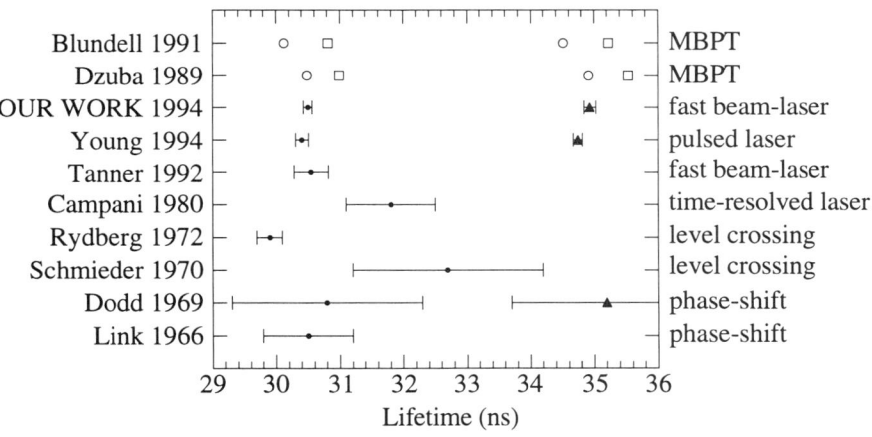

Figure 4. Graphical comparison between recently reported lifetime measurements, previous experimental work, and many-body theoretical values: (○) *ab initio* matrix elements [52, 53] combined with experimental energy levels, (□) *ab initio* matrix elements combined with *ab initio* energy levels [52, 53], (●) $6p\,^2P_{3/2}$ experimental values [12, 45-51], (▲) $6p\,^2P_{1/2}$ experimental values [12, 51, 46].

Two accurate *ab initio* calculations exist for the 6s-6p radial matrix elements [52, 53]. Comparisons of our results with these calculations are shown also in Figure 4. For the heavy alkali atom cesium, the reduced matrix elements for the two fine structure transitions, 6s-6p $^2P_{1/2, 3/2}$, have only a small inherent difference of 0.4% that comes from slight differences in the excited-state wave functions. For the lithium 2s-2p transition, this difference is negligible. Comparison of calculated matrix elements with experimental lifetimes requires a conversion factor of the transition wavelength cubed as indicated by Equation 12. The difference in the transition wavelengths accounts for the majority of the difference between the 6p $^2P_{1/2}$ and 6p $^2P_{3/2}$ state lifetimes. Using theoretical energy intervals and matrix elements, *ab initio* lifetimes are obtained and indicated by the open squares in Figure 4. The *ab initio* energy intervals calculated in References [52, 53] differ from the accurately known experimental transition energies [54] by only 0.5-0.8%, but when the transition energies are cubed this results in a 2% shift in the 6p $^2P_{1/2,\,3/2}$ state lifetimes. Experimental energies are more accurate for testing the calculated matrix elements alone. Using experimental transition energies with the *ab initio* matrix elements, we obtained the energy-corrected theoretical values indicated by the open circles in Figure 4.

Our measured lifetimes for both fine structure states happen to show excellent agreement with the energy-corrected values derived from Reference [53], but are more than 1% longer than the values derived from Reference [52], a 5σ difference from our experimental results. However, agreement between experiment and these energy-corrected theoretical lifetimes at the precision of our experiment is not necessarily to be expected. Both apply relativistic many-body perturbation theory (MBPT), but the calculations are inherently different. Blundell *et al.* use time independent perturbation theory, and Dzuba *et al.* use a time dependent approach. Although both calculations include most of the same effects due to infinite sums of many-body diagrams, certain contributions are not included in one or both formulations, as discussed in Reference [52]. The resulting *ab initio* values for the transition matrix elements differ from each other by 0.4% for 6s→6p $^2P_{1/2}$ and by 0.3% for 6s→6p $^2P_{3/2}$. Therefore, the corresponding *ab initio* lifetimes, proportional to the matrix elements squared, differ by 0.8% and 0.6%, respectively as shown in Figure 4. One source of inaccuracy in both theoretical calculations of transition matrix elements arises from the omission of high-order terms due to inexact electron binding energies. Estimates of the omitted contributions are given in References [52, 53] using experimental energy-scaling considerations. These estimated corrections increase the calculated lifetime values of Reference [53] by about 0.2% and those of Reference [52] by about 0.7%, resulting in a 2σ agreement between our measured values and both many-body calculations.

CONCLUSION

The measurement of lifetimes has found a role in many fields of physics, providing fundamental information about interactions or wave functions. In atomic physics, there has been and still is an incredibly large body of experimental work applied to measuring lifetimes of excited states of atoms and ions. The motivations for this work range from providing information for the interpretation of astrophysical data, to testing atomic theory. There are a variety of techniques available, and many have been applied to molecular systems. In many instances, only one technique many be practical. In neutral alkali systems, several techniques have been applied to lifetimes of the same state where the systematic effects and uncertainties can be compared. Two techniques have been applied successfully at levels of precision below 1%. Briefly described, these are the techniques of laser excitation of a fast atomic beam, and delayed coincidence detection with pulsed-laser excitation.

The lithium 2p state lifetime is an example of a lifetime that plays an important role in understanding the origins of the universe and testing the most sophisticated atomic structure calculations. A long standing discrepancy between experiment and theory in this simple atom has stimulated recent theoretical work. Agreement between a variety of new theoretical approaches calls for a renewal in experimental interest. The experimental atomic physics group at the University of Notre Dame has initiated a new measurement using the fast-beam laser approach.

Precision measurements of lifetimes and hyperfine structure are important to the interpretation of parity nonconservation experiments. The heavy alkali atom cesium is the subject of a new PNC experiment and interpretation of these results in terms of the weak interaction will require the accuracy of transition matrix elements to reach the 0.2% level. Lifetimes of the 6p $^2P_{1/2}$ and 6p $^2P_{3/2}$ states of neutral cesium are used to test many-body perturbation-theory approaches to the calculation of atomic structure in this many electron system. We have measured the lifetimes of the 6p $^2P_{1/2}$ and 6p $^2P_{3/2}$ fine structure states of atomic cesium with sufficient precision to determine the 6s $^2S_{1/2} \rightarrow$ 6p $^2P_{1/2,3/2}$ transition matrix elements to better than ±0.14%. Our lifetime results suggest that the accuracies of existing relativistic many-body calculations of these matrix elements in cesium are of order 0.3-0.6%. This new level of experimental precision is expected to motivate improved theoretical calculations of transition matrix elements for atomic cesium as well as for lighter alkali atoms.

ACKNOWLEDGEMENTS

I would like to thank my collaborators A. Eugene Livingston, Robert J. Rafac, H. Gordon Berry, Charles A. Kurtz, Kris W. Kukla, and Diana DiBerardino for helpful conversations and their dedication to experimental needs during the preparation of this manuscript. Partial support is acknowledged for C.E.T. from the National Institute of Standards and Technology, and the Luce Foundation.

REFERENCES

[1] S. D. Bergeson and J. E. Lawler, "Abstracts of Contributed Papers," ICAP-14, Univ. of Colorado-Boulder, July 31-August 5, 1994, 1K-5.
[2] R. Hutton, S. Huldt, B. Nyström, I. Martinson, K. Ando, T. Kambara, Y. Kanai, Y. Nakai, Y. Awaya, and J. Sugar, "Abstracts of Contributed Papers," ICAP-14, Univ. of Colorado-Boulder, July 31-August 5, 1994, 2K-2.
[3] V. Escalante, A. Góngora-T., and G. A. Victor, "Abstracts of Contributed Papers," ICAP-14, Univ. of Colorado-Boulder, July 31-August 5, 1994, 1K-6.
[4] J. Sapirstein, these same proceedings.
[5] J. Olsen, A. Godefroid, P. Jönsson, P. A. Malmquist, and C. Froese Fischer, "Abstracts of Contributed Papers," ICAP-14, Univ. of Colorado-Boulder, July 31-August 5, 1994, 1K-3.
[6] D. R. Plante, W. R. Johnson, and J. Sapirstein, "Abstracts of Contributed Papers," ICAP-14, Univ. of Colorado-Boulder, July 31-August 5, 1994, 1K-4.
[7] E. Biémont, F. Delahaye, and C. J. Zeippen, "Abstracts of Contributed Papers," ICAP-14, Univ. of Colorado-Boulder, July 31-August 5, 1994, 1K-7.
[8] Sy-Sang Liaw, "Abstracts of Contributed Papers," ICAP-14, Univ. of Colorado-Boulder, July 31-August 5, 1994, 2K-3.
[9] Jian Jin and D. A. Church, Phys. Rev. Letts. 70, 3213-3216 (1993).

[10] M. D. Ray and G. P. Lafyatis, "Abstracts of Contributed Papers," ICAP-14, Univ. of Colorado-Boulder, July 31-August 5, 1994, 2K-7.
[11] R. J. Rafac and C. E. Tanner, "Abstracts of Contributed Papers," ICAP-14, Univ. of Colorado-Boulder, July 31-August 5, 1994, 2K-5.
[12] R. J. Rafac, C. E. Tanner, A. E. Livingston, K. W. Kukla, H. G. Berry, C. A. Kurtz, Phys. Rev. A **50**, R1976 (1994).
[13] R. E. Imhof and F. H. Read, Rep. Prog. Phys. **40**, 1 (1977).
[14] R. G. DeVoe and R. G. Brewer, "Abstracts of Contributed Papers," ICAP-14, Univ. of Colorado-Boulder, July 31-August 5, 1994, 2K-1.
[15] L. J. Curtis, H. G. Berry, J. Bromander, Phys. Lett. **34A**, 169 (1971).
[16] E. H. Pinnington and R. W. Berends, "Abstracts of Contributed Papers," ICAP-14, Univ. of Colorado-Boulder, July 31-August 5, 1994, 1K-8.
[17] M. Walhout, A. Witte, and S. L. Rolston, Phys. Rev. Lett. **72**, 2843 (1994).
[18] C. Otes and J. Hall, post deadline contribution, ICAP-14.
[19] Particle Data Group, "Review of Particle Properties," Phys. Rev. D **45**, VI.15 (1992).
[20] G. W. F. Drake, J. Kwela, and A. van Wijngaarden, Phys. Rev. A **46**, 113 (1992).
[21] J. A. Thorburn, L. M. Hobbs, C. P. Deliyannis, M. H. Pinsolleault, Astro. J. **415**, 150 (1993).
[22] A. Gaupp, P. Kuske, and H. J. Andrä, Phys. Rev. A **26**, 3351 (1982).
[23] I. Carlsson and L. Sturesson, Atoms, Molecules and Clusters **14**, 281-287 (1989).
[24] J. Heldt, G. Leuchs, Z. Phys. A - Atoms and Nuclei **291**, 11 (1979).
[25] W. Hansen, J. Phys. B **16**, 933 (1983).
[26] W. Nagourney, W. Happer, Phys. Rev. A **17**, 1394 (1978).
[27] W. S. Bickel, I. Martinson, I. Lundin, R. Buchta, J. Bromander, I. Bergström, J. Opt. Soc. Am. **59**, 830 (1969).
[28] T. Andersen, K. A. Jessen, O. Sorensen, Phys. Lett. **29A**, 384 (1969).
[29] J. P. Buchet, A. Denis, J. Desesquelles, M. Dufay, C. R. Acad. Sci. Ser. B **265**, 471 (1967).
[30] K. C. Brog, T. G. Eck, and H. Wieder, Phys. Rev. **153**, 91 (1967).
[31] T. Brage and C. F. Fischer, Phys. Rev. A**49**, 2181-2184 (1994).
[32] K. T. Chung, in "The Physics of Highly Charged Ions," edited by P. Richard, M. Stockli, C. L. Cocke, and C. D. Lin, AIP Conf. Proc. No. 274 (AIP, New York, 1993).
[33] J. Pipin and D. M. Bishop, Phys. Rev. A **45**, 2736 (1992).
[34] A. W. Weiss, Can J. Chem. **70**, 456 (1992).
[35] A.-M. Mårtensson-Pendrill and A. Ynnerman, Phys. Scr. **41**, 329 (1990).
[36] S. A. Blundell, W. R. Johnson, Z. W. Liu, and J. Sapirstein, Phys. Rev. A **40**, 2233 (1989).
[37] C. E. Theodosiou, L. J. Curtis, and M. El-Mekki, Phys. Rev. A **44**, 7144 (1991).
[38] W. L. Wiese, Physica Scripta. **35**, 846 (1987).
[39] E. R. I. Abraham, W. I. McAlexander, N. W. M. Ritchie, C. Williams, and R. G. Hulet, "Abstracts of Contributed Papers," ICAP-14, Univ. of Colorado-Boulder, July 31-August 5, 1994, 1K-2.
[40] C. E. Tanner, R. J. Rafac, A. E. Livingston, K. W. Kukla, H. G. Berry, and C. A. Kurtz, "Abstracts of Contributed Papers," ICAP-14, Univ. of Colorado-Boulder, July 31-August 5, 1994, 2K-6.
[41] M. C. Noecker, B. P. Masterson, and C. E. Wieman, Phys. Rev. Lett. **61**, 310 (1988).
[42] W. J. Marciano and J. L. Rosner, Phys. Rev. Lett. **65**, 2963 (1990).
[43] C. S. Wood, D. Cho, S. C. Bennett, C. E. Wieman, "Abstracts of Contributed Papers," ICAP-14, Univ. of Colorado-Boulder, July 31-August 5, 1994, 1D-1.
[44] S. A. Blundell, W. R. Johnson, and J. Sapirstein, Phys. Rev. Lett. **65**, 1411 (1990).
[45] C. E. Tanner, A. E. Livingston, R. J. Rafac, F. G. Serpa, and K. W. Kukla, H. G. Berry, L. Young, and C. A. Kurtz, Phys. Rev. Lett. **69**, 2765 (1992).
[46] J. N. Dodd and A. Gallagher, Jour. of Chem. Phys. **50**, 4838 (1969).
[47] E. Campani, R. DeSalvo, and G. Gorini, Nuovo Cimento Lett. **29**, 485 (1980).
[48] S. Rydberg and S. Svanberg, Phys. Scr. **5**, 209 (1972).
[49] R. W. Schmieder, A. Lurio, W. Happer, and A. Khadjavi, Phys. Rev. A **2**, 1216 (1970).
[50] J. K. Link, J. Opt. Soc. Am. **56**, 1195-1199 (1966).

[51] L. Young, W. T. Hill, S. J. Sibener, S. D. Price, C. E. Tanner, C. E. Wieman, S. R. Leone, Phys. Rev. A **50**, 2174 (1994).
[52] S. A. Blundell, W. R. Johnson, and J. Sapirstein, Phys. Rev. A **43**, 3407 (1991).
[53] V. A. Dzuba, V. V. Flambaum, and O. P. Sushkov, Phys. Lett. A **142**, 373 (1989).
[54] C. E. Moore, National Bureau of Standards, "Atomic Energy Levels", Vol. III, 125 (1958), and Avila *et al.*, Metrologia **22**, 111 (1986).

PRECISE MEASUREMENTS OF ATOMIC MASSES

Accurate Atomic Mass Measurements from Penning Trap Mass Comparisons of Individual Ions

F. DiFilippo, V. Natarajan[†], M. Bradley, F. Palmer, and D.E. Pritchard

Research Laboratory of Electronics, Department of Physics, Massachusetts Institute of Technology, Cambridge, MA 02139, USA and [†]AT&T Bell Laboratories, Murray Hill, NJ 07974 USA.

Abstract. We report measurements of mass ratios of 20 pairs of molecular ions with a single ion Penning trap mass spectrometer having an accuracy exceeding one part in 10^{10}. The dominant source of error is random magnetic field fluctuations which cause a 2.6×10^{-10} rms scatter in measurements of the cyclotron frequency. Robust statistical analysis of the data ensures that nongaussian outliers are weighted less heavily in a smooth and consistent manner. Systematic errors are estimated to be 2×10^{-11} or below for doublet mass comparisons. The ratios form an overdetermined set, such that the atomic masses of nine isotopes can be derived from at least two independent groups of ion mass ratios, providing many consistency checks for systematic errors at the 10^{-10} level. At this level of precision, certain mass measurements have important implications in fundamental metrology. Results presented here are essential for defining a practical atomic standard of mass, for calibrating γ-ray wavelengths, and for determining the molar Planck constant and the fine structure constant.

1. INTRODUCTION

The precision at which mass comparisons can be made has steadily improved over the years. Recently, the Penning trap has emerged as the most accurate instrument for mass spectrometry. A precision exceeding 10^{-10} is routinely attained in our experiment at MIT (1). The major purpose of this paper is to present accurate measurements of twenty different mass ratios that determine ten atomic masses and to present a thorough analysis of the uncertainties in these measurements. We also describe several new metrological implications of our mass comparisons.

The basic advantages of a Penning trap for mass measurement are that the mass is determined from a frequency measurement, that a long time is available to make this measurement, and that the ion is confined to a small spatial region of a highly uniform magnetic field. A Penning trap consists of a strong magnetic field $\vec{B}$ (providing radial confinement) and a weak quadrupole electric field $\vec{E}$ (providing axial confinement) (2). In order to eliminate coulombic perturbations due to other nearby ions and uncertainties due to unspecified internal motion of a cloud of trapped ions, we perform measurements on a single trapped ion.

The physics of a single ion in a Penning trap is well-understood and has been described in detail in the literature (3). The motion is a superposition of three normal modes of oscillation: the "axial" mode, the "trap cyclotron" mode, and the "magnetron" mode. The axial mode is harmonic oscillation along the magnetic field lines at a frequency ω_z that is proportional to the square root of the trap voltage. The trap cyclotron mode (at ω'_c) is similar to ordinary cyclotron motion in the radial plane, with the frequency slightly perturbed by the electric field. The magnetron mode (at ω_m) is a much slower circular motion essentially due to $\vec{E} \times \vec{B}$ drift. The "free-space" cyclotron frequency ω_c, which would be the frequency of cyclotron motion if the electric field were removed, is obtained by adding the three mode frequencies in quadrature:

$$\omega_c = \left(\omega'^2_c + \omega_z^2 + \omega_m^2\right)^{1/2} = \frac{qB}{mc}, \quad (1)$$

where q and m are the charge and mass of the ion (in CGS units), and c is the speed of light. In our trap, $B = 85000$ G, and for an N_2^+ ion (mass 28 u, where u ≡ atomic mass units), the mode frequencies ω'_c, ω_z, and ω_m are 4.6 MHz, 160 kHz, and 2.8 kHz, respectively. Therefore, only the trap cyclotron frequency ω'_c must be measured to the desired precision of ω_c.

For optimum precision, the frequency, phase and amplitude of the ion's axial motion must be accurately extracted from the smallest possible signal. The ion's axial motion is observed by detecting the image current (~10^{-14} A) induced in the endcaps of the trap. A high-Q (~30,000) tuned circuit and an rf SQUID are used to attain a sufficient signal-to-noise ratio (4). The axial signal is a sinusoid that decays as the ion loses its energy into the detector. The simplest method of analysis is to perform a digital Fourier transform on the data. The ion's spectral parameters are then the parameters of the peak bin. However, such an analysis gives biased estimates since it fails to account for the damping.(5) A better procedure is to perform an analog of the Laplace transform with a complex frequency, where the imaginary part is the damping constant. This is implemented simply by pre-multiplying the data with $\exp(-t/\tau)$, where τ is the amplitude decay time, and then taking a fourier transform. The ion's parameters are obtained by finding the peak of the transform. This procedure works better because it weights the data taken at later times progressively less (as the signal damps). We have shown (6) that this procedure gives unbiased estimates and gives errors that are close to theoretical minimum bounds for a given signal to noise ratio.

The radial modes are observed and cooled indirectly by coupling them to the axial mode with a diagonally oriented quadrupole rf field (7). The advantage of this scheme is that the cyclotron mode is not damped by the detector (thus having a nearly zero linewidth) and also does not experience tuned circuit pulling. However, an indirect approach must be taken to measure ω'_c.

We have developed two such approaches to determine ω'_c: the "*p*ulse a*n*d *p*hase" (PNP) method and the "*s*eparated *o*scillatory *f*ields" (SOF) method. Both methods utilize "π-pulses" of the diagonal rf field at $\omega'_c - \omega_z$ to coherently exchange the amplitudes (scaled by $\omega^{1/2}$) and phases of the axial and trap cyclotron modes, allowing the amplitude and phase of the trap cyclotron mode to be determined (8). With the PNP method, the trap cyclotron frequency is measured by exciting the ion to a cyclotron amplitude ρ_c, allowing the ion to evolve "in the dark" for a delay time T, and applying a π-pulse to measure the accumulated phase (9). With the SOF method(10), the ion is excited by a pair of cyclotron pulses separated by a delay time T, so that the final cyclotron amplitude varies sinusoidally with the phase accumulated between the pulses. The amplitude is then measured by a π-pulse. The SOF method is well suited for measurements of non doublets since the cyclotron motion of both ions can be studied with the same electric fields, with the trap voltage then being changed just before the π-pulse in order to bring the axial frequency into resonance with the detector. With both methods, ω'_c is measured to 10^{-10} precision with a series of measurements, the longest having a delay time of ~1 minute, so that proper phase unwrapping is achieved. Related techniques are used to cool the magnetron motion so that the ion is located at the center of the trap before measuring ω'_c.

2. DETERMINING A MASS RATIO

If the magnetic field were known as a function of time, a mass ratio of two different ion species could be determined by comparing the free-space cyclotron frequencies for two ions, measured at times t_1 and t_2,

$$\frac{m_1}{m_2} = \frac{q_1}{q_2} \frac{B(t_1)}{B(t_2)} \frac{\omega_{c2}(t_2)}{\omega_{c1}(t_1)} , \qquad (2)$$

since the ratio q_1/q_2 is a known rational number, and the ratio ω_{c2}/ω_{c1} is measured to high precision. Unfortunately, the magnetic field drifts unpredictably in the time between and during two measurements and does not cancel exactly. The field can change by processes internal or external to the magnet, the major source during the day being external magnetic fields from a nearby subway. Motion of the trap relative to the magnet may also change the field at the trap center because of field gradients. Although other sources of random error (such as trap voltage fluctuation and thermal noise) contribute to temporal variations in repeated measurements of the cyclotron frequency, the magnetic field fluctuations dominate (see section 5). We therefore model all temporal variation of ω_c as if the magnetic field were the only contributor.

Fitting to field drift

In order to account for the effects of temporal drift, measurements are repeated while alternating between the two species being compared. A plot of the cyclotron frequency data for a typical run is shown in Fig. 1. With this scheme, ω_c is measured for each ion at several different times, allowing the drift to be determined. However, the field drift can be determined and corrected for only if it occurs on a time scale longer than the time between measurements of different ions, which is typically 10-20 minutes. Short-term field fluctuations cannot be modeled, and they contribute to the uncertainty in the mass ratio. The time dependence of the magnetic field can be written as:

$$B(t) = B(0)\left(1 + f(t) + \delta B(t)\right), \tag{3}$$

where $f(t)$ is the modeled long-term field drift and $\delta B(t)$ is the unknown short-term field behavior.

The mass ratio is obtained by fitting to the data for both ions simultaneously.

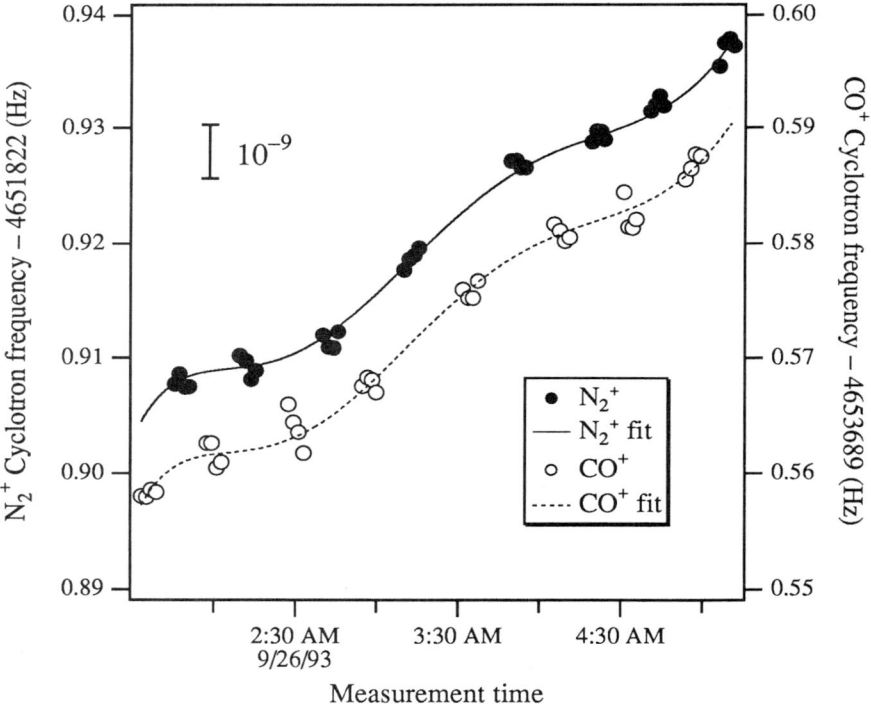

Figure 1. Typical data for an ion mass ratio measurement. The cyclotron frequency is measured alternately for the two ions in order to fit to the field drift and to determine the mass ratio.

If the fitting is done by the least squares method, then the quantity

$$\sum_j \left(\omega_{c1}(t_j) - f(t_j) - \Delta\omega_c\right)^2 / \sigma_j^2 + \sum_k \left(r_0\, \omega_{c2}(t_k) - f(t_k)\right)^2 / \sigma_k^2 \qquad (4)$$

is minimized with respect to the fitting function $f(t)$ and the frequency difference between the two curves $\Delta\omega_c$. Here, the indices j and k indicate individual measurements for the two ions, r_0 is an initial guess of the mass ratio (obtained from existing atomic mass data), and σ is the measurement uncertainty. For a doublet (a pair of ions of nominally the same mass, such as N_2^+ and CO^+), $r_0 \cong 1$. The best value of the frequency difference,

$$\Delta\omega_c = \omega_{c1}(0) - r_0\, \omega_{c2}(0) \qquad (5)$$

leads directly to the mass ratio r,

$$r = \frac{m_1}{m_2} = \left(1 - \frac{\Delta\omega_c}{\omega_{c1}(0)}\right) r_0^{-1} \qquad (6)$$

The uncertainty in r arises from the short-term field fluctuations $\delta B(t)$ and will be discussed in the next section.

The simplest general functional form to use for $f(t)$ is a polynomial. The order of the polynomial is critical. If too few terms are used, the polynomial may not fit the drift adequately. Conversely, if too many terms are used, the polynomial may exaggerate short-term fluctuations, giving unreliable results. Thus there is an optimal order of polynomial to use for a data set.

The data reported here were taken over a period exceeding one year, and the roughly equal earlier and later portions were analyzed in two significantly different ways. For the earlier portion (6) we used our best judgement to determine the number of polynomials, and used conventional least squares fitting algorithms. For the latter portion (11) we used a more conservative statistical test to determine the number of polynomials, and a modification of least square statistics called "robust" statistics which provides improved handling of points (called "outliers" henceforth) that deviate from the mean by several standard deviations.

For the more recent portion of the data, the basis for deciding whether the next higher-order term should be added to the polynomial is whether it produces a significant statistical improvement in the fit. Goodness of fit is characterized by the χ^2 statistic (the quantity minimized in Equation (4)):

$$\chi_n^2 = \sum_{i=1}^N (\Delta_i / \sigma_i)^2, \qquad (7)$$

where Δ is the deviation from the fit, σ is the measurement uncertainty, n is the order of the polynomial fit, and N is the total number of points for both ions. Improvement in the fit due to an additional term (the F-test) (12) is characterized by the relative change in χ_n^2:

$$F_\chi(n) = \frac{\chi_{n-1}^2 - \chi_n^2}{\chi_n^2 / \nu}, \qquad (8)$$

where ν is the number of available degrees of freedom (to be discussed shortly). Assuming purely random fluctuations, the quantity F_χ follows the F distribution, since it is a ratio of χ^2 statistics. The probability P that the term of order n is statistically significant can be determined from tables of F_χ vs v (12). (By coincidence, a probability of 0.5 corresponds to a value of $F_\chi \sim 0.5$ for low orders of n.) Our procedure is to increase the order of the polynomial until two consecutive fits give $P < 0.5$ (because the field may accidently have no variation corresponding to a particular order). The optimal order is chosen to be the last one having $P > 0.5$.

In most statistical analyses, the number of degrees of freedom v is taken to be $N - n - 1$, the number of data points minus the number of terms in the fit (n plus a constant term). This assumes that the data points are all uncorrelated, which is not true in our case, since points within a "cluster" (a group of successive measurements on the same ion – see Fig. 1) are more strongly correlated. We adopted a more conservative estimate for v:

$$v = N^* - n - 2 ,\qquad(9)$$

where N^* is the number of clusters in the data set. (The "2" arises from the fact that $\Delta\omega_c$ is an extra parameter in the fit.) This ensures that only long-term drift affecting more than one cluster is considered and prevents the polynomial from fitting to field jumps on a time scale smaller than the time between clusters. Thus, for example, a minimum of five clusters is needed for a quadratic ($n = 2$) fit.

Figure 2 illustrates the dependence of χ^2, F_χ, and the calculated mass ratio on the order of the polynomial fit, for the CO^+/N_2^+ measurement illustrated in Fig. 1. Fig. 2a shows that χ^2 decreases quickly at first and later stabilizes as the order is increased. In Fig. 2b, F_χ is plotted along with a dotted line corresponding to $P = 0.5$, showing that $n = 6$ is the optimal order in this case. As seen in Fig. 2c, the calculated mass ratio did not vary much, although variations on the order of 1 σ have been observed in other measurements for small n.

A result of this conservative treatment of the number of degrees of freedom has been that some measurements needed to be discarded. For certain runs (apparently when the magnet's liquid helium level was low), the magnetic field was exceedingly erratic. Polynomials whose order approached the number of clusters of data seemed to be necessary in order to fit to the field fluctuations, indicating significant field variation on the time scale required to change ions. The above definition for v does not allow such a high-order fit, and we decided to discard such data completely. These data included our earlier measurement of the $^{14}N/CH_2^+$ ratio (13), which was 1.5 sigma higher than the value of this ratio determined from our more recent measurements of $^{14}N_2/C_2H_4^+$.

3. MAGNETIC FIELD NOISE

The above procedure describes how we model long-term field drift and extract the mass ratio from the data. The unknown component of Equation (3), the short-term field fluctuation $\delta B(t)$, appears as scatter about the fit and leads to uncertainty in the mass ratio. This field noise is the dominant source of error in our

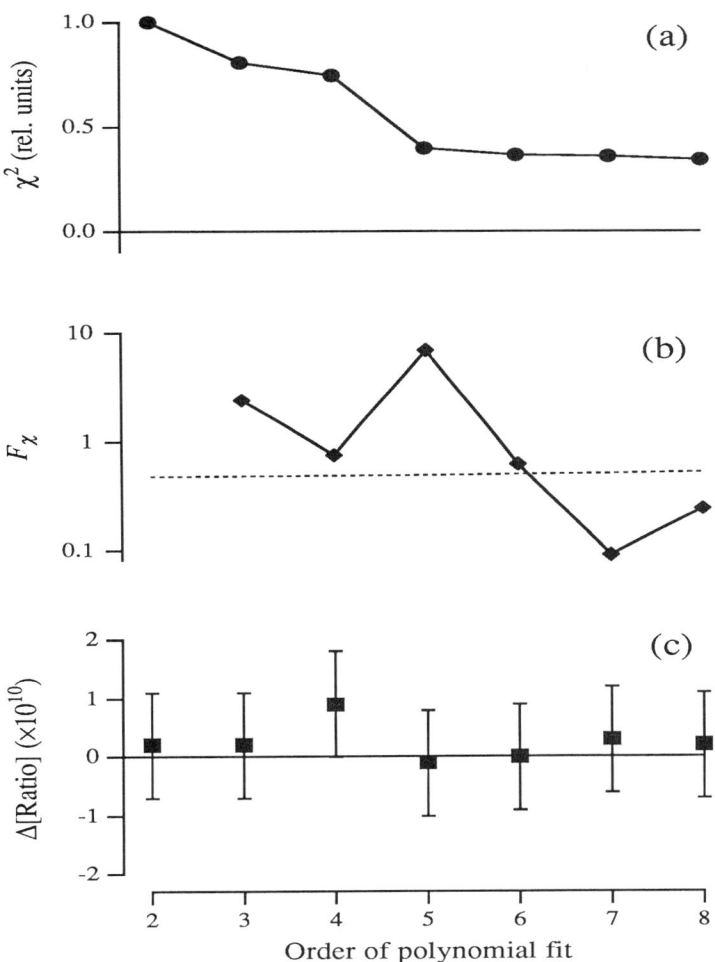

Figure 2. Choosing the order of the polynomial fit. The χ^2 statistic (a), the F_χ statistic (b), and the calculated mass ratio (c) are plotted as a function of the order of the polynomial v for the data in Fig. 1. The dotted line in (b) indicates the value of F_χ corresponding to a 50% statistical probability that the observed value would be exceeded if the data were normally distributed. (Note that it increases slightly with v.) In this case, a 6th order polynomial was determined to be optimal.

experiment, and its statistics must be well understood. A detailed analysis of the magnetic field noise and its effect on mass ratios is presented in this section. Other sources of error, both random and systematic, are considered in the next section and shown to be insignificant.

Traditional statistics is based on the assumption that random variables follow the normal (gaussian) distribution. This assumption often does not apply to real experiments, where some of the fluctuations may be due to a less frequent but more intense external noise source, or where the fluctuations may not be normally distributed ($1/f$ noise for example). A qualitative model of such an observed noise distribution P is a superposition of a dominant gaussian component P_G along with a small component P_N that is much wider: (14)

$$P = (1-\varepsilon)P_G + \varepsilon P_N \qquad (10)$$

where $\varepsilon \ll 1$. The effect of P_N is to increase the probability of observing a large fluctuation. Although a gaussian distribution predicts that variations larger than 3 σ should only occur 0.3% of the time, they occur much more often in most real experiments. Thus, P is a nearly gaussian distribution for small variations, but with tails that approach zero more slowly.

The magnetic field noise distribution $P(\delta B)$ for all the data in the second part of our experiment has been observed to follow this model. Experimentally, $P(\delta B)$ is measured by compiling a histogram of deviations from curve fits to the data from many runs. The histogram shown in Fig. 3a consists of about 1000 measurements and appears to be gaussian near the center with standard deviation $\sigma = 2.6 \times 10^{-10}$. When the histogram is plotted on a semi-logarithmic scale (Fig. 3b), the gaussian central portion of the distribution is parabolic, and the extra outliers are readily apparent.

These excess outliers can have adverse effects on data analysis, especially if the least squares method from traditional statistics is used. Since the goal is to minimize the sum of squares of the deviations from the fit, outliers are heavily weighted and could significantly pull the mean. Data rejection methods exist which attempt to identify and eliminate non-gaussian outliers, but these methods have been shown to be biased for varying degrees of noise contamination (14).

Robust statistical analysis

Our approach has been to use robust statistics (14), which maintains the least squares philosophy while accounting for outliers in a smooth and consistent manner. A class of robust statistics called "M-estimates" is a generalization of least squares statistics and is easily implemented in nonlinear regression. Near the center, a robust distribution approximates a gaussian distribution. Data with larger deviations from the fit Δ are weighted less, and the corresponding probability distribution $P(\Delta)$ has larger tails than a gaussian to account for extra outliers. This distribution can be expressed in terms of an "estimator" $\psi(\Delta)$:

$$P(\Delta) = \exp\left(-\int_0^\Delta \psi(\Delta')d\Delta'\right) \qquad (11)$$

Since the maximum likelihood estimates of the fit parameters are obtained by solving $\sum_i \psi(\Delta_i) = 0$, the estimator $\psi(\Delta)$ is proportional to the effect of a data

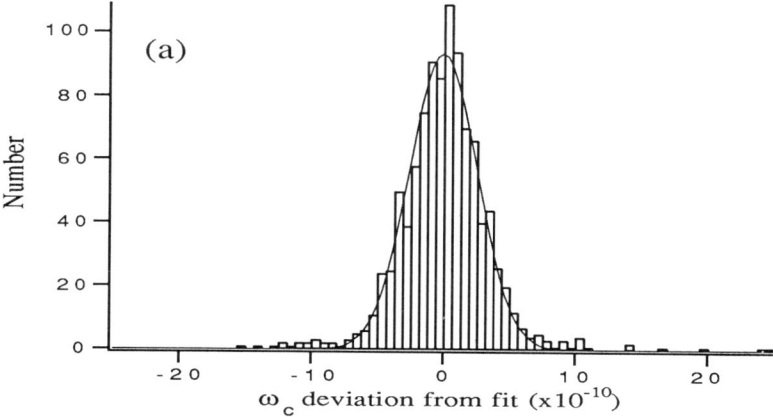

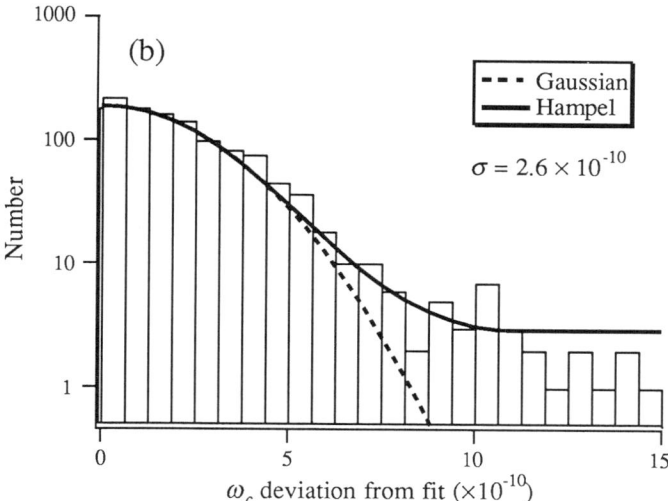

Figure 3. Cyclotron frequency noise. In (a), a histogram of the differences between the measured ω_c's and the fit to the field drift is plotted for ~1000 measurements, along with a gaussian fit with $\sigma = 2.6 \times 10^{-10}$. The excess number of outliers is apparent in (b), where the histogram is folded and plotted on a logarithmic scale. The distribution corresponding to the Hampel estimator in Fig. 4 represents the observed noise more accurately than the conventional gaussian (least-squares) distribution.

point with deviation Δ on the fit. The least squares case is given by $\psi(\Delta) = \Delta$, in which all data points are weighted evenly and affect the fit in linear proportion to their deviation from the fit, and probability distribution is the familiar gaussian: $P(\Delta) = \exp(-\Delta^2/2)$. A robust estimator, on the other hand, has $\psi(\Delta) = \Delta$ for points with small deviations (where gaussian statistics is valid) and $\psi(\Delta) < \Delta$ for points with large deviations (thus diminishing the importance of the outliers).

We use a three-part descending Hampel estimator $\psi(x)$ (Fig. 4) to fit to the field noise. (Many statisticians consider this estimator to be well representative of actual physical data (15).)

$$\psi(x) = \begin{cases} x & \text{for } |x| < a \\ a\,\text{sgn}(x) & \text{for } a < |x| < b \\ a(c - |x|)/(c - b) & \text{for } b < |x| < c \\ 0 & \text{for } |x| > c \end{cases} \quad (12)$$

where $x = \Delta/\sigma$, and where a, b, and c are parameters, chosen to be 1.6, 2.5, and 4.3, respectively. These parameter values were selected to accurately reflect the observed probability distribution in Fig. 3. Points with deviations larger than $a\sigma$ have reduced weight, and those with deviations larger than $c\sigma$ are completely rejected from the fit (i.e., $\psi \to 0$). Although only six points were seen outside the range of Fig. 3(b), others may have been missed because their phase error exceeded π, resulting in misassignment of the phase or in rejection of that datum due to uncertainty in unwrapping the phase.

Fitting with robust statistics is equivalent to performing a weighted least squares fit (14). The assigned error in each measurement is weighted by a factor of $w = \psi(\Delta)/\Delta$. Points with small Δ thus receive full consideration by the fit

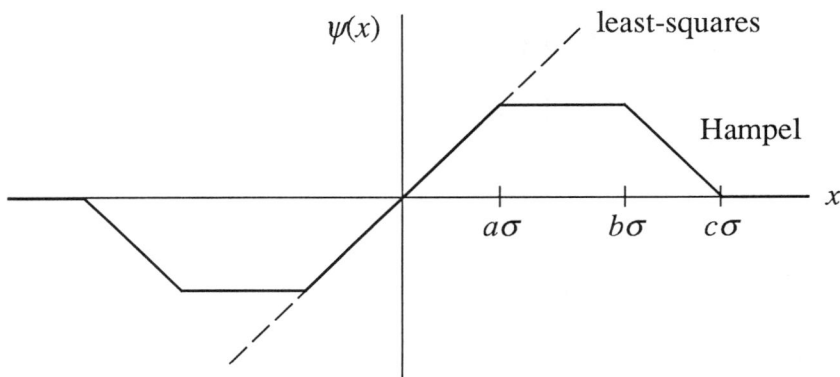

Figure 4. Hampel estimator. The three-part descending Hampel estimator (12) used for modeling fluctuations in ω_c (Fig. 3) is shown, along with the conventional least-squares estimator. Points with large deviations ($> a\sigma$) are weighted less and have a smaller effect on fitting to field drift.

($w=1$), and points with large Δ receive reduced consideration ($w<1$). Least squares fitting is done iteratively (since Δ changes after each attempt) until the fit parameters converge. The resultant uncertainty in the difference frequency $\Delta\omega_c$ is the standard statistical error (from χ^2 curvature) from assigning an uncertainty of $2.6\times 10^{-10}/w$ to each point. For a nighttime run, the statistical uncertainty in $\Delta\omega_c$ and therefore in the associated mass ratio is typically 1×10^{-10}.

Ultimately, the use of robust statistics has made relatively little difference in our final results. When robust and standard techniques were applied to the same data, the results were usually within $\sigma/2$ of each other. The standard deviations observed in the later set of runs were generally close to 2.6×10^{-10} whereas standard deviations of the earlier set of runs occasionally were as high as 4×10^{-10}. We believe this is primarily due to better shimming of the field (so that movement of the trap vacuum system in its dewar does not affect the results so much, but it results partly from the fact that the robust statistics deweight the outliers somewhat. As described below (see Consistency checks), comparisons of final results from the two different sets of runs showed no statistically significant deviations.

4. ATOMIC MASSES

Mass ratios were measured for a wide variety of molecular ions, and the results are listed in Table 1. The ion species were selected so that atomic masses of the neutral isotopes could be determined in terms of $m[^{12}C]$, the basis of the atomic mass scale. In this section, the procedure for converting the ratios into atomic masses is described.

The first step in converting ion mass ratios into masses of neutral atoms in the ground state is to account for chemical binding energies and ionization energies:

$$m[A_nB_k^+] = m[nA_{(g)} + kB_{(g)} - e^-] + \Delta E/c^2 \tag{13}$$

The energy ΔE required to form a molecular ion from the neutral atoms is calculated from the standard heats of formation of the species in the gaseous state at 0 K temperature (16). Afterwards, the mass ratio r can be expressed in terms of the individual atomic masses, as in this example:

$$r = \frac{m[A_nB_k^+]}{m[C_pD_q^+]} = \frac{m[nA + kB - e^-] + \Delta E_1/c^2}{m[pC + qD - e^-] + \Delta E_2/c^2} \tag{14}$$

For doublet measurements, $(1-r) \sim O(10^{-3})$, and the mass ratio should be considered as a determination of a mass difference:

$$m[pC + qD - nA - kB] = (1-r)\bigl(m[pC + qD - e^-]\bigr) + \Delta E'/c^2 \tag{15}$$

Table 1. Measured mass ratios and corresponding mass differences (in atomic mass units). The numbers in parentheses indicate the uncertainty in the last decimal places. Several ratio measurements were repeated in order to reduce statistical error.

Ratio	Value	Runs	Mass Difference	Value [u]
$N_2^+ / C_2H_4^+$	0.999 102 696 201 (55)	4	$C + 2H - N$	0.012 576 059 8 (8)
$^{15}N_2^+ / C_2D_2H_2^+$	0.998 547 569 780 (50)	4	$C + D + H - {}^{15}N$	0.021 817 911 9 (8)
CO^+ / N_2^+	0.999 598 887 572 (77)	3	$2N - C - O$	0.011 233 390 9 (22)
$CO^+ / C_2H_4^+$	0.998 701 943 805 (66)	2	$C + 4H - O$	0.036 385 507 3 (19)
$^{13}C_2H_4^+ / C_2D_2H_2^+$	0.999 805 486 870 (77)	2	$C + D - {}^{13}C - H$	0.002 921 908 6 (12)
Ar^{++} / Ne^+	0.999 437 341 275 (106)	2	$2Ne - Ar$	0.022 497 224 5 (42)
$Ar^+ / C_3H_4^+$	0.998 278 399 350 (88)	2	$3C + 4H - Ar$	0.068 917 005 3 (35)
$CD_3^+ / CD_2H_2^+$	0.999 914 190 780 (100)	2	$2H - D$	0.001 548 283 6 (18)
Ne^+ / CD_4^+	0.996 810 562 610 (130)	1	$C + 4D - Ne$	0.063 966 932 9 (26)
Ar^{++} / CD_4^+	0.996 249 698 100 (100)	1	$2C + 8D - Ar$	0.150 431 104 5 (40)
CD_4^+ / C^+	1.671 397 950 390 (310)	1	D	2.014 101 778 5 (9)
CD_3^+ / C^+	1.503 548 462 350 (200)	1	D	2.014 101 777 6 (6)
Ar^+ / Ne^+	1.998 902 121 050 (300)	1	$Ar - Ne$	19.969 942 947 9 (60)
O^+ / CH_4^+	0.997 730 269 420 (80)	1	$C + 4H - O$	0.036 385 506 2 (13)
CH_4^+ / C^+	1.335 957 033 780 (230)	1	H	1.007 825 031 7 (7)
$SiH_2^+ / C_2D_2H_2^+$	0.998 293 230 200 (80)	1	$2C + 2D - Si$	0.051 277 022 4 (24)
$SiH_2^+ / {}^{15}N_2^+$	0.999 745 290 400 (80)	1	$2{}^{15}N - Si - 2H$	0.007 641 200 7 (24)
$^{15}N^+ / CH_3^+$	0.998 444 631 990 (110)	1	$C + 3H - {}^{15}N$	0.023 366 197 9 (17)
$CO_2^+ / C_3H_8^+$	0.998 348 443 160 (100)	1	$C + 4H - O$	0.036 385 506 0 (22)
$^{13}CH_4^+ / CDH_3^+$	0.999 828 496 650 (90)	1	$C + D - {}^{13}C - H$	0.002 921 907 4 (15)

Although C and D also appear on the right side of the equation, they are multiplied by the small factor $(1-r)$ and do not need to be known *a priori* to high precision. Note that species which appear in both the numerator and the denominator cancel to first order in the mass difference. (For example, the ratio $m[CD_3^+]/m[CD_2H_2^+]$ determines the mass difference $2H - D$.)

The mass differences and their uncertainties, also listed in Table 1, can be expressed in matrix form:

$$xM = y \pm \sigma \quad (16)$$

where x is a $P \times Q$ matrix of the coefficients, M is a column vector of the Q atomic masses, and $y \pm \sigma$ is a column vector of the P mass differences and uncertainties. (The reference atom ^{12}C, which is defined to have a mass of exactly 12 u, is included on the right side of the equation.) The best values of the atomic masses are obtained from a global least squares fit:

$$M = X^{-1}Y \quad (17)$$

where

$$X_{jk} = \sum_{i=1}^{P} x_{ij} x_{ik} / \sigma_i^2 \quad , \quad Y_j = \sum_{i=1}^{P} x_{ij} y_i / \sigma_i^2 \quad (18)$$

The inverted matrix X^{-1} is the covariance matrix, the diagonal elements of which are the uncertainties in the atomic masses.

The ability to invert the X matrix depends on whether each individual species can be directly related to ^{12}C. This is rather difficult to accomplish solely with doublets in the mass range in which we operate (10-50 u). (For example, the three ratios N^+/CH_2^+, O^+/CH_4^+, and CO^+/N_2^+ would seem sufficient to determine the atomic masses of N, O, and H, but the resultant matrix is singular; *i.e.*, the mass differences from these ratios using (15) are linearly dependent.) We have measured one set of doublets ($Ar^+/C_3H_4^+$, Ar^{++}/CD_4^+, and $CD_3^+/CD_2H_2^+$) which determines H and D directly in terms of C and breaks the singularity in the matrix. We have also developed techniques to make non-doublet measurements (10), such as CH_4^+/C^+, which are very useful for providing links to invert the matrix.

Atomic mass table

An atomic mass table (Table 2) of nine isotopes and the neutron is obtained by fitting to the entire set of our mass ratio measurements. For comparison, the best values from conventional mass spectrometry (the 1983 atomic mass evaluation (17)) are also listed, showing that Penning trap measurments have achieved an improvement in precision of a factor of 10-1000. The latest (1993) atomic mass evaluation (13) contains some data from Penning trap experiments, including preliminary values of some results reported here. Our latest results are consistent with our earlier (6) results, except for the 1.5 σ adjustment in the ^{14}N mass that followed the new data analysis techniques described previously.

The neutron mass follows from the atomic masses of ^1H, ^2H, and the deuteron binding energy from:

$$^1H + n \rightarrow {}^2H + \gamma \quad (2.2 \text{ MeV}) , \quad (19)$$

Table 2a. Atomic mass table. The masses (in u) of nine isotopes and the neutron are listed as determined from this experiment, from the 1983 atomic mass evaluation [15], and from other single-ion Penning trap experiments [17,19]. The numbers in parentheses indicate the error in the rightmost figures. For the purpose of comparison, zeros have been added so that the numbers of digits are equal. The uncertainty in the neutron mass from this experiment is limited by the error in the deuteron binding energy [16].

Species	Mass (this work)	Non-Penning trap values [15]	Other Penning trap values
^{1}H	1.007 825 031 6 (5)	1.007 825 035 0 (120)	1.007 825 032 6 (10) [17]
n	1.008 664 923 5 (23)	1.008 664 919 0 (140)	1.008 664 918 7 (26) [17]
^{2}H	2.014 101 777 9 (5)	2.014 101 779 0 (240)	2.014 101 776 9 (11) [17]
^{13}C	13.003 354 838 1 (10)	13.003 354 826 0 (170)	13.003 354 840 4 (41) [17]
^{14}N	14.003 074 004 0 (12)	14.003 074 002 0 (260)	14.003 074 005 6 (18) [17]
^{15}N	15.000 108 897 7 (11)	15.000 108 970 0 (400)	
^{16}O	15.994 914 619 5 (21)	15.994 914 630 0 (500)	15.994 914 626 3 (30) [17]
^{20}Ne	19.992 440 175 4 (23)	19.992 435 600 0 (22000)	
^{28}Si	27.976 926 532 4 (20)	27.976 927 100 0 (7000)	27.976 926 575 0 (380) [19]
^{40}Ar	39.962 383 122 0 (33)	39.962 383 700 0 (14000)	

Table 2b. Atomic mass differences measured by the MIT Penning trap experiment and the Ohio State FT-ICR experiment.

Mass difference	MIT value [u]	Ohio State value [u]
$CH_2^+ - N^+$	0.012 576 046 4 (6)	0.012 576 039 0 (190) [20]
$H_2O^+ - DO^+$	0.001 548 279 4 (9)	0.001 548 296 0 (120) [21]
$D_2O^+ - {}^{20}Ne^+$	0.030 677 980 2 (25)	0.030 677 480 0 (670) [22]

The most accurate measurement of the γ-ray wavelength (18) is accurate to 1×10^{-6} and limits the precision of our determination of the neutron mass to 2×10^{-9}.

Other Penning trap measurements

Table 2 also compares our measurements with other Penning trap experiments. The results are in good agreement with our more precise values, except for a two sigma difference for ^{16}O. Van Dyck *et. al.* (19) have written a detailed review of high precision mass measurements in Penning traps. Their group has measured the masses of several light isotopes (19), most notably ^{3}He and ^{3}H (20), with uncertainties typically a factor of 2 larger than ours. They use a trap similar to ours, but having a small anharmonicity such that the axial frequency is a weak function of the cyclotron amplitude. Measurements are made by monitoring the axial frequency while a drive is swept through the (trap) cyclotron frequency. They have solved the matrix inversion problem by measuring q/m doublets containing highly ionized species such as $^{12}C^{3+}$ and $^4He^+$. The Stockholm-Mainz group has measured the mass of ^{28}Si with 10^{-9} precision (21). Their technique is similar to one used for unstable nuclei (22). Ions are ejected from the trap through a hole in the endcap. As they leave the magnetic field, their cyclotron energy is converted to additional axial velocity, which is monitored by time of flight. Fourier transform ion cyclrometers are widely used in analytical and physical chemistry. (23) These devices have a ring that is divided into four quadrants, with two quadrants used to excite the cyclotron motion and two used for detection. The group at Ohio State has determined several mass differences by FT-ICR spectrometry on ion clouds with uncertainties about 20 times larger than our corresponding values (24-26).

5. UNCERTAINTIES

In the previous discussion, it was assumed that random field fluctuations dominated the uncertainty in the experiment. There are several other sources of error below 10^{-10} that prove negligible but still should be considered. In this section, the various uncertainties, both random and systematic, are summarized and estimated. In Table 3, the uncertainties contributed from the various sources are listed for a single measurement of ω_c as well as for a measurement of the mass ratio during a nighttime run. For the purpose of calculation, errors are estimated for a typical ion of mass 30 u; Table 3 also lists how each error scales with the mass of the ion being measured.

Random errors

The observed noise histogram in Fig. 3 is a composite of all short-term fluctuations which contribute to the random error of the experiment. The rms error is 2.6×10^{-10} per measurement of ω_c, and the resultant error in the mass ratio is typically 1×10^{-10} for a nighttime run of ~60 measurements with ~15 alternations of the ions. Variations in the magnetic field at the site of the ion (from both internal and external processes) are primarily responsible for this random error.

Electric field fluctuations also contribute to the random error, but to a lesser extent. The axial frequency ω_z varies as the square root of the trap voltage, which

Table 3. Summary of estimated random [R] and systematic [S] uncertainties for doublet (d) and non-doublet (n) comparisons, in parts per trillion (10^{-12}). The first column lists the error contributed from each source per measurement of the cyclotron frequency ω_c, and the second column lists the resultant error in the mass ratio for a single nighttime run. Errors are calculated for mass 30 u ions, assuming a cyclotron amplitude $\rho_c = 0.025$ cm, a magnetic bottle $B_2/B_0 < 10^{-6}$ cm^{-2}, and a higher-order electric field $C_4 = 2 \times 10^{-4}$, among other assumptions described in the text. The last column indicates how each uncertainty scales with the mass m of the ion.

Source of uncertainty	Error in ω_c (ppt)	Error in ratio (ppt)	Scaling
Magnetic field fluctuations	260 [R]	100 [R]	—
Electric field fluctuations	90 [R]	30 [R]	$(m/30)^2$
Thermal noise	30 [R]	10 [R]	$(30/m)^{3/2}$
Spectral estimation (ω_z)	90 [R]	30 [R]	$m/30$
Spectral estimation (phase)	100 [R]	30 [R]	—
r_c imbalance (relativity)	10(d), 20(n) [S]	10(d), 20(n) [S]	$30/m$
r_c imbalance (B_2)	10(d), 30(n) [S]	10(d), 30(n) [S]	$m/30$
r_c imbalance (C_4)	20(d), 60(n) [S]	20(d), 60(n) [R]	$(m/30)^3$
C_4 imbalance	20 [S]	20 [R]	$30/m$
Surface patch charges	4(d), 0(n) [S]	4(d), 0(n) [S]	$30/m$
Trap tilt (magnetron)	0(d), 40(n) [S]	0(d), 40(n) [S]	$(m/30)^4$
Tuned circuit pulling	2 [S]	2 [R]	$m/30$

is subject to electrical noise and drift. Since the trap cyclotron frequency ω_c' is measured over a ~1 min integration time, and ω_z is measured immediately afterwards, the trap voltage fluctuations do not cancel in Equation (1) and appear as random fluctuations in ω_c. The shot-to-shot variation in ω_z due to trap voltage instability has been observed to be ~10 mHz. For an ion of mass 30 u, this corresponds to a 9×10^{-11} error in ω_c per shot, resulting in a 3×10^{-11} random error in the mass ratio from a typical run.

Thermal noise in the cyclotron amplitude also causes random error in ω_c. Since the ion is cooled by a resistive detector, there is thermal uncertainty in the ion's initial location in phase space. The axial temperature T_z has been measured to be ~10 K, which is slightly higher than the 4 K liquid He bath because of additional noise from the rf SQUID detector. The cyclotron mode is thermally coupled to the axial mode using sideband cooling (7), resulting in a cyclotron temperature T_c equal to $T_z(\omega_c'/\omega_z)$, which is ~300 K for a mass 30 u ion, corresponding to a thermal rms amplitude of 0.0015 cm. During a measurement, this thermal amplitude adds vectorially to a mean amplitude of typically 0.025 cm from the applied cyclotron excitation pulse, causing variations in the cyclotron amplitude ρ_c from measurement to measurement.

Due to anharmonicities from relativity and higher-order field imperfections, these variations in ρ_c lead to variations in the measured cyclotron frequency ω_c. To lowest order (3),

$$\Delta\left(\frac{\delta\omega_c}{\omega_c}\right) = \left(-\frac{\omega_c^2}{2c^2} - \frac{B_2}{2B_0} + \frac{3\omega_m C_4}{2\omega_c d^2}\right)\Delta(\rho_c^2) \tag{20}$$

where B_2/B_0 is the magnetic bottle, C_4 is the fourth-order electric field coefficient, and $d^2 = 0.3$ cm^2 is the characteristic trap size. We have made the field flaws B_2/B_0 and C_4 smaller than 10^{-6} cm^{-2} and 2×10^{-4}, respectively, by adjusting the magnet's shim coils and the trap's guard ring electrodes. After shimming, the relativistic correction is the dominant term, particularly for light ions. Therefore thermal noise, in conjunction with these anharmonicities, causes a random error of 3×10^{-11} per run for a mass 30 u ion and 1.6×10^{-10} per run for a mass 10 u ion. Although this error is insignificant now, it may be dominant in two-ion mass spectrometry (27), in which both magnetic and electric field fluctuations cancel. We have proposed classical squeezing methods able to reduce the effects of thermal noise by about a factor of 5 (28). Recently, we have demonstrated the ability to squeeze the thermal noise and to reduce amplitude fluctuations by parametric amplification at $2\omega_z$ (29).

Another source of random error is detector noise, limiting the ability to extract frequency and phase information from the detected signal. The axial frequency ω_z can be determined to ~10 mHz. This adds to the random error in the same way as trap voltage fluctuations, contributing an uncertainty of 9×10^{-11} in a typical mass ratio. The phase of the ion's axial motion can be determined to ~10 degrees, out of a total phase which is accumulated over an integration time T of typically 1 minute. For a mass 30 u ion, this phase error contributes an uncertainty of ~1×10^{-10} per measurement of ω_c. This error depends inversely on T, and T is chosen long enough so that the magnetic field fluctuations are dominant.

Systematic errors

The anharmonicities in Equation (20) could also lead to systematic errors in the mass ratio. Since ω_c is measured with a nonzero cyclotron amplitude ($\rho_c \sim 0.02$ cm), nonlinear terms cause a frequency shift of $\sim 10^{-9}$. For doublets, this shift cancels to lowest order since the ions are pulsed to nearly the same amplitude. However, a systematic imbalance in ρ_c would cause a systematic error in the mass ratio. One source of imbalance in ρ_c is the transfer function of the cyclotron drive electronics. The transfer function was characterized from π-pulse data, and the upper limit for ρ_c imbalance was found to be 2% (11). This limit leads to upper bounds on systematic error of 1×10^{-11} from both relativity and the magnetic bottle B_2. The error from the higher-order electric field C_4 in conjunction with an imbalance in ρ_c is below 2×10^{-11}.

It is also possible to have an imbalance in C_4 between the two ions, if the trap is tuned differently for each. Assuming a 50 µV difference in the scaled guard ring potential, the resultant error in the mass ratio would be 2×10^{-11}. (This estimate is based on an experimentally observed shift of 8×10^{-10} for a mass 40 u ion with a 2.4 mV offset (6).) This error is only significant for measurements on heavy ions (>20 u), for which the present apparatus requires that the guard ring potentials be set manually for both ions. (In an earlier measurement of the CO^+/N_2^+ ratio, the guard ring potentials were set improperly, and a 2×10^{-10} correction subsequently had to be made in the ratio reported in (30).) Although the errors involving C_4 are systematic over the course of one run, they are random for ratio measurements on different nights because the trap is retuned (changing C_4) before each run.

For non-doublets, these anharmonic shifts do not cancel to lowest order. Instead, the cyclotron amplitudes are controlled so that the relativistic shifts cancel, and corrections are made for the B_2 and C_4 shifts (10). The resultant systematic errors are calculated to be $\sim 6 \times 10^{-11}$, small compared to the random error from field noise.

Other sources of error contribute at the 10^{-12} level. Surface charge patches on the trap electrodes cause a shift in the ion's equilibrium position, which is different for the two ions because the trap voltage is different. Because of magnetic field gradients, the two ions experience different fields, causing a systematic error of 4×10^{-12} for doublets. (Surface patches would not cause such an error in non-doublet measurements, since the two ions are measured with the same trap voltage (10).)

The magnetron frequency ω_m is not measured during the run; it does not need to be known very accurately, since the effect of an error in ω_m is scaled down by a factor of $(\omega_m/\omega_c)^2$. In an ideal trap, ω_m can be determined from $\omega_z^2/2\omega_c'$; however this relation is perturbed by a factor of $\left(1 + (9/4)\sin^2\theta\right)$ when the trap is tilted with an angle θ with respect to the magnetic field (3). The magnetron frequency is measured once per run by observing the avoided crossing with an rf coupling drive (8) to determine the trap tilt (0.66(2)°, in our case), so that ω_m can be deduced from subsequent measurements of ω_z and ω_c'. The correction factor $(9/4)\sin^2\theta$ can be determined to about 5%. The resultant error for a doublet measurement is completely negligible — a few parts in 10^{14}. For non-doublets, the error is a few parts in 10^{11}, still considerably smaller than the random error.

Tuned circuit pulling also is a potential cause of systematic error. Since the ion's axial mode is coupled to a high-Q tuned circuit, ω_z is perturbed. Assuming an unlikely systematic difference of 50 mHz in the two ions' axial frequencies during one run, the resultant uncertainty in ω_c is 2×10^{-12}.

In summary, the magnetic field noise causes a random error of $\sim10^{-10}$ in one mass ratio run, and other sources of random and systematic errors are calculated to be about an order of magnitude smaller. In the next section, it is shown that the quoted uncertainties are verified by a series of checks.

Consistency Checks

In a precision experiment, the reported uncertainty is just as important as the reported result, and there is a need to check for unknown errors to ensure that none have been overlooked. The fact that we measure mass ratios of molecular ions, which involve various combinations of atomic species, affords an opportunity for self-consistent checks of systematic and random errors. We have done many such checks to ensure that our uncertainties are accurate. These checks may be classed in several categories: repeated measurements, closed loops of ratios, repeated measurements after a complete reshimming of our magnet, redundant ratios, non-doublet measurements, and the overall agreement of the global fit to all ratios.

Repeated measurements. In many cases, measurements were repeated on the same pair of ions on several nighttime runs. The field fluctuations typically were different for each run, therefore testing the method of fitting to the field drift. There were 13 repeated measurements, having a reduced chi-square of $\chi_\nu^2 = 0.75$. This test indicates that the random error from field fluctuations, as determined by the histogram in Fig. 3, has not been underestimated.

Closed loops. Another check involves "closed loops" of ratios. Given three ions A^+, B^+, and C^+, there are three possible doublets that can be measured: A^+/B^+, B^+/C^+, and C^+/A^+. If the ratios are multiplied together, the product should be equal to one, within experimental error. Closed loops are basically a check on the field fitting uncertainties, like repeated measurements, except that they are also sensitive to any systematic errors which are nonlinear with respect to the difference in mass. Three such closed loops were measured, having a reduced chi-square of $\chi_\nu^2 = 1.53$. (Statistically, a reduced chi-square of this value or higher should arise in 20% of the cases, so $\chi_\nu^2 = 1.53$ is not anomalously high.)

Magnet reshimming. In the midst of our measurements, our superconducting magnet accidentally quenched. This divided the data into "earlier" and "later" categories which were analysed differently as discussed in sections 2 and 3. The magnet was rebuilt, reenergized, and reshimmed, changing the higher-order inhomogeneities in the magnetic field. In this process, the trap was thermally cycled between 300 K and 4 K, changing the surface patch charges on the trap electrodes which contribute to higher-order terms in the electric field. Comparison of measurements done before and after the magnet rebuild therefore checks systematic errors resulting from field imperfections. There were three such measurements, having a reduced chi-square of $\chi_\nu^2 = 0.26$.

Redundant ratios. A particularly powerful check for systematics is provided by redundant ratios, measurements that determine the same mass difference using different molecular ions. For example, the ratios O^+/CH_4^+, $CO^+/C_2H_4^+$, and $CO_2^+/C_3H_8^+$ all determine the mass difference $C + 4H - O$, but the measurements are made at mass 16, 28, and 44 u, respectively. Such redundant ratios check for virtually all systematic errors, since the ratios are measured under widely different experimental conditions. For example, different trap voltages test for errors from surface patch charge effects, different cyclotron frequencies test for errors in the phase-coherent cyclotron mode coupling techniques, and different chemical energies test for errors in the method of calculating atomic masses from ratio measurements. The ratios that were measured contain a total of four redundancies with $\chi_\nu^2 = 0.39$.

Non-doublet ratios. We have developed techniques (10) to measure ratios of non-doublets, which are pairs of ions with greatly different mass. Non-doublet ratios were compared with measurements of doublet ratios, providing a test of possible systematic errors arising from differences in the measurement techniques. Anharmonic frequency shifts which cancel to lowest order for doublets do not naturally cancel for non-doublets. The cyclotron modes are intentionally driven to different amplitudes in order to cancel the relativistic shift, but a magnetic bottle shift remains. Three doublet / non-doublet redundancies resulted in $\chi_\nu^2 = 0.16$, indicating that such systematic errors are insignificant at the 10^{-10} level of precision.

An additional test of the non-doublet comparison method is by the measurement of the known ratios N_2^+/N^+ and Ar^+/Ar^{++} (10). In each case, the ratio is about equal to two, except for corrections from the electron mass and the chemical energies. Since N and Ar are compared against themselves, their atomic masses with respect to C cancel to lowest order, and the ratios can be calculated from existing mass data with an accuracy of $\sim 10^{-12}$. The measured ratios were found to agree with the calculated values, adding further confidence to our non-doublet measurement technique.

Table 4. Summary of consistency checks. The number of excess independent measurements ν and the reduced chi-square χ_ν^2 are listed. The last column lists the statistical probability P of exceeding the observed value of χ_ν^2 for each check.

Check	ν	χ_ν^2	P
Repeated measurements	13	0.75	71%
Closed loops	3	1.53	20%
Magnet rebuild	3	0.26	85%
Redundant ratios	4	0.39	82%
Doublet / non-doublet	3	0.16	92%
Overall	24	0.74	81%

Overall agreement. The final consistency check is the overall agreement of all the results in the global least squares fit. There are a total of 33 ratio measurements and 9 atomic masses, and therefore 24 degrees of freedom. The reduced chi-square for the fit was $\chi_v^2 = 0.74$. Table 4 lists all the reduced chi-squares from the different types of checks, as well as the statistical probability for exceeding the observed value of χ_v^2. In all cases, $\chi_v^2 \sim 1$, and the reported uncertainties can be considered to be consistent with the data. These consistency checks therefore are compelling evidence that the errors in the mass ratios are dominated by the observed statistical noise. The checks also imply that if any systematic error had been overlooked, it would have to be smaller than 10^{-10}.

Every isotope in our mass table is derived from at least two independent sets of ratios (except for the neutron, which depends on a single gamma-ray experiment). This not only provides the same checks as the redundant ratios described above, but also ensures that non-canceling calculational and measurement errors have been avoided.

6. APPLICATIONS TO METROLOGY

At the 10^{-10} level of precision, certain mass measurements have important implications for fundamental metrology (1). In this section, we discuss the contributions of our measurements to defining an atom-based mass standard, calibrating γ-ray wavelengths, and determining fundamental constants.

Atomic mass standard

Our demonstrated ability to compare atomic masses at the 10^{-10} level establishes comparison of atomic masses as a more precise operation than comparison of macroscopic masses, which is limited to a relative precision of $\sim 10^{-9}$, especially for masses of different density (31). This suggests the wisdom of an atomic definition of mass, as might be achieved by defining the Avogadro constant, N_A. Finding an accurate way to realize such a definition would have the additional advantage of replacing the last artifact standard, the kilogram.

The S.I. unit of mass, the kilogram, is defined to be the mass of the prototype platinum-iridium cylinder at Bureau International des Poids et Measures. Besides being unique, such an artifact mass standard has many disadvantages, including the possibility of long-term drift and damage due to mishandling (31). (To guard against mishandling, the prototype kilogram has been compared to secondary standards only three times this century.) A more desirable standard would be based on an atomic mass, such as the mass of a ^{28}Si atom, which avoids these disadvantages (32). However, replacing the artifact mass standard depends on the ability to realize the kilogram (*i.e.*, to develop a practical macroscopic mass standard from this atomic definition).

One promising method for realizing an atomic kilogram is to accurately measure the lattice constant and the mass density of a highly-pure silicon crystal (33). With the present mass standard, this experiment determines the Avogadro constant N_A; with an atomic mass standard based on a defined value of N_A, the crystal becomes a mass density standard which would lead to a macroscopic realization of the kilogram. A precision in N_A of 1×10^{-6} has been attained so far (33), and it is anticipated that modifications (including the use of a crystal isotopically enriched with ^{28}Si) will allow N_A to be measured to 10^{-8} in the future.

(Recently, a measurement of the silicon lattice constant d_{220} has been reported to 3×10^{-8} (34).) Realizing the kilogram with 10^{-8} accuracy would at the very least provide a check on the long-term drift of the artifact mass standard. The previous (non-Penning-trap) value of $M(^{28}Si)$ was accurate to 2.5×10^{-8} and would have been a limitation in the accuracy of N_A. The value from our experiment, accurate to 7×10^{-11} and confirmed to 10^{-9} (21), removes this limitation.

γ-ray calibration

Another application of precision mass spectrometry in the field of metrology is to "weigh" γ-rays. By Einstein's principle, $\Delta E = \Delta mc^2$, the energy released in a nuclear process in the form of γ-rays can be measured as a difference in the mass of the initial and final nuclei. If the γ-ray energy is in the form of a single photon with effective wavelength λ^*, after correcting for nuclear recoil, then the energy balance equation is:

$$E_\gamma = hc/\lambda^* = \Delta mc^2 \tag{21}$$

Absolute measurements of γ-ray wavelengths are often imprecise. For this reason, neutron separation energies determined by mass spectrometry are used to calibrate γ-ray wavelengths, particularly in the 2-13 MeV range (35).

The neutron capture reactions $^{14}N(n, \gamma)$ and $^{12}C(n, \gamma)$ are two processes that are attractive for γ-ray wavelength calibration:

$$^{14}N + n \rightarrow {}^{15}N + \gamma_1 + \gamma_2 \quad (10.8 \text{ MeV}) \tag{22}$$

$$^{12}C + n \rightarrow {}^{13}C + \gamma \quad (4.9 \text{ MeV})$$

When combined with $^1H(n, \gamma)$:

$$^1H + n \rightarrow {}^2H + \gamma \quad (2.2 \text{ MeV}), \tag{23}$$

the neutron mass cancels, yielding the energy balance equations:

$$m[^{14}N + {}^2H - {}^{15}N - {}^1H]c^2 = hc/\lambda_1^* \tag{24}$$

$$m[^{12}C + {}^2H - {}^{13}C - {}^1H]c^2 = hc/\lambda_2^*$$

Thus, precise measurements of the mass differences $^{14}N+{}^2H-{}^{15}N-{}^1H$ and $^{12}C+{}^2H-{}^{13}C-{}^1H$ are valuable for γ-ray spectroscopy.

Our best values for these mass differences (Table 5) are accurate to 1×10^{-7} and 4×10^{-7}, respectively. By selecting molecular ions which have optimal correlation among the individual atoms, the final uncertainties in ΔM are minimized. The mass difference $^{14}N+{}^2H-{}^{15}N-{}^1H$ would be most directly measured from $^{15}NH_3^+/NDH_2^+$, but technical difficulties prevented us from loading ammonia ions into the trap. Instead, the ratios $N_2^+/C_2H_4^+$ and $^{15}N_2^+/C_2D_2H_2^+$ were measured, leading to a value of 9 241 852.1 (1.1) nu for ΔM. As a redundancy check, a value of 9 241 853.7 (1.7) nu was obtained independently from all other ratios ($^{15}N^+/CH_3^+$, N_2^+/CO^+, *etc.*) and is in agreement. The mass difference $^{12}C+{}^2H-{}^{13}C-{}^1H$ was also measured in two ways to verify its precision. The ratios $^{13}CH_4^+/CDH_3^+$ and $^{13}C_2H_4^+/C_2D_2H_2^+$ determined this difference to be

2 921 907.4 (1.5) nu and 2 921 908.6 (1.2) nu, respectively, and are also in agreement. Our best values for the mass differences, 9 241 852.7 (0.9) nu and 2 921 908.2 (1.1) nu, result from combining these values obtained by independent routes. It is important to note that we have pushed the errors of these critical mass differences down by averaging several runs, with the result that these important ratios have been measured to 5×10^{-11}, an accuracy roughly a factor of two beyond that at which our consistency checks indicate freedom from systematic error. Thus the quoted error depends on our theoretical analysis of systematics.

The previously accepted values of these mass differences from conventional mass spectrometry (36) are also listed in Table 5. Our values for ΔM are about a factor of 10 more accurate than the prior values. The $^{12}C(n,\gamma)$ mass differences are in good agreement; however, the $^{14}N(n,\gamma)$ mass differences do not agree, differing by nine times the reported uncertainty in the prior value. The current γ-ray energy calibration (35) is based on the inconsistent prior value of $^{14}N+^2H-^{15}N-^1H$. Unlike our redundant Penning trap measurements, the earlier result was based on a single mass comparison. The improved mass difference obtained by Penning trap mass spectrometry considerably increases the accuracy of the energy calibration and suggests that an 8 ppm revision of this calibration is necessary. This is consistent with recent high precision measurements of γ-ray energies with a Ge detector (37).

Fundamental constants

A collaboration of researchers using the High Flux Reactor in Grenoble, France is undertaking precision experiments to measure absolute γ-ray wavelengths corresponding to the above reactions $^{14}N(n,\gamma)$, $^{12}C(n,\gamma)$, and $^1H(n,\gamma)$ (38). A precision of $\sim 2\times10^{-7}$ is expected for the effective wavelengths λ^* (after taking into account nuclear recoil) corresponding to the mass differences $^{14}N+^2H-^{15}N-^1H$ and $^{12}C+^2H-^{13}C-^1H$ (39). When this work is completed, the mass differences would not be needed for calibration purposes. Instead, the precise masses and wavelengths could be combined to determine the fundamental constants $N_A h$ and α (40-41).

The molar Planck constant $N_A h$ follows from the fact that the neutron separation energies are measured in different systems of units. We measure the

Table 5. Mass differences for determining α and $N_A h$. The mass differences associated with the neutron capture reactions $^{12}C(n,\gamma)$ and $^{14}N(n,\gamma)$ determined by this experiment and by conventional mass spectrometry (33) are listed. The new results are a factor of 10 more accurate and show considerable discrepancy with the previous value of $^{14}N+^2H-^{15}N-^1H$.

Mass difference	This work [nu]	Ref. [33] [nu]
$^{14}N+^2H-^{15}N-^1H$	9 241 852.7 (0.9)	9 241 780 (8)
$^{12}C+^2H-^{13}C-^1H$	2 921 908.2 (1.1)	2 921 911 (12)

mass defect ΔM in (microscopically defined) atomic mass units (u), while the effective γ-ray wavelength λ^* is measured in S.I. units (m). Equating the energies in Equation (21) leads to:

$$N_A h = \lambda^* \Delta M c \times 10^{-3} \tag{25}$$

where the Avogadro constant $N_A = 10^{-3} \Delta M/\Delta m$ is needed to convert ΔM into the mass difference Δm (in kg), required in Equation (21).

Another route for the accurate determination of $N_A h$ is the precise determination of h/m from measurements of the velocity, v, and deBroglie wavelength of a particle, λ_{dB}. This has been accomplished for the neutron at the 10^{-7} level(42), and for Cs at the 10^{-6} level (43). $N_A h$ is determined from these measurements as follows:

$$N_A h = \frac{M}{m} h = M \frac{h}{m} = M \lambda_{dB} v \tag{26}$$

where the last equality follows from the relation, $\lambda_{dB} = h/mv$.

The fine structure constant α can also be determined from $N_A h$, and therefore from a measurement of λ^* and ΔM:

$$\alpha^2 = \frac{2R_\infty}{c} \left(\frac{m_p}{m_e}\right) \frac{N_A h}{M_p} \times 10^3 \tag{27}$$

$$= 2R_\infty \left(\frac{m_p}{m_e}\right) \frac{\Delta M}{M_p} \lambda^*$$

The Rydberg constant R_∞, the proton-electron mass ratio m_p/m_e, and the proton atomic mass M_p are known to 4×10^{-11} (44), 3×10^{-9} (45), and 5×10^{-10} (10), respectively. Therefore, measuring λ^* and ΔM with a relative accuracy of $\sim 2 \times 10^{-7}$ also would determine α to 10^{-7}. Although α has been determined by other experiments with accuracy as high as 10^{-8}, this measurement would be valuable as an independent check and also to verify the consistency of physical theories (46), especially QED, which currently gives the best value of α (47).

ACKNOWLEDGEMENTS

This work was supported by the National Science Foundation (Grant No. PHY-9222768) and the Joint Services Electronics Program (Grant No. DAAL03-92-C-0001). F.D. also acknowledges additional support from an N.S.F. Graduate Fellowship. We are grateful to K. Boyce, E. Cornell, R. Flanagan, G. Lafyatis, and R. Weisskoff for earlier work on this experiment, and to S. Rusinkiewicz for technical assistance. We also would like to thank M. Matthews for helpful discussions on fitting to the field drift.

REFERENCES

1. F. DiFilippo, V. Natarajan, K.R. Boyce, and D.E. Pritchard, *Phys. Rev. Lett.*. **73**, 1481 (1994).
2. F.M. Penning, *Physica* **3**, 873 (1936).
3. L. Brown and G. Gabrielse, *Rev. Mod. Phys.* **58**, 233 (1986).
4. R.M. Weisskoff et. al., *J. Appl. Phys.* **63**, 4599 (1988).

5. R. Kumaresan and D.W. Tufts, *IEEE Trans. Acous., Speech, and Sig. Proc.* **ASSP-30**, 833 (1982).
6. V. Natarajan, Ph.D. thesis (M.I.T., unpublished, 1993).
7. D.J. Wineland and H.G. Dehmelt, *Int. J. Mass Spec. Ion Proc.* **16**, 338 (1975).
8. E.A. Cornell, R.M. Weisskoff, K.R. Boyce, and D.E. Pritchard, *Phys. Rev. A* **41**, 312 (1990).
9. E.A. Cornell *et. al.*, *Phys. Rev. Lett.* **63**, 1674 (1989).
10. V. Natarajan, K.R. Boyce, F. DiFilippo, and D.E. Pritchard, *Phys. Rev. Lett.* **71**, 1998 (1993).
11. F. DiFilippo, Ph.D. thesis (M.I.T., unpublished, 1994).
12. P.R Bevington and D.K. Robinson, *Data Reduction and Error Analysis for the Physical Sciences*, Second Edition (McGraw-Hill, Inc., New York, 1992).
13. G. Audi and A.H. Wapstra, *Nucl. Phys.* **A565**, 1 (1993).
14. P.J. Huber, *Robust Statistics*, (John Wiley & Sons, New York, 1981).
15. D.F. Andrews *et. al.*, *Robust Estimates of Location: Survey and Advances*, (Princeton University Press, Princeton, NJ, 1972).
16. S.G. Lias *et. al.*, *J. Phys. Chem. Ref. Data* **17**, Suppl. 1 (1988).
17. A.H. Wapstra and G. Audi, *Nucl. Phys.* **A432**, 1 (1985).
18. G.L. Greene, E.G. Kessler, Jr., R.D. Deslattes, and H. Börner, *Phys. Rev. Lett.* **56**, 819 (1986).
19. R.S. Van Dyck, Jr., D.L. Farnham, and P.B. Schwinberg, *J. Mod. Opt.* **39**, 243 (1992).
20. R.S. Van Dyck, Jr., D.L. Farnham, and P.B. Schwinberg, *Phys. Rev. Lett.* **70**, 2888 (1993).
21. R. Jertz *et. al.*, *Physica Scripta* **48**, 399 (1993).
22. H.-J. Kluge, *Phys. Scrip.* **T22**, 85 (1988).
23. M. Comisarow and A. Marshall, *Int. J. Mass Spect. Ion Proc.* **118** 37 (1992).
24. M.V. Gorshkov *et. al.*, *J. Am. Soc. Mass Spec.* **4**, 855 (1993).
25. M.V. Gorshkov *et. al.*, *Phys. Rev. A* **47**, 3433 (1993).
26. M.V. Gorshkov *et. al.*, *Int. J. Mass Spect. Ion Proc.* **128**, 47 (1993).
27. E.A. Cornell, K.R. Boyce, D.L.K. Fygenson, and D.E. Pritchard, *Phys. Rev. A* **45**, 3049 (1992).
28. F. DiFilippo, V. Natarajan, K.R. Boyce, and D.E. Pritchard, *Phys. Rev. Lett.* **68**, 2859 (1992).
29. V. Natarajan, F. DiFilippo, and D.E. Pritchard, to be published.
30. V. Natarajan, K.R. Boyce, F. DiFilippo, and D.E. Pritchard, in *Nuclei Far From Stability / Atomic Masses and Fundamental Constants 1992*, R. Neugart and A. Wöhr, Eds., p. 13 (1992).
31. T.J. Quinn, *IEEE Trans. Instr. Meas.* **40**, 81 (1991).
32. B.N. Taylor, *IEEE Trans. Instr. Meas.* **40**, 86 (1991).
34. P. Seyfried *et. al.*, *Z. Phys. B* **87**, 289 (1992).
34. G. Basile *et. al.*, *Phys. Rev. Lett.* **72**, 3133 (1994).
35. A.H. Wapstra, *Nucl. Instr. Meth. Phys. Res.* **A292**, 671 (1990).
36. L.G. Smith and A.H. Wapstra, *Phys. Rev. C* **11**, 1392 (1975).
37. S. Raman, private communication.
38. M.S. Dewey *et. al.*, *Nucl. Instr. Meth. Phys. Res.* **A284**, 151 (1989).
39. G.L. Greene and E.G. Kessler, Jr., private communications.
40. R.D. Deslattes and E.G. Kessler, Jr., in *Atomic Masses and Fundamental Constants – 6*, J.A. Nolen, Jr. and W. Benenson, Eds., (Plenum Press, New York, 1979), p. 203.
41. W.H. Johnson, in *Precision Measurements and Fundamental Constants II*, B.N. Taylor and W.D. Phillips, Eds., Natl. Bur. Stand. (U.S.), Spec. Publ. 617 (U.S. GPO, Washington D.C., 1984), p. 335.
42. E. Krüger, W. Nistler, and Weirauch, "Determination of the fine-structure constant by measuring the quotient of the Planck constant and the neutron mass" presented at the Conference on Precision Electromagnetic Measurements, Boulder CO, June 27 - July 1 (1994).

43. D. S. Weiss, B. C. Young, and S. Chu, *Phys. Rev. Lett* **70**, 2706 (1993).
44. T. Andreae et. al., *Phys. Rev. Lett.* **69**, 1923 (1992).
45. R. S. Van Dyck, Jr., D. L. Farnham, and P. B. Schwinberg, "Proton/ electron mass ratio and the electron's 'atomic mass'", presented at the Conference on Precision Electromagnetic Measurements, Boulder CO, June 27 - July 1 (1994).
46. G.L. Greene, M.S. Dewey, E.G. Kessler, Jr., and E. Fischbach, *Phys. Rev. D* **44**, R2216 (1991).
47. T. Kinoshita, *Metrologia* **25**, 233 (1988).

Mass Measurements of Short-Lived Isotopes in Traps and Storage Rings

H.-Jürgen Kluge*,§, Georg Bollen†, and Bernhard Franzke*

* Gesellschaft für Schwerionenforchung mbH,
D-64220 Darmstadt, Germany
§ Physikalisches Institut, Universität Heidelberg,
D-61290 Heidelberg, Germany
† Institut für Physik, Universität Mainz,
D-55099 Mainz, Germany

Abstract. High-accuracy mass measurements can be performed on short-lived radionuclides with half-lives $T_{1/2} \geq 1 ms$ by storing them in ion traps or storage rings. The masses of about 80 isotopes have been determined with a resolving power of typically one million and an accuracy of $\delta m/m = 10^{-7}$ with the tandem Penning trap mass spectrometer ISOLTRAP installed at the on-line mass separator ISOLDE/CERN. Schottky and time-of-flight mass measurements will soon start at the storage ring ESR at GSI/Darmstadt on highly charged ions delivered at relativistic energies from the fragment separator. Further developments and projects are discussed.

INTRODUCTION

The mass of a quantum mechanical system, i.e. the sum of the rest masses of its constituents minus the binding energy is it's most fundamental property since all forces acting in such a system are involved in it. In nuclear physics, precise masses are especially important since an exact theory of strong interaction is still missing. A large variety of nuclear models has been developed in the course of the last decades. The comparison of the predictions of state of the art models with experimental masses has led to the development of collective models like the liquid-drop model or single-particle models like the (spherical) shell model and the (deformed) Nilsson model. More accurate mass values as well as the extension of the data set to those of isotopes far from the valley of nuclear stability have stimulated and still stimulate an on-going refinement of those models.

Today the masses of around 2000 isotopes are known: stable isotopes typically with an accuracy of about $\delta m/m = 10^{-8}$, deteriorating to about 10^{-7} for long-lived isotopes and to about 10^{-6} for short-lived ones. This decrease in accuracy is due to the fact that almost all the masses of radioisotopes are determined by Q-value measurements in nuclear decays or reactions.

Hence, the mass of an isotope far from stability is linked by many mass differences to the well known mass of a stable isotope. As a consequence, the errors in the mass differences accumulate. In addition, Q-value measurements require the knowledge of the initial and final nuclear state, quite often a difficult task in case of complex or unknown nuclear level schemes and a source of many wrong entries in the tables of nuclear masses.

About 4500 - 6000 nuclei are assumed to exist as bound systems between the hydrogen atom and the very heavy elements (with Z=109, meitnerium, being the last one detected) and the neutron and proton drip lines (where neutrons and protons become unbound, respectively). The masses of neutron-rich isotopes far from the valley of nuclear stability are especially important, since these nuclei were involved in the nucleosynthesis (r-process) of the elements and the one-neutron separation energy (S_{1n}) of these isotopes enters exponentially in the rate equations modelling the element cooking process. Because the r-process path runs so far away from known isotopes, nuclear models or mass formulae have to be used to calculate S_{1n}. The accuracy presently achieved for such models is $\delta m/m \simeq 10^{-5}$, corresponding to about 1MeV or 10% of S_{1n} for medium heavy nuclei. In order to improve the predictive power of the models, high-accuracy mass data are required of radionuclides as far as possible from the valley of nuclear stability.

THE GOAL

Short-lived isotopes are produced by nuclear reactions of accelerated particles or neutrons with target atoms. The mass of the resulting radionuclide, quite often extremely rare, should be determined with

- high resolving power (in order to resolve the ground-state mass from that of the isomer),

- high accuracy ($\delta m/m \leq 10^{-6}$),

- in a short time period (at least comparable to the nuclear half-life $T_{1/2}$),

- with high efficiency and,

- in a direct or even absolute way (relative to the mass of a freely chosen reference isotope or relative to the mass of ^{12}C establishing the atomic mass unit on the microscopic scale).

In the following, two new approaches to accurate mass measurements of short-lived isotopes are described which both make use of storage techniques, i.e. the observation of the cyclotron frequency of ions confined in Penning traps or storage rings (Fig. 1).

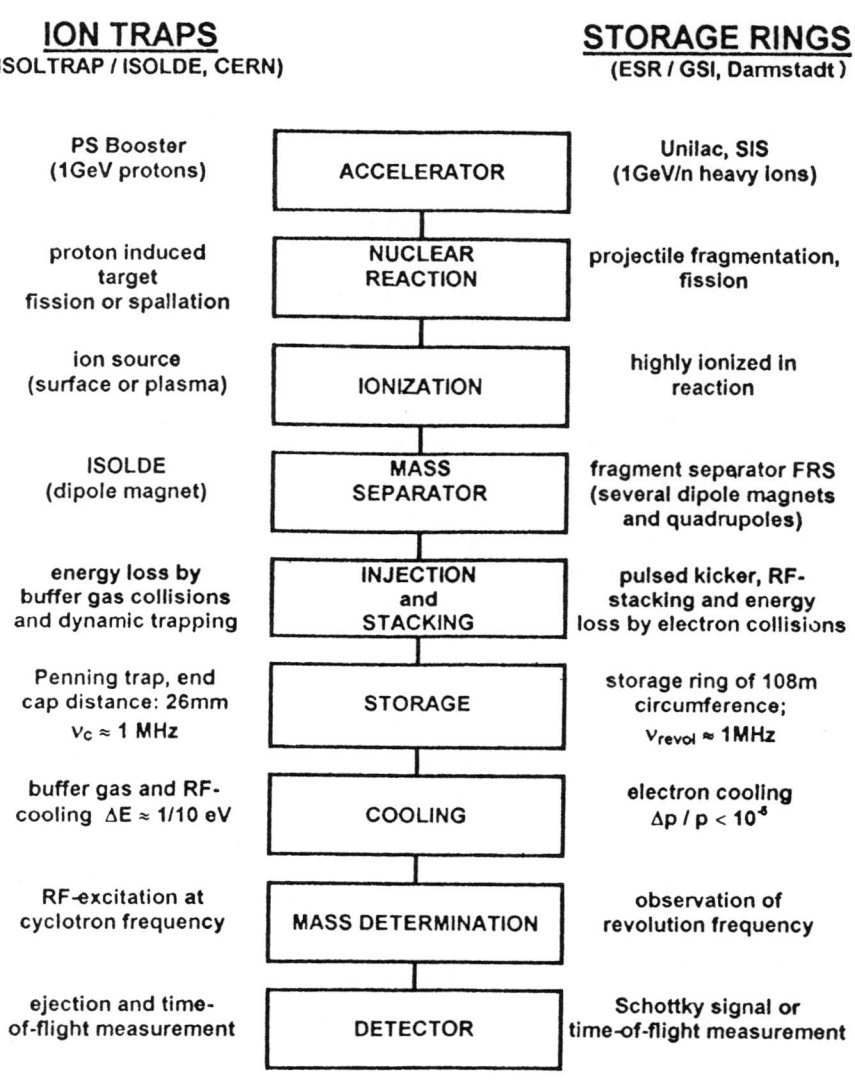

FIGURE 1. Principle of mass determinations of short-lived isotopes in a Penning trap and storage ring.

PENNING TRAP MASS SPECTROMETRY OF SHORT-LIVED ISOTOPES

Penning traps have been proven to be very effective instruments for high-accuracy mass determinations via the relation

$$\omega_c = \frac{qe}{m} B, \qquad (1)$$

where ω_c is the cyclotron frequency, q the charge state and m the mass of the stored ion, e the elementary charge and B the magnetic field. With these devices a number of stable or long-lived particles, for example the electron and positron, electron and proton, proton and antiproton, helium-3 and tritium, and light ions and molecules have been determined to very high accuracy [1–6]. However, the current methods used in these experiments are not readily adapted to mass measurements of radioactive isotopes. In the following the special requirements which are imposed in the case of mass measurements of short-lived isotopes will therefore be discussed.

Resolving Power

Due to the superposition of an electrostatic quadrupole field and a homogenous magnetic field the ion motion in a Penning trap is not a pure cyclotron motion with frequency ω_c according to (1) but consists of three independent harmonic eigen-motions [7]: a reduced cyclotron motion with frequency ω_+ modified from the pure cyclotron motion by the presence of the electric quadrupole field, an oscillation along the magnetic field lines with frequency ω_z, and a slow drift around the trap axis, called magnetron motion, with frequency ω_-. Generally,

$$\omega_+ \gg \omega_z \gg \omega_- . \qquad (2)$$

The energy of an ion is therefore given as the sum of the energies of three harmonic oscillators by

$$E = \hbar\omega_+(n_+ + 1/2) + \hbar\omega_z(n_- + 1/2) - \hbar\omega_-(n_- + 1/2). \qquad (3)$$

Note that the magnetron motion is unstable. The eigenfrequencies ω_+, ω_-, and ω_z can be determined by driving the corresponding motions with oscillating dipole fields ($n \to n+1$) and detecting the change of the motional amplitudes $\varrho \sim \sqrt{n+1}$. However, the most direct approach is the direct measurement of the sum frequency [8]

$$\omega_c = \omega_+ + \omega_- . \qquad (4)$$

The motion of a trapped ion can be directly excited at this frequency by an azimuthally quadrupolar RF-field [8] which leads to a simultaneous change in quantum numbers by

$$n_+ \rightarrow n_+ + 1; \quad n_- \rightarrow n_- - 1 \ . \tag{5}$$

This excitation couples the magnetron and reduced cyclotron motion so that a periodic conversion from one motion into the other can be achieved. After a particular excitation time T_{RF}, an initially pure magnetron motion is transformed into a pure cyclotron motion which, according to (3), results at the same time in an increase of radial energy. The width of the cyclotron resonance curve $\Delta \nu_c$ is determined by the Fourier limit of the driving RF-field switched on for a time T_{RF}. From [8] the relationship is

$$\Delta \nu_c (\text{FWHM}) \simeq 0.8/T_{RF} \ . \tag{6}$$

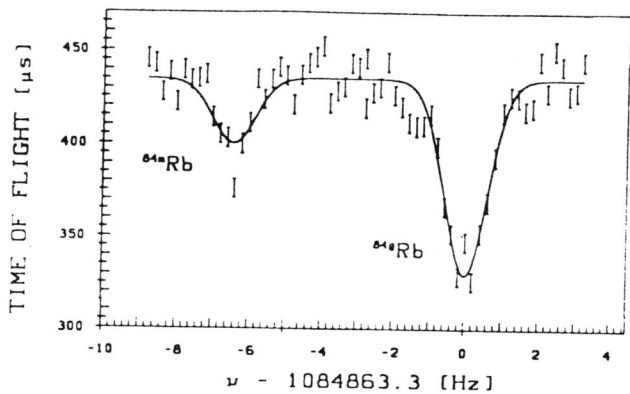

FIGURE 2. Cyclotron resonances of both the isomeric and ground state of ^{84}Rb. The resolving power is about $R = 800000$. The mass difference measured by weighing corresponds to $\Delta E_{mass} = 464(7)\text{keV}$ in full agreement with the value of $\Delta E_\gamma = 463.7(1.0)\text{keV}$ as determined by γ-spectroscopy. Due to the limited statistics obtained in the on-line run, the data points were fitted by a Gaussian (full line) instead of the correct line profile (see Figure 3).

For example, a singly charged ion of mass number $A = 100$ in a field $B = 6\text{T}$ has $\nu_c \simeq 1\text{MHz}$. Exciting its motion for $T_{RF} = 800\text{ms}$ yields a line width of $\Delta \nu_c \simeq 1\text{Hz}$, corresponding to a resolving power of $R = 10^6$. This is more than sufficient to discriminate between different isotopes, more than adequate to distinguish between isobars further away from the valley of nuclear stability, and even suited to resolve ground and isomeric states of an $A \approx 100$ nucleus with an excitation energy of $\Delta E \geq 100$keV. As an example, Fig. 2 shows the cyclotron resonance of both the ground state ($T_{1/2} = 32.9$d) and the isomeric state ($T_{1/2} = 20.5$min) of ^{84}Rb [9]. Note, that nuclei either in the ground or in the isomeric state can be eliminated by RF-excitation at the appropriate frequency ω_+. In such a way an isomer separator can be realized resulting, for example, in an inverted population of a nuclear system.

Accuracy and Calibration

The line profile of the cyclotron resonance curve is the Fourier transform of the rectangular RF-pulse of duration T_{RF} modified by the periodic conversion of magnetron into cyclotron motion and by the transformation of radial energy into a change of time of flight from the trap to the detector when the ions are ejected (see below). As can be seen from Fig. 3, the resonance profile is very well understood. Hence, statistical errors well below $\delta \nu_c / \nu_c = 10^{-8}$ can be obtained. In order to determine the mass of the investigated isotope via (1), the magnetic field has to be known. This can be achieved by loading a stable isotope into the trap the mass of which is precisely known. In the future, it is planned to use ^{12}C cluster ions for calibrating the magnetic field which would enable to perform for the first time absolute mass measurements of short-lived isotopes.

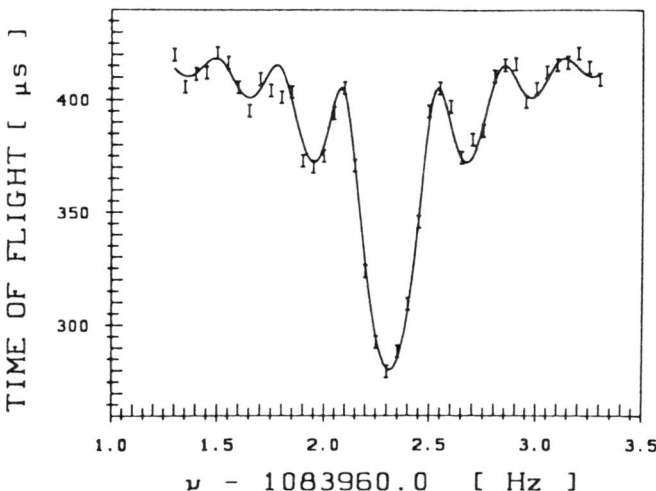

FIGURE 3. Cyclotron resonance for ^{85}Rb ($T_{RF} = 3.6$s, $R = 4 \cdot 10^6$). The solid line represents a fit by the theoretical line shape.

THE ISOLTRAP TANDEM PENNING TRAP MASS SPECTROMETER

Figure 4 shows the experimental set-up of the ISOLTRAP spectrometer [10] at the on-line mass separator ISOLDE, a powerful facility for the production of radioactive isotopes. ISOLTRAP consists of two main parts, the lower Penning trap in an electro magnet acting as an ion beam accumulator, cooler and buncher (trap 1) and the upper high-precision trap (trap 2) in a superconducting coil in which the mass determination takes place.

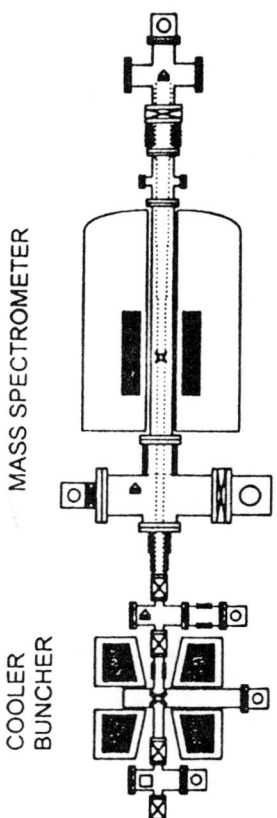

FIGURE 4. The ISOLTRAP tandem Penning trap mass spectrometer installed at the on-line isotope separator ISOLDE at CERN/Geneva. The mass separated ion beam from ISOLDE enters from the bottom.

The continuous 60keV ISOLDE ion beam is collected on a rhenium foil mounted in the lower endcap electrode of trap 1. After a certain collection time the foil is rotated by 180° and heated so as to surface-ionize the accumulated material. The ions released from the foil are mass selectively cooled, and centered in trap 1 by He buffer gas collisions ($p_{He} = 10^{-3} - 10^{-4}$ mbar) which damp the axial as well as the cyclotron amplitudes. The increase in magnetron radius is counteracted by RF-excitation of the ω_c-resonance of the ion under investigation (see Eq. (5)). A fraction of about 10^{-3} to 10^{-4} of the ions accumulated on the foil is centered by this new cooling technique [11]. After cooling to room temperature, the ion cloud is ejected, accelerated to 1keV, retarded again in front of trap 2 and captured there in flight. The transfer between both traps can be realized with an efficiency close to 100% [12].

The high-precision trap is placed in the 5.7T field of the superconducting magnet. Deviations from the ideal quadrupole field are compensated by correction electrodes installed between the ring electrode and the end electrodes and at the holes in the endcaps required for injection and ejection of ions. As in trap 1 the ring electrode is split into four quadrants to create the azimuthal quadrupole field for ω_c-excitation.

A destructive technique is used to detect the increase in radial energy (Eqs. (3) and (5)) as a function of the applied RF-frequency [13]. After excitation for a period T_{RF}, the ions are gently pulled from the trap as short pulses, leaving the homogeneous part of the magnetic field and drifting along the magnetic field lines to an ion detector placed in the weak fringe field of the magnet. As they drift, the radial energy gained in resonance by the RF-excitation is converted into longitudinal energy in the inhomogeneous part of the magnetic field. Therefore, the measurement of the drift time as a function of the applied radio frequency yields a resonance curve with a minimum at $\omega_c = \omega_+ + \omega_-$ corresponding to a maximum in radial energy.

RESULTS

Up to the present, cyclotron resonances have been observed for about 80 different isotopes and isomers (Table 1). Three of these isotopes (^{39}K, ^{85}Rb, and ^{133}Cs) were used as reference isotopes. Results are published in Refs. [10,14-17].

Typically, the statistical accuracy in on-line runs is of the order of some parts in 10^8. A systematic error of $1 \cdot 10^{-7}$ is added which takes care of possible calibration errors (in first order proportional to the mass difference of the isotope under investigation and the reference isotope), jumps in the magnetic field, and the accuracy with which the various parameters can be controlled under the condition of limited beam time.

One remarkable feature of the ISOLTRAP mass data is that the size of the error is almost constant being in a first approximation only proportional to the mass of the isotope under investigation (dashed lines in Fig. 5.). The excellent agreement with earlier data [18] of superior accuracy in the mass region $84 \leq A \leq 88$ demonstrates the reliability of the trap technique. For isotopes far away from stability, Fig. 5 shows impressively the superiority of direct mass measurements over measurements of mass differences.

TABLE 1. Isotopes for which cyclotron resonances have been determined by use of the ISOLTRAP mass spectrometer.

Element	Mass Number
K	39
Rb	75-89, 90m, 91 - 94
Sr	78 - 83, 91 - 95
Cs	118 - 121, 122+122m, 122m, 123 - 140, 142
Ba	124, 126, 128, 138 - 144
Fr	209 - 212, 221, 222
Ra	226, 230

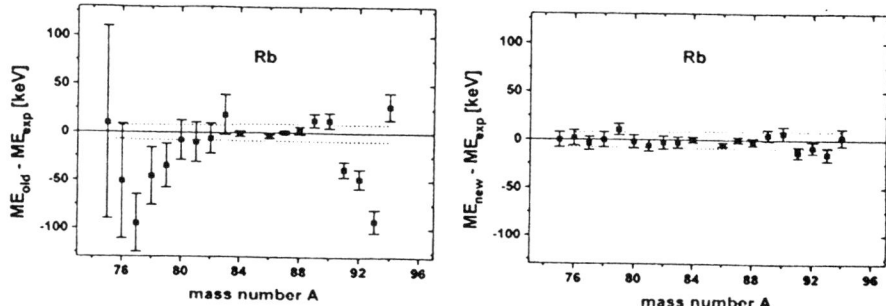

FIGURE 5. Rubidium masses as a function of mass number. Left: the difference is plotted between the Penning trap data (zero line with experimental errors given by the dashed line) and a mass adjustment of all masses known up to now except the trap data (squares). Right: difference between the Penning trap data (zero line) and a new adjustment of all known mass data.

FURTHER DEVELOPMENTS

Presently, the lower trap (Fig. 4) is being replaced by a cylindrical trap in a superconducting magnet. An increase of the overall efficiency by about two orders of magnitude to 10^{-2} is expected. In addition, the high resolving power of this trap of $R = 10^5$ under cooling conditions will enable to eliminate already in trap 1 contaminations like isobars before the ion cloud is transfered to the precision trap. Extension to isotopes of those elements not surface ionizable will be possible by laser desorption of the accumulated sample and resonance ionization. Incorporated with this system will be a laser desorption source for the creation of ionized carbon clusters over a wide mass range. These clusters will be used to calibrate the magnetic field for absolute mass measurements.

A further approach is to extend the tandem configuration to a triple trap system by addition of a Paul trap which collects the 60keV ISOLDE ion beam in-flight and cools the ions by buffer gas collisions [19]. Such a system is being developed at McGill, Montreal. This trap will have a very large phase space volume for ion collection allowing bunching efficiencies considerable in excess of $2 \cdot 10^{-3}$ obtained in a first test by use of a prototype [20].

THE EXPERIMENTAL STORAGE RING (ESR) AT GSI AS A MASS SPECTROMETER

Recently, the accelerator complex at GSI/Darmstadt has been extended by the heavy ion synchrotron SIS to a relativistic heavy-ion facility delivering ion beams up to uranium with energies in excess of 1GeV/u. Radioactive beams might be produced by use of the fragment separator FRS and injected at energies up to 550MeV/u into the storage ring ESR (Fig. 6).

Except for its size (e.g., circumference = 108m), such a storage ring has features very similar to those of an ion trap (Fig. 1). Cooling of the stored ions (to low momentum spread but not to rest like in traps!) is achieved by Coulomb interactions with "cold" electrons. These electrons are produced from a hot cathode in the electron cooler (Fig. 6) and accelerated to 170keV for 300MeV/u uranium ions. Since the energy spread ($\Delta E \approx 0.1 eV$) is a constant of motion, the velocity spread of the electrons in longitudinal direction is reduced when thermal even at the velocity is increased to $\beta = v/c \approx 0.65$. For bare uranium ions at 300MeV/u, with an electron beam current of 200mA, and for an interaction region of 2m length, cooling times of the order of τ_{cool} (U^{92+}) = 0.1 s and an momentum spread of the stored ions of $\Delta p/p \leq 10^{-6}$ were obtained. Due to the small emittance of the cooled ion beam and the ultra-high vacuum ($\leq 10^{-10}$mbar), the storage time τ_{store} is as long as 1h for U^{92+}.

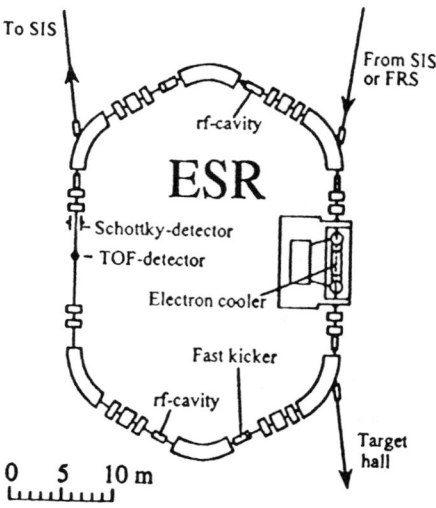

FIGURE 6. Experimental Storage Ring (ESR) at GSI/Darmstadt.

Since the velocity of the cooled ion beam is equal to the velocity of the electrons in the cooler which can easily be kept constant, a measurement of the ratio of the revolution frequencies of two ion species ($\nu_R = 1.7$ MHz for 300 MeV/u U^{92+}) can be used for the determination of the mass ratio of those species very similar to the Penning trap technique.

Mass Determination via the Schottky Signal

Pick-up electrodes mounted in the beam tube of the ESR are used to observe the image currents induced by the circulating beam particles and, after Fourier transformation from the time to the frequency domain, to determine the momentum spread of the stored ions. This tool for diagnosis, the so-called Schottky signal, can also be used to measure the revolution frequency with high precision [21]. In a pilot experiment [22], the mass ratio of the mass doublet $^{18}F^{9+}/^{20}Ne^{10+}$ was determined to agree with the known value within $3\cdot10^{-7}$ at a level of an accuracy of $4\cdot10^{-6}$.

Figure 7 shows the potential of this technique. In the course of the (successful) search for the bound β-decay of $^{163}Dy^{66+}$, an argon target in the ESR was used for detection of the β-decay daughter nucleus ^{163}Ho [23]. By nuclear (p;xp,yn)-reactions, neighbouring isotopes are produced which show up as extremely narrow peaks in the Schottky spectrum (Fig. 7). The widths of these signals correspond to a momentum spread of $\Delta p/p \approx 10^{-6}$.

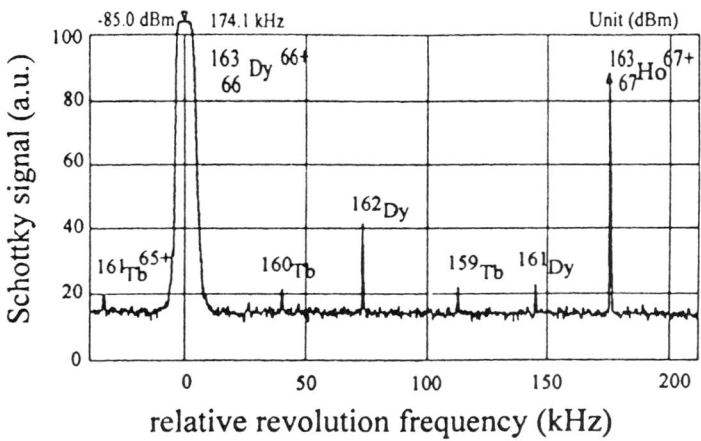

FIGURE 7. Schottky spectrum of stored $^{163}Dy^{66+}$ and the signals of $^{163}Ho^{67}$ (bound β-decay) and other nuclides produced by nuclear reactions with an argon target.

Hence, a precision in frequency ratios of better than $\delta\nu_R/\nu_R = 10^{-6}$ is obtained [24]. How this transforms into accuracy of mass ratios is presently under investigation.

It should be noted that mass measurements at storage rings provide the mass of relativistic, highly charged or even fully stripped ions but not the mass of the neutral atom. The measured mass has to be corrected for relativistic effects, the mass of the stripped electrons and their binding energies. This is of no concern in the case of the anticipated accuracy of $\delta m/m \lesssim 10^{-6}$. On the contrary, measuring the revolution frequency of an ion in different charge states presents an excellent tool to search for non-linearities in the mass-frequency relation.

Isotopes with half-lives down to about $T_{1/2} \approx 1s$ can be investigated with the Schottky technique. This is due to the time required to prepare a cooled ion beam after injection from the FRS. In the case of highly charged, heavy ions the detection limit is of the order of 1 to 10 ions of the same isotope and charge state stored at a time in the ESR.

Mass Determination via Time-of-Flight Measurement

Another approach to mass determinations in the ESR is the measurement of the time for one revolution in the ring [25,26]. In this case, the ESR is operated in an isochronous mode and the measurements are performed on "uncooled" ions immediately after injection from the FRS. Mass measurements of single ions with half lives of $T \geq 1$ms are planned.

The time-of-flight detector is similar to the one used in the TOFI-experiment at Los Alamos [27]. When the stored ion is transversing a thin carbon foil, secondary electrons are ejected. They are detected by a multichannel plate detector with a time resolution of $\delta t \approx 150$ ps. This corresponds to a time resolution per turn of $3 \cdot 10^{-4}$ for a typical revolution time of 500 ns. Since up to 100 revolutions should be observable per injected ion, the expected resolving power is of the order of $t/\Delta t = 3 \cdot 10^5$.

Still it has to be shown how to inject into a storage ring operated in the isochronous mode and how many turns an ion will survive. First mass measurements are planned for the second half of 1995.

Mass Measurements à la the RF Smith Spectrometer

There might be still other possibilities to perform mass measurements in a storage ring. Consider two resonators producing an electric field transversal ("stochastic") or longitudinal (RF-Smith) to the path of the ion in the storage ring. If those resonators are placed at exactly two opposite positions in the ring and are out of phase by exactly 180° after a half turn of the ion in the storage ring, the transversal (longitudinal) energy picked up in the first resonator will exactly be compensated in the second one as long as the resonators are operated at the revolution frequency. Off resonance, the ion will pick up transversal (longitudinal) energy and get finally lost.

An ion stored in the ring will perform many turns in contrast to the two revolutions in the RF-Smith spectrometer. This will increase the resolving power. A further increase will be obtained by operating the resonators at higher harmonics just as in the Smith spectrometer.

Quantitative calculations have still to be performed but the proposed technique might have advantages in respect to resolution because it avoids the timing detector and possible changes in the charge state of the ion traversing the foil.

CONCLUSIONS

Powerful techniques for mass measurements of short-lived isotopes have been developed. A first Penning trap mass spectrometer (ISOLTRAP) is operational at the on-line mass separator ISOLDE/CERN which has a resolving power of about one million and an accuracy of $\delta m/m = 10^{-7}$. The half life limit at present is about $T_{1/2} = 1$s. A similar spectrometer is under construction by a Chalk-River–Manitoba–Montreal collaboration [28]. A RF-Smith spectrometer is presently prepared by an Orsay group for mass measurements on isotopes with $T_{1/2} < 1$s [29]. Soon, mass measurements will also start at the storage ring ESR at GSI on highly charged ions. First, the Schottky technique will be applied for mass measurements of isotopes with $T_{1/2} \gtrsim 1$s. Later the time-of-flight method will be used for isotopes with half-lives in the region of less than one second down to 1ms. At Stockholm, the SMILETRAP [30] is coupled to the electron beam ion source CRYSIS of the storage ring CRYRING. Here, mass measurements of stable, highly charged ions up to the completely stripped nucleus are measured with an accuracy of some parts in 10^9.

Obviously, storing techniques have led to a breakthrough in mass spectrometry.

REFERENCES

1. Schwinberg, P.B., Van Dyck, R.S., Jr, and Dehmelt, H.G., Phys. Lett. *A81* (1981) 119
2. Van Dyck et al., Bull Am. Phys. Soc. *31* (1986) 244
3. Gabrielse et al., Phys. Rev. Lett. *65* (1990) 1317
4. Van Dyck, R.S., Jr, Farnham, D.L., and Schwinberg, P.B., J. Mod. Optics *39* (1992) 243

5. Cornell, E.A. et al., Phys. Rev. Lett. *63* (1989) 1674
6. Natarajan, V. et al., Phys. Rev. Lett. *71* (1993) 1998
7. Brown, L.S., and Gabrielse, G., Rev. Mod. Phys. *58* (1986) 233
8. Bollen, G. et al., J. Appl. Phy. *68* (1990) 4355
9. Bollen, G. et al., Phys. Rev. *C46* (1992) R2140
10. Stolzenberg, H. et al., Phys. Rev. Lett. *65* (1990) 3104, and to be published
11. Savard, G. et al., Phys. Lett. *A158* (1991) 247
12. Schnatz, H. et al., Nucl. Instrum. Methods *A251* (1936) 17
13. Gräff, G., Kalinowsky, H., and Traut, J., Z. Phys. *A297* (1980) 35
14. Bollen, G. et al., Physica Scripta *46* (1992) 581
15. Bollen, G. et al., J. Mod. Optics *39* (1992) 257
16. Otto, T. et al., Nucl. Phys. *A567* (1994) 281
17. Bollen, G., Kluge, H.-J., Otto, T., Savard, G., Schweikhard, L., Stolzenberg, H., Audi, G., Moore, R.B., Rouleau, G., Szerypo, J., Patyk, Z. and The ISOLDE Collaboration, Habilitationsschrift von Bollen, G., Mainz 1994, to be submitted to Nucl. Phys. *A* (1994)
18. Wapstra, A.H., Audi, G., and Hoekstra, R., ed. Haustein, P., Atom. Data Nucl. Data Tables *39* (1988) 290
19. Moore, R.B., Hyperfine Interact. *81* (1993) 45
20. Moore, R.B., and Rouleau, G., J. Mod. Optics *39* (1992) 361
21. Bechert, K. et al., Proc. of 2nd Europ. Part. Accel. Conf., Nice 1990, Vol.1, p. 777 (Edition Frontieres, Gif-sur Yvette 1990)
22. H. Geissel et al., Phys. Rev. Lett. *68* (1992) 3412
23. Jung, M. et al., Phys. Rev. Lett. *69* (1992) 2164
24. Franzke, B. et al., Proc. of 2nd Europ. Part. Accel. Conf., Berlin 1990, Vol.1, p.444 (Edition Frontieres, Gif-sur Yvette 1992)
25. Fujita, Y. et al., Adv. in Mass Spectrometry *11a* (1989) 640
26. Balog, K.S. et al., Nucl. Instr. Meth. *B70* (1992) 459
27. Vaziri, K. et al., Nucl. Instr. Meth. *B26* (1987) 280
28. Sharma, K.S et al., Jour. Mod. Opt. *39* (1992) 349
29. Coc, A. et al., Nucl. Instr. Meth. *A305* (1991) 143
30. Borgenstrand, H., Carlberg, C., Rouleau, G., Schuch, R., Söderberg, F., Beebe, E., Bergström, I., Paal, A., Liljeby, L., Bollen, G., Hartmann, H., Jertz, R., Senne, P., Schwarz, T., Kluge, H.-J., and Mann, R., submitted to Physica Scripta (1994)

ADVANCES IN LASER COOLING AND TRAPPING OF ATOMS; ATOM OPTICS

Recent Advances in Subrecoil Laser Cooling

John Lawall[a], François Bardou[a], Jean-Philippe Bouchaud[b],
Bruno Saubamea[a], Nick Bigelow[c], Michèle Leduc[a],
Alain Aspect[d] and Claude Cohen-Tannoudji[a]

a - Laboratoire Kastler Brossel * et Collège de France,
24 rue Lhomond, 75231 Paris Cedex 05, France
b - Service de Physique de l'Etat Condensé, CEA - Saclay,
Orme des Merisiers, 91191 Gif-sur-Yvette, France
c - Department of Physics and Astronomy and the Laboratory for Laser
Energetics, University of Rochester, Rochester, NY 14627, USA
d - Institut d'Optique Théorique et Appliquée,
B.P. 147, 91403 Orsay Cedex, France

Abstract. This paper presents new experimental realizations of subrecoil laser cooling by velocity selective coherent population trapping (VSCPT). Starting from a cloud of trapped and precooled metastable helium atoms, it has been possible to achieve a VSCPT interaction time of 500 μs. This has enabled the momentum distribution to be compressed along one or two orthogonal axes to better than $\delta p = \hbar k/4$ where $\hbar k$ is the photon momentum. The corresponding temperature is at least 16 times smaller than the single photon recoil limit, and the de Broglie wavelength of the atoms is at least 4 times larger than the wavelength of the laser used to cool the atoms. Recent theoretical developments are also described, establishing a connection between subrecoil cooling and Lévy flights and allowing one to get analytical expressions for the proportion of cooled atoms in the long time limit. Finally, the possibility of increasing the efficiency of VSCPT by Sisyphus precooling is briefly mentioned.

INTRODUCTION

Reducing the velocity spread δv of an ensemble of atoms, which amounts to cooling their translational degrees of freedom, opens the way to various interesting applications in atomic physics : longer observation times allowing more precise measurements, longer de Broglie wavelengths which can be useful in new research fields like atomic interferometry, and the search for quantum statistical effects. Figure 1 summarizes a few important steps which have been achieved in this domain during the last 25 years.

The advent of tunable lasers in the early seventies gave access to the homogeneous width Γ of an atomic transition, by selecting a group of atoms having

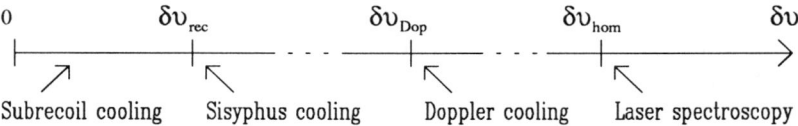

FIGURE 1: A few important landmarks in the velocity scale corresponding to various velocity selection or cooling mechanisms.

a velocity spread δv_{hom} such that the Doppler effect associated with δv_{hom} is equal to Γ :

$$k\delta v_{\text{hom}} = \Gamma \tag{1}$$

For example, a monochromatic laser light "burns" a hole with a width δv_{hom} in the Doppler profile of an atomic vapor. Other sub-Doppler schemes in laser spectroscopy use nonlinear effects such as saturated absorption or two photon absorption processes from two counterpropagating laser beams (1). Note that all these methods use a velocity selection mechanism and do not compress the atomic velocity distribution.

The Doppler cooling method, proposed in the mid seventies (2), produces a compression of the velocity distribution by introducing a velocity damping force due to a Doppler induced imbalance between the radiation pressure forces exerted by three sets of counterpropagating laser beams. The laser beams are detuned slightly to the red of the atomic transition, so that the detuning :

$$\delta = \omega_L - \omega_A \tag{2}$$

between the laser frequency ω_L and the atomic frequency ω_A is negative. An atom subjected to such a set of laser beams is said to be in an optical molasses, due to the strong viscous damping forces it experiences. The theory of Doppler cooling (3) shows that there is a lower limit T_D to the temperature which can be achieved by such a method, whose order of magnitude is given by :

$$k_B T_D \simeq \hbar\Gamma \tag{3}$$

(k_B : Boltzmann constant), this minimum temperature being reached when $\delta = -\Gamma/2$. The corresponding velocity spread δv_{Dop} is thus given by :

$$M\delta v_{\text{Dop}}^2 \simeq \hbar\Gamma \tag{4}$$

where we have dropped factors of two (M is the mass of the atom). Equation (4) may be written :

$$\delta v_{\text{Dop}} \simeq \sqrt{\frac{\hbar \Gamma}{M}} \simeq \sqrt{\frac{\hbar^2 k^2/2M}{\hbar \Gamma}} \frac{\Gamma}{k} = \sqrt{\frac{E_R}{\hbar \Gamma}} \delta v_{\text{hom}} \tag{5}$$

We have used (1) and introduced the recoil kinetic energy:

$$E_R = \frac{\hbar^2 k^2}{2M} \tag{6}$$

of an atom absorbing or emitting a single photon with momentum $\hbar k$. For most optical lines, the recoil energy E_R is much smaller than $\hbar \Gamma$, so that:

$$\varepsilon = \frac{E_R}{\hbar \Gamma} \ll 1 \tag{7}$$

We conclude that δv_{Dop} is much smaller than δv_{hom}:

$$\delta v_{\text{Dop}} \simeq \sqrt{\varepsilon} \delta v_{\text{hom}} \ll \delta v_{\text{hom}} \tag{8}$$

In 1988, it became clear that other cooling mechanisms, more efficient than Doppler cooling, were operating in optical molasses (4). Among the new cooling mechanisms, using optical pumping, light shifts and laser polarization gradients, a particularly efficient one is the so-called Sisyphus cooling mechanism, where the moving atom is running up potential hills more frequently than down (5). The quantum theory of Sisyphus cooling (6) shows that there is a lower limit to the velocity spread δv which can be achieved by Sisyphus cooling. One find that:

$$\delta v > \delta v_{\text{rec}} = \hbar k/M \tag{9}$$

Using (5), (6), (7) and (9), one gets:

$$\delta v_{\text{rec}} = \frac{\hbar k}{M} = \sqrt{\frac{\hbar^2 k^2/M}{\hbar \Gamma}} \sqrt{\frac{\hbar \Gamma}{M}} \simeq \sqrt{\varepsilon} \delta v_{\text{Dop}} \ll \delta v_{\text{Dop}} \tag{10}$$

The same reduction factor, $\sqrt{\varepsilon}$, appears for δv when one goes from Doppler cooling to Sisyphus cooling as when one goes from velocity selection in laser spectroscopy to Doppler cooling.

$$\delta v_{\text{rec}} \simeq \sqrt{\varepsilon} \delta v_{\text{Dop}} \simeq \varepsilon \delta v_{\text{hom}} \tag{11}$$

Subrecoil cooling corresponds to a situation where:

$$\delta v < \delta v_{\text{rec}} \tag{12}$$

The present paper is devoted to the description of recent experimental and theoretical advances in this field. After a brief review of subrecoil cooling, we describe a new generation of experiments using velocity selective coherent population trapping (VSCPT), which lead to values of δv significantly smaller than $\delta v_{\rm rec}$ in one and two dimensions. We then review new theoretical approaches well adapted to the long time limit where the standard methods of quantum optics (optical Bloch equations) become inappropriate. Finally, we briefly discuss the possibility of combining VSCPT and Sisyphus cooling.

BRIEF REVIEW OF SUBRECOIL COOLING

General Considerations

We first note that subrecoil cooling results in a delocalization of atoms in the laser wave. Condition (12), which according to (9) can also be written :

$$\delta p < \hbar k \tag{13}$$

where $p = Mv$ is the atomic momentum, is equivalent to :

$$\lambda_{DB} = \frac{h}{\delta p} > \frac{h}{\hbar k} = \lambda_L \tag{14}$$

Subrecoil cooling thus corresponds to a situation where the de Broglie wavelength λ_{DB} of the atoms is larger than the wavelength λ_L of the laser used to cool them. The spatial extent of the wave packets describing the center of mass of the atom can no longer be neglected and a full quantum treatment of atomic motion is needed.

A second important consequence of equation (13) is that spontaneous emission must be avoided for atoms cooled below the single photon recoil limit, because spontaneous emission, which occurs in random directions, would communicate to the atoms a random recoil δp on the order of $\hbar k$. In other words, ultracold atoms must be prevented from absorbing light.

Up to now, two subrecoil schemes (7) (8) have been proposed and demonstrated (9). The first one, which uses VSCPT (7) (10), is based on a combination of two effects : (i) the existence of certain atomic states $\left|\Psi_p^{NC}\right\rangle$ which are not coupled to the lasers and which, for p small enough, are perfect traps in momentum space, and (ii) a random walk of atoms in momentum space due to exchange of momentum with photons during fluorescence cycles, which allows atoms to diffuse from non trapping states with $p \neq 0$ to trapping states $p \simeq 0$ where they accumulate. The second scheme (8) uses an appropriate sequence of stimulated Raman and optical pumping pulses tailored in such a way that atoms are pushed in momentum space towards the zone $p \simeq 0$ where no resonant light can excite them.

General Expression of the Trapping State for a $J_g = 1 \longleftrightarrow J_e = 1$ Transition

In this paper, we focus on VSCPT. We describe new experimental realizations of VSCPT in one (11) and two (12) dimensions, using the transition $2^3S_1 \longleftrightarrow 2^3P_1$ of Helium. It will thus be useful to give an expression of the trapping state for a $J_g = 1 \longleftrightarrow J_e = 1$ transition, valid in any dimension. Other transitions have been considered in the literature (13).

We will follow here the method of Ol'shanii and Minogin (14). In the lower state g, the atom can be considered as a spin-1 particle, because $J_g = 1$. Consequently, its state is described by a wave function $\boldsymbol{\Psi}(\mathbf{r})$, which is a vector field. Similarly, the state of the atom in the excited state e (with $J_e = 1$) is described by a vector field $\boldsymbol{\Phi}(\mathbf{r})$. Finally, there is in the atom-laser interaction Hamiltonian V_{AL} a third vector field, the laser electric field $\mathbf{E}_L(\mathbf{r})$. Let $\mathbf{E}_L^+(\mathbf{r})$ and $\mathbf{E}_L^-(\mathbf{r})$ be the positive and negative frequency components of $\mathbf{E}_L(\mathbf{r})$, respectively. The transition amplitude induced by V_{AL} between g and e may be shown to be proportional to the following integral:

$$\int d^3 r \; \boldsymbol{\Phi}^*(\mathbf{r}) \cdot \left[\mathbf{E}_L^+(\mathbf{r}) \times \boldsymbol{\Psi}(\mathbf{r}) \right] \tag{15}$$

which is in fact the only scalar which can be constructed from the three vector fields $\boldsymbol{\Phi}^*(\mathbf{r})$, $\mathbf{E}_L^+(\mathbf{r})$, $\boldsymbol{\Psi}(\mathbf{r})$. Such a result can also be directly checked by expanding $\boldsymbol{\Psi}(\mathbf{r})$ and $\boldsymbol{\Phi}^*(\mathbf{r})$ on an orthornormal basis of Zeeman sublevels and by using the Clebsch-Gordan coefficients of the transition $J_g = 1 \longleftrightarrow J_e = 1$.

Suppose now that we take:

$$\boldsymbol{\Psi}(\mathbf{r}) = \alpha(\mathbf{r}) \mathbf{E}_L^+(\mathbf{r}) \tag{16}$$

where $\alpha(\mathbf{r})$ is any scalar function of $\mathbf{r}$. It is then clear that the integral of (15) vanishes for all $\boldsymbol{\Phi}^*(\mathbf{r})$. This means that the states (16) are not coupled by the laser to the excited state. Expanding $\alpha(\mathbf{r})$ in plane waves $\exp(i\mathbf{p}\cdot\mathbf{r}/\hbar)$, one gets a set of non coupled states labelled by $\mathbf{p}$:

$$\boldsymbol{\Psi}_\mathbf{p}^{NC}(\mathbf{r}) = \mathbf{E}_L^+(\mathbf{r}) \exp(i\mathbf{p}\cdot\mathbf{r}/\hbar) \tag{17}$$

If we take $\mathbf{p} = 0$ in (17), we get a particularly important state:

$$\boldsymbol{\Psi}^T(\mathbf{r}) = \boldsymbol{\Psi}_0^{NC}(\mathbf{r}) = \mathbf{E}_L^+(\mathbf{r}) \tag{18}$$

Because of the monochromaticity of the laser field, all wave vectors $\mathbf{k}_i$ appearing in the plane wave expansion of $\mathbf{E}_L^+(\mathbf{r})$:

$$\mathbf{E}_L^+(\mathbf{r}) = \sum_{i=1}^{N} \mathcal{E}_{oi}\hat{\epsilon}_i \exp(i\mathbf{k}_i \cdot \mathbf{r}) \qquad (19)$$

have the same modulus $|\mathbf{k}_i| = \omega_L/c = k$. It follows that the state (18) is not only a non coupled state as (17), but is also an eigenstate of the kinetic energy operator $\mathbf{P}^2/2M$. This ensures that $\mathbf{\Psi}^T$ will not be destabilized by a motional coupling induced by $\mathbf{P}^2/2M$ between $\mathbf{\Psi}^T$ and other ground states coupled to e. The state $\mathbf{\Psi}^T$ is therefore a perfect trap for atoms, sometimes called a "dark" state.

Replacing in (17) $\mathbf{E}_L^+(\mathbf{r})$ by (19), one sees that the states (17) with $\mathbf{p} \neq \mathbf{0}$ are linear superpositions of plane waves with wave vectors $\mathbf{k}_i + \mathbf{p}/\hbar$. In general, these wave vectors do not have the same modulus, and (17) is not an eigenstate of $\mathbf{P}^2/2M$. There are motional couplings, proportional to $\mathbf{k}_i \cdot \mathbf{p}/M$, which destabilize the states (17) and introduce a photon absorption rate $\Gamma'_{NC}(\mathbf{p})$ from these states, proportional to $\mathbf{p}^2$ (if $|\mathbf{p}|$ is small enough). Such an absorption is then followed by a spontaneous emission process, which introduces a random change of momentum and allows atoms to diffuse in momentum space. The smaller $|\mathbf{p}|$, the smaller is the diffusion rate. Atoms thus progressively accumulate in a set of states $\mathbf{\Psi}_\mathbf{p}^{NC}(\mathbf{r})$, with $|\mathbf{p}|$ distributed over a range δp around the value $\mathbf{p} = \mathbf{0}$ corresponding to the perfectly dark state (18). Arguments similar to those used in (10) show that δp decrease as $1/\sqrt{\Theta}$ where Θ is the laser-atom interaction time.

In fact, several perfectly dark states can exist for a given laser configuration. The conditions for having a single dark state, which are discussed in reference 14, are fulfilled for all the experiments described in this paper.

NEW GENERATION OF EXPERIMENTS

The new experimental scheme (11) (12) has been radically altered from the initial one (7), in order to achieve much longer atom-laser interaction times and to confine atoms in a smaller volume. Instead of applying the VSCPT laser beams to a supersonic beam of metastable helium atoms, we now start with atoms precooled to $\sim 200\mu K$ in a magneto-optical trap (15) (16). The trap is loaded from a cryogenic (6K) beam of He* in the 2^3S_1 state, decelerated by radiation pressure using a Zeeman slowing technique (17). The trap contains $\sim 10^5$ He* atoms in a volume of $\sim 1mm^3$, forming a well localized source of slow atoms upon which to perform further cooling. The trap is shut off, and the beams for the VSCPT cooling process, tuned to the $2^3S_1 \longleftrightarrow 2^3P_1$ transition, are pulsed on. All of the VSCPT beams are derived from the same laser, thus ensuring phase coherence. During the time of the VSCPT cooling, the atoms move less than 1 mm, after which they follow ballistic trajectories under the influence of gravity. Atoms are detected 5 cm below the trap by means of a microchannel plate detector. The initial temperature of the trapped cloud of atoms is measured by switching off the trap and observing the time of flight distribution as the atoms fall. The observed distribution, in which both the initial velocity and gravity play important roles, peaks around 45 ms. The corresponding initial rms velocity in 60 cm/s, corresponding to a temperature of 180 μK. High spatial resolution (0.5 mm) is obtained by accelerating the output of the microchannel plate toward a phosphor screen, and the resulting blips of light are recorded with a triggered CCD camera which provides temporal resolution. For more experimental details, see references (11) and (12).

1D Experiments

The laser configuration is formed by two counterpropagating beams along the x-axis with σ^+ and σ^- polarizations. According to equation (18), the trapping state is then a linear superposition of two de Broglie waves with wave vectors $\pm k$ along the x-axis. One thus expects that, after the VSCPT cooling process, atoms have been pumped into a linear superposition of two wave packets moving with momenta $\pm \hbar k$ along the x-axis. Since no cooling takes place along the y and z axes, the velocity spread along these two axes is the same as in the magneto-optical trap. Atoms are detected on the microchannel plate detector a time τ_f (30-80 ms) after the VSCPT cooling beams are turned off, within a temporal window τ. By varying τ_f, one can probe the (uncooled) vertical velocity distribution, and by varying τ, one can select the time resolution. Images from the CCD camera are digitized in a PC and the entire process is repeated. The images are averaged in software.

The left part of figure 2 gives an example of atomic position distribution obtained after a single release from the trap. Each dot corresponds to a single He* atom. The right part of the figure is obtained after averaging over 80 single releases. One clearly sees the double band structure which is the signature of VSCPT cooling along the laser axis (x-axis). Note that there is no cooling along the y-axis. Here the VSCPT interaction time was $\Theta=400$ μs, about one order of magnitude longer than the interaction time in the first experimental realization of VSCPT (7).

The observed width of each band of figure 2 is manifestly smaller than the spacing between the bands which corresponds to $2\hbar k$. This is a clear indication of cooling below the recoil limit. The intensity at the center of each band is increased by a factor 5 when the VSCPT beams are applied, a sign of real cooling (increase of the density in momentum space). The width of each band reflects, in addition to the final VSCPT momentum distribution, contributions due to the size of the cloud of trapped atoms, the size of the blips of light emitted by the phosphor, and the dispersion of atom arrival

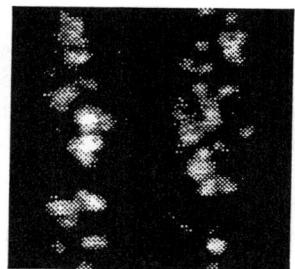 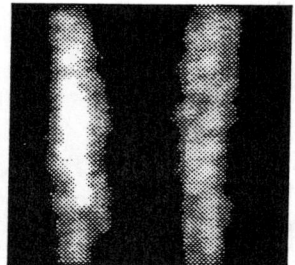

FIGURE 2: Atomic distribution on the detector obtained by 1D-VSCPT. (Interaction time $\Theta=400\mu$s, Rabi frequency $\Omega \simeq \Gamma$, $\tau_f = 45$ ms, $\tau=10$ ms). The left part is obtained after a single release from the trap. The right part is obtained after averaging over 80 single releases. The separation of the two bands is 0.9 cm. For larger values of τ_f, it can reach 1.4 cm.

times during the observation time τ. As explained in reference (11), one can extract the contribution of the velocity spread of the cooled atoms along the x-axis. One finds $\delta v = \delta p/M \simeq 2$ cms^{-1} (half width at $1/\sqrt{e}$), which corresponds to $\delta p/\hbar k = 1/4.5$. Converting this into an effective temperature by $k_B T/2 = (\delta p)^2/2M$, one finds $T \simeq T_R/20 = 200nK$, where T_R is the recoil temperature defined by $k_B T_R/2 = E_R = \hbar^2 k^2/2M$. Note that we introduce here a temperature T as a convenient indicator of the width of the peaks of the momentum distribution, rather than in a strict thermodynamic sense.

2D Experiments

The laser configuration now consists of four counterpropagating beams along the x and y axes, with σ^+ and σ^- polarizations (Fig. 3). In equation (19), $N = 4$, $\mathbf{k}_1 = +k\hat{x}$, $\mathbf{k}_2 = -k\hat{x}$, $\mathbf{k}_3 = +k\hat{y}$, $\mathbf{k}_4 = -k\hat{y}$. One thus expects that, after such a laser configuration has been applied for a time Θ to an atom, the state of this atom will be a linear superposition of four wave packets with mean momenta $\pm\hbar k\hat{x}$, $\pm\hbar k\hat{y}$. Thus, on the detector plane, one should observe four spots separated by a distance $2\hbar k\tau_f/M$, where τ_f is the flight time to the detector, the width of these spots decreasing as $1/\sqrt{\Theta}$.

An example of atomic position distribution detected on the microchannel plate is shown in figure 4. The figure is obtained by averaging over 25 consecutive single releases from the trap, for each of which the camera was exposed from 45 ms to 65 ms after the VSCPT interaction. The four peaks are clearly

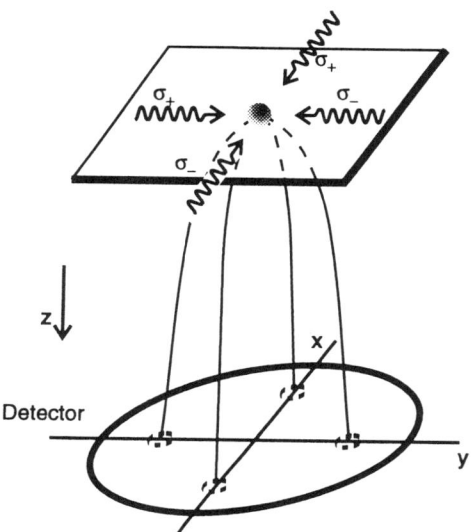

FIGURE 3: Principle of the 2D-VSCPT experiment. By interacting with the four VSCPT cooling beams, atoms are pumped into a coherent superposition of four wave packets whose centers follow ballistic trajectories to the position sensitive detector located 5 cm below.

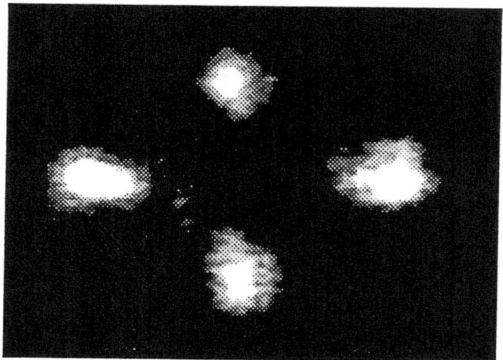

FIGURE 4: Atomic position distribution on the detector obtained by 2D-VSCPT (Interaction time $\Theta=500\mu s$, Rabi frequency $\Omega=0.8\Gamma$, detuning $\delta \simeq 0.5\Gamma$, $\tau_f = 55$ ms, $\tau=20$ ms). The image is obtained after averaging over 25 single releases.

resolved. As for the 1D experiment, the detected peaks contain instrumental broadenings in addition to the atomic momentum spread (size of the cloud of trapped atoms, size of the blips of light, dispersion of arrival times). An upper bound to the momentum spread is obtained by neglecting the initial cloud size and the imperfect detector resolution. From the width of the peaks, we deduce a momentum spread (half width at $1/\sqrt{e}$) $\delta p \simeq \hbar k/4$. The corresponding effective temperature is $T \simeq T_R/16$. These values of $\delta p/\hbar k$ and T/T_R are the lowest ever achieved in 2D subrecoil cooling (18).

An important issue is whether the VSCPT process actually increases the density in momentum space or merely acts to select atoms within a small velocity group. The answer is dependent on the laser parameters. Figures 5a and 5b describe how the momentum distribution varies with the laser power and the detuning. The heavy lines represent a profile of one of the spots of Fig. 4 taken in the direction perpendicular to the recoil momentum. In this direction, the broadening due to the dispersion of arrival times is absent. The thin lines correspond to the uncooled ditribution.

It appears in figure 5a that the efficiency of VSCPT increases with the laser power. When Ω increases, the peak of the cooled distribution becomes higher than the uncooled distribution (which is a signature of cooling), while the width of the peak increases, in agreement with theoretical predictions (10).

Figure 5b shows that the efficiency of 2D-VSCPT is higher for a blue detuning ($\delta = \omega_L - \omega_A > 0$)) than for a red one ($\delta = \omega_L - \omega_A < 0$). We will come back to this point in the last section of this paper devoted to the discussion of possible Sisyphus-type friction mechanisms.

A Few Prospects

The previous results demonstrate that VSCPT is a pratical means for achieving a significant subrecoil laser cooling in one or two dimensions. The final temperature T, expressed in units of the recoil temperature T_R, is about the same in both cases : $T/T_R \simeq 1/20$ or $1/16$. One can hope to reduce this temperature by another order of magnitude by increasing the interaction time to a few ms. Further study is required to thoroughly understand the limitations imposed, for example, by the residual stray magnetic field, imperfect laser beam polarization, imperfect vacuum, and multiple scattering of resonant light.

An interesting feature of VSCPT is that atoms are prepared in a coherent superposition of wave packets whose centers, in our work, are separated by macroscopic distances, on the order of 1 cm. An obvious challenge is to recombine the two stripes of figure 2 or the four beams of figure 3 in order to observe interference and thus demonstrate the coherence.

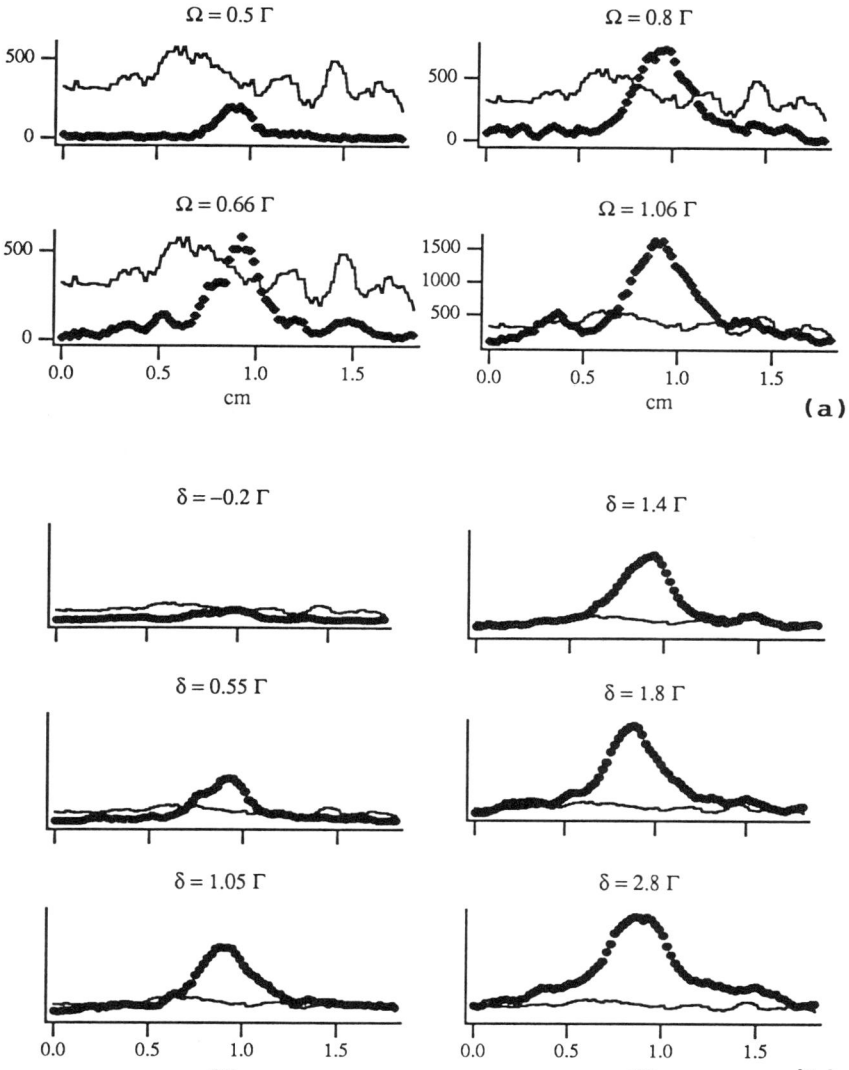

FIGURE 5: a - Influence of the laser power on VSCPT efficiency. The detuning is fixed at $\delta = 0.55\Gamma$ and the interaction time at $\Theta = 500~\mu s$. Each curve corresponds to a different value of the Rabi frequency Ω. b - Influence of the detuning on VSCPT efficiency. The Rabi frequency is fixed at $\Omega=0.8\Gamma$ and the interaction time at $\Theta = 500~\mu s$. Each curve corresponds to a different value of the detuning $\delta = \omega_L - \omega_A$. For both figures, the heavy and thin lines correspond to the cooled and uncooled momentum distributions, respectively.

The experiments described here can be extended to three dimensions. During an interaction time of 1 ms, the velocity change due to gravity is 1 cms^{-1}, which remains small compared to the recoil velocity of He* (9.2 cms^{-1}). One can therefore neglect gravity during a VSCPT cooling process lasting for less than 1 ms. Generalizing to six laser beams with counterpropagating σ^+ and σ^- beams along each axis, one expects to get a linear superposition of six wave packets falling with well defined initial velocities. Since there is now no uncooled velocity component, it is no longer necessary to trigger the detector and one expects to observe wave packets arriving at different times (and different positions) because they start with different initial velocities. Experiments of this type are under way and encouraging preliminary results have been obtained.

NEW THEORETICAL DEVELOPMENTS

Connection with Lévy Flights

Making quantitative predictions of the efficiency of VSCPT in the long time limit ($\Theta \longrightarrow \infty$) seems rather difficult. Because atoms are delocalized in the laser wave when $\delta p < \hbar k$, all atomic degrees of freedom must be treated quantum mechanically. All of the usual treatments (19), leading to a Fokker Planck equation description of atomic motion through an expansion of the density matrix elements in powers of $\hbar k/\delta p$, cannot be applied here because $\hbar k/\delta p$ is not a small parameter. Furthermore, the fact that no steady state exists for VSCPT complicates the search for a numerical solution of the optical Bloch equations in the limit $\Theta \longrightarrow \infty$.

During the last few years, new theoretical approaches have been developed for circumventing these difficulties. Monte Carlo simulations of the time evolution of a single atom in VSCPT have been made (20). Such a time evolution consists of a sequence of quantum jumps occurring at random times and associated with spontaneous emission processes. Between two successive quantum jumps, a coherent evolution period takes place, associated with absorptions and stimulated emissions of laser photons. The simulation uses the so called "delay function" which gives the distribution of the time intervals between two successive spontaneous emissions (21). As in the Wave Funtion Monte Carlo approach (22), the description of the atomic state by a wave function rather than by a density matrix simplifies the numerical calculations which can be extended to much longer times. Furthermore, such Monte Carlo simulations provide a better physical understanding of VSCPT. They clearly show that the smaller the atomic momentum p, the longer the delay τ_d between two successive spontaneous emission jumps, which is the principle of VSCPT.

There is another striking feature of the Monte Carlo simulations of VSCPT [see for example figure 1 of reference (23)] which suggested a completely new statistical approach for such a cooling scheme. The random sequence of time intervals τ_d is dominated by a few terms, the longest ones, which are on the order of the total observation time. This uncommon domination of a random sequence by rare events is a signature of "Lévy flights" and "broad" distributions (24) (25), in sharp contrast with the usual Brownian motion statistics encountered in other cooling schemes. In fact, one can show (23) that VSCPT provides simple examples of Lévy flights. Using the statistical properties of Lévy flights, one can then derive new analytical results for the asymptotic properties of VSCPT in the limit $\Theta \longrightarrow \infty$ (23).

More precisely, one can define in the neighbourhood of $p = 0$ a very narrow trapping zone $|p| < p_{\text{trap}}$. When the atomic momentum lies in this zone, the atom is considered as being trapped. Then, after a certain time, the atom leaves the trap, diffuses out of the trap before returning to the trap, and so on. The temporal evolution of the atom thus appears as a sequence of trapping periods where $|p| \leq p_{\text{trap}}$, with duration τ_1, τ_2..., alternating with diffusion periods where $|p| > p_{\text{trap}}$, with durations $\hat{\tau}_1$, $\hat{\tau}_2$... The $\hat{\tau}_i's$ are actually "first return times" in the trap. Consider $2N$ successive alternating trapping and diffusion periods, with $N \gg 1$, and let $T(N) = \sum_{i=1}^{N} \tau_i$ be the total trapping time, and $\hat{T}(N) = \sum_{i=1}^{N} \hat{\tau}_i$ the total escape time. Understanding how $T(N)$ and $\hat{T}(N)$ grow with N is important for predicting the proportion of cooled atoms in the limit $\Theta \longrightarrow \infty$. Since the $\tau_i's$ are independent random variables, as well as the $\hat{\tau}_i's$, one needs only to find their probability distributions $P(\tau)$ and $\hat{P}(\hat{\tau})$.

In fact, from the physics of VSCPT, more precisely from the $p-$dependence of the photon absorption rate for $p \longrightarrow 0$ and $p \longrightarrow \infty$, one can determine the asymptotic behaviour of $P(\tau)$ and $\hat{P}(\hat{\tau})$ at large τ and $\hat{\tau}$. The important point is that these distributions are broad. In several cases, they behave as $\tau^{-(1+\mu)}$ for $P(\tau)$ and as $\hat{\tau}^{-(1+\hat{\mu})}$ for $\hat{P}(\hat{\tau})$. For example, for 1D-VSCPT one finds $\mu=1/2$ and $\hat{\mu} = 1/4$. The distributions $P(\tau)$ and $\hat{P}(\hat{\tau})$ are then so broad that $\langle \tau \rangle$ and $\langle \hat{\tau} \rangle$ are infinite. Consequently, the central limit theorem (CLT) does not apply to the sums $T(N)$ and $\hat{T}(N)$. It must be replaced by a generalized CLT, established by Lévy and Guedenko [see, e.g., (25) for a concise account]. If $0 < \mu, \hat{\mu} < 1$, one finds that $T(N)$ and $\hat{T}(N)$ do not grow as N for large N, but rather as $N^{1/\mu}$ and $N^{1/\hat{\mu}}$. If $\hat{\mu} < \mu$, the total time spent by the atom outside the trap predominates over the total time spent in the trap at large N, and one expects that the proportion f of cooled atoms tends to 0 when $\Theta \longrightarrow \infty$. In other cases, in which $\mu = \hat{\mu}$, $T(N)$ and $\hat{T}(N)$ have the same $N-$dependence, so that f is expected to tends towards a constant when $\Theta \longrightarrow \infty$. Finally, it may happen, for example in the presence of a friction mechanism which pushes the atom towards $p = 0$ in momentum space, that

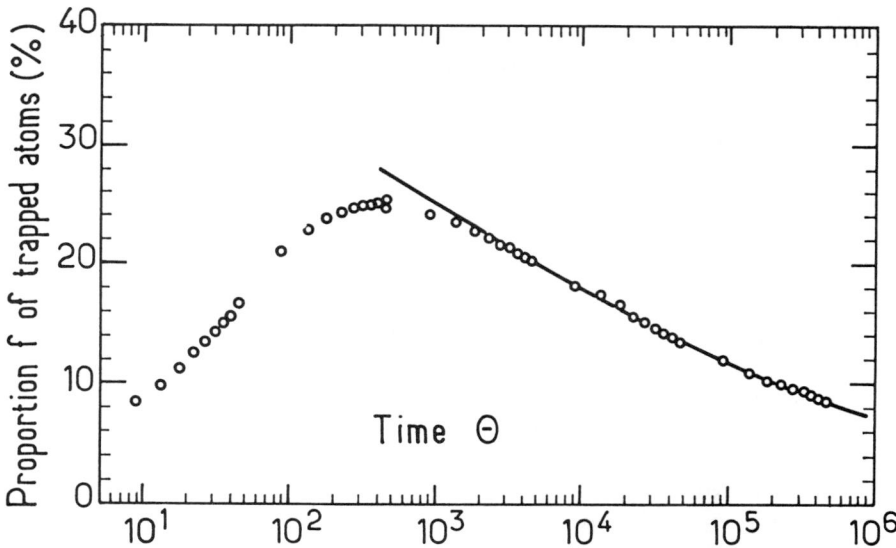

FIGURE 6: Proportion f of cooled atoms versus the interaction time Θ (in units of $1/\Gamma'$, where $\Gamma' = \Gamma s$, s being the saturation parameter). The full line represents the analytical asymptotic prediction given by the Lévy flight approach. The circles represent the results of Monte Carlo simulations at intermediate times.

only $P(\tau)$ is a broad distribution, $\hat{P}(\hat{\tau})$ being a narrow distribution leading to a "normal" linear growth ($\sim N$) of $\hat{T}(N)$. In such a case, $T(N)$ predominates over $\hat{T}(N)$ and $f \longrightarrow 1$ if $\Theta \longrightarrow \infty$. In fact, more precise calculations can be done, using the convolution of two Lévy laws, and one can derive analytical expressions for f in the limit $\Theta \longrightarrow \infty$. These analytical predictions have been quantitatively checked by comparison with the results of Monte Carlo simulations at intermediate times. For example, for 1D-VSCPT, one predicts that f should vary as $\left[A + B\Theta^{1/4}\right]^{-1}$ when $\Theta \longrightarrow \infty$, where A and B are constants. Such a prediction (full line of Fig. 6) is in very good agreement with the results of Monte Carlo numerical calculations (circles of Fig. 6).

The Lévy flight approach can also give analytical predictions for the momentum distribution and for the influence of the dimensionality d. One finds that, when d increases, $P(\tau)$ narrows whereas $\hat{P}(\hat{\tau})$ broadens. One thus expects that for pure VSCPT, where the return of atoms in the trap is only due to momentum diffusion, the cooling efficiency should rapidly decrease when d increases. The fact that the 2D experiment described above gives a good signal is therefore an indication that an additional friction mechanism exists.

Combining VSCPT and Sisyphus Cooling

The laser configurations used in the first experimental realizations of VSCPT (7) consisted of two counterpropagating beams with σ^+ and σ^- polarizations. The resulting laser electric field then has an intensity which does not vary in space. One can easily show that, in such a laser configuration, the light shifts of the various ground state Zeeman sublevels are position independent. Plotting these light shifts as a function of the position of the atom, one gets flat "adiabatic" potential curves, the light shift of the non-coupled state being equal to zero everywhere. There are therefore no potential hills and no possibility of Sisyphus cooling. Furthermore, and contrary to what happens for a $J_g \longleftrightarrow J_e = J_g + 1$ transition, there is no polarization gradient force for a $J_g = 1 \longleftrightarrow J_e = 1$ transition in a $\sigma^+ - \sigma^-$ laser configuration (19).

Several authors have recently mentioned that other laser configurations in one and two dimensions could lead to a coexistence of VSCPT and Sisyphus cooling [see references (26) to (30)]. For a $J_g = 1 \longleftrightarrow J_e = 1$ transition, there is always a non-coupled state whose light shift is zero everywhere, giving rise to a perfectly flat adiabatic potential. But for the laser configurations discussed in (26) to (30), the total laser intensity varies in space and the other eigenvalues of the light shift operator (other than zero) are in general position-dependent, giving rise to adiabatic potential curves with potential wells and potential valleys. For a moving atom, there are nonadiabatic couplings which can transfer the atom from the non-coupled state to such a position dependent potential curve, and a Sisyphus cooling can occur if the detuning δ is positive (when δ is positive, light shifts are positive and the position-dependent potential curves are above the flat line corresponding to the non-coupled state ; furthermore, the transfer rate by optical pumping from a coupled state to the non coupled one is maximum at the tops of the potential hills). Such a semiclassical picture of Sisyphus cooling becomes questionable in the quantum regime where atoms are delocalized in the laser wave. More precise treatments using quantum Monte Carlo methods and confirming the existence of a Sisyphus type cooling may be found in references (27) and (28).

The coexistence of VSCPT and Sisyphus friction forces may be very attractive at higher dimensions, because the random walk process of pure VSCPT becomes less and less efficient at bringing atoms back towards $p = 0$. Studying how the cooling efficiency depends on the detuning δ can provide useful informations. For pure VSCPT, the cooling efficiency should not be very sensitive to the detuning (10). The fact that, in the 2D experiment described above, the signal is much better for $\delta > 0$ than for $\delta < 0$ (see Fig. 5b) seems to indicate the existence of a Sisyphus cooling improving the efficiency of 2D-VSCPT when $\delta > 0$. Further experimental work is needed to confirm such a result.

CONCLUSION

In conclusion, important advances have been achieved in 1D and 2D subrecoil cooling of atoms. Temperatures significantly lower than the single photon recoil limit (by a factor 16 at least) have been observed. The de Broglie wavelength of the cooled atoms now reaches values of the order of 4.5 μm. One can thus hope to put several atoms in a volume λ_{DB}^3 while keeping these atoms separated by distances large compared to the optical wavelength λ_L which determines the range of radiative interactions between atoms. This could be important for reducing the limitations associated with atom-atom interactions in the search for quantum statistical effects. A better understanding of the long time limit of VSCPT has been obtained with the development of new statistical approaches inspired by the Lévy flight description of anomalous random walks. By studying the competition between trapping and escape processes, determined by the distribution of trapping times and first return times in the trap, one can predict in a quantitative way how the efficiency of pure VSCPT varies in the long time limit and how it depends on the dimensionality. Theoretical and experimental studies indicate that VSCPT could be improved by a Sisyphus precooling. This is important for future developments because there is still room for an increase of the interaction time and for a corresponding decrease of the temperature.

REFERENCES

* Laboratoire associé au CNRS et à l'Université Pierre et Marie Curie.

1. See for example the Proceedings of the 2nd Laser Spectroscopy Conference, ed. by Haroche, S., Pebay-Peyroula, J. C., Hänsch, T. W., and Harris, S. H., Springer, Berlin, 1975.

2. Hänsch, T. W., and Schawlow, A. L., *Opt. Commun.*, **13**, 68 (1975).
   Wineland, D. J., and Dehmelt, H. G., *Bull. Am. Phys. Soc.*, **20**, 637 (1975).

3. Wineland, D. J., and Itano, W., *Phys. Rev.* **A20**, 1521 (1979)
   Gordon, J. P., and Ashkin, A., *Phys. Rev.* **A21**, 1606 (1980).

4. For a review on these developments, see Cohen-Tannoudji, C., and Phillips, W. D., *Physics Today* **43** (10), 33 (1990). See also the papers of the Gaithersburg, Paris and Stanford groups in "Laser Cooling and Trapping of Atoms", Special Issue, ed. by Chu, S., and Wieman, C., *J. Opt. Soc. Am.* **B6**, 2019 (1989).

5. Dalibard, J., and Cohen-Tannoudji, C., *J. Opt. Soc. Am.* **B6**, 2023 (1989).

6. Castin, Y., Dalibard, J., and Cohen-Tannoudji, C., in *Proceedings of Light Induced Kinetic Effects*, ed. by Moi, L., et al., Pisa, ETS Editrice, 1991.

7. Aspect, A., Arimondo, E., Kaiser, R., Vansteenkiste, N., and Cohen-Tannoudji, C., *Phys. Rev. Lett.* **61**, 826 (1988).

8. Kasevich, M., and Chu, S., *Phys. Rev. Lett.* **69**, 1741 (1992).

9. See also the proposals described in Pritchard, D. E., Hermerson, K., Bagnato, V. S., Lafyatis, P., and Martin, A. G., in *Laser Spectroscopy*VIII, ed. by Persson, W., and Svanberg, S., Berlin, Springer, 1987 ; Wallis, H., and Ertmer, W., *J. Opt. Soc. Am.* **B6**, 2211 (1989) ; Molmer, K., *Phys. Rev. Lett* **66**, 2301 (1991).

10. Aspect, A., Arimondo, E., Kaiser, R., Vansteenkiste, N., and Cohen-Tannoudji, C., *J. Opt. Soc. Am.* **B6**, 2112 (1989).

11. Bardou, F., Saubamea, B., Lawall, J., Shimizu, K., Emile, O., Westbrook, C., Aspect, A., and Cohen-Tannoudji, C., *C. R. Acad. Sci.***318**, Série II, 877 (1994).

12. Lawall, J., Bardou, F., Saubamea, B., Shimizu, K., Leduc, M., Aspect, A., and Cohen-Tannoudji, C., to appear in *Phys. Rev. Lett.*, 1994.

13. See for example the review paper of Arimondo, E., in *Laser Manipulation of Atoms and Ions*, ed. by Arimondo, E., Phillips, W. D., and Strumia, F., Amsterdam, North-Holland, 1992, and references therein.

14. Ol'shanii, M. A., and Minogin, V. G., in *Proceedings of Light Induced Kinetic Effects*, ed. by Moi, L., et al., Pisa, ETS Editrice, 1991 ; Ol'shanii, M. A., and Minogin, V. G., *Opt. Comm.* **89**, 393 (1992).

15. Raab, E. L., Prentiss, M., Cable, A., Chu, S., and Pritchard, D. E., *Phys. Rev. Lett.* **59**, 2631 (1987).

16. Bardou, F., Emile, O., Courty, J.-M., Westbrook, C. I., and Aspect, A., *Europhys. Lett.* **20**, 681 (1992).

17. Phillips, W. D., and Metcalf, H., *Phys. Rev. Lett.* **48**, 596 (1982).

18. The Raman cooling scheme of reference (8) has been generalized to 2 and 3 dimensions : Davidson, N., Lee, H. J., Kasevich, M., and Chu, S., *Phys. Rev. Lett.* **72**, 3158 (1994). At two dimensions, and with our definition of δp, $\delta p/\hbar k$ is just below the recoil limit.

19. For an overview on laser cooling, see Cohen-Tannoudji, C., in "Fundamental Systems in Quantum Optics", Les Houches, session LIII, ed. by Dalibard, J., Raimond, J.-M., and Zinn-Justin, J., Amsterdam, North-Holland, 1992, and references therein.

20. Cohen-Tannoudji, C., Bardou, F., and Aspect, A., in *Laser Spectroscopy* **X**, ed. by Ducloy, M., Giacobino, E., and Camy, G., Singapore, World Scientific, 1992.

21. Cohen-Tannoudji, C., and Dalibard, J., *Europhys. Lett.* **1**, 441 (1986) ; Zoller, P., Marte, M., and Walls, D. F., *Phys. Rev.* **A35**, 198 (1987).

22. For an overview on this approach, see Castin, Y., Dalibard, J., and Mølmer, K., in *Atomic Physics* **13**, ed. by Walther, H., Hänsch, T. W., and Neizert, B., New York, AIP, 1993, and references therein.

23. Bardou, F., Bouchaud, J.-P., Emile, O., Aspect, A., and Cohen-Tannoudji, C., *Phys. Rev. Lett.* **72**, 203 (1994).

24. Montroll, E. W., and Schlesinger, M. F., in *Statistical Mechanics*, ed. by Lebowitz, J., and Montroll, E. W., Amsterdam, North-Holland, 1984 ; Schlesinger, M. F., Zaslavsky, G. M., and Klafter, J., *Nature*, **363**, 31 (1993).

25. Bouchaud, J.-P., and Georges, A., *Phys. Rep.* **195**, 125 (1990).

26. Shahriar, M. S., Hemmer, P. R., Prentiss, M. G., Marte, P., Mervis, J., Katz, D. P., Bigelow, N. P., and Cai, T., *Phys. Rev.* **A48**, R4035 (1993).

27. Marte, P., Dum, R., Taïeb, R., Zoller, P., Shahriar, M. S., and Prentiss, M., *Phys. Rev.* **A49**, 4826 (1994).

28. Dum, P., Marte, P. and Zoller, P., *Technical Digest of IQEC'94*, **9**, 238 (1994).

29. Shahriar, M. S., Widmer, M. T., Bellanea, M. J., Vredenbregt, E., and Metcalf, H. J., *Technical Digest of IQEC'94*, **9**, 238 (1994).

Colder and Better: Some New Developments in Laser Cooling, Trapping and Manipulation of Atoms

William D. Phillips

National Institute of Standards and Technology, PHY A167, Gaithersburg, MD 20899, USA

Abstract. This paper is a short summary of a presentation given at the Fourteenth International Conference on Atomic Physics in Boulder CO, August 1994. It describes several recent advances in the field of laser cooling, trapping and manipulation of neutral atoms: adiabatic transfer of linear momentum from light to atoms; trapping and cooling atoms in optical lattices; evaporative cooling of alkali atoms held in magnetic traps; cyclic cooling of atoms in a magnetic trap; and optical shielding of laser-cooled and trapped atoms from collisions with other cold atoms. Some of the work described was done in my own laboratory at NIST in Gaithersburg, and some in other laboratories. I have tried to choose topics complementary to those treated by other speakers at the conference. This paper is intended more as a guide to the literature on the subjects treated than a complete treatment in itself.

ADIABATIC MOMENTUM TRANSFER

The transfer of momentum to atoms is a fundamental aspect of atom optics and interferometry [1], being involved in the reflection, focusing, re-direction, and beamsplitting of atomic beams. Atom interferometry requires that the momentum transfer be coherent, i.e., free from dissipative processes that would reduce the contrast of any interference fringes. In particular, coherent momentum transfer must avoid spontaneous emission of photons by the atoms. Two recent experiments [2,3] have demonstrated a new technique whereby momentum is coherently transferred to atoms by adiabatic passage in a time-varying resonant laser field.

The experiments avoid spontaneous emission by keeping the atoms in a "dark state" with coherent population trapping. Such dark states were first described [4] as coherent superpositions of the internal states of an atom that cannot absorb from a particular, albeit resonant, light field. A trivial example of a dark state (which in this basis has only one component) would be an atom in the $m_J = 1$ Zeeman sublevel of a $J = 1$ ground level, irradiated by σ^+ light resonant for a $J = 1 \rightarrow J' = 1$ transition. More generally, in this $J = 1 \rightarrow J' = 1$ system, for any polarization of light, there is a dark state, a coherent superposition of the ground sublevels, that cannot absorb that light. Dark states were later generalized [5] to include the state of the atom's center-of-mass motion and the momentum of the light field. Here the dark state represents a specific superposition of momentum states of the atom as well as of its internal states, and one must consider the direction as well as polarization of the light fields.

The suggestion [6] that such velocity selective dark states could be used to coherently transfer momentum to atoms followed the study and demonstration of adiabatic transfer of population between internal states in Raman systems [7]. The basic idea, continuing from the example given above, would be to start with an atom in m = -1, with momentum $-\hbar k$ (a photon momentum), illuminated with σ^- light propagating in the negative direction. This state, $|-1,-\hbar k\rangle$, is dark to that light. If a σ^+ light field, propagating in the positive direction is slowly turned on while the σ^- light is slowly turned off, the nature of the light field changes continuously from σ^+ to σ^- and the state which is dark changes continuously from $|-1,-\hbar k\rangle$ to $|+1,+\hbar k\rangle$. The remarkable thing is that if the change is slow enough to be adiabatic, the state of the atom adiabatically follows in the dark state, and gains two photon momenta while changing from $|-1,-\hbar k\rangle$ to $|+1,+\hbar k\rangle$, without absorbing light or being in the excited state. The requirement for adiabaticity is that the frequency corresponding to the light shift of the non-dark states be large compared to the rate at which the nature of the light field changes.

The Harvard experiments [2] performed transfer on metastable He, using a transition as described above. The transfer was transverse to a thermal atomic beam that passed through partially overlapped laser beams orthogonal to the atomic beam. Multiple passes allowed the transfer of six photon momenta ($6\hbar k$). The NIST experiments [3] used laser cooled Cs, with light resonant on a F = 4 → F' = 4 transition. In this case, $8\hbar k$ was transferred in a single adiabatic passage. In more recent experiments at Stanford [8], adiabatic transfer in Cs between non-degenerate ground states was used to demonstrate atom interference and in multiple passages more than 140 $\hbar k$ was transferred.

OPTICAL LATTICES

When multiple, intersecting light beams are used for laser cooling, as for example in optical molasses [9,10], they create a periodic interference pattern of varying intensity and polarization, producing a pattern of light-shift potentials that can confine or trap atoms. This was demonstrated in some early experiments [11,12,13]. More recently the term "optical lattice" has been applied to the trapping of atoms in such periodic structures where the nature of the interference pattern is well defined and time-invariant, and where the polarization of the light at the points where atoms are trapped is such that the atoms are optically pumped into the most strongly trapped sublevel. These features lead to long trapping times for the atoms and the ability to observed quantization of the center-of-mass motion of the trapped atoms.

The first such observations involved 1D lattices for Cs [14] and Rb [15]. The lattices were formed from counterpropagating laser beams with orthogonal linear polarization (lin ⊥ lin). The configuration produces a polarization that varies periodically in space from σ^+ through linear to σ^-, and is the classic configuration for Sisyphus laser cooling [16]. At the $\sigma^\pm$ positions atoms are pumped into $m_F = \pm F$, the state which is most deeply trapped at that location. Different, complementary techniques were used to study the motion of atoms trapped at the lattice sites. The Paris group [14] looked at the absorption or gain of a probe laser directed through the optical lattice and saw features corresponding to transitions between quantized vibrational levels in the optical lattice potential wells. The NIST group [15] measured the spectrum of fluorescence emitted by the atoms in the lattice and observed sidebands, corresponding to the same kind of transitions, on either side of the elastically scattered light. The position, width, relative magnitude and symmetry of the spectral features in these experiments give information about the vibrational level spacing, the degree of localization, and the temperature of the trapped atoms. The observations are in excellent agreement with theoretical calculations [17]. Using different techniques a group at SUNY has also seen the effects of quantized motion in a 1D lattice [18]

More recently a number of groups have extended the study of optical lattices to 2 and 3D. In 2 or 3D one faces the issue of stability of the interference pattern. For example, a typical 3D optical molasses with three pairs of counterpropagating laser beams produces an interference pattern whose character depends on the relative phases between the laser beams. On the other hand, a 3D configuration of 4 beams produces an interference pattern whose position, but not character, depends on the relative phases [19]. One can therefore insure a well defined and stable interference pattern by using 4 beams in 3D (3 beams in 2D or 2 beams in 1D) or by actively stabilizing the relative phase of the laser beams. The former technique is used by the Paris group [19, 20, 21], the NIST group [22] and the Stanford group [23], while the Munich group uses the latter technique [24,25]. The motion of the trapped atoms is studied by the probe laser technique in Paris and Munich and by the fluorescence spectrum technique at NIST.

Spectra obtained in 2 and 3D lattices are similar to those obtained in 1D. An important additional feature is that the lattices may not be isotropic so that different or multiple oscillation frequencies (sideband frequencies) are seen along different directions of observation. Another important feature is that in 3D, there is trapping and cooling in all directions, while in 1D, the transverse directions are heated and untrapped. This leads to residual Doppler broadening of the spectra observed in 1D, a broadening that is absent in 3D. The Munich group has observed spectra containing a series of sub-harmonics of the vibrational spacing corresponding to multiphoton transitions between adjacent vibrational levels [25]. The Paris group has applied a uniform static magnetic field to their lattice and seen spectral changes corresponding to paramagnetism involving the orientation of the atomic angular momentum [21].

Using time-of-flight techniques [26] the NIST group has measured the temperature of Cs atoms held in a 4-beam optical lattice detuned 25 linewidths from resonance. Temperatures as low as 1 µK were achieved [22], significantly lower than the 2.5 µK reported for Cs in a 6-beam optical molasses [27]. Presumably this lower temperature is due to the stable lattice of light-shift potentials in the 4-beam configuration compared to the uncontrolled interference pattern in the 6-beam molasses. The atoms trapped in the lattice can be cooled even further by a process of adiabatic expansion. (Such adiabatic cooling has previously been observed in 1D for lithium atoms in an atomic beam [28].) By slowly reducing the intensity of the trapping laser beams, the depth of the potential wells and hence the oscillation frequency of the atoms is reduced. This allows an atom to expand and cool within the well where it is trapped. The adiabatic condition requires that the normalized rate of intensity reduction be small compared to the atomic oscillation frequency, but the rate must be fast enough to avoid heating from spontaneous emission. We found that for characteristic times from a few tens to a few hundreds of microseconds the cooling produced the same final temperature. By observing the ballistic spreading of the atoms after cooling, we measured final temperatures as low as 700 nK along three orthogonal axes. This is the lowest 3D, steady state kinetic temperature yet observed, although lower or comparable 1D [29,30], 2D [31] and transient [32] kinetic temperatures have been reported for laser cooled atoms, and much lower spin temperatures have been observed in solids.

EVAPORATIVE COOLING

One of the outstanding challenges of low temperature and atomic physics is to achieve Bose-Einstein condensation (BEC) of a dilute, weakly interacting gas [33]. The expectation is that when a sample of weakly interacting bosonic atoms is cold and dense enough that the thermal deBroglie wavelength is comparable to the average interatomic spacing, a macroscopic fraction of the sample will condense into the lowest available momentum state. BEC in spin-polarized atomic hydrogen has been pursued for a number of years, and recently there has been considerable interest in pursuing BEC of alkali atoms, assisted by laser cooling and electromagnetic trapping techniques.

One of the most important techniques used by the spin polarized hydrogen researchers is evaporative cooling [34]. Hydrogen atoms held in a magneto-static trap may occasionally escape when individual atoms have energy higher than the trap depth. Since only those atoms with higher than average energy escape, the average energy of the remaining atoms is reduced, these atoms re-thermalize, producing more atoms in the high energy tail of the distribution, and the evaporation effectively cools the sample. In this way the MIT spin polarized hydrogen group has achieved conditions close to what is required for BEC [35].

Although laser cooling of alkali and other atoms can achieve very low temperatures, collisions involving excited atoms limit the density of laser cooled samples. It seems likely that achieving BEC in such laser cooled samples will require the final stages of cooling and compression to be accomplished with ground state atoms. Thus, evaporative cooling is an attractive technique even for atoms than can be effectively laser cooled. Recent experiments [36,37] have demonstrated such evaporative cooling for alkali atoms that were originally laser cooled and then held in a static magnetic trap. At MIT sodium atoms [36] and at JILA rubidium atoms [37] were laser trapped, laser cooled, and then transferred to a spherical quadrupole magnetic trap [38]. A radio frequency field coupled atoms in a trapped spin state to an untrapped state at a specific value of magnetic field and therefore at a specific locus of positions in the trap. This allowed the effective depth of the trap to be adjusted so as to control the rate of evaporation. As of the date of the conference these groups had achieved cooling and compression of the atomic samples corresponding to an increase of phase space density of about a factor of 4 or 5. Just after the ICAP, a group at Stanford [39] reported achieving evaporative cooling of sodium atoms trapped in a Far-Off-Resonant laser dipole Trap (FORT) [40,41].

CYCLIC COOLING

Sisyphus laser cooling [16] extracts kinetic energy from atoms by converting the kinetic energy to light-shift potential energy and then dissipating the potential energy through optical pumping among the differently light-shifted levels. It is possible to accomplish the same thing in a magnetic or other macroscopic trap by converting kinetic energy to trap potential energy and dissipating that energy by optical pumping among the differently Zeeman-shifted levels. This process was proposed [42] even before the discovery of Sisyphus laser cooling. The original proposal involves starting with an energetic atom in a deeply bound potential, allowing it to move to its turning point where it is coherently transferred by radio frequency coupling to a state that is only weakly bound. The atom then accelerates slowly to the bottom of the weak potential where it is optically pumped to the more deeply bound state, having completed a cycle and lost considerable kinetic energy.

In a recent experiment at JILA [43] a slightly modified version of this scheme was demonstrated using magnetically trapped Rb atoms. Instead of two potential curves with very different depths, the JILA experiment uses two similar potentials whose equilibrium points were displaced from each other. The displacement results from the fact that the magnetic trapping alone is insufficient to support the Rb atoms against gravity, so an additional, linear gradient magnetic field is applied to levitate the atoms. Because the two relevant states have slightly different magnetic moments, the equilibrium points where gravitational and magnetic forces balance are different [42b]. This leads to two displaced potential wells with nearly the same curvature. Cooling occurs when an energetic atom oscillates to the turning point of its potential, is transferred to the other, displaced potential and oscillates to the opposite turning point of that potential, which is near the bottom of the first potential. Optical pumping returns the atom to a point near the bottom of the first potential, completing the cycle and removing kinetic energy. In this way, Rb atoms were cooled to a temperature of 1.5 μK [43], considerably lower than the 6 μK reported for laser cooling of Rb in optical molasses [44].

OPTICAL SHIELDING OF COLLISIONS

Inelastic collisions between laser cooled, trapped atoms are often a major factor in limiting the density that can be achieved for a sample of trapped atoms. Collisions where one of the collision partners is optically excited are generally the most destructive. Atoms approaching each other on an attractive potential accelerate and may decay to the relatively flat ground state potential after having gained enough kinetic energy to leave the trap. Alternately, the atoms may undergo a non-adiabatic curve crossing resulting in an exothermic change of state, acquiring enough energy to leave the trap. Collisions between ground state atoms can also be destructive. Exothermic changes in a ground hyperfine state can eject atoms from a trap. In the case of magnetically trapped atoms, spin-flip collisions can put the atom into an untrapped state. In the case of trapped metastable rare gas atoms, collisions between atoms in the lowest lying metastable state (the effective ground state) can lead to destructive Penning ionization. Avoiding destructive collisions can allow higher density to be achieved. One strategy for doing this is to reduce or avoid optical excitation of the trapped atoms as in a dark MOT [45], a FORT [40,41], a magnetic trap [38] or a microwave trap [46]. Another alternative is to protect the atoms from collisions with other atoms by application of an appropriate laser field.

The basic idea of such optical shielding of atomic collisions is to tune a laser field to be resonant with a transition between the relatively flat ground molecular potential and a repulsive, excited molecular potential. The resonance should occur at a large interatomic separation; ideally, the separation should be larger than the separation at which any other laser fields are resonant for coupling to any attractive potentials. Such ideas have been discussed for some time [47,48], and recent experiments [49] have demonstrated reduction of collisional loss from ground-excited state collisions in a magneto-optical trap (MOT).

In experiments at NIST in Gaithersburg [50] we have used optical shielding to reduce Penning ionization collisions between trapped metastable xenon atoms. This optical shielding was effective both for collisions between Xe atoms in the lowest metastable state, $6s[3/2]_2$ ($^3P_2$), and for collisions involving excitation on the laser cooling transition to the $6p[5/2]_3$ state. Shielding was provided by a strong laser tuned as much as a few hundred MHz to the blue of the laser cooling transition. When all the atoms were in the lowest metastable state (laser cooling and trapping light turned off) the reduction of Penning ionization was more than a factor of five. With the cooling and trapping light on, the reduction of the (larger) ionization rate was as much as a factor of 30. The reduction of Penning ionization collisions produces a significant increase in the collision-limited lifetime of atoms in a dense trap, as well as a significant increase in the achievable density of trapped atoms. Similar shielding of Penning ionization collisions has been reported for metastable Kr atoms in experiments in Tokyo [51].

ACKNOWLEDGMENTS

I thank my colleagues in the Laser Cooled and Trapped Atoms Group at NIST-Gaithersburg for their substantial help in preparing this presentation. I also thank G. Grynberg, T. Hänsch, W. Ketterle, E. Cornell, S. Chu and P. Julienne for providing me with the latest information about their work. Work at NIST-Gaithersburg is supported in part by the U. S. Office of Naval Research and by the National Science Foundation.

REFERENCES

1. Appl. Phys. B**54**, 319-485 (1992), special issue on optics and interferometry with atoms. edited by J. Mlynek, V. Balykin, and P. Meystre.
2. J. Lawall and M. Prentiss, Phys. Rev. Lett. **72**, 993-996 (1994).
3. L. Goldner, C. Gerz, R. Spreeuw, S. Rolston, C. Westbrook, W. Phillips, P. Marte, and P. Zoller, Phys. Rev. Lett. **72**, 997-1000 (1994).
4. G. Alzetta, A. Gozzini, L. Moi, and G. Orriols, Nuovo Cimento **36B**, 5 (1976); E. Arimondo and G. Orriols, Lett. Nuovo Cimento **17**, 333 (1976); H. Gray, R. Whitley, and C. Stroud, Opt. Lett. **3**, 218 (1978).
5. A. Aspect, E. Arimondo, R. Kaiser, N. Vansteenkiste, and C. Cohen-Tannoudji, J. Opt. Soc. Am. B **6**, 2112-2124 (1989), and Phys. Rev. Lett. **61**, 826-829 (1988).
6. P. Marte, P. Zoller and J. Hall, Phys. Rev. A **44**, R4118-R4121 (1991).
7. U. Gaubatz, P. Rudecki, M. Becker, S. Schiemann, M. Külz, and K. Bergmann, Chem. Phys. Lett. **149**, 463 (1988); J. Kuklinski, U. Gaubatz, F. Hioe, and K. Bergmann, Phys. Rev. A **40**, 6741 (1989); U. Gaubatz, P. Rudecki, S. Schiemann, and K. Bergmann, J. Chem. Phys. **92**, 5363 (1990); B. W. Shore, K. Bergmann, J. Oreg, and S. Rosenwaks, Phys. Rev. A **44**, 7442 (1991); P. Pillet, C. Valentin, R.-L. Yuan, and J. Yu, Phys. Rev. A **48**, 845 (1993).
8. M. Weitz, B. Young, and S. Chu, to be published; S. Chu, "Progress in atom manipulation and polymer experiments with single molecules of DNA," presented at the 14th International Conference on Atomic Physics, Boulder CO 1994.

9. S. Chu, L. Hollberg, J. Bjorkholm, A. Cable, and A. Ashkin, Phys. Rev. Lett. **55**, 48-51 (1985).
10. P. Lett, W. Phillips, S. Rolston, C. Tanner, R. Watts, and C. Westbrook, J. Opt. Soc. Am. B **6**, 2084-2107 (1989).
11. M. Prentiss and S. Ezekiel, Phys. Rev. Lett. **56**, 46-49 (1986).
12. C. Salomon, J. Dalibard, A. Aspect, H. Metcalf, and C. Cohen-Tannoudji, Phys. Rev. Lett. **59**, 1659-1662 (1987).
13. C. Westbrook, R. Watts, C. Tanner, S. Rolston, W. Phillips, P. Lett, and P. Gould, Phys. Rev. Lett. **65**, 33-36 (1990).
14. P. Verkerk, B. Lounis, C. Salomon, and C. Cohen-Tannoudji, Phys. Rev. Lett. **68**, 3861-3864 (1992).
15. P. Jessen, C. Gerz, P. Lett, W. Phillips, S. Rolston, R. Spreeuw, and C. Westbrook, Phys. Rev. Lett. **69**, 49-52 (1992).
16. J. Dalibard and C. Cohen-Tannoudji, J. Opt. Soc. Am. B **6**, 2023-2045 (1989).
17. P. Marte, R. Dum, R. Taïeb, P. Lett and P. Zoller, Phys. Rev. Lett. **71**, 1335-1338 (1993).
18. R. Gupta, S. Padua, T. Bergeman, and H. Metcalf, "Search for motional quantization of laser-cooled atoms," in *Laser Manipulation of Atoms and Ions* (Proceedings of the International School of Physics "Enrico Fermi", Course CXVIII), E. Arimondo, W. Phillips, and F. Strumia, editors (North Holland, Amsterdam, 1992) pp. 345-360.
19. P. Verkerk, D. Meacher, A. Coates, J.-Y. Courtois, S. Guibal, B. Lounis, C. Salomon and G. Grynberg, Europhys. Lett. **26**, 171-176 (1994).
20. G. Grynberg, B. Lounis, P. Verkerk, J.-Y. Courtois, and C. Salomon, Phys. Rev. Lett. **70**, 2249-2252 (1993).
21. D. Meacher, S. Guibal, C. Mennerat, K. Petas, J-Y. Courtois, and G. Grynberg, "Paramagnetism and spin temperature in a caesium optical lattice," to be published.
22. A. Kastberg, W. Phillips, S. Rolston, R. Spreeuw, and P. Jessen, "Adiabatic cooling of cesium to 700 nK in an optical lattice," to be published.
23. B. Anderson, T. Gustavson and M. Kasevich, presented at the International Conference on Quantum Electronics, Anaheim, CA, 1994.
24. A. Hemmerich and T. Hänsch, Phys. Rev. Lett. **70**, 410-413 (1993); A. Hemmerich, C. Zimmermann, and T. Hänsch, Europhys. Lett. **22**, 89-95 (1993).
25. A. Hemmerich, C. Zimmermann, and T. Hänsch, Phys. Rev. Lett. **72**, 625-628 (1994).
26. P. Lett, R. Watts, C. Westbrook, W. Phillips, P. Gould and H. Metcalf, Phys. Rev. Lett. **61**, 169 (1988).
27. C. Salomon, J. Dalibard, W. Phillips, A. Clairon, and S. Guellati., Europhys. Lett. **12**, 683-688 (1990).
28. J. Chen, J. Story, J. Tollett, and R. Hulet, Phys. Rev. Lett. **69**, 1344-1347 (1992).
29. M. Kasevich and S. Chu, Phys. Rev. Lett. **69**, 1741-1744 (1992); N. Davidson, H.-J. Lee, M. Kasevich, and S. Chu, Phys. Rev. Lett. **72**, 3158-3161 (1994).
30. J. Reichel, O. Morice, G. Tino, and C. Salomon, "Sub-recoil Raman cooling of cesium atoms," to be published.
31. J. Lawall, F. Bardou, B. Saubamea, K. Shimizu, M. Leduc, A. Aspect, and C. Cohen-Tannoudji, "Two-dimension subrecoil laser cooling," to be published; also, C. Cohen-Tannoudji, presentation at the 14th International Conference on Atomic Physics, Boulder CO, 1994.
32. C. Monroe, W. Swann, H. Robinson, and C. Wieman, Phys. Rev. Lett., **65**, 1571-1574 (1990).
33. I. Silvera and J. Walraven, "Spin-polarized atomic hydrogen," in *Progress in Low Temperature Physics*, Volume X, " (Elsevier, Amsterdam, 1986) chap. 3, pp. 139-370; T. Greytak and D. Kleppner, "Lectures on Spin-Polarized Hydrogen," in *New Trends in Atomic*

Physics, Vol. 2, edited by G. Grynberg and R. Stora, (North Holland, Amsterdam 1984) pp. 1127-1230; T. Greytak, "Prospects for Bose-Einstein condensation in magnetically trapped atomic hydrogen," in *Bose-Einstein Condensation*, edited by A. Griffin, D. Snoke, and S. Stringari (Cambridge University Press, 1994)
34. H. Hess, Phys. Rev. B **34**, 3476 (1986).
35. J. Doyle, J. Sandberg, I. Yu, C. Cesar, D. Kleppner, and T. Greytak, Phys. Rev. Lett. **67**, 603-606 (1991).
36. K. Davis, M.-O. Mewes, M. Joffe and W. Ketterle "Evaporative cooling of sodium atoms," poster presentation at the 14th International Conference on Atomic Physics, Boulder CO, 1994, and private communications.
37. W. Petrich, M. Anderson, J. Ensher and E. Cornell, "Evaporative cooling of rubidium in a magnetic trap," poster presentation at the 14th International Conference on Atomic Physics, Boulder CO, 1994, and private communications.
38. A. Migdall, J. Prodan, W. Phillips, T. Bergeman, and H. Metcalf, Phys. Rev. Lett. **54**, 2596-2599 (1985).
39. S. Chu, private communication, August 1994.
40. W. Phillips, "Laser cooling and trapping of neutral atoms," in *Laser Manipulation of Atoms and Ions* (Proceedings of the International School of Physics "Enrico Fermi," Course CXVIII) E. Arimondo, W. Phillips and F. Strumia, editors (North Holland, Amsterdam, 1992) pp. 289-343
41. J. Miller, R. Cline and D. Heinzen, Phys. Rev. Lett. **71**, 2204-2207 (1993); Phys. Rev. A **47**, R4567-R4570 (1993). .
42. D. Pritchard, Phys. Rev. Lett. **51**, 1336-1339 (1983); D. Pritchard and W. Ketterle, "Atom traps and atom optics," in *Laser Manipulation of Atoms and Ions* (Proceedings of the International School of Physics "Enrico Fermi," Course CXVIII) E. Arimondo, W. Phillips and F. Strumia, editors (North Holland, Amsterdam, 1992) pp. 473-496.
43. N. Newbury, C. Myatt, E. Cornell and C. Wieman, poster presentation at the 14th International Conference on Atomic Physics, Boulder CO, 1994, and private communications.
44. C. Gerz, T. Hodapp, P. Jessen, K. Jones, W. Phillips, C. Westbrook, and K. Mølmer, Europhys. Lett. **21**, 661-666 (1993).
45. W. Ketterle, K. Davis, M. Joffe, A. Martin, and D. Pritchard, Phys. Rev. Lett. **70**, 2253-2256 (1993).
46. R. Spreeuw, C. Gerz, L. Goldner, W. Phillips, S. Rolston, C. Westbrook, M. Reynolds, and I. Silvera, Phys. Rev. Lett. **72**, 3162-3165 (1994).
47. Private communications with: J. Dalibard, K. Mølmer, and P. Julienne; P. Gould; and G. Shlyapnikov.
48. K.-A. Suominen and P. Julienne, "Optical shielding of cold collisions," to be published.
49. L. Marcassa, S. Muniz, E. de Queiroz, S. Zilio, V. Bagnato, J. Weiner, P. Julienne, and K.-A. Suominen, "Optical suppression of photoassociative ionization in a magneto-optical trap," to be published.
50. M. Walhout, U. Sterr, C. Orzel, M. Hoogerland, and S. Rolston, "Optical control of ultracold collisions in metastable xenon," to be published.
51. H. Katori and F. Shimizu, "Laser induced ionization-collision rate of ultra-cold krypton gas in the $1s_5$ metastable state," poster presentation at the 14th International Conference on Atomic Physics, Boulder CO, 1994.

Spectroscopy and Quantum Optics with Stored Ions

Rainer Blatt

*Drittes Physikalisches Institut, Universität Göttingen,
Bürgerstraße 42 – 44, D-37073 Göttingen, Germany*

Abstract: New developments in spectroscopy and quantum optics with stored ions are summarized and discussed. Currently used ion trap devices are briefly described, and newly proposed cooling techniques are indicated. Recent results in precision spectroscopy are reviewed with particular emphasis on time and frequency standards. New applications of stored ions in the field of quantum optics are proposed such as the preparation of nonclassical states of motion in an ion trap. These can be detected by observation of the quantum collapse and revival phenomenon.

1. INTRODUCTION

Ion storage in combination with laser cooling has proved to be a very valuable tool for precision spectroscopy and quantum optics for many years now [1–4]. Trapping and cooling of ions are currently routinely used in a number of laboratories both for investigating the fundamental aspects of the interaction between matter and radiation as well as for applications such as mass spectrometry and for time and frequency standards.

In particular, during the last few years precision spectroscopy on clouds of trapped ions has been improved in such a way that frequency standards based on this technique are currently available which show long-term stability superior to current primary Cs standards [4, 5]. More recently, first precision measurements on single laser-cooled ions have demonstrated that future clocks may well be realized with such experiments [6–16].

Aside from these applications, spectroscopy on single stored and laser-cooled ions is utilized for a variety of fundamental experiments in the field of quantum optics [3]. For example, several years ago quantum jumps in the fluorescence light of single ions were observed for the first time [17–19]. Currently such measurements are used for a detailed investigation of the quantum mechanical measurement process, and they form the basis of measuring very narrow clock transitions. Also, the nonclassical antibunching property of the resonance fluorescence radiation has been observed with single trapped ions [20, 21], and currently several experiments are under way to further investigate nonclassical characteristics of resonance fluorescence [22, 23].

In this contribution several new developments in stored ion research are summarized and discussed with an emphasis on experiments using Paul and rf traps. Recent results with Penning traps, especially in the field of precision mass spectrometry, are summarized in the papers by D. Pritchard and J. Kluge in these proceedings.

2. STORAGE DEVICES

Ion traps, especially Paul and Penning traps have long been used for precision spectroscopy, for the investigation of nonneutral plasmas, and for time and frequency standard applications. Whereas Penning traps are most often applied to plasma research and precision mass spectrometry, Paul traps are mostly used for precision laser spectroscopy and in quantum optics.

One of the decisive drawbacks of conventional quadrupole Paul traps is that clouds of more than a few tens of ions cannot be efficiently optically cooled [4]. Due to Coulomb repulsion in an ion cloud there are always many ions away from the trap center undergoing a strong micromotion at the frequency of the trapping field. Thus the kinetic energy of an ion cloud in a Paul trap cannot be reduced below a value determined by the trap parameters and the extension of the ion cloud. In order to overcome this limitation, various rf-traps have been designed which allow the trapping of many ions with little or no micromotion. The simplest approach to such a device is a trap variant of the Paul mass filter [24–26]. Here the trap consists of a linear quadrupole potential created by four rods with additional endcaps for axial confinement (linear rf trap). Thus all ions located along the trap axis undergo very little or no micromotion, and the mean kinetic energy for ion clouds in these traps is generally much lower. Laser cooling of ions then can result in rather cold samples of ions arranged like a string of pearls along the trap axis. This concept of minimizing the micromotion has also been pursued with the use of an rf-ring trap where the electrodes of a linear quadrupole are bent so that a closed quadrupole ring is formed ("race-track" trap) [27, 28]. In this way all ions located on the central ring axis are free of micromotion.

A different concept for minimizing the micromotion was pursued with the development of rf-traps with a potential of higher multipolarity, as, e.g., the rf-octupole trap [29]. Here the effective trap potential and thus the amplitude of the micromotion increase with the third power of the ions' distance from the trap center, and therefore all those ions near the center of such a trap undergo a much reduced micromotion. Consequently, reduced kinetic energies have been observed for ion clouds in an rf-octupole trap, and laser cooling of such clouds seems possible [29].

For experiments with single laser-cooled ions conventional Paul traps with sizes on the order of mm are routinely employed [30–32, 21, 17, 33, 35, 34]. For better laser cooling (see below) and for applications in quantum optics, miniature traps (with sizes of ≈ 0.01 to 0.1 mm) would be desirable. This is difficult to achieve with the Paul trap consisting of a ring electrode and two endcaps. With laser beams applied in the gap between endcap and ring, stray scattered light would seriously hamper the experiments. Recent approaches to miniaturizing the Paul trap therefore consist in using electrodes of simple planar geometry which, e.g., permit precise photolithographic fabrication of a microtrap [35]. The simplest example of such a trap is a one-hole trap, i.e., a hole in a metallic sheet to which the trapping potential is applied. This is a Paul trap where the endcaps have been moved to infinity. This trap was originally demonstrated by H. Straubel [36] and is known as the Paul-Straubel trap [37, 38]. Similarly the trapping potential can be applied to two endcaps with the ring-electrode moved to infinity resulting in the endcap trap (or inverted Paul-Straubel trap) [39]. Such storage devices allow an unobstructed access to the trap center, which is imperative for laser cooling of single ions and its efficient detection via resonance fluorescence. In first experiments, it was found that the voltages for an endcap trap are about 2 times higher than for a comparable Paul trap whereas for the Paul-Straubel trap 8 times higher voltages are required to generate a similar trap potential [39].

Miniature Paul traps consisting of three rings or three holes have also been devised, and it is currently being investigated in how far whole arrays of such traps can be produced by photolithograpy [35]. Such arrays were first considered by Major [40] and could be used to provide many single trapped and laser-cooled ions simultaneously similar to a string of stored ions in a linear trap. However, with such an array all ions would be independent whereas in a linear trap there is always a coupling due to the mutual Coulomb repulsion.

3. COOLING TECHNIQUES

Laser cooling of single atomic particles has been of increasing interest for more than a decade [1–4]. Proposed by Hänsch and Dehmelt [41] and by Wineland and Dehmelt in 1975 [42], laser cooling was first observed by Neuhauser, Hohenstatt, Toschek, and Dehmelt [30] and by Wineland, Drullinger, and Walls in 1978 [43]. This started an investigation of the fundamental interaction between electromagnetic radiation and matter with particular emphasis on the coupling between internal (atomic) and external (motional) degrees of freedom. This interest is due to the fact that a single stored and cooled ion provides a quantum system close to theoretical models in quantum optics. Many aspects of laser cooling of a single trapped ion are well understood [30, 43–47]. For two-level systems, as long as the trap frequency ν is smaller than the natural linewidth Γ of the optical transition for laser cooling (i.e., the weak-binding limit), the final temperature is limited by $T \approx \hbar\Gamma/2k_B$ where k_B is Boltzmann's constant (Doppler limit). On the other hand, for trap frequencies $\nu > \Gamma$ (i.e., the strong binding limit), the trapped particle develops well-resolved absorption sidebands at the trap frequency and can be optically pumped to its lowest vibrational state by exciting selectively on the lower sideband (sideband cooling). Experimentally, the limiting temperature in this case (given by the probability of not being in the lowest state) can be very low. Both the Doppler limit and the sideband limit have been observed in experiments and agree well with the theoretical predictions [32].

During the past few years significant progress has been made in our understanding of laser cooling of multilevel systems, particularly as applied to free atoms [48]: aside from the usual cooling by scattering forces, many cooling schemes have been investigated and proved successful, in particular those concerned with dipole forces in a standing wave [49]. Extremely low temperatures have been achieved with polarization gradient cooling and dark state cooling [50]. These results have led to an investigation of similar cooling techniques for stored ions.

Recently, the cooling mechanisms of a trapped ion in a standing wave were investigated [52, 51]. Since individual stored ions can be spatially localized to dimensions smaller than an optical wavelength (Lamb-Dicke limit), cooling can be studied for different positions of the ion within the standing wave. Thus it was found [51] that for a two-level system located at the node of the standing wave, the final temperature is a factor of 2 lower than the limit for a traveling wave, and more important, the cooling rates do not saturate with the laser intensity. At the point of maximum (intensity) gradient of the standing wave, cooling is obtained by the Sisyphus mechanism [49] for positive detunings with respect to the resonance. With this technique very low temperatures (corresponding to very low mean excitation numbers $n \approx 1$ of the oscillatory ion motion) may be obtained when applied with three-level and multilevel configurations. This can be achieved even for the weak-binding case where otherwise only the Doppler limit (with $n \ll 1$) can be reached [52, 51].

Aside from using intensity gradients, as with an ion placed at the point of maximum gradient in a standing wave, an ion can similarly be subjected to a laser beam configuration leading to polarization gradients [53, 54]. As the results show [53], laser cooling of a stored ion using polarization gradients leads to final quantum numbers $n \approx 1$, i.e., to final energies $E = \hbar\nu(\langle n \rangle + 1/2) \ll \hbar\Gamma/2$, and as it turns out, the cooling results are quite insensitive to either the precise localization of the ion or the laser detuning. Thus, for weak confinement, laser cooling with polarization gradients seems superior to other cooling techniques in traps, such as Doppler cooling, cooling at the node of a standing wave, or even Sisyphus cooling with pure intensity gradients.

First experimental results for laser cooling of trapped ions with polarization gradients were recently obtained by Birkl et al. [55]. In their experiment a single Mg$^+$ ion was trapped in a one-dimensional trap along the axis of a quadrupole-ring trap and was cooled to 0.4 mK, which is 2.5 times lower than the Doppler limit with hints of possibly lower temperatures. Since in this experiment the ion was not within the Lamb-Dicke regime (i.e., its oscillation amplitude was larger than a wavelength of the cooling laser), the results should be compared with the theory for free atom cooling and qualitative agreement was obtained.

The ultimate goal of laser cooling is to prepare an ion in the ground state of the trapping potential. This can be achieved by sideband cooling which requires the condition $\nu > \Gamma$. For typical dipole allowed transitions the spontaneous width Γ is usually of the order of a few tens of MHz, while typical trap frequencies ν are of the order of a few MHz. Thus the sideband cooling conditions are usually not satisfied. In order to reach the sideband regime two strategies are currently being pursued: (i) decreasing the "effective" spontaneous decay rate of the atom or (ii) increasing the trap frequency. The first demonstration of sideband cooling was accomplished by the NIST Boulder group [32] using a weak quadrupole transition in Hg$^+$. Since the excited state in this experiment has a lifetime of about 100 ms, the achievable cooling rate is rather small. Therefore, after absorption on the sideband an additional laser was applied and tuned to resonance with a strongly allowed transition for fast optical pumping of the metastable state population back to the ground state [32].

As an alternative way to reach the sideband cooling conditions, a reduction of the effective spontaneous transition rate of the two-level system can be obtained, e.g., with Raman transitions coupling two atomic states (Raman-cooling) [47] with a sufficiently long lifetime. This can be realized, e.g., with the S-P-D transitions in three-level ions such as Ba$^+$, Sr$^+$, Ca$^+$, and Yb$^+$ or with atomic ground states exploiting Zeeman structure or hyperfine splitting. In this configuration a three-level Λ system driven by two counterpropagating laser beams is considered. Tuning the two-photon Raman transition to the lower motional sideband, one obtains a two-level atom involving the two atomic ground states where the optical pumping rate Γ' plays the role of an effective spontaneous decay rate. For sufficiently low laser power one obtains $\Gamma' \ll \Gamma$, i.e., the sideband cooling regime $\Gamma' < \nu$ can be reached. Increasing the trap frequency can be accomplished with microtrap design (see section II above) where trap frequencies on the order of a few tens of MHz may be achieved [34].

More recently, a generalization of the approach by Diedrich et al. [32]

was proposed to achieve sideband cooling on a metastable transition with a very long lifetime where the metastable state is quenched (or "dressed") by a nonresonant laser [56]. The purpose of the dressing laser is to mix the metastable state with a third atomic state which has a dipole allowed (fast) decay to the ground state, thus providing an effective spontaneous decay rate Γ' in the two-level system which—by adjusting the intensity and detuning of the strong laser—can be "designed" and optimized for sideband cooling, i.e., $\nu > \Gamma'$. The advantage of this scheme is that it can be applied to a broad range of ion level configurations. In the case of Ba^+, for example, simultaneous interaction on the weak $S_{1/2} - D_{3/2}$ transition and the strongly allowed $S_{1/2} - P_{1/2}$ dipole transition would allow one to reach the sideband limit in a three-level V configuration. It is expected that this proposed cooling scheme will be of utmost importance for precision experiments as, e.g., in frequency and time standard applications. Owing to decay rates that are usually several tens of MHz, the strong confinement regime is experimentally very hard to achieve; hence, in almost all applications so far only the Doppler limit could be reached. With the proposed technique it should be easily possible to reach the sideband limit since in most spectroscopic applications a weak transition is available anyway. Only optical transitions have been considered so far for such a scheme. However, Zeeman and hyperfine states could also serve for weakly coupled transitions, and these levels are available in any ion. In this case, howeever, the Lamb-Dicke parameters $\eta = \hbar k^2/2m\nu$ (here k denotes the k-vector of the applied radiation, m is the mass of the ion and ν is the trap frequency) for these transitions become very small, and correspondingly one has to apply high radiation intensities to achieve the required transition rates. Nevertheless, in the rf and the microwave domain, this could be possible, and the required parameters can be readily derived from expressions given in [56].

The standard theory of laser cooling of trapped ions assumes the motion of a laser-driven ion in a (static) one-dimensional harmonic oscillator trapping potential [44–47]. Almost all experiments with single laser-cooled ions, on the other hand, have been performed with Paul traps where the trapping potential is explicitly time dependent. In general, the ion motion in a Paul trap is governed by a fast oscillation at the driving frequency Ω (micromotion), superimposed on a slow secular motion (macromotion) at frequency ν [1]. To the extent that the frequency of the macromotion is much smaller than that of the micromotion, adiabatic elimination of the fast time scales $1/\Omega$ allows one to describe the ion dynamics as motion in an effective harmonic oscillator potential (pseudopotential). Thus standard laser cooling theory is based on the assumption that the time scale of the rf field is much faster than all other time scales of the problem. In experiments, the effects of the micromotion are clearly visible as additional resonances in excitation spectra [57], and it appears necessary to investigate in which way the time-dependent trapping field influences the cooling dynamics, the cooling rates and the final temperatures.

When discussing the influence of micromotion, the notion of temperature has to be reconsidered. In a harmonic trap, laser cooling theories predict a Boltzmann distribution of the occupation of the harmonic oscillator states, which allows the assignment of a temperature. However, with the micromotion present, the Hamiltonian that describes the motion of the ion in the trap is time dependent, and therefore the concept of time-independent eigenstates of the trap Hamiltonian fails [58–61]. Nevertheless, it is always possible to define the kinetic energy via the expectation value of the squared momentum $\langle P(t)^2 \rangle$ which is explicitly time dependent. For a comparison with experimental results the mean kinetic energy is defined as a time averaged value $\overline{\langle P^2 \rangle}/(2m)$ where the time average is taken over one period $2\pi/\Omega$ of the micromotion [62]. This mean kinetic energy can be compared to the kinetic energy (as given by the temperature) obtained for the harmonic pseudopotential approximation, and thus one can study the influence of the micromotion on the dynamics of laser cooling of a single ion in Paul traps.

It turns out that the treatment of laser cooling with harmonic traps describes the cooling dynamics sufficiently well as long as the laser-atom detuning Δ is small compared to the micromotion frequency Ω, more precisely, for $|\Delta| < \Omega - \nu - \Gamma$ [63]. When this condition is not fulfilled, cooling may arise for additional detunings, and heating may appear where cooling was expected.

4. PRECISION SPECTROSCOPY

Storing ions in an ultrahigh vacuum provides samples which are ideally suited for precision spectroscopy. The absence of collisions with walls or a buffer gas and the virtually infinite interaction times make these samples a good choice for applications with time and frequency standards. The potential instability of clocks is determined by the Q value of a clock transition, i.e., $Q = \omega/\Delta\omega$ (where $\Delta\omega$ denotes the linewidth and ω is the center frequency) and the signal-to-noise ratio with which the center frequency can be determined. Since it is currently still very hard to lock optical frequencies by means of a frequency chain to the Cs standard, there is continuing interest in the search for improved frequency standards in the radio frequency domain.

For this reason, rf-optical double resonance spectroscopy of hyperfine splittings in ion clouds is performed in many laboratories [65–68, 76, 5, 70]. Several candidate ions have been investigated in the past, all of them with similar level schemes as a neutral atom. Most promising results have been obtained with measurements of the ground-state hyperfine splitting in ^{199}Hg$^+$ and ^{171}Yb$^+$ ions. As reported by Maleki et al., the most advanced ion trap standard today with a cloud of trapped ^{199}Hg$^+$ ions in a linear quadrupole trap yields a stability figure of $\sigma_y(\tau) = 7 \cdot 10^{-14} \tau^{-1/2}$ with a long term stability of about $5 \cdot 10^{-16}$ for averaging times longer than 10^5s [5]. The latter has been measured by comparing two trapped ion standards, and it is expected that a long-term stability of at least $1 \cdot 10^{-16}$ can be reached with the current setup. The accuracy, however, is limited essentially by the correction for the 2nd order Doppler shift and is expected to achieve 10^{-14} with fieldable, room temperature and lamp based systems.

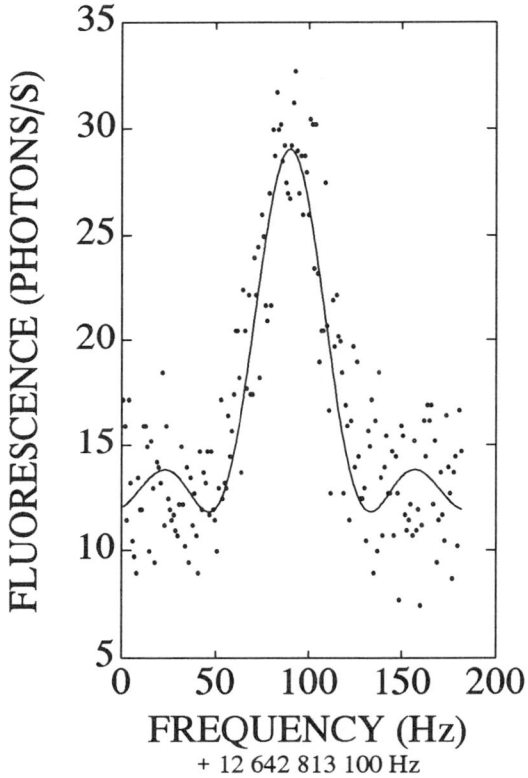

Figure 1: Ground-state hyperfine resonance in a single laser-cooled ^{171}Yb$^+$ ion, from Ref. [7].

The search for ions that can used as rf standards and can be optically pumped and prepared by means of efficient solid state and diode laser sources led to an investigation of the ^{171}Yb$^+$ ion [71]. Measurements with ^{171}Yb$^+$ ions are currently performed in several laboratories [65–67], and its ground-state hyperfine splitting has been determined to be $\nu_{hfs}=$ 12 642 812 118.471

(9) Hz [67]. The uncertainty in this measurement by Tamm et al. is almost entirely due to the corrections for the 2nd order Doppler shift and a residual pressure shift. Nevertheless, the measured linewidth of about 15 mHz and the signal-to-noise ratio yield a clock performance almost comparable to primary Cs standards [67]. These results were obtained with a conventional Paul trap; more recently P. Fisk et al. obtained even higher Q values, i.e., $Q = 1.5 \cdot 10^{13}$ at a lower temperature of the ion cloud [66].

All of these results, however, indicate that for ultimate accuracy the ions should be laser cooled, and that requires the use either of a single ion with a conventional Paul trap or very few ions along the axis of a linear Paul trap. Thus, very recently, the first rf-optical double resonance spectra were reported from single laser-cooled ^{199}Hg$^+$ [6] and ^{171}Yb$^+$ [7]. As an example, a measurement of the ground-state hyperfine transition observed in a single laser-cooled ^{171}Yb$^+$ ion is shown in Fig. 1. Since the ion was cooled to the Lamb-Dicke regime, residual systematic shifts such as Doppler effect and Stark effect contribute only at the 10^{-16} level. Therefore, a single stored and laser-cooled ion seems to be a good choice for future frequency standards.

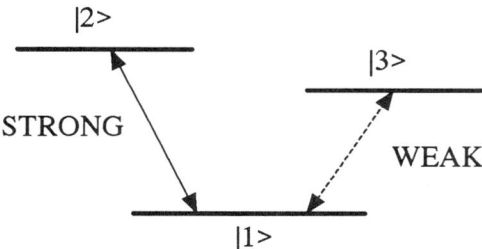

Figure 2: Level scheme for absorption measurements on a weakly coupled transition.

For optical frequency standards, a variety of narrow clock transitions is currently under investigation in several ions. In order to observe the absorption on a weak transition with a single stored ion, usually "V" type three-level ions are considered with a strongly allowed dipole transition for cooling and monitoring and with a narrow clock transition, as shown in Fig. 2.

Absorption on the weak transition causes the monitored fluorescence on the strong transition to cease such that a digital ("on-off") scattered intensity is observed. In this way absorption of a single photon can be detected, and the spectroscopic line shapes are obtained by recording the absorption rate as a function of the detuning on the clock transition. This technique, known as *spectroscopy with quantized fluorescence* and based on Dehmelt's *electron shelving* scheme [72], is now routinely employed in many laboratories. Very promising results have been obtained with the $S_{1/2}-D_{5/2}$ quadrupole transition in a single Hg$^+$ ion at 282 nm where a linewidth of approximately 100 Hz was observed [8] and with the $D_{3/2} - D_{5/2}$ magnetic dipole transition in a single Ba$^+$ ion at 12.5 μm [9]. The latter transition is of interest since the radiation of a ^{15}NH$_3$ laser near 24 THz is directly compared to a Cs standard and thus allows the determination of the absolute frequency of the single ion transition in the far-infrared regime.

Similar forbidden transitions are currently being investigated in Sr$^+$ ions near 674 nm [10], in Yb$^+$ ions near 411 nm [11] and 467 nm [12], in Ca$^+$ ions at 729 nm [13–15], and in In$^+$ ions at 236 nm [16]. It is expected that with further development of solid state and diode lasers the observation of narrow optical clock transitions from single trapped ions can be routinely used for frequency standard applications.

5. ORDERED STRUCTURES

Ordered structures of ions in Paul and Penning traps were first observed a few years ago [74]. When laser cooled to less than 1 K ions in Paul traps were seen to form orderly, rigid arrangements such as rings and hollow shells. In the Penning trap ions form concentric shells [73], however, individual ions are usually not distinguished because of their rapid rotation about the symmetry axis.

Recently, very elongated ion structures have been observed by Birkl et al. in a "race-track" trap [28, 75] and by Raizen et al. [76] in a linear trap. In the ring trap thousands of Mg$^+$ ions were seen to form a single strand or more complex shapes like multiple layers of intertwined helices. In the linear trap similar structures were obtained with Hg$^+$ ions, forming linear strings, zigzags, and helices. The NIST researchers found that the laser light scattered by a two-ion crystal and projected onto a screen formed alternating light and dark bands [77]. The bands result from the interference of the emitted fluorescence radiation. The distance between the two ions could be deduced from the spacing of the bands. In the future, the arrangements of ions in more complicated crystals will be deduced from the patterns of Bragg-scattered light, just as the arrangements of atoms in ordinary crystals are deduced by the patterns of scattered x-rays or neutrons. The first observation of Bragg scattering of laser light off a cold ion cloud in a Penning trap was recently reported by Tan et al. [78].

6. QUANTUM OPTICS

A single trapped and laser-cooled ion provides an almost ideal quantum system and is thus a preferred choice for quantum optical experiments. In particular, the resonance fluorescence of a single ion can be observed without

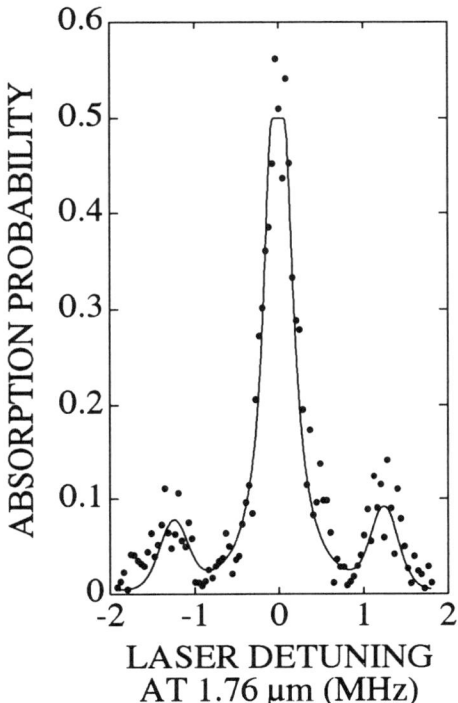

Figure 3: Quantized absorption spectrum of the $S_{1/2}-D_{5/2}$ quadrupole transition in a single laser-cooled Ba$^+$ ion showing motional sidebands, from Ref. [64].

any perturbation by the presence of other ions or the interaction with surrounding walls. Confinement to the Lamb-Dicke regime, moreover, allows one to position an ion at a fixed location in space and with respect to optical fields. These features have been exploited for the observation of the nonclassical antibunching property of the resonance fluorescence from a single ion [20, 21], and they are paramount for the observation of other nonclassical properties, such as squeezing of the resonance fluorescence [22, 23].

During recent years, spectacular advances in laser cooling and trapping have enabled the experimental observation of *nonclassical motion* of trapped ions and atoms [32, 79], in confirmation of theoretical predictions. The quantized motion of the particles is indicated by the presence of asymmetric motional sidebands in probe field absorption spectra or in resonance fluorescence spectra. As an example, Fig. 3 shows the absorption on the $S_{1/2} - D_{3/2}$ transition in a single Ba$^+$ ion which was obtained by Appasamy et al. [64] employing the quantized fluorescence technique as described above. From the asymmetry in the height of the sidebands a mean excitation number of $\langle n \rangle = 5$ could be derived which corresponds to an ion temperature of about 0.33 mK. Such an asymmetry is a clear proof of the quantum nature of the ion motion in a trap. As is well known [46], Doppler cooling as well as sideband cooling leave the ion oscillation in a thermal distribution, however this access to quantized states of motion invited further investigation of the possibility of generating novel and interesting nonclassical states, such as Fock states and squeezed states of the ion motion, in analogy with the quantum theory of light.

The interaction of a harmonic oscillator coupled to a two-level system is generally investigated in quantum optics by studying the Jaynes-Cummings Hamiltonian in the rotating wave approximation ($\hbar = 1$),

$$H = \omega_f a^\dagger a + \frac{1}{2}\omega_0 \sigma_z + g(\sigma_+ a + a^\dagger \sigma_-), \qquad (1)$$

where $a^\dagger$ and a are creation and annihilation operators for the harmonic oscillator oscillating at frequency ω_f, and $\sigma_{\pm,z}$ are the Pauli spin matrices describing a two-level system with transition frequency ω_0. The last term in (1) describes the coupling of the two-level system to the harmonic oscillator with a coupling strength g.

The dynamics of a two-level ion in a standing wave laser field which moves in a harmonic trapping potential is described by the Hamiltonian [80]

$$H = \nu a^\dagger a + \frac{1}{2}\Delta \sigma_z + \frac{\Omega}{2}\eta(\sigma_+ a + a^\dagger \sigma_-). \qquad (2)$$

Here ν is the trap frequency, $\Delta (\approx -\nu)$ is the detuning of the laser frequency from the two-level transition, and Ω is the laser Rabi frequency; $\eta = 2\pi a_0/\lambda$ is the Lamb-Dicke parameter with a_0 the amplitude of the ground state of the trap and λ the optical wavelength.

Therefore this leads to an alternative realization of the Jaynes-Cummings model (JCM) [80–82] with the associations

$$\begin{aligned} \Delta &\leftrightarrow \omega_0, \\ \nu &\leftrightarrow \omega_f, \\ g &\leftrightarrow \eta\Omega/2. \end{aligned} \qquad (3)$$

The conditions for such a realization are that the vibrational amplitude of the ion motion is much less than the wavelength of the light (Lamb-Dicke limit), and the trap frequency ν is much larger than the atomic spontaneous decay rate Γ (strong confinement limit). The main advantages in the trapped ion case, as compared to conventional cavity QED experiments, are (i) the harmonic oscillator damping is negligible and (ii) the coupling is proportional to the laser amplitude, i.e., it can be made very strong simply by increasing the laser power. Such straightforward control of the coupling strength is usually not possible in cavity QED experiments.

An alternative realization of the Jaynes-Cummings–model coupling with trapped ions has been proposed by Wineland et al. [83, 84]. In this work ions having an unpaired outer electron are considered to be trapped along the axis of a linear trap. The ions are subjected to a homogeneous magnetic field which quantizes the spins, and a superimposed inhomogeneous field gradient is applied whose value averages to zero over the ions' orbits. However, as the ions oscillate, they experience a motional oscillating field which tends to flip the spin and reduces or increases the ions' oscillator quantum number according to a Jaynes-Cummings coupling. This form of coupling has already been realized to couple the spin and cyclotron motion of a single electron in the classic g-2 experiments of Dehmelt and his collaborators [85].

Taking advantage of the analogy to the JCM, use of a single trapped ion has been proposed for the preparation of Fock states in a strongly coupled atom-oscillator system [80]. This proposal is based on the observation of quantum jumps from the JCM manifold of dressed energy levels to a third weakly coupled atomic level as is shown in Fig. 2. For a strongly driven $|1\rangle - |2\rangle$ transition the atom develops the well-known JCM energy levels whereas the weakly coupled state $|3\rangle$ shows only the manifold introduced by the harmonic oscillator, i.e., the trap levels. Excitation to any one of these weakly coupled states is observed by the cessation of the fluorescence on the strong transition, i.e., by observation of a quantum jump. Since the detuning of the exciting laser (at the $|1\rangle - |3\rangle$ transition) is known, and the system is projected to one of the excited states only, the quantum number of the ion oscillation is precisely known, i.e., a Fock state of the ion motion is prepared. In this configuration, preparation of a Fock state corresponds to preparation of a nonclassical state of motion of the trapped ion with fixed energy $n\hbar\omega$ whereas in cavity quantum electrodynamics (CQED) the Fock state corresponds to a nonclassical state of light with no intensity fluctuations and undetermined phase. The significance of the use of the trapped ion configuration is that it demonstrates the potential of the well-established field of ion trapping for investigations of features of the JCM which have thus far been studied only in the context of CQED.

An obvious disadvantage of the Fock state created by the aforementioned procedure is the fact that the excited state $|3\rangle$ can spontaneously decay to the ground state. However, the final motional state will closely approximate a Fock state provided η is small, since the probability that the spontaneous decay is accompanied by a change in the motional quantum number n is proportional to η. Another scheme for the preparation of nonclassical states of motion such as Fock states and superpositions of Fock states is based on adiabatic passage along dressed energy levels of the strongly coupled ion-trap system by varying the laser frequency [86]. For this purpose it is assumed that a single two-level ion has been optically pumped by sideband cooling to its lowest vibrational level. Then a slow variation of the laser frequency can be used in such a way that the ion motion is left in a nonclassical state whereas the internal atomic state is the ground state. This adiabatic passage is coherent if one chooses, e.g., an electric dipole–forbidden transition. After the frequency chirp cycle the ion will be left in its internal ground state; therefore the prepared motional state persists owing to the absence of spontaneous emission. It has been shown that such a procedure can be easily extended from one to two or three dimensions and is not restricted to the Lamb-Dicke limit [86]. Finally, this scheme is simpler to realize experimentally and is fairly insensitive to uncertainties in experimental parameters. However, it requires sideband cooling to the lowest vibrational state at the outset.

In addition to Fock states, squeezed states of ion motion can be produced as follows [81]: Consider an ion to be located at a common node of two (different frequency) standing-wave laser fields. When the beat frequency between the light fields is equal to twice the trap frequency, the steady state of the system is a pure state described by a product of the ground internal state of the ion and a squeezed state of the quantized motion. Hence the generation of a squeezed state is indicated by a cessation of the fluorescence emitted by the ion, in this case by a so-called "dark-state." Related dark states, created by coherent population trapping in nonabsorbing atomic states, have already been studied in detail [87]. Alternative schemes for the preparation of oscillator squeezed states in an ion trap have also been proposed involving either a nonadiabatic change in, or a parametric driving of, the voltage between the electrodes of the ion trap [88].

Similar to squeezed states, a coherent state of the trap motion may be produced. This is achieved with a single trapped ion at the node of a standing-wave field with a frequency detuned to the lower sideband and an additional traveling wave on resonance. Also, starting after sideband cooling with an ion in state $n = 0$, a coherent state could be produced by suddenly shifting the center position of the ion's well or by driving the oscillator with a classical resonant excitation [88, 84].

Experimentally the various oscillator states should be measured by the observation of quantum collapse and revival as in CQED [89, 90]. With the alternative realization of the JCM such an experiment could be performed as follows: Consider a single trapped ion with an internal level scheme as indicated in Fig. 2 trapped at the node of a standing-wave laser field exciting the two-level transition $|1\rangle - |3\rangle$. This is a weakly coupled transition (usually a forbidden transition), for which the spontaneous decay time is long compared to the observation time τ and the trap period ν^{-1}. It is assumed that this transition is driven by a very strong standing-wave field in such a way that the effective coupling constant $\eta\Omega/2$ leads to several Rabi oscillations during the interaction time, but it still satisfies the conditions $\nu, |\Delta| \gg \eta\Omega, \Gamma, |\Delta - \nu|$, and $\eta \ll 1$. The third level $|2\rangle$ is used for optical cooling of the ion (initial preparation) and for a measurement of the population inversion (on the $|1\rangle - |3\rangle$ transition). Measuring the population inversion as a function of the interaction time on the $|1\rangle - |3\rangle$ then reveals the collapse and revival phenomenon whose shape provides a signature of the statistical distribution of the initially prepared oscillator state [90].

Finally, it should be noted that for an experiment with a single trapped ion the localization at the node of the standing wave is only required to suppress unwanted transitions $|1, n\rangle \rightarrow |2, n\rangle$ (n denotes the quantum number of the ions oscillatory motion) which may appear at the exact resonance frequency. Placing an ion at the node of the standing wave field allows interaction only with the motional sidebands corresponding to transitions $|g, n\rangle \rightarrow |e, n \pm 1\rangle$. For a highly forbidden transition such as, for example, that used in the quantum jump experiments [17–19], a traveling wave detuned to the lower sideband (i.e., $\omega_L = \omega_0 - \nu$) would be sufficient for the observation of these phenomena. Thus, an experimental realization of the proposed technique seems to be readily feasible with single trapped Ba^+, Ca^+, Sr^+, Hg^+, In^+, and Yb^+ ions.

7. SUMMARY

Ion storage has become a routine technique in many laboratories. It is especially suited for applications in precision spectroscopy and for frequency and time standards. Recent results have been summarized and discussed. New laser cooling techniques have been proposed and reviewed, and applications of a single laser-cooled trapped ion for experiments in quantum optics have been suggested. In particular, it was shown that a single ion trapped at the node of a standing-wave laser field represents an alternative object for a study of the Jaynes-Cummings model and allows for the preparation of nonclassical states of the quantized ion motion in a trap. With this it is perceived that the well-established field of ion trapping provides an alternative testing ground for strongly coupled cavity QED.

References

[1] D.J. Wineland, W. M. Itano, and R.S. VanDyck, Jr., Adv. At. Mol. Phys. **19**, 135 (1983); R. C. Thompson, *ibid.* **31**, 63 (1993)

[2] P. E. Toschek, in *New Trends in Atomic Physics Vol.I*, Proceedings of the Les Houches Summer School, Session XVIII, edited by G. Grynberg and R. Stora (North-Holland, Amsterdam 1984), p. 381

[3] R. Blatt, in *Fundamental Systems in quantum Optics*, Proceedings of the Les Houches Summer School, Session LIII, edited by J. Dalibard, J. M. Raymond, and J. Zinn-Justin (Elsevier Science Publishers B. V. 1992), p. 253

[4] See, e.g., special issue on *The Physics of Trapped Ions*, edited by R. Blatt, P. Gill, and R.C. Thompson, J. Mod. Optics **39**, 192 (1992)

[5] R. L. Tjoelker, J. D. Prestage, G. J. Dick, and L. Maleki, in *Proceedings of the 1993 IEEE International Frequency Control Symposium*, IEEE Catalog No. 93CH3244-1, p.144 (1993)

[6] W. M. Itano, J. C. Bergquist, J.J. Bollinger, J. M. Gilligan, D. J. Heinzen, F. L. Moore, M. G. Raizen, and D. J. Wineland, Phys. Rev. A **47**, 3554 (1993)

[7] V. Enders, Ph. Courteille, L. S. Ma, W. Neuhauser, R. Blatt, and P. E. Toschek, Europhys. Lett. **24**, 325 (1993)

[8] J. Bergquist, F. Diedrich, W. M. Itano, and D. J. Wineland, in *Laser Spectroscopy IX*, edited by M. S. Feld, J. E. Thomas, and A. Mooradian (Academic, San Diego 1989), p. 274

[9] A. A. Madej, K. J. Siemsen, J. D. Sankey, R. F. Clark, and J. Vanier, IEEE Trans. Instr. Meas. **42**, 234 (1993); A. A. Madej, private communication 1994

[10] G. P. Barwood, C. S. Edwards, P. Gill, H. A. Klein, and W. R. C. Rowley, Opt. Lett. **18**, 732 (1993)

[11] H. A. Klein, A. S. Bell, G. P. Barwood, P. Gill, and W. R. C. Rowley, IEEE Trans. Instr. Meas. **40**, 129 (1992)

[12] H. Lehmitz, J. Hattendorf-Ledwoch, R. Blatt, and H. Harde, Phys. Rev. Lett. **62**, 2108 (1989)

[13] S. Urabe, M. Watanabe, H. Imajo, K. Hayasaka, Opt. Lett. **17**, 1140 (1992)

[14] F. Arbes, M. Benzing, T. Gudjons, F. Kurth, and G. Werth, Z. Phys. D **29**, 159 (1994)

[15] M. Knoop, M. Vedel, and F. Vedel, to be published

[16] E. Peik, G. Hollemann, and H. Walther, Phys. Rev. A **49**, 402 (1994)

[17] W. Nagourney, J. Sandberg, and H. Dehmelt, Phys. Rev. Lett. **56**, 2797 (1986)

[18] Th. Sauter, W. Neuhauser, R. Blatt, and P. E. Toschek, Phys. Rev. Lett. **57**, 1696 (1986)

[19] J. C. Bergquist et al., Phys. Rev. Lett. **57**, 1699 (1986)

[20] F. Diedrich and H. Walther, Phys. Rev. Lett. **58**, 203 (1987)

[21] M. Schubert, I. Siemers, R. Blatt, W.. Neuhauser, and P. E. Toschek Phys. Rev. Lett. **68**, 3016 (1992)

[22] W. Vogel and R. Blatt, Phys. Rev. A **45**, 3319 (1992)

[23] R. Blatt, M. Schubert, I. Siemers, W. Neuhauser, and W. Vogel, in *Fundamentals of Quantum Optics III*, edited by F. Ehlotzky (Springer-Verlag, Berlin 1993), p. 156

[24] H. G. Dehmelt, in *Frequency Standards and Metrology*, edited by A. DeMarchi (Springer-Verlag Berlin 1989), p. 286

[25] J. D. Prestage, G. J. Dick, and L. Maleki, J. Appl. Phys. **66**, 1013 (1989)

[26] W. Paul, O. Osberghaus, und E. Fischer, Forschungsberichte des des Wirtschaftsministeriums Nordrhein-Westfalen, Nr. 415, Köln und Opladen 1958

[27] J. Drees and W. Paul, Z. Phys. **180**, 340 (1964); D. A. Church, J. Appl. Phys. **40**, 3127 (1969); B. I. Deutch et al., Phys. Scr. **T22**, 248 (1988)

[28] I. Waki, S. Kassner, G. Birkl, and H. Walther, Phys. Rev. Lett. **68**, 2007 (1992)

[29] J. Walz, I. Siemers, M. Schubert, W. Neuhauser, R. Blatt, and E. Teloy, to be published in Phys. Rev. A **50** (1994)

[30] W. Neuhauser, M. Hohenstatt, P. Toschek, and H. Dehmelt, Phys. Rev. Lett. **41**, 233 (1978)

[31] W. Neuhauser, M. Hohenstatt, P. Toschek, and H. Dehmelt, Phys. Rev. A **22**, 1137 (1980)

[32] F. Diedrich, J. C. Bergquist, W. M. Itano, and D. J. Wineland, Phys. Rev. Lett. **62**, 483 (1989)

[33] G. Janik, W. Nagourney, and H. Dehmelt, J. Opt. Soc. Am. **B 2**, 1251 (1985)

[34] S. R. Jefferts, C. Monroe, A. S. Barton, and D. J. Wineland, Proc. Conf. Prec. Electromagnetic Meas., Boulder, CO, June 1994; to be published in special issue of IEEE Trans. Instrum. Meas.

[35] R. G. Brewer, R. G. DeVoe, and R. Kallenbach, Phys. Rev. A **46**, R6781 (1992)

[36] H. Straubel, Die Naturwissenschaften **18**, 506 (1955)

[37] N. Yu, H. Dehmelt, and W. Nagourney, Proc. Natl. Acad. Sci. USA **86**, 5672 (1989)

[38] N. Yu, W. Nagourney, and H. Dehmelt, J. Appl. Phys. **69**, 3779 (1990)

[39] C. A. Schrama, E. Peik, W. W. Smith, and H. Walther, Opt. Commun. **101**, 32 (1993)

[40] F. G. Major, J. Physique Lett. **38**, L-221 (1977)

[41] T. Hänsch and A. Schawlow, Opt. Commun. **13**, 68 (1975)

[42] D. Wineland and H. G. Dehmelt, Bull. Am. Phys. Soc. **20**, 637 (1975)

[43] D. J. Wineland, R. E. Drullinger, and F. L. Walls, Phys. Rev. A **40**, 1639 (1978)

[44] D. J. Wineland and W. M. Itano, Phys. Rev. **A 20**, 1521 (1979); W. M. Itano and D. J. Wineland, *ibid.* **A 25**, 35 (1982)

[45] M. Lindberg and S. Stenholm, J. Phys. **B 17**, 3375 (1985)

[46] S. Stenholm, Rev. Mod. Phys. **58**, 699 (1086) and references therein.

[47] M. Lindberg and J. Javanainen, J. Opt. Soc. Am. **B 3**, 1008 (1986)

[48] See articles in *Laser Cooling and Trapping*, edited by S. Chu and C. Wieman, special issue of J. Opt. Soc. Am. B **6** (11) (1989)

[49] J. Dalibard and C. Cohen-Tannoudji, J. Opt. Soc. Am. B **2**, 1707 (1985)

[50] P. D. Lett, R. N. Watts, C. I. Westbrook, W. D. Phillips, P. L. Gould, and H. J. Metcalf, Phys. Rev. Lett. **61**, 1069 (1988); J. Dalibard and C. Cohen-Tannoudji, J. Opt. Soc. Am. B **6**, 2023 (1989); P. J. Ungar, D. S. Weiss, E. Riis, and S. Chu, *ibid.* **6**, 2058 (1989)

[51] J. I. Cirac, R. Blatt, P. Zoller, and W. D. Phillips, Phys. Rev. A **46**, 2668 (1992).

[52] D. J. Wineland, J. Dalibard, and C. Cohen-Tannoudji, J. Opt. Soc. Am. B **9**, 32 (1992)

[53] J. I. Cirac, R. Blatt, A. S. Parkins, and P. Zoller, Phys. Rev. A **48**, 1434 (1993)

[54] Sung Mi Yoo and J. Javanainen, Phys. Rev. A **48**, R30 (1993); J. Javanainen and Sung Mi Yoo, Phys. Rev. A **48**, 3776 (1993)

[55] G. Birkl, J. A. Yeazell, R. Rückerl, and H. Walther, to be published

[56] I. Marzoli, J. I. Cirac, R. Blatt, and P. Zoller, Phys. Rev. A **49**, 2771 (1994)

[57] Th. Sauter, R. Blatt, W. Neuhauser, and P.E. Toschek, Physica Scripta **T 22**, 128 (1988)

[58] R. J. Cook, D. G. Shankland, and A. L. Wells, Phys. Rev. A **31**, 564 (1985)

[59] M. Combescure, Ann. Inst. Henri Poincaré **44**, 293 (1986)

[60] L. S. Brown, Phys. Rev. Lett. **66**, 527 (1991)

[61] R. Glauber, in *Proceedings of the International School of Physics "Enrico Fermi,"* Course CXVIII, edited by E. Arimondo, W. D. Phillips, and F. Strumia (1992), p. 643

[62] R. Blatt, P. Zoller, G. Holzmüller, and I.Siemers, Z. Phys. **D 4**, 121 (1986)

[63] J. I. Cirac, L. J. Garay, R. Blatt, A. S. Parkins, and P. Zoller, Phys. Rev. A **49**, 421 (1994)

[64] B. Appasamy, I. Siemers, Y. Stalgies, J. Eschner, R. Blatt, W. Neuhauser, and P. E. Toschek, to be published in Appl. Phys. **B** (1994)

[65] R. Casdorff, V. Enders, R. Blatt, W. Neuhauser, And P. E. Toschek, Ann. Physik **48**, 41 (1991)

[66] P. T. H. Fisk, M. A. Lawn, and C. Coles, in *Proceedings of the 1993 IEEE International Frequency Control Symposium*, IEEE Catalog No. 93CH3244-1, p. 139 (1993); P. T. H. Fisk, M. A. Lawn, and C. Coles, Appl. Phys. **B 57**, 287 (1993)

[67] C. Tamm, D. Schnier, and A. Bauch, to be published in Appl. Phys. B (1994)

[68] J. J. Bollinger, J. D. Prestage, W. M. Itano, and D. J. Wineland, Phys. Rev. Lett. **54**, 1000 (1985)

[69] M. G. Raizen, J. M. Gilligan, J. C. Bergquist, W. M. Itano, and D. J. Wineland, Phys. Rev. A **45**, 6943 (1992)

[70] X. Feng, G. Z. Li, and G. Werth, Phys. Rev. A **46**, 327 (1992); X. Feng, G. Z. Li, and G. Werth, Phys. Rev. A **46**, 2959 (1992); F. Arbes, M. Benzing, Th. Gudjons, F. Kurth, and G. Werth, to be published in Z. Physik D (1994)

[71] R. Blatt, R. Casdorff, V. Enders, W. Neuhauser, and P.E. Toschek, in *Proc. Frequency Standards and Metrology*, edited by A. DeMarchi (Springer-Verlag, Berlin 1989), p. 306

[72] H. G. Dehmelt, Bull. Am. Phys. Soc. **20**, 60 (1975); H. G. Dehmelt, IEEE Trans. Instr. Meas. **31**, 83 (1982)

[73] S. L. Gilbert, J. J. Bollinger, and D. J. Wineland, Phys. Rev. Lett. **60**, 2022 (1988)

[74] F. Diedrich, E. Peik, J. M. Chen, W. Quint, and H. Walther, Phys. Rev. Lett. **59**, 2931 (1987); D. J. Wineland, J. C. Bergquist, W. M. Itano, J. J. Bollinger, and C. H. Manney, Phys. Rev. Lett. **59**, 2935 (1987); B. G. Levi, Phys. Today **41**, 17 (9)(1988)

[75] G. Birkl, S. Kassner, and H. Walther, Europhys. News **23**, 143 (1992)

[76] M. G. Raizen, J. M. Gilligan, J. C. Bergquist, W. M. Itano, and D. J. Wineland, Phys. Rev. A **45**, 6493 (1992)

[77] U. Eichmann, J. C. Bergquist, J. J. Bollinger, J. M. Gilligan, W. M. Itano, D. J. Wineland, and M. G. Raizen, Phys. Rev. Lett **70**, 2359 (1993)

[78] J. Tan, J. J. Bollinger, and D. J. Wineland, private communication (1994)

[79] P. Verkerk et al., Phys. Rev. Lett. **68**, 3861 (1992); P. S. Jessen et al., Phys. Rev. Lett. **69**, 49 (1992); A. Hemmerich and T. W. Hänsch, Phys. Rev. Lett. **70**, 410 (1993); G. Grynberg et al., Phys. Rev. Lett. **70**, 2249 (1993)

[80] J. I. Cirac, R. Blatt, A. S. Parkins, and P. Zoller, Phys. Rev. Lett. **70**, 762 (1993)

[81] J. I. Cirac, A. S. Parkins, R. Blatt, and P. Zoller, Phys. Rev. Lett. **70**, 556 (1993)

[82] J. I. Cirac, A. S. Parkins, R. Blatt, and P. Zoller, Opt. Comm. **97**, 353 (1993)

[83] D. J. Wineland, J. J. Bollinger, W. M. Itano, F. L. Moore, and D. J. Heinzen, Phys. Rev. A **46**, R6797 (1992)

[84] D. J. Wineland, J. J. Bollinger, W. M. Itano, and D. J. Heinzen, Phys. Rev. A **50**, 67 (1994)

[85] H. Dehmelt, Science **247**, 539 (1990)

[86] J. I. Cirac, R. Blatt, and P. Zoller, Phys. Rev. A **49**, R3174 (1994)

[87] G. Alzetta et al., Nuovo Cimento Soc. Ital. Fis. **36B**, 5 (1976); P. M. Radmore and P. L. Knight, J. Phys. B **15**, 561 (1981); I. Siemers, M. Schubert, R. Blatt, W. Neuhauser, and P. E. Toschek, Europhys. Lett. **18**, 139 (1992)

[88] D.J. Heinzen and D.J. Wineland, Phys. Rev. A **42**, 2977 (1990)

[89] C. A. Blockley, D. F. Walls, and H. Risken, Europhys. Lett. **17**, 509 (1992)

[90] J. I. Cirac, R. Blatt, A. S. Parkins, and P. Zoller, Phys. Rev. A **49**, 1202 (1994)

Atom Optics and Interferometry with Laser Cooled Atoms

J.H. Müller, D. Bettermann, V. Rieger, F. Ruschewitz,
K. Sengstock, U. Sterr, M. Christ,
M. Schiffer, A. Scholz, and W. Ertmer

*Institut für Angewandte Physik, Universität Bonn,
Wegelerstr. 8, D-53115 Bonn, Germany*

Abstract: A short review about recent developments of coherent atom optics and atom interferometry is given. The paper mainly concentrates on light fields as optical elements for de Broglie waves. Ramsey type interferometers for measurements of fundamental constants as well as for the study of pure quantum mechanical phase shifts are discussed in more detail. As an example measurements of the scalar Aharonov-Bohm phase shift on laser trapped magnesium atoms are presented. For future atom cavities and large area interferometers evanescent light fields acting as mirrors seem to be best suited. Recent experimental demonstrations of the reflection and diffraction of atoms with evanescent light fields are discussed. The specific advantages in using laser manipulated and cold atoms in the field of atom optics and matter-wave interferometry are described.

INTRODUCTION

The recent realization of atom interferometers has revived many ideas for possible experiments with matter-wave interferometers and some experiments have already been realized (1-9). Matter-wave interferometers offer the unique possibility to measure phase shifts of the wave function. Due to this access to the purely quantum mechanical behaviour of "particles", fundamental tests of quantum mechanics can be performed. E.g., the whole area of topological (geometrical) phases like the Aharonov-Bohm effect, Berry's phase and related effects can be explored. The question of non-locality of quantum mechanics and the principle of causality often studied via "Gedanken"-experiments can be addressed now by laboratory-scale experiments for massive particles. Though, a lot of experiments to very important questions have yet been studied with neutrons (10), atoms offers the unique possibility of the manipulation of the internal degrees of freedom during the interference process. Furthermore, with the increased interaction time using slow atoms, precision tests of charge neutrality, parity violation, sensitive measurement of gravity and highest precision spectroscopy can be envisaged. With increasing density in phase space the influence of the different kinds of quantum statistics will be monitored and will lead to completely new effects in nonlinear atom optics.

The output of an interferometer depends on its reference frame, i.e. it can be used as an inertial sensor. For optical interferometers this has led to the development of laser gyros as high-sensitivity rotation sensors. For an atom interferometer using thermal atoms the Sagnac effect already has been demonstrated (3). The extrapolation of the achieved sensitivity to a device using cold atoms shows values far superior to existing laser gyros.

With the Stanford Raman interferometer measurements of the gravitational acceleration have been performed already rivaling the best optical methods in resolution (8). For this experimental scheme considerable improvement in sensitivity is also possible, so that a gravimeter based on atom interferometry will be the most sensitive device in the near future.

The high sensitivity of interferometric measurements also allows to study the properties and interactions of the interfering particles itself. Besides measurements of polarizabilities and transition dipole moments, interaction of the interfering particles with each other via cold collisions or the interaction with nearby surfaces can be monitored. From a metrological point of view a precision determination of the recoil shift of an optical transition constitutes an independent way to measure the atomic mass. A determination of the Cs atom rest mass accurate to 10^{-7} has been performed with a cold atom interferometer (9).

All those experiments described above demand for coherent optics for matter waves: The key elements to build an atom interferometer are coherence preserving beam splitters and mirrors, atomic sources with high brightness and convenient and efficient detectors for the interfering particles.

Mirrors for atoms can be realized in different ways. Efficient reflection for rare gas atoms has been achieved in the early work of Stern and Estermann with diffraction from single crystal surfaces (11). Unfortunately only ground state atoms are reflected and the possible interaction with surface phonons leads to a loss of coherence. The up to now most promising mirror configuration makes use of the strong dipole force an atom experiences in the steep intensity gradient of an evanescent light field (12). With an evanescent light field mirror atoms can be reflected considerably distant from the substrate surface so that surface quality is not a demand as critical as for direct surface diffraction. Evanescent wave mirrors have been demonstrated so far for several alkalies and metastable rare gases (13-16).

A second class of deflecting devices make use of the discrete transfer of photon momenta during Rabi cycling. The basic idea is applicable to two-level systems as in the case of optical Ramsey-Bordé interferometry and also by Raman transitions to atoms with multiple ground states (4, 17-20). The beam splitting property of the light matter interaction becomes apparent by considering fractional Rabi cycles, e.g. $\pi/2$-pulses. These kind of beam splitters have the additional advantage of producing an entanglement between internal and external state of the atom. In atom interferometers based on these mirrors and beamsplitters like the Ramsey-Bordé interferometer (3, 7) and the Raman-echo interferometer (8) the interference process takes place in real space and is also labelled by different internal states in the spinor space of the atoms.

Transmitting and reflecting gratings as beam splitters based on the dipole force have been realized with standing light fields and also with evanescent waves. A problem of these beam splitters is the low efficiency so that up to now no interferometer of this type has been realized. Nevertheless, the use of optical potentials e.g. for the realization of lenses and for atom lithography is promising.

The progress made in microfabrication has allowed to build material gratings, slits, zone plates, etc. with dimensions suitable to act as diffractive elements of de Broglie waves. The successful operation of the MIT three-grating interferometer (2) and the demonstration of Young's double slit experiment for matter waves in Konstanz (1) emphasize the suitability of microstructures for atom or to be more general de Broglie wave optics and interferometry.

Laser cooling of atoms allows to prepare ensembles for atom optics and interferometry at low temperature with high density in phase space. Furthermore, due to the possibility to prepare cooled beams at arbitrary velocity the de Broglie wavelength can be tuned from 10^{-11} m at thermal velocities to 10^{-6} m and beyond for atoms below the recoil limit. Laser-slowed and compressed atomic beams with brightness $B = 10^{14}$ atoms sterad^{-1} cm^{-2} s^{-1} at a total flux between 10^8-10^9 s^{-1} have been reported (21-23). Although supersonic expansion sources for rare gases may easily exceed this value by several orders of magnitude the real advantage of a laser prepared atomic beam is the low velocity spread and the density in phase space. Most atom optics devices show large chromatic aberrations which favors the use of low-velocity-spread beams. Low absolute velocity also allows for larger splitting angles for beam splitters because the ratio between the fixed maximum transverse momentum transfer and the longitudinal momentum becomes better. For atoms dropped from a magneto-optical trap bouncing from an evanescent wave mirror has been demonstrated (24, 25) so that a resonator for de Broglie waves with one reflecting mirror and the use of gravity is possible (26).

Also the sensitivity of atom interferometry benefits in two ways from lower velocity atoms (27). First the larger splitting angles allow usually for a larger enclosed area of the interferometer. This way the sensitivity to perturbations that enter the Hamiltonian as a vector potential, e.g. rotations, is enhanced. Second the lower velocity allows for longer interaction time with perturbing scalar potentials like gravity so that a larger phase shift can be sampled.

To resume the short overview we state that microfabricated structures are at present the most efficient optical elements for thermal atoms and - recently demonstrated at MIT - molecules. Introducing the techniques of laser cooling and laser manipulation of atoms light fields as optical elements become an attractive alternative and furthermore cold atom interferometers on "trapped" ensembles, nearly at rest in the laboratory frame, can be considerably more sensitive than their thermal counterparts.

In the next sections we will stress in more detail the advantages of cold atoms for interferometry and atom optics. Some different cold atom interferometers will be described, the application of cold atom interferometry to a possible optical frequency standard will be discussed and the measurement of a topological phase akin to the scalar Aharonov-Bohm effect will be presented. The use of evanescent waves as optical elements for cold atoms is discussed in the last section. Finally a short outlook to future developments is given.

COLD ATOM INTERFEROMETRY

Overview

Cold atom interferometers have been used so far to study the effect of gravity, to measure atomic properties, to prepare a highly accurate optical frequency reference and to detect a topological phase.

An ingenious, conceptually simple interference experiment using cold atoms has been done by Shimizu et al. with metastable Ne atoms freely falling from a magneto-optical trap through a double-slit structure on a microchannelplate detector (5). The distance between the source, the slits and the detector was 76 mm and 113 mm respectively. With the 6 μm spacing of the slits a typical spacing of the interference fringes of 200 μm has been achieved. To ensure coherent illumination of the slits the diffractive spreading from the source had to be larger than the slit separation. The low initial velocity of the atoms allowed for a source diameter of more than 20 μm determined by the size of an optical pumping beam. Using a gated detection method interference fringes from different initial velocities and thus different de Broglie wavelength have been observed. The measured fringe spacing matches well the theoretical exspectation. Because the fringe spacing depends on the de Broglie wavelength and thus on the atomic inertial mass, while the energy gain in the gravitational field depends on the gravitational mass, this experiment tests the weak equivalence principle.

A precision measurement of the gravitational acceleration has been done by Kasevich et al. using a Raman-echo interferometer for sodium atoms in an atomic fountain (8). In difference to the experiment described above the phase shift due to gravitation here is independent of the atomic mass. Sodium atoms are cooled and trapped in a magneto-optical trap, cooled to subdoppler temperatures in polarization gradient molasses and launched vertically by a moving molasses. A pulsed standing lightwave containing appropriate sidebands drives a velocity selective Raman transition between the two hyperfine levels (F = 3, F = 4) of the $^2S_{1/2}$ ground state of the sodium atoms. A $\pi/2$-π-$\pi/2$ pulse sequence splits, redirects and recombines the wavefunction of an atom. The interference signal can be read out by monitoring the population of the hyperfine levels due to the entanglement between external and internal state of the atom. The fringe phase of the output signal of the interferometer depends on the total phase difference of the Raman pulses at the interaction points with the atom:

$$\Phi_{out} = \phi_1 - 2\phi_2 + \phi_3. \qquad (1)$$

The interference fringes can be scanned just by scanning the phase of one pulse. Note that a variation of the pulse phases linear in time e.g. due to a constant Doppler shift cancels. The quadratic phase shift due to the free falling of the atoms against the beamsplitters can be measured by a compensation method. For a total interaction time of 70 ms the acquired phase shift corresponds to about 10^5 fringes. A phase resolution better than $3 \cdot 10^{-3}$ rad ensured a sensitivity of $3 \cdot 10^{-8}$ g in the measurements at Stanford. Further increase in sensitivity just by choosing longer interaction times requires careful stabilization of the reference frame.

In a similar experiment Weiss et al. performed a precision determination of the atomic mass of Cs by measuring the recoil velocity with a Raman-Ramsey interferometer (9). Here Cs atoms are prepared in an atomic fountain. Near the turning point of their trajectory the atoms are exposed to a $\pi/2$-$\pi/2$-π-...-π-$\pi/2$-$\pi/2$ Raman pulse sequence inducing transitions between the hyperfine levels of the ground state. Without the central sequence of π-pulses just as in the case of optical Ramsey interferometry discussed below the interference signal consists of two recoil components separated in frequency space by $2\hbar k^2/m$. The effect of the sandwiched π-pulses is to enlarge the recoil splitting by the number of transferred photon momenta. This way the recoil shift has been measured to a precision of 10^{-7} within 2 hours of integration time. With the known photon momentum the ratio $\hbar/m$ can be derived.

The Ramsey-Bordé Interferometer with Mg Atoms

Instead of using a Raman transition the four pulse Ramsey sequence can be applied readily to a narrow optical transition. Using a Ramsey-Bordé interferometer (28) in the time domain we investigate the possibility of an optical frequency standard based on the $^1S_0$-$^3P_1$ intercombination transition of Mg. Mg atoms are ideally suited for this application as the lifetime of the $^3P_1$ state of 5.1 ms allows for long coherent interaction time with light fields and the closed

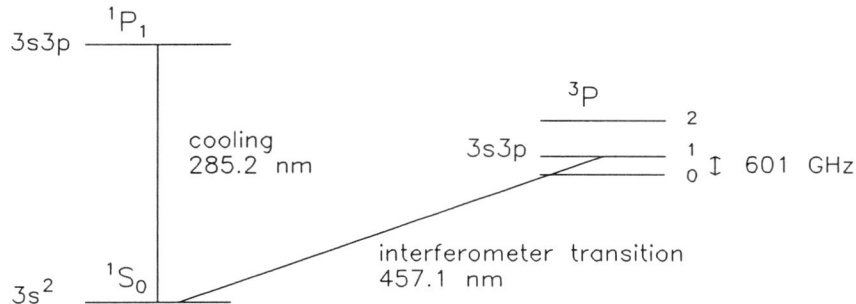

Figure 1: Level scheme of the Mg atom (only levels relevant for the experiment are shown)

$^1S_0$-$^1P_1$ transition with an upper level lifetime of only 2 ns can be used for efficient laser manipulation of the atomic motion (Fig. 1). The basic idea of pulsed interferometry is sketched as a recoil diagram in Figure 2. Two counterpropagating pairs of parallel laser pulses resonant with the intercombination transition irradiate the free atoms. Among the different paths of the split atomic wave packets due to the momentum exchange with the pulses there are two pairs forming two closed trapezoids. The probability amplitudes for a packet to move along the different paths of a trapezoid interfere at the last beamsplitter pulse. Thus again the interference can be read out by monitoring the

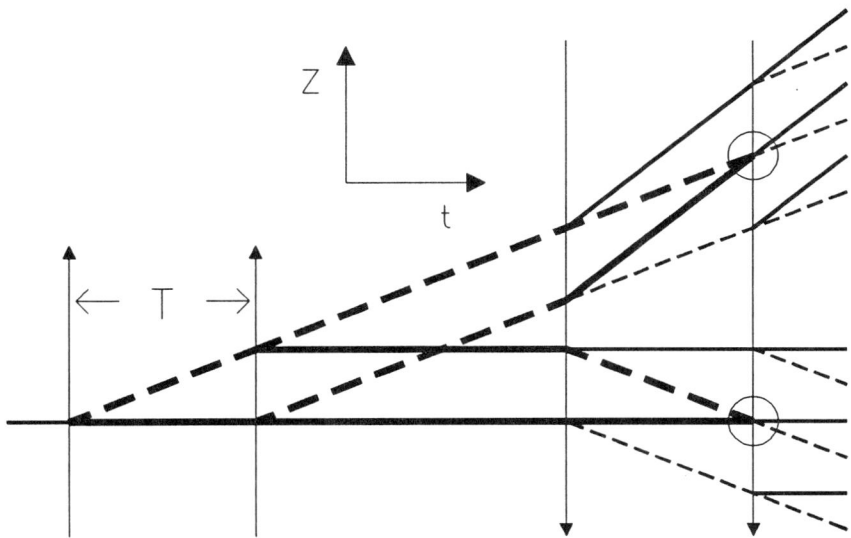

Figure 2: Recoil diagram for pulsed Ramsey-Bordé interferometry. The recoil induced displacement of the wavepacket is shown versus time.

population of the atomic states. Evaluating the phase difference between the two arms of a closed trapezoid for an atom finally in the excited state renders (29):

$$\Phi_{out} = 2(\Delta \pm \delta)T + (\phi_2 - \phi_1 + \phi_4 - \phi_3) + \pi \qquad (2)$$

Here ϕ_i denote the phases of the beamsplitter pulses, T denotes the delay between copropagating pulses, Δ denotes the detuning between the atomic resonance and the laser frequency and δ is the recoil shift. The signal thus consists of two sinusoidal fringe systems separated in frequency space by two recoil shifts. Of course also the open paths of the recoil diagram give rise to interference terms, but their separation at the last beamsplitter is usually larger than the coherence length so that their contribution is negligible. The squared probability amplitude for the open paths sums up to the incoherent Doppler and saturation dip background of the interference signal.

The fringe position depends explicitly on the laser frequency. This fact has been used for many years to perform ultrahigh resolution spectroscopy on the narrow lines in the infrared and visible wavelength region (30, 31). With the use of a pulsed scheme on trapped atoms considerable progress in accuracy is achieved compared to a conventional beam setup. The two main sources of systematic uncertainties 2nd order Doppler shift and laser phase error due to imperfect optics are inherently suppressed because of the low absolute velocity of the atoms and the possibility to cut the pulse pairs from single laser beams. A estimate of all known systematic sources of error shows a potential accuracy of $\delta\nu/\nu = 2 \cdot 10^{-15}$ (7).

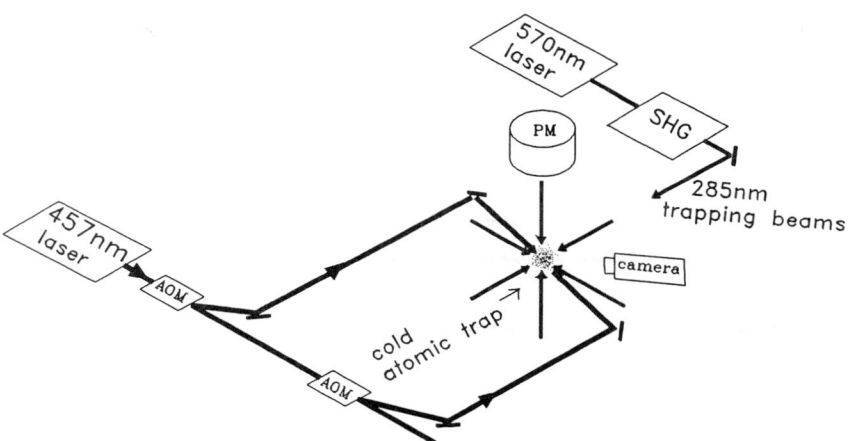

Figure 3: Sketch of the experimental setup for pulsed interferometry with trapped Mg atoms

The experimental setup is sketched in Figure 3. Mg atoms either from a thermal or a laser slowed atomic beam are stored in a magneto-optical trap operating on the strong $^1S_0$-$^1P_1$ transition at 285.2 nm. A frequency-doubled dye laser at 570 nm provides the necessary UV radiation. Typical values for the number of trapped atoms in a first experimental run were 10^3 from a thermal beam and $4 \cdot 10^5$ from the laser slowed beam. The temperature of the atoms measured with a time-of-flight technique is close to the Doppler limit of the pure two level system corresponding to a rms velocity of $\sigma_v = 0.81$ m/s. The trap fluorescence decays linearly when the loading is switched from the slowed source to the thermal beam with a time constant between $\tau = 0.2$-0.8 s depending on the laser intensity and the detuning from the trap transition. The linear behavior of the trap population enables the use of the trap dynamics for a signal amplification of the interference signal as will be discussed below.

The light pulses for interferometry are delivered by an ultrastable dye laser spectrometer at 457 nm. The linewidth of the laser frequency, relative to a reference cavity, as derived from the electronic error signal from the frequency control circuit is at the Hz level. The true linewidth of the radiation irradiating the atoms is of course larger due to the imperfect isolation and drift of the reference cavity, systematic errors in the locking circuitry, the Doppler effect of vibrating mirrors etc. From the observed contrast of the interference fringes for different dark times T we can state a linewidth surely below 2 kHz. The pulses for the interferometry are cut from single beams with acousto-optic modulators.

The pulsed scheme demands a periodic experiment. A cycle begins with trap loading for about 10 ms. To avoid lineshifts and destruction of coherence the trapping light and the quadrupole magnetic field are switched off within 20 μs before the interferometry pulses irradiate the freely expanding atomic cloud. The total trap off time is 300 μs and the time T between copropagating pulses was chosen between 6.3 μs and 100 μs in the experiments described here. After the last of the four pulses which closes the interferometers the trap is switched on again and its fluorescence is monitored. The decrease in the trap fluorescence is a quantum amplified measure of the number of transitions to the metastable state. Typically 1% of the trapped atoms are excited to the metastable state in a single cycle. Due to the long lifetime of the $^3P_1$ level the probability for an excited atom to leave the capture range of the trap before decaying to the ground state is high. Thus the excitation to the metastable state can be thought of as an additional loss mechanism. Repeating the cycle many times during the lifetime of the undisturbed trap the losses are accumulated and the number of trapped atoms tends to a new equilibrium value. Solving the rate equation for the number of trapped atoms with typical experimental parameters the above mentioned 1% excitation probability per cycle leads to a 30% decrease of the stationary fluorescence.

Figure 4 shows interference signals for different time delays T corresponding to fringe periodicities between 80 kHz and 5 kHz. The integration time per point is 4 s. The resolved line Q is $\nu/\delta\nu = 5 \cdot 10^{11}$. The noise of the signal is

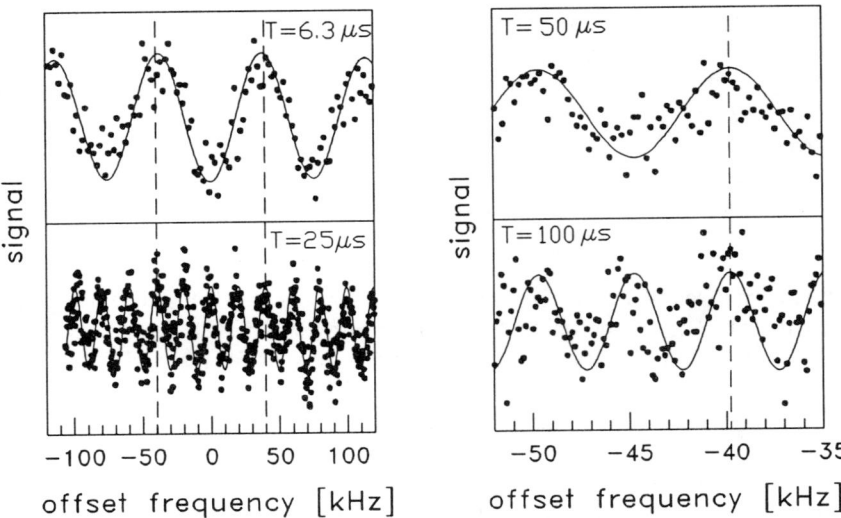

Figure 4: Ramsey fringes obtained with trapped atoms for different pulse separations T. The incoherent background signal has been subtracted.

consistent with the shot noise fluctuations in the number of trapped atoms. Locking the laser frequency to the resolved resonances yields a stability of $\delta\nu/\nu = 8.7 \cdot 10^{-13}$ within an integration time of 20 s (32). Since the data shown here have been achieved with the trap filled from the thermal beam improving the S/N is straightforward by working with the laser slowed beam if additional noise sources, which hindered higher resolution in first experiments, can be kept under control.

Measurement of the Scalar Aharonov-Bohm Effect

In a first application of our cold atom interferometer we used the interferometer to detect a topological phase shift akin to the scalar Aharonov-Bohm effect (33-35). Consider a particle under the influence of a spatially constant but time varying potential U(t). Evaluating the extra phase via the action integral regarding that momentum is conserved because gradU = 0 by definition yields:

$$\delta\phi = -\frac{2\pi}{h}\delta\int_\Gamma dL = -\frac{2\pi}{h}\int U(t)\,dt \qquad (3)$$

In order to observe this extra phase a reference wave not experiencing the extra potential is needed for interference. For a particle without internal structure this is not possible in a simple connected space-time region because the gradient of the potential cannot vanish everywhere between the particle and the reference. In their original proposal for an electron interacting with a pulsed electrostatic potential Aharonov and Bohm used suitable temporal and spatial barriers to prevent the electrons from piercing the regions of a non-vanishing gradient of the potential.

Due to the entanglement of internal and external states in the pulsed Ramsey interferometer the original idea of the Aharonov-Bohm can be transferred without the need of an excluded spatial region, if a pulsed and state selective potential is applied. Here the internal structure of the atom shelves part of the atom from the potential. A convenient choice for the state selective potential is the light shift generated by an expanded laser beam detuned from the strong $^1S_0$-$^1P_1$ transition (36). The additional potential is applied during the dark time between the pulses of a copropagating pair of interferometer pulses. The intensity I, detuning D and rise time of the shifting laser pulse are suitably chosen to ensure adiabatic evolution and minimization of excited state admixture in order to avoid coherence destruction by spontaneous emission. In the limit of large detuning and low intensity the phase shift due to a pulse of duration T' is proportional to:

$$\delta\Phi \propto \frac{I}{D}T' \qquad (4)$$

Figure 5 shows typical interference signals with and without the additional potential. The phase shift of nearly π is clearly visible. The two signals are recorded in parallel at every frequency step in order to minimize the influence of the drift of the dye laser spectrometer. Figure 6 shows detected phase shifts for different experimental parameters versus $I/D \cdot T'$. The phase shift is only dependent on the mentioned combination of experimental parameters and independent of the direction of the shifting beam or the chosen periodicity of the Ramsey fringes and agrees well with the expected linear dependency. The topological nature of the effect is proven by the contrast of the Ramsey fringe envelope and by laser beam reversal and will be discussed in a forthcoming publication.

250 Atom Optics and Interferometry

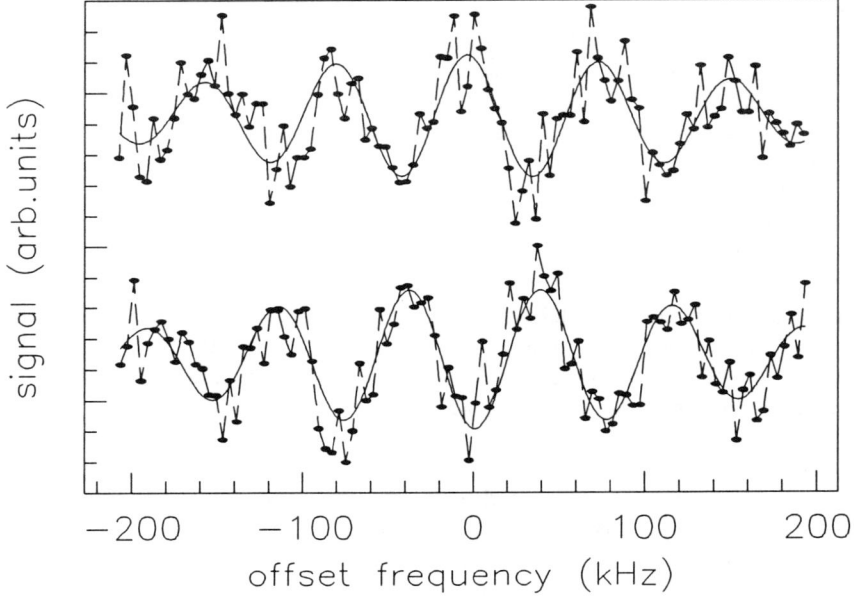

Figure 5: Simultaneously detected Ramsey fringes showing the Aharonov-Bohm phase shift. Upper curve with and lower curve without the additional light shift potential.

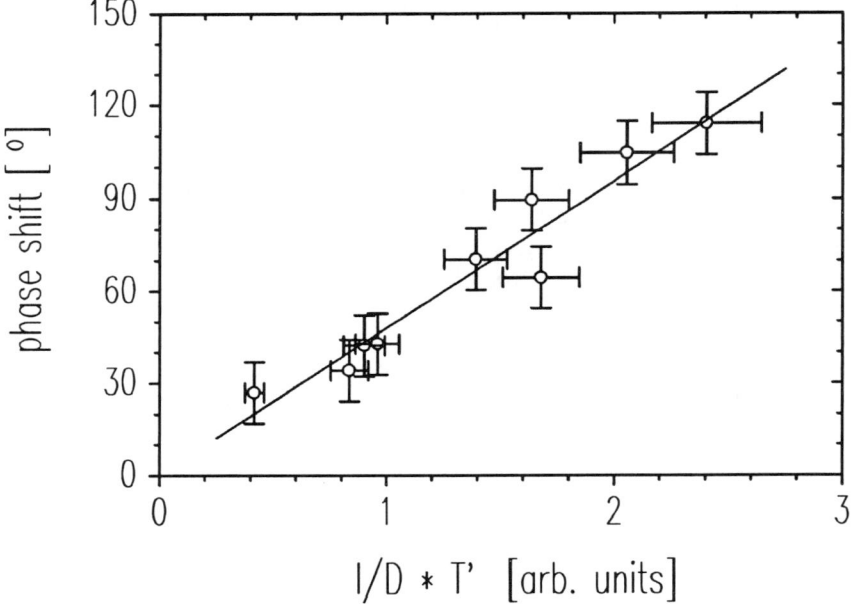

Figure 6: Measured phase shift vs. I/D T' for different parameters of the additional potential. The error bars contain the fit error for the phase shift (y) and the fluctuations in the intensity of the light shift beam (x).

COLD ATOM OPTICS

Overview

For interference experiments requiring true spatial separation of the beams and large enclosed areas as well as for atomic cavities there is urgent need for large angle mirrors and beam splitters. As already mentioned in the introduction for slow atoms the use of evanescent waves of experimentally achievable intensity allows steep potentials capable to invert velocities of several m/s.

The principal setup for the reflection of a two-level atom consists of an evanescent wave detuned to the blue side of the atomic resonance travelling along a glass-vacuum interface. The instantaneous dipole force expels an incoming atom from the region of high intensity near the surface. A large detuning ensures adiabatic evolution of the internal state of the atom while the atom moves towards the exponential potential well. Neglecting the direct atom-surface interaction the height of the potential well is determined by the light shift at the interface. To ensure coherence preserving reflection the probability for spontaneous emission has to be minimized. This favors the use of a light field with large detuning but high power which achieves just the necessary barrier height. To follow this idea, a lot of attempt was done to develop surface layers to increase the power in the evanescent wave. Enhancement factors of several hunderts could recently be demonstrated with wave guide layers (37).

The extension to a beam splitter is possible by the use of a standing evanescent wave where diffractive reflection takes place. The periodic potential allows multiple coherent exchange of grating momenta $n\hbar Q$ where Q denotes the wavevector of the standing wave along the surface. The smooth exponential envelope of the evanescent wave normal to the surface leads to the reflecting barrier. For a theoretical description of the diffraction process to each diffraction order labeled by the net number of exchanged photon momenta one adiabatic potential is assigned (38). Far from the surface and in the limit of grazing incidence the potentials are separated by a Doppler energy $\hbar\delta_{Dop} = \hbar Q v_\|$ where $v_\|$ denotes the velocity component parallel to Q. In the vincinity of the surface the potentials are shifted due to the strong atom-field coupling. For significant large enough shifts considerably higher than the Doppler energy avoided crossings of the adiabatic potentials appear which give rise to motion induced non-adiabatic transitions between the different potentials hence to diffraction (38).

The total interaction time with the potential may be of the order of the excited state lifetime.

The avoided crossings of the potential curves act as beamsplitters for the wavepackets. Since for reflection the splitted wavepackets may meet again at the same crossing rich interference phenomena occur in the population of the accessible diffraction orders when the initial kinetic energy of the atom is varied.

Reflection and Diffraction of Atoms by an Evanescent Light Field

An experiment to explore the possibilities of an evanescent grating beam splitter is performed with a laser-prepared beam of metastable Ne atoms (39). A reduced level scheme of the Ne atom showing the relevant transitions for cooling and reflection is shown in Figure 7. An overview of the whole experimental setup is given in Figure 8. Metastable Ne atoms in the $3s[3/2]_2$ state from a DC-discharge supersonic expansion source are decelerated in a Zeeman slower by counterpropagating laser light at 640 nm to a final velocity of 70 m/s. To increase the phase space density and to further reduce the longitudinal velocity the atoms pass a tilted two dimensional magneto-optical funnel. The funnel axis is tilted by 68° versus the original atomic beam axis and seperates also the metastable atoms from ground state atoms in the atomic beam. The metastable atoms are deflected and compressed to a beam of diameter d = 150 μm with a mean longitudinal velocity of v_L = 25 m/s. The spreads (FWHM) of the longitudinal and the transverse velocity distributions are Δv_L = 12 m/s and Δv_T = 0.75 m/s respectively.

The evanescent wave for reflection of the atoms is prepared by total internal reflection of a laser beam at 594 nm at the base of a quartz prism. For the diffraction experiments a standing evanescent wave is prepared by modematched retroreflection of the beam leaving the prism. The position of the prism and the angle between the prism base and the atomic beam can be adjusted independently.

After interaction with the evanescent wave the metastable atoms are detected by a two-stage MCP with an attached phosphoric screen. With a CCD camera monitoring the screen two dimensional profiles of the beam are recorded.

The upper state of the transition nearly resonant with the 594 nm radiation decays spontaneously to the ground state of the Ne atom. In this way an unwanted background signal due to atoms which undergo spontaneous emission during the interaction with the evanescent wave is avoided.

We observed reflection of the atoms for detuning values between δ = 0.5 GHz and δ = 2 GHz with an optimum value of δ = 0.9 GHz at a laser power of 120 mW and the laser radiation polarized parallel to the plane of incidence. In Figure 9 vertical cuts of the beam profile at the detector for different reflection angles between 30 mrad and 74 mrad at optimized detuning and polarization are shown. The large peaks at 10 mrad stem from atoms missing the prism. For increasing reflection angle their number is reduced because they hit the downstream edge of the prism. The decrease of the peak height for the reflected atoms with increasing angle is caused by the decrease of the effective mirror size i.e. the area where the light shift potential is sufficiently high to invert the perpendicular motion. The maximum observed reflection angle of 74 mrad corresponds to a velocity vertical to the prism base of $v_\perp$ = 1.9 m/s.

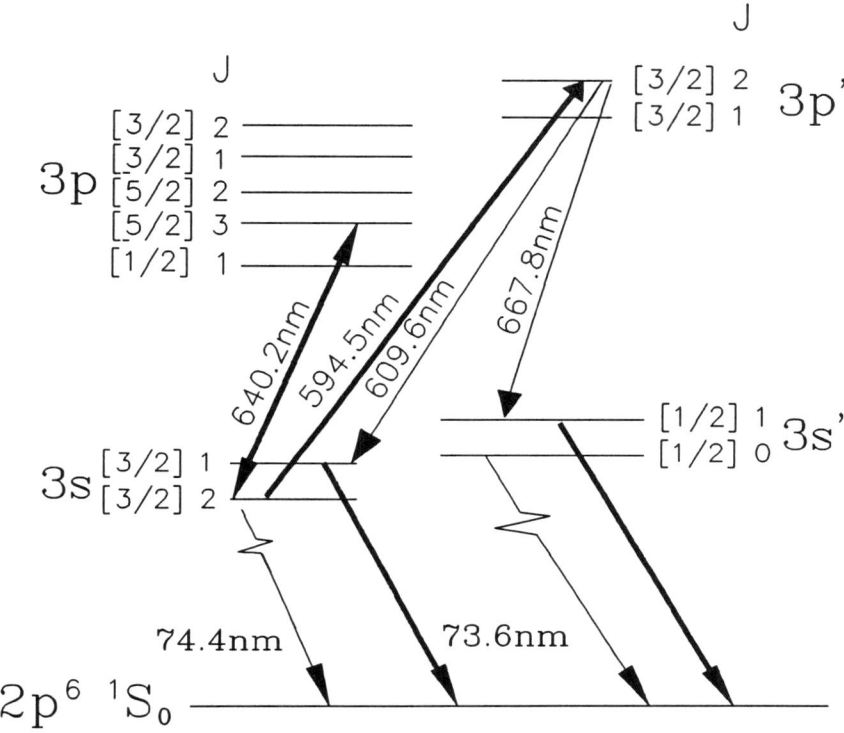

Figure 7: Level scheme of the Ne atom (only energy levels relevant to the experiment are shown)

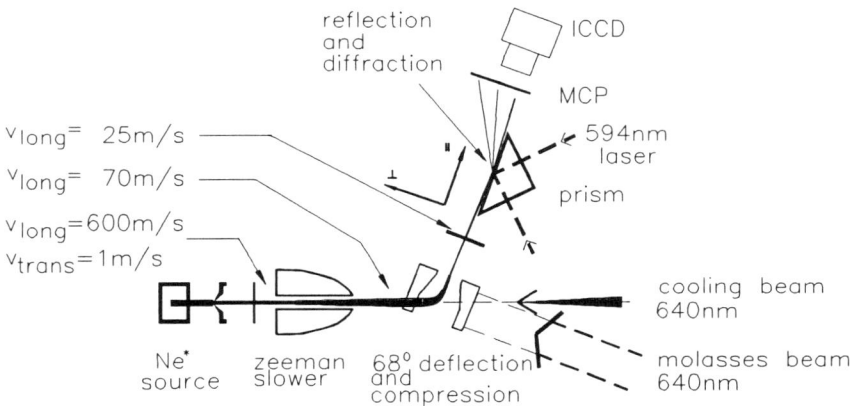

Figure 8: Sketch of the experimental setup for reflection and diffraction of cooled metastable Ne atoms from evanescent waves.

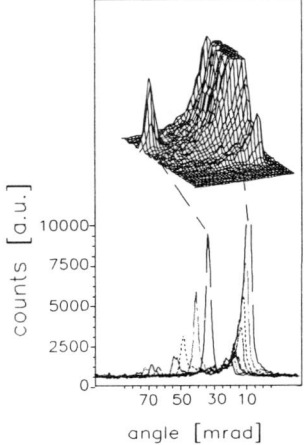

Figure 9: Experimental results for atomic reflection. The upper graph shows part of a 2-D profile of the beam. Below vertical cuts of the profiles for different incident angles are shown.

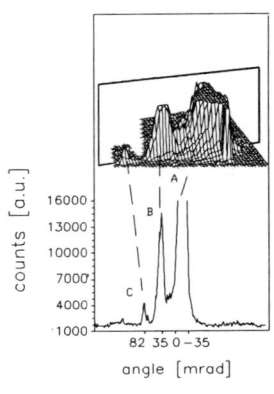

Figure 10: Experimental result for diffraction of the beam by two counterpropagating evanescent waves. The marked peaks in the vertical cut originate from uninfluenced (A), reflected (B) and diffracted (C) atoms.

Diffraction of the atoms was observed for the grating configuration of the light field. The intensity ratio of the two running waves was used as an additional experimental parameter. Optimum diffraction was observed for a detuning $\delta = 900$ MHz, an incident laser power of 150 mW and an intensity ratio between incoming and retroreflected beam of 1.64. Figure 10 shows the two dimensional profile together with a vertical cut for an incident angle of $\alpha = 36$ mrad. The peaks marked in the vertical cut correspond to atoms missing the prism (A), the specular reflected atoms at 36 mrad (B) and to atoms diffracted to the $n = -2$ diffraction order at 84 mrad (C). Normalized to the flux hitting the effective mirror surface 70 % of the atoms show up in the specularly reflected peak while 1.5-3 % of the atoms are detected in the $n = -2$ diffraction order. Figure 11 shows the observed diffraction angles versus different incident angles. The comparison with the theoretical values for a beam velocity of 25 m/s shows excellent agreement. The population of higher diffraction orders was not detectable because of the too low laser power for the evanescent field.

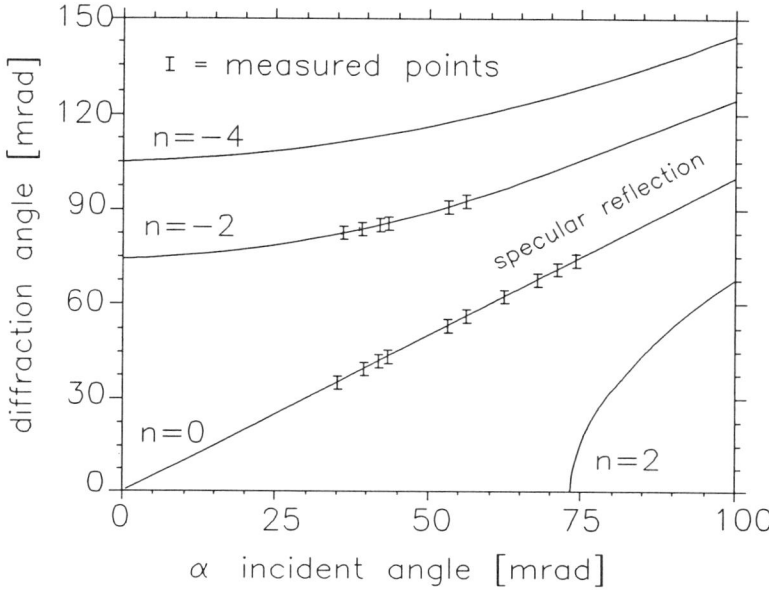

Figure 11: Measured angular position of the n = 0 and the n = -2 diffraction orders for different incident angles of the beam. The solid lines mark the theoretical positions of the diffraction peaks calculated for a beam velocity of 25 m/s.

OUTLOOK

As described in this article, up to now several schemes of atom deflection, reflection and interferometry were demonstrated in the last few years, together with first important applications.

The technical standardisation and simplification of atom manipulation setups (MOT's, funnels, fountains, ...) together with more sophisticated coherent optic arrangements will strongly improve further experiments.

The combination of elements for manipulating and cooling with mirrors and beam splitters will improve the sensitivity of Mach-Zehnder type interferometers by several orders of magnitude.

Like in light optics, several different matter wave devices, not yet demonstrated (Michelson interferometers, multiple beam interferences, phase contrast interferencies, ...) optimized for special applications may lead to standard setups, like the electron microscop being in electron optics.

The potential for precise measurements of gravitational effects, the gravitational constant G, the mass of atoms, and of symmetry and conservation laws in physics is given and was partially yet demonstrated.

Studies of the fermionic and bosonic statistical properties of atoms will reconnect the field of atom physics with that of statistical physics and thermodynamics and will help to classify yet open questions in quantum statistics of weakly interacting particles.

The fields of cold collisions and the scattering of atoms from surfaces strongly gain in their information capabilities and precision by the ability to read out the amplitude *and* phase shifts in interferometric measurements.

The multiple particle interference ("beating of atoms") and Hanbury-Brown-Twiss like experiments for atoms will be further important steps in quantum physics.

REFERENCES

1. Carnal, O., and Mlynek, J., *Phys. Rev. Lett.* **66**, 2689 (1991).
2. Keith, D.W., Ekstrom, C.R., Turchette, Q.A., and Pritchard, D.E., *Phys. Rev. Lett.* **66**, 2693 (1991).
3. Riehle, F., Kisters, Th., Witte, A., Helmcke, J., and Bordé, Ch. J., *Phys. Rev. Lett.* **67**, 177 (1991).
4. Kasevich, M., and Chu, S., *Phys. Rev. Lett.* **67**, 181 (1991).
5. Shimizu, F., Shimizu, K.,and Takuma, H., *Phys. Rev.* **A46**, R17 (1992).
6. Robert, J., Miniatura, Ch., Gorceix, O., Le Boiteux, S., Lorent, V., Reinhardt, J., and Baudon, J., *J. Phys. II France* **2**, 601 (1992).
7. Sengstock, K., Sterr, U., Hennig, G., Bettermann, D., Müller, J.H., and Ertmer, W., *Opt. Comm.* **103**, 73 (1993).
8. Kasevich, M., and Chu, S., *Appl. Phys. B* **54**, 321 (1992).
9. Weiss, D.S., Young, B. N., and Chu, S., *Phys. Rev. Lett.* **70**, 2706 (1993).
10. see for example *Physica B* **151**, special issue on 'Matter Wave Interferometry' (1988)
11. Estermann, I., Stern, O., *Zeits. f. Phys.* **53**, 95 (1930).
12. Cook, R.J., and Hill, R.K., *Opt. Comm.* **43**, 258 (1982).
13. Balykin, V.I., Lethokov, V.S., Ovchinnikov, Y.B., and Sidoron, A.I., *Phys. Rev. Lett.* **60**, 2137 (1988).
14. Hajnal, J.V., Baldwin, K.G.H., Fisk, P.T.H., Bachor, H.A., and Opat, G.I., *Opt. Comm.* **73**, 331 (1989).
15. Esslinger, T., Weidemüller, M., Hemmerich, A., and Hänsch, ,T.W., *Opt. Lett.* **18**, 450 (1993).
16. Seifert, W., Adams, C.S., Balykin, V.I., Heine, C., Ovchinnikov, Y., and Mlynek, J., *Phys. Rev. A* **49**, 3814 (1994).
17. Marte, P., Zoller, P., and Hall, J. L., *Phys. Rev.* **A 44**, R4118 (1991).
18. Goldner, L. S., Gertz, Ch., Spreeuw, J. C., Rolston, S. L., Westbrook, Ch., Phillips, W. D., Marte, P., and Zoller, P., *Phys. Rev. Lett.* **72**, 997 (1994).
19. Lawall, J., and Prentiss, M., *Phys. Rev. Lett.* **72**, 993 (1994).
20. Pfau, T., Adams, C.S., Sigel, M., and Mlynek, J., *Phys. Rev. Lett.* **71**, 3427 (1993).
21. Nellessen, J., Werner, J., and Ertmer, W., *Opt. Comm.* **78**, 300 (1990).
22. Riis, E., Weiss, D. S., Moler, K. A., and Chu, S., *Phys. Rev. Lett.* **64**, 1658 (1990).

23. Scholz, A., Christ, M., Doll, D. Ludwig, J., and Ertmer, W., acc. for publication in *Opt. Comm.* (1994).
24. Kasevich, M. A., Weiss, D. S., and Chu, S., *Opt. Lett.* **15**, 607 (1990).
25. Aminoff, C. G., Steane, A. M., Bouyer, P., Desbiolles, P., Dalibard, J., and Cohen-Tannoudji, C., *Phys. Rev. Lett.* **71**, 3083 (1993).
26. Wallis, H., Dalibard, J., and Cohen-Tannoudji, C., *Appl. Phys.* B **54**, 407 (1992)
27. Clauser, J.F., *Physica* B **151**, 262 (1988).
28. Bordé, Ch. J., *Phys. Lett.* A **140**, 10 (1989).
29. Bordé, Ch. J., Salomon, Ch., Avrillier, S., Van Lerberghe, A., Bréant, Ch., Bassi, D., and Scoles, G., *Phys. Rev.* A **30**, 1836 (1984).
30. Bergquist, J. C., Lee, S. A., and Hall, J. L., *Phys. Rev. Lett.* **38**, 159 (1977)
31. Helmcke, J., Zevgolis, D., and Yen, B. Ü., *Appl. Phys.* B **28**, 82 (1982).
32. Sengstock, K., Sterr, U., Müller, J.H., Bettermann, D., Rieger, V., and Ertmer, W., *accepted for publication in Appl. Phys.* B (1994).
33. Aharonov, Y., and Bohm, D., *Phys. Rev.* **115**, 485 (1959).
34. Cimmino, A., Opat, G.I., Klein, A.G., Kaiser, H., Werner, S.A., Arif, M., and Clothier, R., *Phys. Rev. Lett.* **63**, 380 (1989).
35. Sangster, K., Hinds, E.A., Barnett, S.M., and Riis, E., *Phys. Rev. Lett.* **71**, 3641 (1993).
36. Sterr, U., Sengstock, K., Müller, J.H., Bettermann, D., and Ertmer, W., *Appl. Phys.* B **54**, 341 (1992).
37. Kaiser, R., Lévy, Y., Vansteenkiste, N., Aspect, A., Seifert, W., Leipold, D., Mlynek, J., *Opt. Comm.* **104**, 234 (1994).
38. Deutschmann, R., Wallis, H., and Ertmer, W., *Phys. Rev.* A **47**, 2169 (1993).
39. Christ, M., Scholz, A., Schiffer, M., Deutschmann, R., and Ertmer, W., *Opt. Comm.* **107**, 211 (1994).

Dipole Trapping, Cooling in Traps, and Long Coherence Times

Heun-Jin Lee, Charles Adams, Nir Davidson,
Brent Young, Martin Weitz, Mark Kasevich,
and Steven Chu

Physics Department, Stanford University, Stanford, CA 94305

Abstract. Conditions for trapping sodium atoms in far detuned dipole traps are explored with the goal of maximizing phase space density and coherence times. With red detuned traps, we have been able to achieve densities as high as 2×10^{12} atoms/cm^3 at temperatures on the order of $(11\hbar k)^2/2M$. We also report Raman cooling in a red de-tuned trap to a one dimensional "temperature" of $0.7(\hbar k)^2/2M$. With a blue detuned trap, we have been able to observe Ramsey interference fringes between the ground states of sodium for over 4 seconds. Finally, we report an atom interferometer based on adiabatic transfer of ground state populations.

Laser cooling and trapping techniques[1] have enabled one to dramatically increase the phase space density of a vapor of atoms. Beginning with atomic beams at a temperature of hundreds of Kelvin and a densities on the order of 10^{10} atoms/cm^3 or less, one can routinely cool alkali atoms to temperatures on the order of $4(\hbar k)^2/2M$ (microKelvin temperatures) at densities in excess of 10^{11} atoms/cm^3 with magneto-optic traps (MOT).[2] Despite attempts to increase the density of atoms in a MOT, no one has been able to achieve atomic densities in excess of 10^{12} atoms/cm^3. One of the effects that limit the density is the repulsive force generated by radiatively trapped photons.[3]

Recently, we have begun to study methods of increasing the density while reducing the temperature of atoms in optical traps. Concurrently, we have also begun to explore methods of increasing the coherence times of atoms in traps.

We have concentrated on studying optical traps based on induced dipole forces[4], especially dipole traps tuned far off resonance[5]. In these traps, the probability of scattering a photon is greatly reduced so that the trap approaches a conservative potential, and the ground state hyperfine and Zeeman levels of the atom experience nearly equal light shifts. This second virtue has permitted us to apply Raman cooling to trapped atoms and establish long ground state coherence times. Finally, optical traps allows one certain flexibility not allowed with magnetic traps.

In this paper we report the results of sodium atoms held in red de-tuned dipole traps fashioned out of light from the red lines of a krypton laser (647 and 676 nm), the 1.06 μm line of a Nd:YAG laser and in blue de-tuned traps made with the blue (488 nm) and green (514 nm) lines of an argon laser.

LOADING DIPOLE TRAPS

Both the red and blue detuned traps were loaded from a MOT by overlapping the Mot trap with the dipole traps. The MOT, in turn was loaded from a beam of thermal Na atoms that was slowed by a counterpropagating, frequency chirped laser beam. A detailed description of our apparatus can be found elsewhere.[6] After a loading time that was varied between 0.5 to 2.5 seconds, the magnetic field, atomic beam, slowing beam and MOT beams were shut off, leaving only the dipole trap beams present. During the last 10-30 msec of the MOT loading, the intensity of the $3S_{1/2}, F=1 \rightarrow 3P_{3/2}, F=2$ repumping sideband of the MOT beams was reduced to approximately one fifth of the initial value, increasing the number of atoms loaded into the trap by an order of magnitude. The reduced sideband intensity optically pumped atoms into the $F=1$ state, which, in turn reduced the trap loss due to ground state hyperfine changing collisions.[7] If an $F=2$ atom collides with an $F=1$ atom and relaxes to the $F=1$ state, the added kinetic energy of $2\pi\hbar \times 1.77$ GHz is sufficient to eject both atoms out of the trap. Finally, the atoms were optically pumped into the $3S_{1/2}, F=1$ state by shutting off the repumping sideband in the last 0.5 msec. This method of loading the trap is a version of the temporal "dark MOT" scheme of Ketterle, et al.[8]

DIPOLE TRAPPING WITH A KRYPTON LASER

Our first red detuned dipole trap was made with the red lines from a Krypton laser. A single beam of 2 watts was focused to waist sizes between w_0 of 5 μm to 36 μm. We were able to trap over 10^5 atoms. The lifetimes of the atoms in the 5 μm trap is plotted in Fig. 1. After a rapid decay in the first second of trapping, the lifetime was governed by the background pressure in the vacuum can.

In general, the 5 μm trap did not perform well. Atoms only loaded at the edges and not the center of the trap, presumably due to the large light shifts of the ground and excited state energy levels. Furthermore, the large ac Stark shifts prevented optical pumping needed for Raman cooling. As one went to more weakly focused traps, the temperature of the trapped atoms decreased, the total number of trapped atoms increased, and the storage time increased. In the weakest focused traps the atoms were extended in space over several millimeters along the axial dimension.

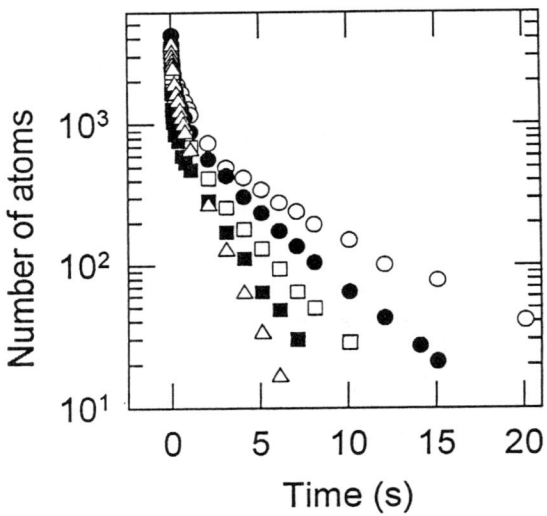

Fig. 1. Lifetime of atoms in a 5 μm krypton laser dipole trap as a function of background pressure of the vacuum chamber. The triangles, filled squares, open squares, filled circles and open circles correspond to background pressures of 11, 7, 5, 2.9, and 1×10^{-10} torr lifetimes, respectively.

Our attempts to Raman cool[9] in traps with $w_0 = 12$ μm only yielded a modest build-up of the phase space density. Apart from gravitational acceleration out of the $v \sim 0$ states, Raman cooling works best under free space conditions. The more tightly focused dipole traps have too high an axial oscillation frequency for the Raman cooling scheme to work well. With a dipole trap of $w_0 = 36$μm, the calculated axial oscillation frequency of the atoms was on the order of 10 Hz while the transverse frequency is roughly 2.3 KHz. Fig. 2 show the results of cooling for 5 milliseconds. The axial temperature was reduced from 30 μK to 0.8 μK. After cooling the atoms settled into the bottom of the trap, and the axial confinement of the atoms decreased from several mm to less than 0.2 mm. Very roughly, the density of the atoms in the trap increased to 5×10^{10} atoms/cm^3 after cooling. Based on the estimated collision rate for hyperfine changing collisions[7], we would expect a collision limited trap lifetime of 1 second at a density of 10^{11} atoms/cm^3.

Cooling along the transverse dimensions was not as rewarding, but the 3 dimensional cooling results were more promising for the 36um trap. Before we were able to optimize simultaneous axial and transverse cooling in the 36 μm focused trap, the Krypton laser tube failed. In our preliminary results with the 12 μm trap, the axial and transverse velocities were reduced to rms velocities of 1.5 and 2.5 photon recoils respectively, but these results were obtained before we achieved 0.8 photon recoil cooling along the axial dimension for the 36 μm trap. Since we could not afford the high operating costs of a krypton laser, we decided to trap sodium atoms with a Nd:YAG laser operating at 1.06 μm.

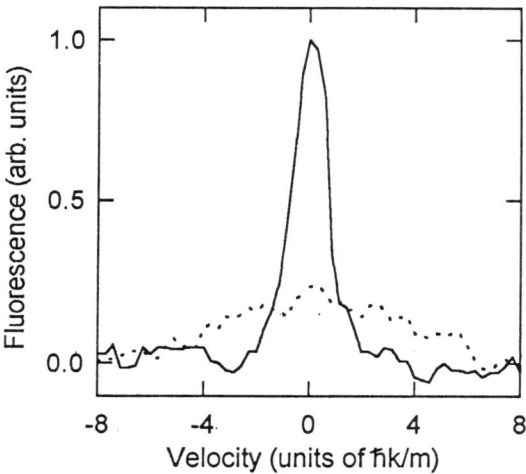

Fig. 2. Raman cooling in a re-detuned single beam dipole trap. Two watts of the red lines of a krypton laser were focused to 36 μm. After a cooling time of 50 ms, the atoms were cooled along the axial dimension from an initial temperature of 30 μK to 0.8 μK.

CROSSED DIPOLE TRAPS WITH A YAG LASER

In order to equalize the axial and transverse trap oscillation frequencies, we constructed a dipole trap from a pair of laser beams focused to $w_0 = 18\mu$m. The two beams crossed at right angles and were orthogonally polarized to avoid standing wave effects. For parallel polarizations, the number of trapped atoms decreased by a factor of two while the temperature remained the same. Up to 4 watts per beam, adjusted with an AO modulator, could be delivered in each laser beam.

The trapping conditions were adjusted in order to maximize the fluorescence from atoms captured in the trap, measured by turning off the trapping light and switching on the MOT beams for a short period of time. The fluorescence was also monitored with a microscope imaging system based on a f 2.5 "objective" lens with a resolution (installed inside the vacuum can) of less than 5 µm. This upper limit was established by imaging a tightly focused (8 µm diameter beam) dipole trap onto a CCD camera. The imaging system had a magnification of 4.5x and looked down onto the crossed trap with horizontally propagating beams. The measured resolution of the optical system sitting on a laser table was 2µm. The spatial dimensions of the trapped atoms could be imaged by pulsing on the MOT light for times between 5 and 50 µsec immediately after the trapping light was turned off. (During the detection time, the atom cloud expanded.) The objective lens was mounted on a uhv micromanipulator and susceptible to mechanical vibrations that degraded the image by ~3 microns. To avoid the effect of lens vibrations, the most critical size measurements of the trapped atom cloud were taken by averaging the widths of several single shot measurements.

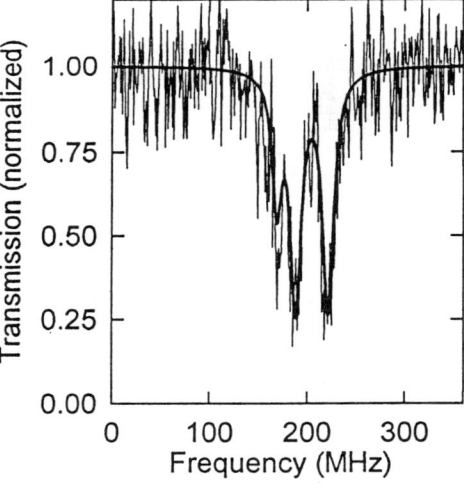

Fig. 3. The transmitted intensity as a function of frequency of a probe beam frequency passing through the center of the crossed region of the trap. The solid curve is the theoretical calculation of the $3S_{1/2}, F=1$ to $3P_{3/2}, F=0,1,2$ absorption lines assuming an unsaturated Lorentzian lineshape for each transition.

The atom density in the trap was measured with a 1 nanowatt probe beam focused to a waist of 7 µm. The duration of the probe beam was typically 2 µs in order to avoid effects of ballistic expansion of the atoms after the trapping light was turned off. Fig. 3 shows the transmitted light as a function of probe light detuning. The equal absorption of the $F=1$ and $F=2$ transitions indicates that the probe light intensity was low enough to avoid saturation and optical pumping effects. From a peak absorption of 75% and the measurement of the spatial extent of the trapped atoms, the peak density was calculated to be 2×10^{12} atoms/cm^3.

The lifetime of the atoms in the trap fit well to a single exponential decay and was independent of the number of atoms loaded into the trap. The loss rate does, however, depend on the intensity of the trapping light. For a 8 watt trap, the number of atoms decayed with a single exponential lifetime of 0.8 second, while the decay rates for a 4 and 2 watt trap were 1.5 and 2.7 seconds. The mechanism for this decay has yet to be determined: however, since this loss was independent of the trap density, it could not be due to the interaction between two trapped atoms and a photon.

The temperature of atoms trapped in the crossed region was measured by a time of flight technique. The detection pulse was delayed by a few hundred microseconds after the trap light was turned off. In this time, the size of atom cloud expanded by roughly an order of magnitude. For our deepest trap (480 μK deep with 8 watts), atoms were localized to 14 μm (FWHM) and had an initial temperature of 140μK. If one assumes equipartition in a harmonic oscillator potential well, the expected trap size would be 9μm. For the 8 watt trap, the temperature of the atoms was seen to increase at a rate equivalent to scattering one 1.06 μm photon every 800 ms and consistent with the expected photon scattering rate. Fig. 4a,b shows the temperature and density of the atoms as function of the time spent in the trap.

For shallower traps made with 2 and 4 watt beams, no heating was observed for observation times up to several seconds as shown in Fig. 4a. In fact, after the first second of trapping, the density increased while the temperature decreased, indicating evaporative cooling.[10] The phase space density increased by a factor of 2 before loss due to collisions with residual background atoms (the pressure was 6×10^{-10} torr) began to decrease the number of trapped atoms.

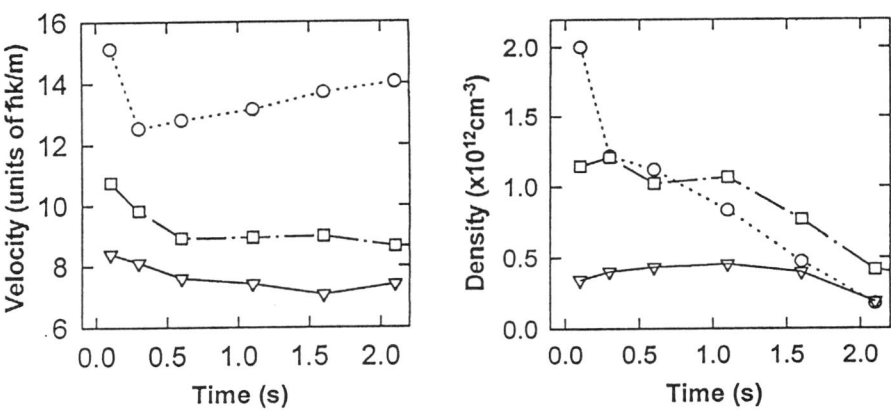

Fig. 4. (a) Temperature of the atoms as a function of the time spent in the trap. The trap heated atoms at the highest trap intensities, but for lower intensities, the atoms cooled for the first second after trapping. (b) Density of the atoms in the central region of the trap as a function of trapping time for a 2, 4, and 8 W trap.

Effective evaporative cooling requires that the depth of the trapping potential and the volume of the trap be independently varied in order to maintain the high density of atoms in the trap. At present, we can only evaporatively cool by turning down the light intensity, and after an initial stage of evaporation, the decrease in the atom density would reduce the evaporation rate considerably.

We intend to explore a dipole trap based on sharply rising, repulsive walls of light. A blue trap could be made where the volume could be easily changed by more than an order of magnitude on the order of 0.1 second. It would also be much more suitable for both Raman cooling and precision spectroscopy since atoms confined in such a trap would spend most of their time under free fall conditions.

BLUE DE-TUNED DIPOLE TRAPS

We report in this section the study of a gavito-optic (atomic trampoline) trap formed by constructing a curved, repulsive barrier of blue-detuned light and gravity. This work is also the precursor to a blue trap with walls suitable for evaporative cooling. A gravito-optic trap was proposed when laser cooled atoms were shown to reflect at normal incidence from an evanescent wave of light[11], and a number of groups have actively studied this type of trap[12,13]. In these traps, atoms spend most of the time in zero potential (except gravity), and for very short intervals are reflected from a very sharp wall of light with characteristic size $\lambda/2\pi$. The resulting phase shifts can be 10^3-10^4 times smaller than for a comparably deep red-detuned trap where the atoms spend most of their time in regions which produce the maximal light shift. Another important property of the evanescent wave potential is that the integrated light shift along the trajectory is only weakly dependent on the energy of the atom, so one can expect even longer coherence times.

In spite of these properties, previously demonstrated atomic trampolines could not be used for precision spectroscopic measurements since the longest 1/e storage times were on the order of 100 msec. The short lifetime and scattering time are mainly due to the difficulties that arise from the requirement of a glass surface: since the atoms have to be dropped from a height of a few mm, a stronger dipole potential is needed to support them against the kinetic energy they gain as they fall onto the surface. The light intensity needed to create the necessary dipole force and simultaneously avoid a high photon scattering rate has prompted a number of groups to study ways of enhancing the light intensity by interferometric techniques with dielectric coatings[14] or by using surface plasmon resonances[15].

We avoided these difficulties by constructing a blue-detuned light trap based on free-propagating laser beams. This trap was placed at the center of the source of cold atoms and loaded with energies of a few $U_{rec}=(\hbar k)^2/2M$ from a MOT by overlapping the two traps in space and time as described in the "trap loading" section. Our trap was based on two linearly polarized cylindrical Gaussian beams (15 μm x 1100 μm, 1/e^2 diameters), co-propagating horizontally and oriented at +45° and -45°. The two beams were overlapped approximately at their 1/e intensity, resulting in a "V" shape that confined the atoms transversely to the beams propagation direction. Along the beam propagation direction, confinement was obtained due to curvature of the beams near their focus. The approximately harmonic dipole potential along the longitudinal direction extends to about the Raleigh range (~700 μm) of the focused beams and has depth of roughly 15% of the transverse directions.

We used the 488 nm and 514.5 nm lines of an argon laser for the two cylindrical beams in order to avoid any intensity or polarization gratings. The average rate for spontaneously scattering photons at such detunings was calculated to be 10^{-3}-10^{-4} sec^{-1}. With 15 W (all lines) from an argon laser, the power of 488 nm and 514.5 nm lines was 4 and 6 W, respectively, corresponding to a maximal dipole wall potential of 190U_{rec} and 90U_{rec}.

The blue trap loading followed the procedure developed for red trap loading as described in the previous section. The lifetime of the traps displayed fast and slow decay times similar to the decays with the Kr$^+$ laser trap shown in Fig. 1. The long time decay is consistent with losses due to collisions with the background gas. There is an additional loss rate on the order of 1 second that may be attributable to either to density dependent loss mechanisms, evaporation, or to initial population of unstable orbits that either "spill over the edge" or tunnel out of the trap. A very high sensitivity of the faster loss rate to the relative alignment between the two cylindrical beams supports either of the latter explanations.

We measured the lifetime of the ground state coherence by performing rf Ramsey spectroscopy on the F=1, m_f=0 and F=2, m_f=0 ground state levels. The transition was excited with ~1.77 GHz rf traveling wave beamed into the vacuum chamber. A bias magnetic field of ~5 mG was applied to separate the magnetic-field sensitive transitions from the m_F=0 → m_F=0 transition.[1] The number of atoms making the transition to the F=2 state was determined by measuring the fluorescence from a short pulse of light resonant with the $3S_{1/2}$, F=2 → $3P_{3/2}$, F=3 transition. The resulting central Ramsey fringes for T=4 sec are shown in Fig. 5. The fringe contrast for this measurement time was 43%. The fringe contrast after T=0.2, 1 and 2 sec was 98%, 90% and 82%, respectively.

---

[1] Note that atoms with an energy of 1 U_{rec} = 25 KHz while the Zeeman sublevels were separated by 7 KHz. At higher densities where collisional de-phasing might be a problem, spin exchange collisions could be prevented by increasing the Zeeman splitting $\mu \cdot B$ to greater than $k_B T$ of the atom cloud.

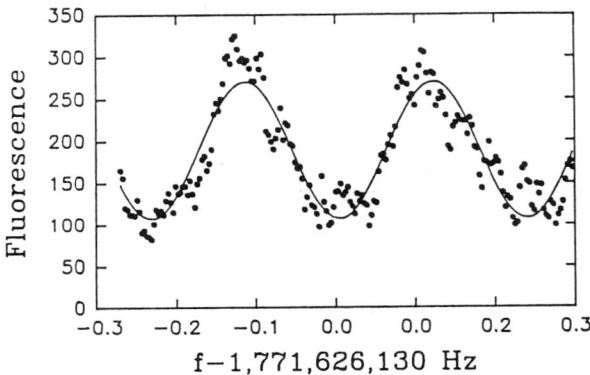

Fig. 5. The central Ramsey fringes of the $F=1, m_f=0$ to $F=2, m_f=0$ transition. The FWHM linewidth is 0.125 Hz corresponding to a fitted measurement time of 3.98 ± 0.04 sec while the actual measurement time was 4.00 seconds. A constant background, present without rf pulses was subtracted from the data. The circles are the data and the line is a fit to $a + b \cdot \cos(2\pi f T + \phi)$.

We feel that the rf dephasing is due mainly to non-uniformities in the light shifts of the dipole potential. To investigate this effect we measured the frequency of the central Ramsey fringe at $T=1$ sec as a function of the trapping laser power P (the 488 and 514.5 nm lines are ~25% and ~40%, respectively, of that power). The measured frequency shift was proportional to the trapping power, with a slope of ~18 mHz/W. Since most of the trap parameters such as density, bounce frequencies, and velocities of the atoms depend on the trapping laser intensity, a simple proportionality is not necessarily expected. The slope corresponds to 270 mHz frequency shift at $P=15$ W.

We calculated the light shifts and coherence time for the atoms in our trap using Monte Carlo simulations based on classical trajectories. The distribution of average light shifts over the atoms that remained in the trap after 1 sec is shown in Fig. 6. The hyperfine light shift distribution (upper scale) is obtained by dividing the frequency scale by $\delta_{argon}/\delta_{hf} \approx 4.5 \ast 10^4$. It yields a coherence decay time of $T_c = 9$ sec, and $\delta f = 70$ mHz frequency shift of the Ramsey fringes[16].

The current trap has "soft" walls along the beam propagation direction. This asymmetric geometry allows atoms to sample wildly different trajectories. A more symmetric trap should give a narrower distribution of phase shifts. Such a trap might consist of "hard" walls made with three intersecting sheets of light, propagating at 45° to gravity and oriented at 90° with respect to each other create an inverted pyramid. The calculated light shift distribution gives a factor of 4 increase in the coherence decay time relative to the two-beam trap while the average Stark shift decreases by only ~35%. In this trap, each atom experiences different collisions with the walls, but the effect of many collisions is to induce an average phase shift that is approximately the same for all atoms.

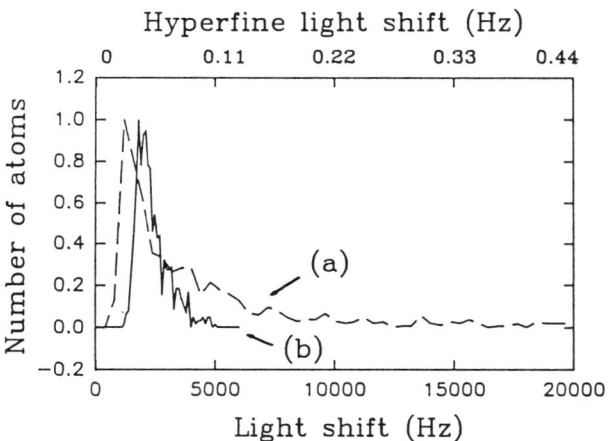

Fig. 6. Light shift distributions among atoms stored in the gravito-optic trap calculated with a classical Monte Carlo simulation. Curve (a) is the simulation of the trap used in the experiment and (b) is the proposed symmetric 3 beam trap.

Similar averaging is seen with atoms confined at higher temperatures by conventional bottles. For example, hydrogen atoms trapped in a teflon coated glass bulb of a hydrogen maser[17] and atoms confined in a cell with buffer gases[18] can undergo a large number of collisions before the relevant atomic coherence is destroyed. Note, that such averaging also plays an essential role in suppressing the first order Doppler shift that would otherwise wash out the fringes completely. One can say that optical atom traps are evolving towards "light bottles".

We also performed rf spectroscopy of Na atoms in a far off resonance, red-detuned dipole trap made by focusing a 10 watt Nd:YAG laser beam ($\lambda = 1.06$ μm) to a 75μm 1/e^2 diameter spot. The depth of the dipole potential ($\sim 150U_{rec}$) was comparable to that of our blue-detuned trap, and the detuning from the Na D lines (2.26*10^{14} Hz) was ~2.5 times larger. The measured coherence times was $T_c \approx$ 15 msec, a factor of ~300 smaller then for the blue-detuned trap.

ATOM INTERFEROMETRY BASED ON ADIABATIC TRANSFER

Longer coherence times will play a crucial role in increasing the sensitivity of atom interferometers. Several groups have recently shown that laser light can coherently split and recombine atoms,[19] with the most precise interferometers using laser cooled atoms. Atom interferometers have been used to measure the acceleration due to gravity[20] with a sensitivity of 3 parts in 10^8 and $\hbar/m_{atom}$ to a resolution of one part in 10^7.[21]

In addition to increasing the coherence time of the interferometers, the precision can also be improved if better methods of coherently changing the momentum of the atom are developed. Our previous experiments used beam splitters and mirrors based on de-tuned stimulated Raman transitions and achieved a transfer efficiency of 85% per $2\hbar k$ of momenta exchanged.

In this section, we report our work on population transfer based on adiabatic passage. As with many laser coherence techniques, adiabatic passage was first introduced in nuclear magnetic studies[22] before being carried over into the optical domain.[23] Population transfer by adiabatic passage using delayed laser pulses was first conceived and observed by Gaubatz, et al.[24] in molecular systems. Marte, et al.[25] then proposed the use of this method for atomic beam splitters, and coherent transfer of $6\hbar k$ momentum using adiabatic following has recently been demonstrated by Lawall and Prentiss,[26] and $8\hbar k$ was achieved in the work of Goldner, et al.[27]

Adiabatic transfer has some experimental advantages. It is experimentally robust since it is not sensitive to changes in parameters such as laser power or the time the light is on. Also, we have shown that adiabatic transfer in a pure three level system performed with light on resonance does not introduce any ac Stark shift even when the transfer is not 100% adiabatic. This fact may also have applications in optical frequency standards since one of the major drawbacks of using a two-photon clock transition is the induced ac Stark shift.

We presented the formal proof in a previously submitted paper,[28] but the absence of a Stark shift also has an intuitive explanation. The two light-coupled eigenstates of the three level system have equal and opposite energy level shifts at zero detuning. If the non-adiabatic transfer from the uncoupled state populates the two coupled states with an equal weight, there will be no net light shift.

Our experiment differs from the previous work in several ways. We used a magnetic field insensitive transition, which is essential for precision interferometry. For example, in our previous $\hbar/m_{atom}$ experiment the positions of the interferometer fringes were measured to a precision of less than 50 mHz while the magnetic precession of a $m_F=\pm F$ state is 1.4 MHz/Gauss. Our adiabatic transfer was between the two cesium hyperfine ground states $6S_{1/2}$, $F=3$, $m_F=0$ and $6S_{1/2}$, $F=4$, $m_F=0$ via the excited state $6P_{1/2}$, $F'=3$ or 4, $m_F=1$. We used two counterpropagating laser beams in a σ^+-σ^+ polarization configuration, and since both beams have the same helicity, atoms that are not transferred adiabatically will be predominantly optically pumped into the $F=4, m_F=+4$ and $F=3, m_F=+3$ states. Thus, the fraction of atoms that are coherently transferred can be measured and the loss of interferometer contrast though non-adiabatic transfers is decreased. Off-resonant excitation limited the observed efficiency to ~80% per exchanged photon pair in the work of Goldner, et al, using the D_2 line if cesium.[27] We used the $6P_{1/2}$ state, which has a 5.8 times larger excited state hyperfine splitting (1.17 GHz) than the $6P_{3/2}$ state, so that the off-resonant excitation to a fourth level is significantly reduced.[29] Also, by shaping the excitation pulses, we were able to improve the transfer efficiency. The transfer of many photon kicks was achieved by reversing the direction of the two laser beams between successive transfers. By incorporating all of the above changes, we have achieved coherent transfer with an efficiency as high as 95% per exchanged photon pair, and up to 140 photon momenta were coherently added.

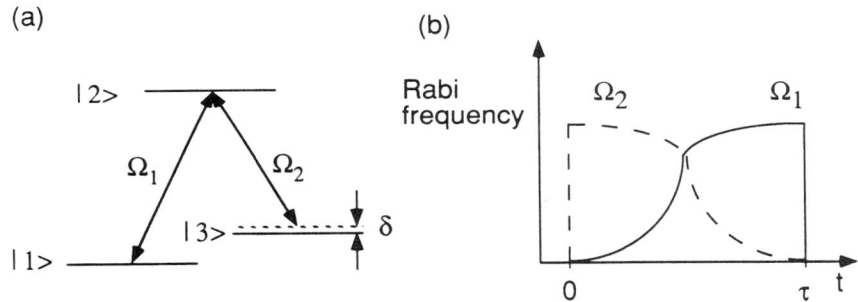

Fig. 7. (a) Level scheme and (b) pulse sequence for complete population transfer for adiabatic following based on delayed laser pulses. The condition for adiabatic following is $\Omega_{max}\tau >> 1$.

Fig. 7a shows the level scheme for adiabatic transfer. Optical beams of frequencies ω_1 and ω_2 connect two ground states $|1>$ and $|3>$ via an excited state $|2>$. With zero detuning and Rabi frequencies Ω_1 at ω_1 and Ω_2 at ω_2, there is a new eigenstate of the atom plus photon field (a coherent superposition of two ground states) in which the amplitudes for absorption into the excited state cancel.[30] By changing the light intensities slowly as shown in Fig. 7b, population can be adiabatically transferred from $|1,p>$ to $|3, p + \hbar k_1 - \hbar k_2>$ while remaining in the uncoupled "dark state". For counterpropagating laser beams two photon momenta are transferred to the atoms.

The adiabatic transfer method can also be used to construct the necessary interferometer components for an interferometer as shown on Fig. 8. A coherent superposition of two states of different momenta is created by turning the intensity of both beams to zero as shown for pulse shape A in Fig. 8. If the light is turned off adiabatically with $\Omega_1 = \Omega_2$, the state immediately after pulse A contains $|1,\mathbf{p}\rangle$ and $|3,\mathbf{p}+\hbar\mathbf{k}_1-\hbar\mathbf{k}_2\rangle$ with equal amplitudes. In pulse B, both Ω_1 and Ω_2 are increased simultaneously. In general, the atom will be found in a superposition of the coupled and non-coupled ground states. As pulse B evolves, the part of the atom in the equal superposition dark state will be adiabatically transferred to the new dark state, $|1,\mathbf{p}\rangle$. Any part of the atom that projects onto the coupled state undergoes spontaneous emission and falls into a number of m_F states. The dotted lines symbolize the loss of atoms out of the dark state. Pulse C takes the part of the atom in the dark state $|1,\mathbf{p}\rangle$ immediately before the pulse and creates another superposition dark state. Finally, pulse D is used to spatially recombine the atom and complete the interferometer. The interference signal is calculated in a straightforward manner, and the details of the calculation are presented in another publication.[31]

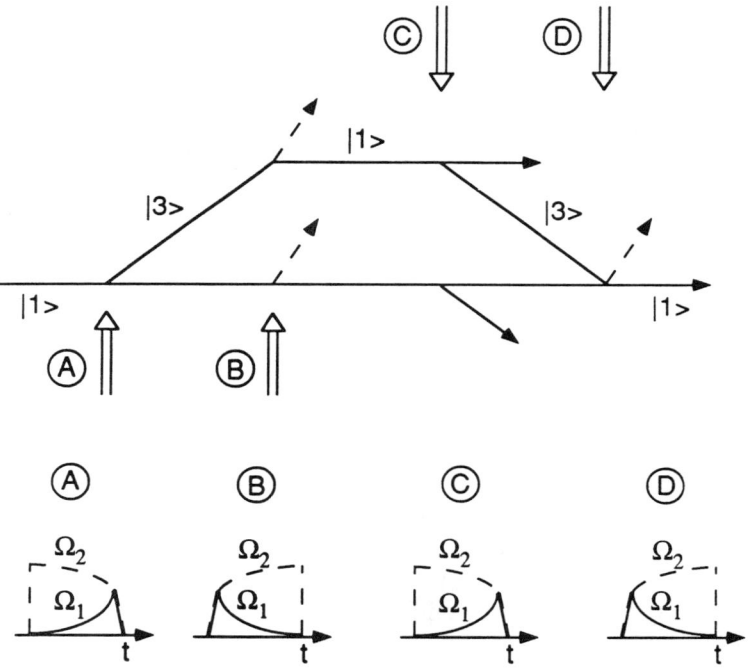

Fig. 8. Pulse sequence for an atomic interferometer using adiabatic following. The double arrows show the propagation axis of the light with frequency w_1 (Rabi frequency Ω_1); light with frequency w_2 (Rabi frequency Ω_2) is directed oppositely. The dashed lines show loss of atoms due to spontaneous emission.

A schematic of the experimental apparatus is shown in Fig. 9. For more details, the reader is referred to our earlier atom interferometer papers.[20,21] A cesium atomic beam, slowed by a chirped laser beam, loaded a magneto-optic trap.[32] The light necessary for cooling and trapping was generated from a Ti:Sapphire laser operating at the cesium D_2 transition $6S_{1/2}$-$6P_{3/2}$. In 0.4 s about 10^9 atoms were loaded, after which the trapping magnetic field was shut off and the atoms were further cooled to 4 μK in polarization-gradient optical molasses.[33] The atoms were then launched in a vertical ballistic trajectory by acousto-optically shifting the molasses beams frequencies to a moving molasses[34] traveling upwards at 2.3 m/s. The molasses beams were then blocked with a mechanical shutter to prevent low level light leaking through the AO switch from destroying the phase coherence. After launching, the atoms were optically pumped into F=4, m_F=0 before entering a magnetically shielded region with a homogeneous 100 mG magnetic bias field oriented parallel to the Raman beams.

The two Raman beams were generated from a second Ti:Sapphire laser locked to the D_1 transition $6S_{1/2}$,F=4 → $6P_{1/2}$,F'=4 or 3 at 894 nm. The second Raman frequency component at $6S_{1/2}$(F=3)-$6P_{1/2}$(F') was generated by directing part of the D_1 light through a 9.2 Ghz electro-optic modulator (EOM) and then through a cavity locked to the upper sideband. The cavity filtered out all optical frequencies by 35 dB except for the desired sideband. Two 40 MHz acousto-

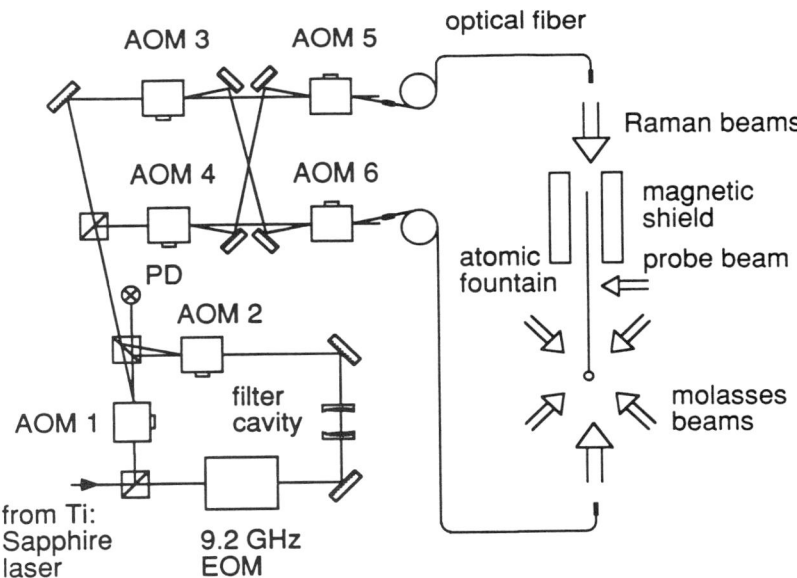

Fig. 9. Experimental setup used for an atomic interferometer based on delayed laser pulses. The optics used to generate the beams for the atomic fountain are not shown and the slowing, clearing and hyperfine pumping beams have been omitted.

optic modulators (AOM's 1 and 2) were used to generate the pulse shapes for adiabatic following.

The undeflected fraction of both acousto-optic modulators were mixed on a fast photodiode and the beat frequency was phase locked to an rf source by controlling the electro-optic modulator drive frequency. This difference frequency was adjusted between each pulse to account for the gravitational acceleration of the atoms. The directions of the Raman beams were switched after each transfer with a set of four acousto-optic modulators. The beams were then coupled into single mode optical fibers and expanded to a 2 cm Gaussian diameter with 6 mW/beam. The beam quality and collimation were measured to be better than $\lambda/10$ over the central 1 cm diameter using an optical sheering interferometer.

We studied the limitations to transfer efficiency. The efficiency is sensitive to the purity of the Raman beam polarizations; if the two beams have different polarizations the dark state will not be a pure $m_F=0$ state. Back reflections from the vacuum can windows also limit the transfer efficiency in the Doppler-sensitive case since the moving atoms would be exposed to light with incorrect Raman difference frequencies. For a quantitative evaluation of how these imperfections degraded the transfer efficiency, we refer the reader to another publication.[31]

After obtaining efficient population transfer, we constructed an atomic interferometer based on adiabatic transfer via the dark state. We operated the interferometer as follows: Adiabatic pulses (typically 150 μs long) as shown in Fig. 9 were applied to the atoms. A laser beam tuned to $6S_{1/2}, F=4 \rightarrow 6P_{3/2}, F'=5$ then pushed away residual population in the $F=4$ state. The population in $6S_{1/2}, F=3, m_F=0$ was measured by first transferring the population to $6S_{1/2}, F=4, m_F=0$ with a microwave π-pulse. After the atoms have fallen back into the probe beam region, light tuned to $6S_{1/2}, F=4 \rightarrow 6P_{3/2}, F'=5$ was pulsed on and the subsequent fluorescence was measured.

Our experimental results for the interferometer is shown in Fig. 10 for a spacing $T=250$ μs between both pairs of pulses. The fringes were recorded by varying the Raman difference frequency of only the final two pulses. The broad envelope in the fringes corresponds to the frequency width of a single pulse. The experimental fringe contrast is 29% when we transferred with the $6P_{1/2}, F'=3$ excited state, while the expected fringe contrast is ~33%. We conjecture that the experimental fringe contrast is less than the theoretically predicted contrast because of imperfect polarization.

We have observed good contrast interferometer fringes with a spacing between pulses of up to $T=1.3$ ms, corresponding to a total interferometer measurement time of ~3.5 ms. Probably, the contrast began to wash out for longer interferometer times because of the frequency jitter on the Raman beams induced by vibrations. By using a vibration isolation system with a feedback circuit, we hope to substantially increase the interferometer drift time.[35]

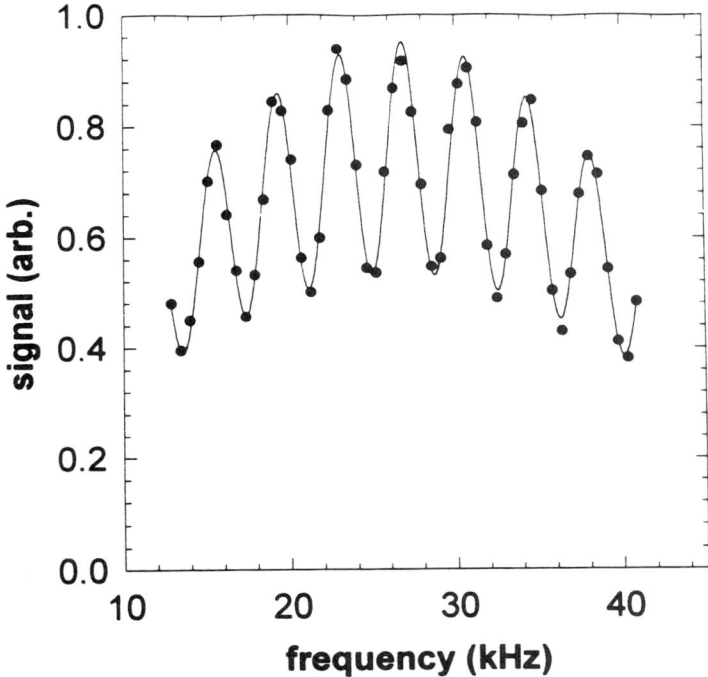

Fig. 10. Interference fringes for an atomic interferometer based on adiabatic following using pulse shapes as shown in Fig. 8. The time between pulses A and B and also C and D was T = 250 μs, while the spacing between pulses B and C was 250 μs. Each point corresponds to 4 launches at a rate of 0.9 launches/s. The solid line is a fit to a cosine function with a Gaussian envelope.

We plan to use an adiabatic transfer interferometer in our next measurement of $\hbar/m_{atom}$.[21] The adiabatic transfer method should allow us to insert more photon recoil momenta than our previous measurement and induce smaller ac Stark shifts. This method can also be used in large area atom interferometers with extremely high sensitivity to inertial effects. However, it is not obvious that an atom interferometer based on adiabatic transfer will be superior to the off-resonant Raman method we had previously introduced, and we plan to try both methods. We will have two different interferometer methods with different systematics for measuring $\hbar/m_{atom}$.

This work was supported in part by grants from the NSF and the AFOSR. CSA was a Lindemann Fellow during the course of this work.

REFERENCES:

1. See the special issue "Laser Cooling and Trapping of Atoms", J. Opt. Soc. Am. **B6**, (1989), eds. S. Chu and C.E. Wieman.

2. E.L. Raab, M. Prentiss, A.E. Cable, S. Chu, and D.E. Pritchard, Phys. Rev. Lett. *59*, 2631 (1987).

3. T. Walker, D. Sesko and C. Wieman, Phys. Rev. Lett. **64**, 408 (1990); D. Sesko T. Walker and C. Wieman, J. Opt. Soc. B**8**, 946 (1991).

4. S. Chu, J.E. Bjorkholm, A. Ashkin, and A. Cable, Phys. Rev. Lett. *57*, 314 (1986).

5. J.D. Miller, R.A. Cline, and D.J. Heizen, Phys. Rev. Lett. A**47**, R4567 (1993); S. Rolston, et al. Proc. SPIE **1726**, 205 (1992).

6. M.A. Kasevich *et. al.*, Phys. Rev. Lett. **66**, 2297 (1991).

7. E. Tiesinga, S.J.M. Kuppens, B.J. Verhaar, and H.T.C. Stoof, Phys. Rev. A**43**, 5188 (1991).

8. W. Ketterle, et al. Phys. Rev. Lett. **70**, 2253 (1993).

9. M. Kasevich and S. Chu, Phys. Rev. Lett. **69**, 1741 (1992).

10. H.F. Hess, G.P. Kochanski, J.M. Doyle, N. Masuhara, D. Kleppner and T.J. Greytak, Phys. Rev. Lett. **59**, 672 (1987); J.M. Doyle, J.C. Sandberg, I.A. Yu, C.L. Cesar, D. Kleppner and T.J. Greytak, Phys. Rev. Lett. **57**, 314 (1991).

11. M.A. Kasevich, D.S. Weiss, and S. Chu, Opt. Lett. **15**, 607 (1990), K. Shimoda, IEEE Trans. Instrum. Meas. **38**, 150 (1989). For grazing incidence see, V. I. Balykin *et al.*, Phys. Rev. Lett. **60**, 2137 (1988).

12. C.G. Aminoff *et al.*, Phys. Rev. Lett. **71**, 3083 (1993).

13. H. Wallis, J. Dalibard, and C. Cohen-Tannoudji, Appl. Phys. B **54**, 407 (1992).

14. R. Kaiser *et al.*, Opt. Comm. Opt. Comm. 104, 234 (1994).

15. T. Esslinger *et al.*, Opt. Lett. **18**, 450 (1993). W. Seifert *et al.*, Phys. Rev. A **49**, 3814 (1994). S. Feron *et al.*, Opt. Comm. 102, 83 (1993).

16. The fringe contrast is calculated by $\{ \int \cos[\pi T(f-\delta f)]N(f)df\}/ \int N(f)df$, where $N(f)$ is the light shift distribution of Fig. 5, and δf is the frequency shift of the Ramsey fringe that is found by maximizing the contrast.

17. H. M. Goldenberg, D. Kleppner, and N. F. Ramsey, Phys. Rev. Lett. **8**, 361 (1960).
18. S. A. Murthy *et al.*, Phys. Rev. Lett. **63**, 965 (1989).
19. See, for example, articles in Appl. Phys. B **54**, 321 ff (1992).
20. M. Kasevich and S. Chu, Phys. Rev. Lett. **67**, 181 (1991).
21. D. S. Weiss, B. C. Young, and S. Chu, Phys. Rev. Lett. **70**, 2706 (1993); D. S. Weiss, B. C. Young, and S. Chu, submitted to Appl. Phys. B.
22. A. Abragam, The Principles of Nuclear Magnetism, (Oxford Univ. Press. London, 1961).
23. D. Grischkowsky, Phys. Rev. Lett. **24**, 866 (1970).
24. U. Gaubatz, P. Rudecker, M. Becker, S. Schiemann, M. Külz, and K. Bergmann, Chem. Phys. Lett. **149**, 463 (1988).
25. P. Marte, P. Zoller, and J. L. Hall, Phys. Rev. A **44**, R4118 (1991).
26. J. Lawall and M. Prentiss, Phys. Rev. Lett. **72**, 993 (1994).
27. L.S. Goldner at al., Phys. Rev. Lett. **72**, 997 (1994).
28. M. Weitz, B. C. Young, and S. Chu, Phys. Rev. A **50**, 2438 (1994).
29. P. Pillet, C. Valentin, R.-L. Yuan, and J. Yu, Phys. Rev. A **48**, 845 (1993).
30. E. Arimondo and G. Orriols, Lett. Nuovo Cimento **17**, 333 (1976).
31. M. Weitz, B.C. Young, and S. Chu, accepted by Phys. Rev. Lett. (1994).
32. E. L. Raab, M. Prentiss, A. Cable, S. Chu, and D. E. Pritchard, Phys. Rev. Lett. **59**, 2631 (1987).
33. J. Dalibard and C. Cohen-Tannoudji, J. Opt. Soc. Am. B**6**, 2023 (1989); P.J. Ungar, D.S. Weiss, E. Riis, and S. Chu, J. Opt. Soc. Am. B**6**, 2058 (1989).
34. D.S. Weiss, E. Riis, M.A. Moler, and S. Chu, in *Light Induced Kinetic Effects on Atoms, Ions, and Molecules*, eds. I. Moi, S. Gozzini, C. Gabbanini, E. Arimondo, and F. Strumia (ETS Editrice, Pisa, 1991), p. 35.
35. A. Peters, J. Hensley and S. Chu, to be published.

QUANTUM OPTICS AND COHERENCE

Measuring the Phase of a Quantum Field

Leonard Mandel

Department of Physics and Astronomy
University of Rochester
Rochester NY 14627

Abstract. The operational approach to the problem of identifying the dynamical variable(s) representing the phase difference between two quantum fields is reviewed. The results of numerous measurements in which the theory has been put to the experimental test all confirm the validity of the approach, according to which the dynamical variables representing the phase depend on the measurement scheme.

INTRODUCTION

The problem of identifying the dynamical variable that properly represents the phase of a quantized electromagnetic field, or more appropriately the phase difference between two fields, is almost as old as quantum mechanics itself. It was first tackled by Dirac in his first paper on field quantization (1), but the phase operator introduced by Dirac was later found to be non-Hermitian. Later Susskind and Glogower (2) and Carruthers and Nieto (3) constructed Hermitian operators $\hat{C}$ and $\hat{S}$ that are analogous to the cosine and sine of the phase. Unfortunately $\hat{C}$ and $\hat{S}$ do not commute, so that there is not one phase but a cosine phase and a sine phase, and this has been widely regarded as an unsatisfactory feature of this theory.

In the following years numerous new theoretical approaches were developed (4), some of which yield the complete probability distribution of the phase (5-13), but no general consensus as to which one is correct emerged. We mention particularly an idea due to Pegg and Barnett (14-16), which has generated a good deal of interest. By working in a truncated Hilbert space in which the photon occupation number has an upper bound s, they succeeded in identifying a Hermitian phase operator. s is allowed to tend to infinity only at the end of the calculation. The results of this approach sometimes agree with the Susskind and Glogower theory (2).

We have approached the phase problem in an entirely different, more operational, manner (17). Instead of looking for an operator with certain mathematical characteristics, we identify operators that correspond to what is usually measured in the laboratory when the phase is determined in classical optics. This leads to the conclusion that there is not a unique phase operator, but rather that different measurement schemes are associated with different operators. The predictions of this operational theory differ, in general, from the predictions of other phase theories. Finally, we show that the operational phase theory is very well confirmed in numerous experiments (17-24).

We start by briefly outlining our operational approach to the phase. We then present the results of several experiments in which the theoretical predictions of this phase theory are tested, and we make some comparisons with other phase theories.

OPERATORS CORRESPONDING TO CERTAIN MEASUREMENTS

In classical optics the phase difference between two light beams is usually measured in some kind of interference or homodyne experiment. A typical arrangement, that we designate as scheme 1, is illustrated in fig. 1. The two input fields, characterized by complex amplitudes a_1, a_2, are mixed by a 50%: 50% beam splitter at 45°, and the two output fields a_3, a_4 are measured with the two photodetectors D3, D4, whose outputs are compared. If ϕ_1, ϕ_2 are the phases of the two incoming fields, then $\sin(\phi_2 - \phi_1)$ is proportional to the difference between the integrated output light intensities registered by detectors D_3 and D_4. This still does not completely determine the phase difference $\phi_2 - \phi_1$, but if it is possible to insert a quarter wave phase shifter in one input, as shown in fig. 1, and remeasure the outputs, now relabeled 5 and 6, and if the light intensities and the phases have not changed significantly, then we can determine $\cos(\phi_2 - \phi_1)$, which is proportional to the difference between the integrated light intensities 5 and 6.

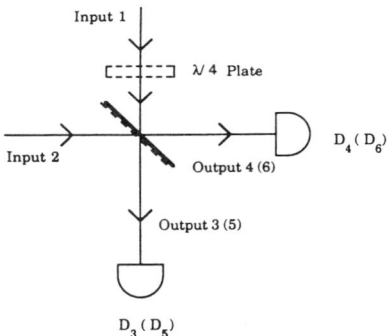

FIGURE 1. Outline of a scheme for measuring the sine or cosine of the phase difference between two optical fields at input ports 1 and 2. [Reproduced from reference 17]

When the field is quantized and each input may be treated as a single mode, $\hat{a}_1, \hat{a}_2$ are photon annihilation operators obeying the usual commutation relations, and so are the output fields $\hat{a}_3, \hat{a}_4$ and $\hat{a}_5, \hat{a}_6$. From the relations between inputs and outputs of the beam splitter we readily obtain (17):

$$\sin(\phi_2 - \phi_1) \propto \hat{n}_3 - \hat{n}_4 \equiv \hat{S} \tag{1}$$

$$\cos(\phi_2 - \phi_1) \propto \hat{n}_5 - \hat{n}_6 \equiv \hat{C}. \tag{2}$$

If we take the operators $\hat{S}$ and $\hat{C}$ to characterize the sine and the cosine of the phase difference, we find that

$$[\hat{S}, \hat{C}] \neq 0, \tag{3}$$

which reflects the fact that the sine and cosine measurements are alternatives and therefore are not mutually compatible, even if the unperturbed fields do not fluctuate between measurements.

There is, however, no need to make the sine and cosine measurements successively. With the help of the more complicated 8-port arrangement shown in fig. 2, (25-28) that we call scheme 2, we can obtain the sine and cosine of the phase difference $\phi_2 - \phi_1$ simultaneously. A simple analysis of the input and output variables leads to the conclusion that (17)

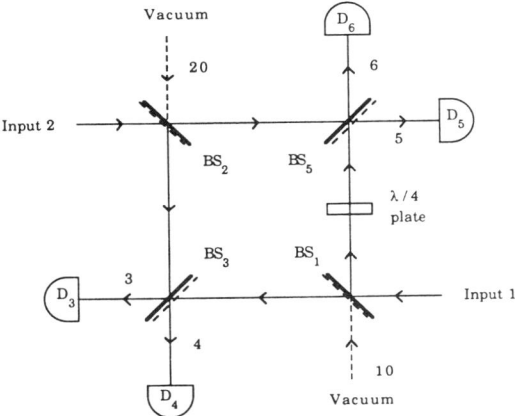

FIGURE 2. Outline of an eight-port scheme for simultaneously measuring the sine and cosine of the phase difference between two optical fields at input ports 1 and 2. [Reproduced from reference 17]

$$\cos(\phi_2 - \phi_1) \propto \hat{n}_4 - \hat{n}_3 = \frac{1}{2}(\hat{a}_1^\dagger - i\hat{a}_{10}^\dagger)(\hat{a}_2 - i\hat{a}_{20}) + \text{hc} \equiv \hat{C} \quad (4)$$

$$\sin(\phi_2 - \phi_1) \propto \hat{n}_6 - \hat{n}_5 = \frac{1}{2}(-i\hat{a}_1^\dagger + \hat{a}_{10}^\dagger)(\hat{a}_2 + i\hat{a}_{20}) + \text{hc} \equiv \hat{S}, \quad (5)$$

where $\hat{a}_{10}$, $\hat{a}_{20}$ represent the vacuum modes at the two unused input ports. Unlike $\hat{C}$ and $\hat{S}$ in Eqs. (1) and (2), these $\hat{C}$, $\hat{S}$ operators commute, which reflects the fact that they are measurable at the same time. In order to ensure compatibility of the sine and cosine we shall take the measured operators to be

$$\hat{C}_m = \hat{C}(\hat{C}^2 + \hat{S})^{-\frac{1}{2}} \quad (6)$$

$$\hat{S}_m = \hat{S}(\hat{C} + \hat{S}^2)^{-\frac{1}{2}}, \quad (7)$$

and then $\hat{C}_m$ and $\hat{S}_m$ commute also. We note that measurement schemes 1 and 2 correspond to different measured sine and cosine operators.

Because of the irrational operator factors in Eqs. (6) and (7), it is worth noting that the expectation of functions of the sine and cosine can be calculated without using irrational operators. In general, for any function $f(\hat{n}_3, \hat{n}_4, \hat{n}_5, \hat{n}_6)$ of the photon numbers, we may write for the quantum expectation

$$\langle f(\hat{n}_3, \hat{n}_4, \hat{n}_5, \hat{n}_6) \rangle = \sum_{\{n\}} f(n_3, n_4, n_5, n_6) P(n_3, n_4, n_5, n_6), \quad (8)$$

where $P(n_3, n_4, n_5, n_6)$ is the joint probability for n_3 photons to be detected at output port 3, n_4 photons at output port 4, n_5 photons at output port 5, n_6 photons at output port 6. With perfect detectors this is given by

$$P(n_3, n_4, n_5, n_6) = \left\langle : \prod_{j=3}^{6} \frac{(\hat{a}_j^\dagger \hat{a}_j)^{n_j} e^{-\hat{a}_j^\dagger \hat{a}_j}}{n_j!} : \right\rangle. \quad (9)$$

With photodetectors of detection efficiency α and surface area S used at normal incidence, the joint probability $P(m_3, m_4, m_5, m_6)$ for m_3 detections by D_3, m_4 detections by D_4, m_5 detections by D_5, and m_6 detections by D_6 during the measurement time T is given by

$$P(m_3,m_4,m_5,m_6) = \prod_{j=3}^{6} \left\langle : \frac{\left[\alpha cS \int_t^{t+T} \hat{I}_j(t')dt'\right]^{m_j}}{m_j!} e^{-\alpha cS \int_t^{t+T} \hat{I}_j(t')dt'} : \right\rangle, \qquad (10)$$

when $\hat{I}_j(t)$ is light intensity expressed in photons per unit volume.

Two different limiting situations are commonly encountered. When the photon counting time T is much shorter than the coherence time T_c of the light, we may approximate the integrals in Eq. (10) by writing

$$\alpha cS \int_t^{t+T} \hat{I}_j(t')dt' \Rightarrow \alpha\left(\frac{cTS}{L^3}\right)\hat{a}_j^\dagger \hat{a}_j = \beta \hat{n}_j. \qquad (11)$$

Here L^3 is the quantization volume, and β is the product of the detection and the effective collection efficiency. Then

$$P(m_3,m_4,m_5,m_6) = \left\langle : \prod_{j=3}^{6} \frac{(\beta \hat{n}_j)^{m_j} e^{-\beta \hat{n}_j}}{m_j!} : \right\rangle. \qquad (12)$$

On the other hand, when the photon counting time T is much longer than T_c, then the time integral can be approximated by the time average multiplied by T. But for an ergodic process the time average equals the average over the ensemble, so that we may write

$$\alpha cS \int_t^{t+T} \hat{I}_j(t')dt' \to \alpha\left(\frac{cTS}{L^3}\right)\langle \hat{n}_j \rangle = \beta \langle \hat{n}_j \rangle \qquad (13)$$

and substitute in Eq. (10). Then

$$P(m_3,m_4,m_5,m_6) = \prod_{j=3}^{6} \frac{(\beta \langle \hat{n}_j \rangle)^{m_j} e^{-\beta \langle \hat{n}_j \rangle}}{m_j!}. \qquad (14)$$

In both cases, the effect of less than perfect detection and collection efficiency is incorporated by replacing $\hat{a}_j^\dagger \hat{a}_j$ by $\beta \hat{a}_j^\dagger \hat{a}_j$.

The experimental outcome $m_3 = m_4$ and $m_5 = m_6$ deserves special mention. For measurement scheme 1 we see from Eqs. (1) and (2) that both $\cos(\phi_2 - \phi_1)$ and $\sin(\phi_2 - \phi_1)$ are then zero, but there are no phase angles whose cosine and sine both vanish. For measurement scheme 2 we find that C_m and S_m are both undefined when $m_3 = m_4$, $m_5 = m_6$. We shall therefore exclude such combinations from the sum in Eq. (8), and renormalize the sum by dividing by $1 - \mathcal{P}$, where $\mathcal{P}$ is the probability for the outcome $m_3 = m_4$, $m_5 = m_6$. By using Eq. (9) we may show that (17)

$$\mathcal{P} = \left\langle :I_0\left\{\beta[(\hat{a}_1^{\dagger 2} - \hat{a}_2^{\dagger 2})(\hat{a}_1^2 - \hat{a}_2^2)/2]^{1/2}\right\} I_0\left\{\beta[(\hat{a}_1^{\dagger 2} + \hat{a}_2^{\dagger 2})(\hat{a}_1^2 + \hat{a}_2^2)/2]^{1/2}\right\} e^{-\beta(\hat{n}_1 + \hat{n}_2)} : \right\rangle \tag{15}$$

where $I_0(z)$ is the modified Bessel function of zero order. Henceforth we shall assume that this renormalization has been carried out. The resulting correction will generally be negligible when $\langle m_1 \rangle, \langle m_2 \rangle$ are large numbers, but it becomes increasingly significant the smaller $\langle m_1 \rangle, \langle m_2 \rangle$. Discarding the all zero outcomes makes it possible to obtain some information about the phase difference $\phi_2 - \phi_1$ in direct measurements of two weak optical fields, as we shall demonstrate below.

As a check on the reasonableness of the resulting formalism, we now calculate some expectations of $\hat{C}_m$ and $\hat{S}_m$. When the two inputs to the interferometer in scheme 2 are in the two-mode state $|\theta_1\rangle|\theta_2\rangle$ of definite phase (14-16), where

$$|\theta\rangle \equiv \frac{1}{(s+1)^{1/2}} \sum_{n=0}^{s} e^{in\theta} |n\rangle, \tag{16}$$

and the limit $s \to \infty$ is to be taken only at the end of the calculation, we find that

$$\left. \begin{array}{l} \langle \hat{C}_m \rangle = \cos(\theta_2 - \theta_1) \\ \langle \hat{S}_m \rangle = \sin(\theta_2 - \theta_1) \\ \langle \hat{C}_m^2 \rangle = \cos^2(\theta_2 - \theta_1) \\ \langle \hat{S}_m^2 \rangle = \sin^2(\theta_2 - \theta_1). \end{array} \right\} \tag{17}$$

This is exactly what is to be expected for a state of well-defined phase.

When the interferometer input is in the 2-mode Fock state $|n_1\rangle|n_2\rangle$, for which no definite phase is to be expected, one finds (19)

$$\langle \hat{C}_m \rangle = 0 = \langle \hat{S}_m \rangle$$
$$\langle \hat{C}_m^2 \rangle = \frac{1}{2} = \langle \hat{S}_m^2 \rangle, \qquad (18)$$

which is typical of a randomly phased system.

Finally, we note that the same formalism allows the probability distribution $P(\phi_2 - \phi_1)$ of the phase difference $\phi_2 - \phi_1$ between two fields to be calculated and compared with measurement (22). For this purpose let us identify the function $f(n_3, n_4, n_5, n_6)$ in Eq. (8) with

$$f(n_3,n_4,n_5,n_6) = \left\{ \frac{(n_4-n_3)+i(n_6-n_5)}{\left[(n_4-n_3)^2+(n_6-n_5)^2\right]^{1/2}} \right\}^x \equiv e^{i\Theta x}, \qquad (19)$$

whose expectation is the characteristic function C(x) of the phase difference $\phi_2 - \phi_1$. Fourier inversion of C(x) then should give the measured probability distribution $P_m(\phi_2 - \phi_1)$,

$$P_m(\phi_2 - \phi_1) = \frac{1}{2\pi} \int_{-\infty}^{\infty} dx\, C(x) e^{-ix(\phi_2-\phi_1)} . \qquad (20)$$

In practice, the situation is complicated by the fact that when there are few photons, there are also few non-zero outcomes of the measurement. For example, when $\langle m_1 \rangle, \langle m_2 \rangle \ll 1$, then non-zero experimental outcomes other than $m_3 = 1$, and others zero; $m_4 = 1$, and others zero; $m_5 = 1$, and others zero; $m_6 = 1$, and others zero are very improbable, and each of these 4 outcomes corresponds to one of 4 possible values of the phase difference (19). No other values are ever encountered. But a simple procedural change is capable of dealing with this problem and distributing the outcomes over the range $-\pi$ to π. We deliberately shift the phase of one of the input fields, say $\hat{a}_2$, by some amount θ, and then reconstruct the conditional probability density $p(\phi_2 - \phi_1|\theta)$. We repeat this procedure for many different values of θ in small equal steps B over the range

$-\pi/2$ to $\pi/2$. This distributes the output photons over all the detectors. Then the desired probability distribution $P(\phi_2 - \phi_1)$ is obtained by averaging all the different $p(\phi_2 - \phi_1|\theta)$ over θ. Thus, if $\langle \ \rangle'$ denotes the quantum expectation in the phase shifted state, then (22)

$$p(\phi_2 - \phi_1|\theta) = \frac{1}{2\pi} \int_{-\infty}^{\infty} \langle e^{ix\Theta n}\rangle' e^{-ix(\phi_2 - \phi_1 - \theta)} dx, \qquad (21)$$

and after averaging over all θ we have in the limit of sufficiently small B,

$$P_m(\phi_2 - \phi_1) = \frac{1}{2\pi} \int_{-\pi}^{\pi} d\theta p(\phi_2 - \phi_1|\theta). \qquad (22)$$

EXPERIMENTS TO MEASURE PHASE

We now turn from theory to experiment and we describe a number of phase measurements which allow the foregoing operational phase theory to be tested. The measurements below are all based on scheme 2, which makes use of the eight-port device to give both the sine and cosine of the phase difference between the two input fields.

For the first experiments the light source was a stable single-mode He:Ne laser whose output beam is split with a 50%:50% beam splitter into two parts serving as inputs 1 and 2 to the eight-port interferometer. (18,19) The beam splitter is mounted on a piezo electric transducer that allows the phase difference between the two inputs to be varied, but the phase difference was held constant for one series of measurements. The mean photon counts at the inputs were varied over a 3000:1 range from 30 down to 10^{-2} by adjustment of the photon counting time T from just over 1msec to 0.5μsec. The interference fringes at the interferometer output generally have visibility above 98%.

Fig. 3 shows the measured values of $\langle \hat{C}_m \rangle$ as a function of the mean detected photon number $\langle m_1 \rangle$, for a fixed ratio (a) $\langle m_2 \rangle / \langle m_1 \rangle = 1$ and (b) $\langle m_2 \rangle / \langle m_1 \rangle = 7$, superimposed on the theoretical curve derived from Eqs. (6) and (12). Good agreement is obtained. Two features are worth noting. The ratio $\langle \hat{C}_m \rangle / \cos(\theta_2 - \theta_1)$ turns out to be independent of the phase difference $\theta_2 - \theta_1$ between the two input beams, and it becomes unity when $\langle m_1 \rangle, \langle m_2 \rangle \to \infty$. Also we observe that, even when $\langle m_1 \rangle, \langle m_2 \rangle \ll 1$, the data still contain some information about the phase difference, which is connected with the fact that all zero outcomes $m_3, m_4, m_5, m_6 = 0$ are discarded, although they are most probable.

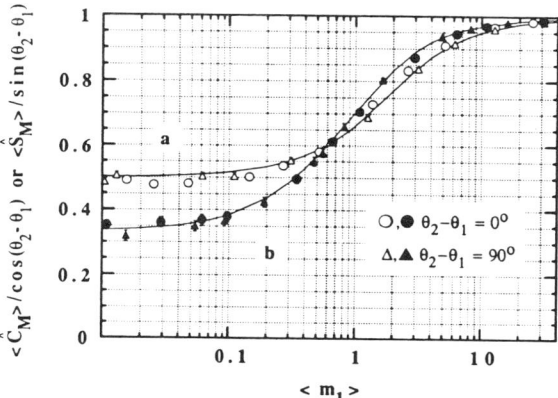

FIGURE 3. Comparison of measured and calculated values of $\langle \hat{C}_m \rangle$ or $\langle \hat{S}_m \rangle$ as function of average detected photon number $\langle m_1 \rangle$ for two coherent fields with (a) $\langle m_2 \rangle / \langle m_1 \rangle = 1$; (b) $\langle m_2 \rangle / \langle m_1 \rangle = 7$. [Reproduced from reference 18]

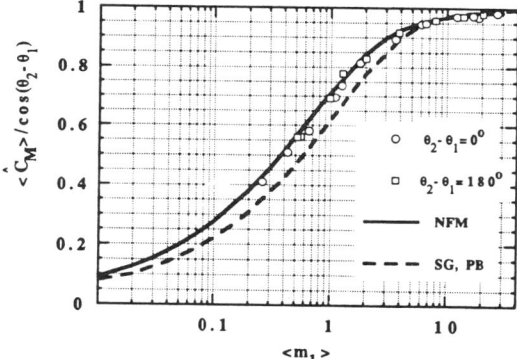

FIGURE 4. Comparison of measured and calculated values of $\langle \hat{C}_m \rangle$ as function of average detected photon number $\langle m_1 \rangle$ for two coherent fields with $\langle m_2 \rangle = 50$. The full curve corresponds to the theoretical prediction of the theory of Noh, Fougères and Mandel, and the dashed curve to the theories of Susskind and Glogower and Pegg and Barnett. [Reproduced from reference 19]

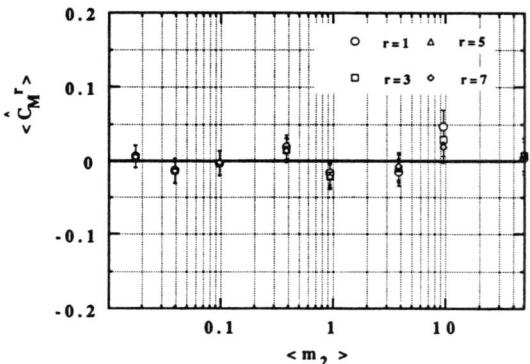

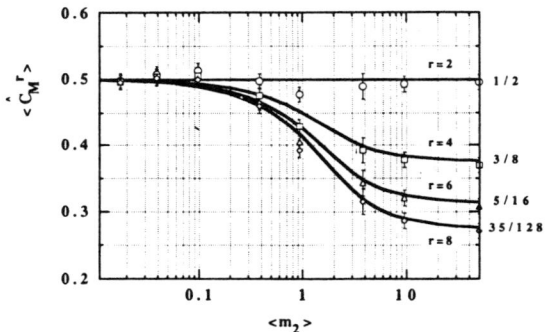

FIGURE 5. Comparison of measured and calculated values of the (a) odd and (b) even moments of $\hat{C}_m$ as function of average detected photon number $\langle m_2 \rangle$ for a vacuum field entering at input port 1 and a coherent field entering at port 2. [Reproduced from reference 19]

Fig. 4 shows the results of similar measurements (19), in which $\langle m_2 \rangle$ is held fixed at $\langle m_2 \rangle = 50$ and $\langle m_1 \rangle$ is varied, superimposed on two different theoretical curves based on our formalism (NFM) and on the Susskind-Glogower (SG) and Pegg-Barnett (PB) theories. Once again the data are in good agreement with the (NFM) theory. Figs. (5a) and (5b) present the results of measurements of the first 8 moments of $\hat{C}_m(\phi_2 - \phi_1)$ in the special case in which input 1 is in the vacuum state, while input 2 is a coherent laser field with $\langle m_2 \rangle = 50$. As the vacuum field is randomly phased, one might expect $\phi_2 - \phi_1$ to be distributed uniformly over the range $-\pi$ to π. However, for the reasons already mentioned in connection with the derivation of $P_m(\phi_2 - \phi_1)$ in Eq. (22), the uniform probability distribution of the phase difference is encountered only in the limit $\langle m_2 \rangle \to \infty$, and the higher moments of $\hat{C}_m$ reflect this fact. (19)

Fig. 6 presents the results (22) of measurements of the complete probability distribution $P_m(\phi_2 - \phi_1)$ of the phase difference between two input fields in coherent states $|v_1\rangle$ and $|v_2\rangle$, with $v_1 = 1$, $v_2 = 2.36$. The experimental results are in the form of a histogram of bin width 18°, because one input was phase shifted externally in steps of 18°, as described above. The black dots give the expected theoretical form of $P_m(\phi_2 - \phi_1)$, and we see that theory and experiment are again in good agreement.

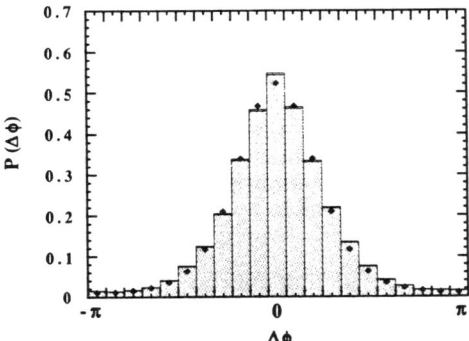

FIGURE 6. Comparison of measured and calculated probability distributions of the phase difference $\Delta\phi$ between two coherent input fields $|v_1\rangle, |v_2\rangle$ with $v_1 = 1, v_2 = 2.36$. The black dots give the predictions of the NFM theory. [Reproduced from reference 22]

So far all the measurements made use of coherent input fields for which the phase difference is effectively constant. But the phase formalism should apply to much more general situations, including those in which the two input fields are only partially coherent. Fig. 7 illustrates the experimental set-up for producing two input fields to the 8-port interferometer whose degree of mutual coherence can be varied. (23) One source is a very stable single-mode He:Ne laser, whereas the other source is a multimode He:Ne laser. The two laser beams are effectively mutually incoherent. By mixing these two beams with the help of beam splitters BS3 and BS5 in various proportions that can be controlled by adjustment of the variable neutral density filters F4 and F6, we generate two new fields $\hat{E}_1^{(+)}$ and $\hat{E}_2^{(+)}$, whose degree of coherence can be varied between 0 and 1. These serve as inputs to the 8-port interferometer in Fig. 7. Variable filter F8 allows the intensity of $\hat{E}_2^{(+)}$ to be controlled relative to $\hat{E}_1^{(+)}$. The rest of the apparatus functions as described previously.

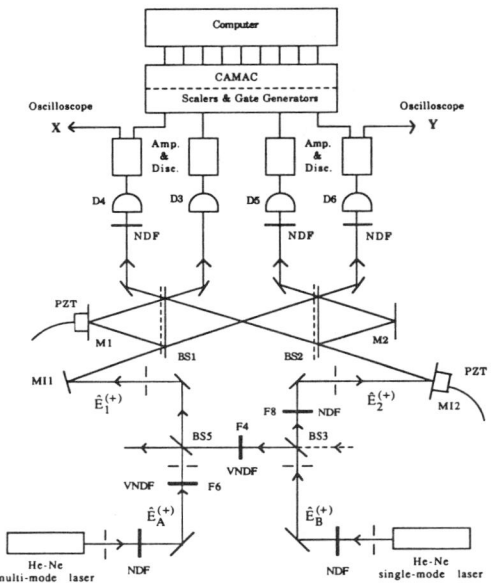

FIGURE 7. Outline of the experimental setup for measuring the probability distribution of the phase difference between two fields with variable degree of coherence. [Reproduced from reference 23]

Fig. 8a shows the results of measurements in which the average photon numbers $\langle m_1 \rangle, \langle m_2 \rangle$ at the input were both unity, for three different values of the degree of coherence. (23) Fig 8b shows the results of similar measurements in which $\langle m_1 \rangle, \langle m_2 \rangle$ were increased to 5. All the histograms are superimposed on the theoretically expected probability distributions given by Eq. (22). Once again there is very good agreement between theory and experiment. As $\langle m_1 \rangle, \langle m_2 \rangle$ decrease from large to small values the probability distribution broadens, and the same is true as the degree of coherence decreases from 1 to 0.

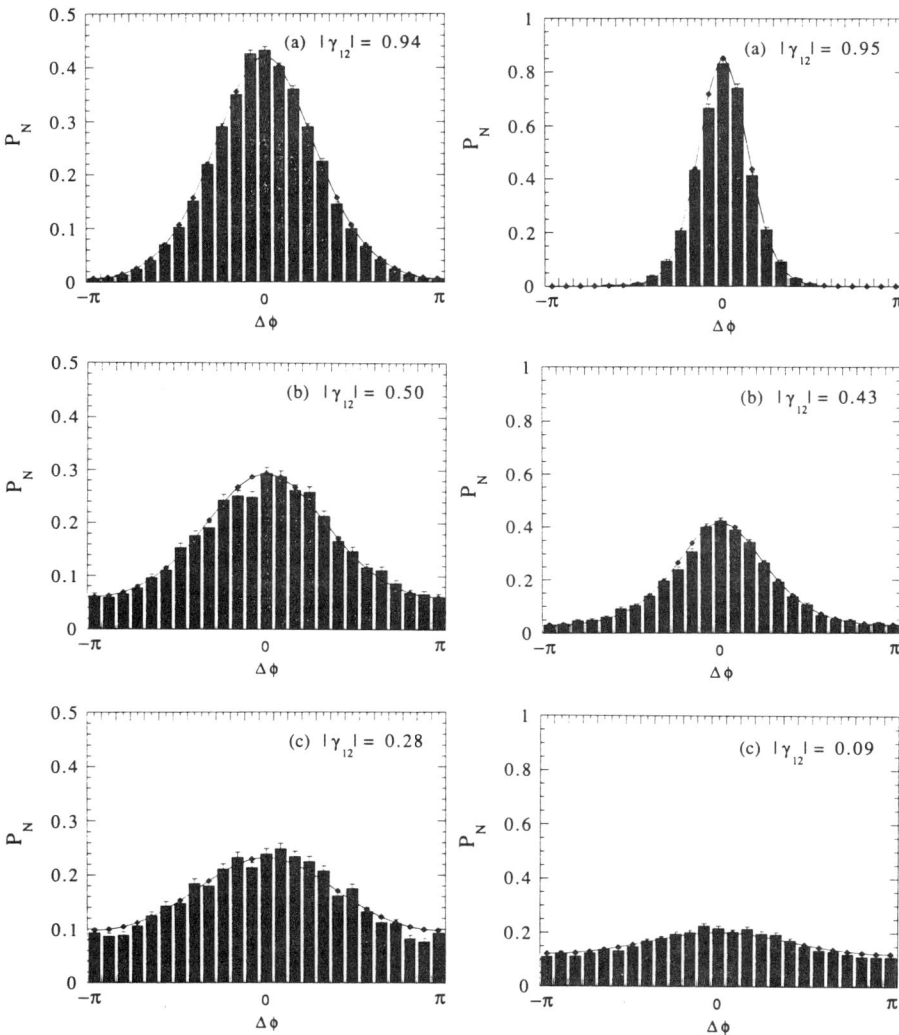

FIGURE 8. Comparison of measured and calculated probability distributions of the phase difference $\Delta\phi$ between two input fields with variable degree of mutual coherence $|\gamma_{12}|$, for $\langle m_1 \rangle = 1 = \langle m_2 \rangle$, left, and $\langle m_1 \rangle = 5 = \langle m_2 \rangle$, right. [Reproduced from reference 23]

292 Measuring the Phase of a Quantum Field

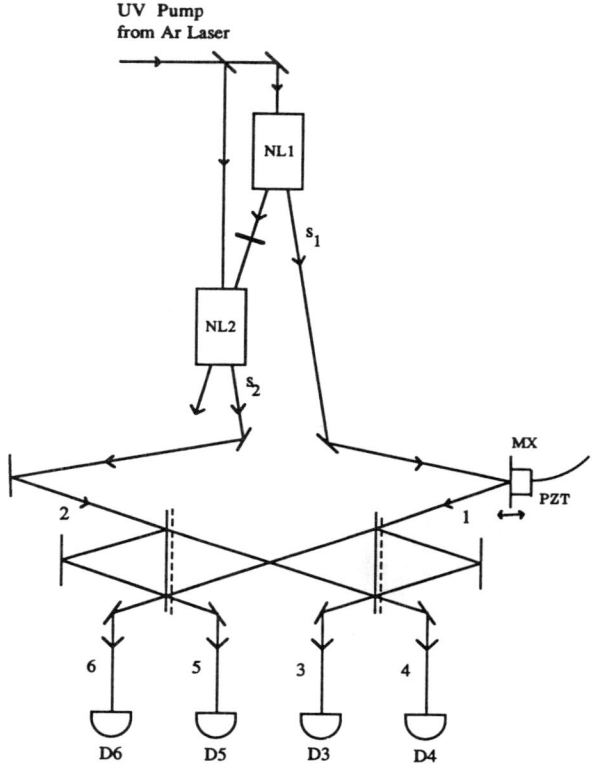

FIGURE 9. Outline of the experimental set up for measuring the probability distribution of the phase difference between two quantum fields produced by downconversion in two non-linear crystals. [Reproduced from reference 24]

So far the good agreement between our formalism for the phase and the results of measurements has been limited to optical fields having a classical character, which could be regarded as ensembles of coherent states. Finally we turn to a situation in which the fields have no classical description, and are purely quantum mechanical in nature. (24) Fig 9 shows the set-up for such an experiment. NL1 and NL2 are two crystals with $\chi^{(2)}$ non-linear susceptibilities that function as parametric downconverters. When they are optically pumped by a UV beam from an argon ion laser oscillating on the 351 nm line, some of the incident pump photons split into two lower frequency photons that are emitted simultaneously, such that either s_1 and i_1 are emitted from NL1 or s_2 and i_2 are emitted from NL2. The two crystals are so oriented that i_1 from NL1 passes through NL2 and is colinear with i_2 emitted from NL2. s_1 and s_2 serve as inputs to the eight-port

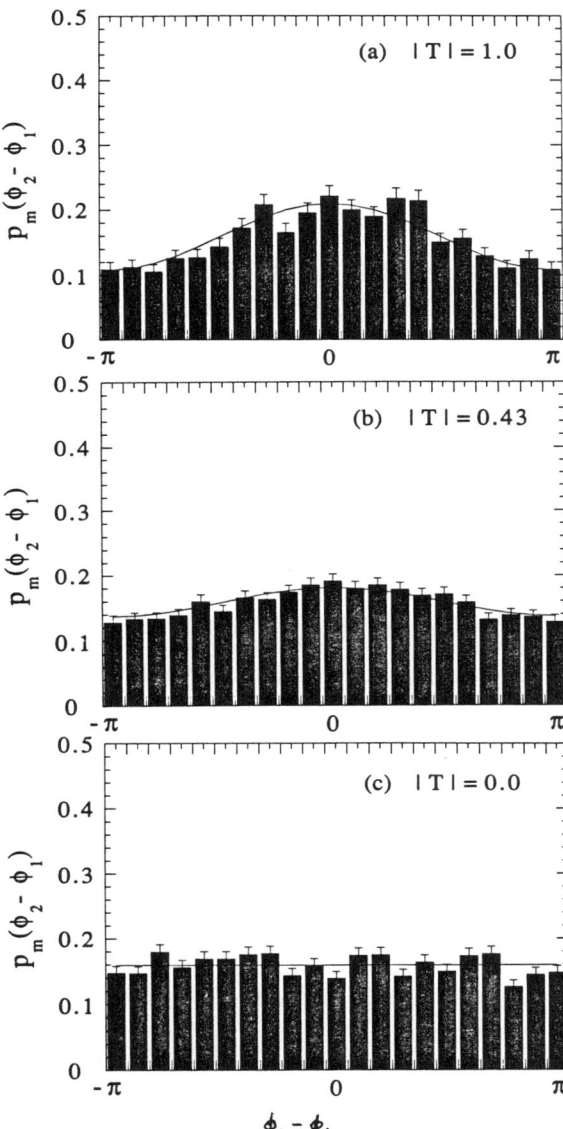

FIGURE 10. Comparison of the measured and calculated probability distributions of the phase difference $\phi_2 - \phi_1$ between two quantum fields with variable degrees of mutual coherence $|\gamma_{12}|$. $|T|$ is the absolute transmissivity of the filter placed in the idler 1 beam. [Reproduced from reference 24]

interferometer, which is configured as before. It has been demonstrated that the degree of coherence $|\gamma_{12}|$ between the s_1 and s_2 can be controlled by the transmissivity $|\mathcal{T}|$ of variable filter F inserted between NL1 and NL2, as shown in Fig. 9. The phenomenon cannot be described classically, but it can be understood in terms of the intrinsic indistinguishability of the s_1 and s_2 photon trajectories. It can be regarded as an example of a process in which i_1 induces coherence between s1 and s2, but without inducing emission. (29)

Our theoretical formalism for the probability distribution of the phase difference leads to the analytic form (24)

$$P_m(\phi_2 - \phi_1) = \frac{1}{2\pi}\left[1 + \beta|\mathcal{T}|\cos(\phi_2 - \phi_1 - \psi)\right], \quad (23)$$

in which β is a real number ($0 \leq \beta \leq 1$) that mainly reflects the lack of complete overlap between i_1 and i_2. In practice we find that $\beta \approx 0.32$. The results of measurements of the probability distribution $P_m(\phi_2 - \phi_1)$ for three different values of the transmissivity $|\mathcal{T}|$ are shown in Fig. 10, in which the probability has been centered somewhat arbitrarily on $\phi_2 - \phi_1 = 0$. Also shown are the theoretically expected forms given by Eq. (23). Once again we find reasonable agreement. We have also calculated the corresponding probability distribution by use of the Pegg and Barnett Hermitian operator formalism, and we find a result of the same general form as that given by Eq. (22), except that the coefficient of the cosine term is about five times smaller. (24) This agrees less well with the data.

It is encouraging that our operational formalism for the quantum phase exhibits good agreement with experiment in every case in which it has been put to the test. This suggests that there may not be one unique operator that represents the phase of a quantized field, perhaps because the concept of phase is intrinsically classical.

ACKNOWLEDGMENTS

The work described in this paper was carried out in collaboration with several former graduate students and coworkers, including A. Fougères, T. Grayson, C. Monken, J. W. Noh, and J. R. Torgerson, whose contributions are gratefully acknowledged. The research was supported by the National Science Foundation and by the U.S. Office of Naval Research.

REFERENCES

1. P. A. M. Dirac, Proc. Roy. Soc. (London) **A114**, 243 (1927).
2. L. Susskind and J. Glogower, J. Physics (N.Y.) **1**, 49 (1964).

3. P. Carruthers and M. M. Nieto, Phys. Rev. Lett. **14**, 387 (1965); Rev. Mod. Phys. **40**, 411 (1968).

4. See for example <u>Quantum Phase and Phase Dependent Measurements</u>, eds. W. P. Schleich and S. M. Barnett, special issue in Physica Scripta **T48**, 1993.

5. W. Vogel and W. Schleich, Phys. Rev. A **44**, 7642 (1991).

6. G. S. Agarwal, S. Chaturvedi, K. Tara and V. Srinivasan, Phys. Rev. A **45**, 4904 (1992).

7. W. Schleich, A. Bandilla and H. Paul, Phys. Rev. A **45**, 6652 (1992).

8. M. Freyberger and W. Schleich, Phys. Rev. A **47**, 1230 (1993).

9. U. Leonhardt and H. Paul, Phys. Rev. A **47**, R 2460 (1993).

10. D. T. Smithey, M. Beck, A. Faridani and M. G. Raymer, Phys. Rev. Lett. **70**, 1244 (1993).

11. D. T. Smithey, M. Beck, J. Cooper, M. G. Raymer and A. Faridani, Phys. Scripta T 48, 35 (1993)

12. M. G. Raymer, J. Cooper and M. Beck, Phys. Rev. A **48**, 4617 (1993).

13. D. T. Smithey, M. Beck, J. Cooper, and M. G. Raymer, Phys. Rev. A **48**, 3159 (1993).

14. D. T. Pegg and S. M. Barnett, Phys. Rev. A **39**, 1665 (1984).

15. S. M. Barnett and D. T. Pegg, J. Mod. Opt. **36**, 7 (1989).

16. S. M. Barnett and D. T. Pegg, Phys. Rev. A **42**, 6713 (1990).

17. J. W. Noh, A. Fougères and L. Mandel, Phys. Rev. A **45**, 424 (1992).

18. J. W. Noh, A. Fougères and L. Mandel, Phys. Rev. Lett. **67**, 1426 (1991).

19. J. W. Noh, A. Fougères and L. Mandel, Phys. Rev A **46**, 2840 (1992).

20. X. Y. Zou, L. J. Wang and L. Mandel, Phys. Rev. Lett. **67**, 318 (1991).

21. L. J. Wang, X. Y. Zou and L. Mandel, Phys. Rev. A **44**, 4614 (1991).

22. J. W. Noh, A. Fougères and L. Mandel, Phys. Rev. Lett **71**, 2579 (1993).

23. A. Fougères, J. R. Torgerson and L. Mandel, Opt. Comm. **105**, 199 (1994).

24. A. Fougères, C. Monken and L. Mandel, Opt. Lett., to be published (1994).

25. N. G. Walker and J. E. Carroll, Opt. Quant. Electron. **18**, 355 (1986).

26. R. Loudon, in <u>Frontiers in Quantum Optics</u>, eds. E. R. Pike and S. Sarkar (Hilger, Bristol, 1986)p. 42.

27. N. G. Walker, J. Mod. Opt. **34**, 15 (1987).

28. B. J. Oliver and C. R. Stroud,Jr., Phys. Lett A **135**, 408 (1989).

29. X. Y. Zou, L. J. Wang and L. Mandel, Phys. Rev. Lett. **67**, 318 (1991).

Measuring and Manipulating Quantum Fields in a Cavity by Atom Interferometry

J.M. Raimond, M. Brune, S. Haroche
F. Schmidt-Kaler, L. Davidovich and N. Zagury[1]

Laboratoire Kastler-Brossel, [2]
Département de Physique, Ecole Normale Supérieure
24 rue Lhomond F-75231 Paris Cedex, France

Abstract. Long lived circular Rydberg atoms interacting with high finesse superconducting cavities offer unique opportunities to perform tests on fundamental quantum mechanics. "Schrödinger cats", multiparticle nonlocal correlations, "quantum teleportation" offer striking illustrations of the less intuitive aspects of quantum mechanics. At the heart of these experiments is the dispersive atom-cavity interaction. We present its experimental manifestations, with the measurement of single photons light shifts and of Lamb shifts due to a single, empty, cavity mode. We then describe some possible extensions of this experiment, some of which being under way in our laboratory.

INTRODUCTION

The advent of new experimental techniques allowing one to prepare and study single quantum systems in a carefully controlled environment has considerably renewed the interest in fundamental experiments on quantum mechanics. The gedankenexperiments of the founders of quantum theory are now accessible. In this field, cavity quantum electrodynamics (CQED) is particularly promising (1, 2, 3, 4). A matter-field system made of a circular Rydberg atom (5) interacting with a single high–quality mode of a superconducting cavity can be understood in terms of the basic quantum mechanics postulates. Observing its behavior allows one to realize extensive tests of our understanding of the quantum measurement theory.

---

[1]Permanent address: Departamento de Fisica, PUC, Caixa Postal 38071, 22452–97 Rio de Janeiro, Brasil
[2]Laboratoire de l'Ecole Normale Supérieure et de l'Université Pierre et Marie Curie, Associé au CNRS (URA 18).

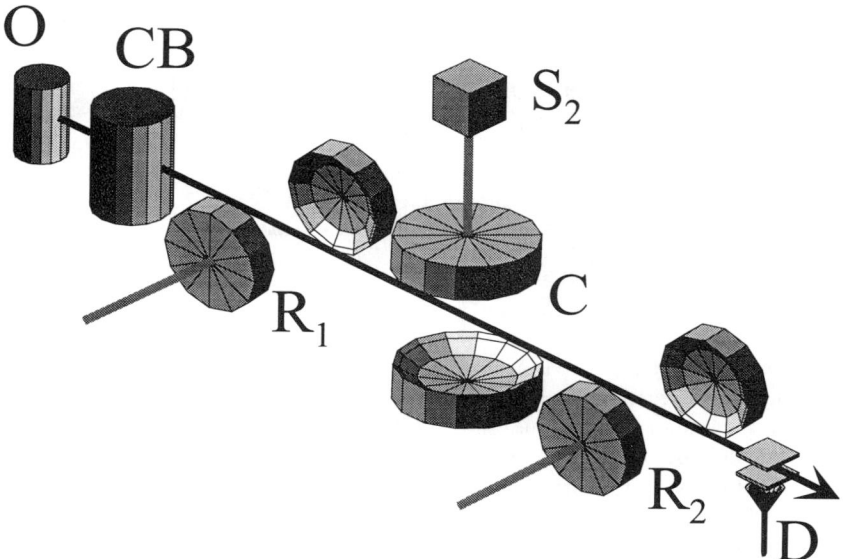

FIGURE 1: Principle of a cavity quantum electrodynamics experiment

The basic principle of a CQED experiment is sketched in Figure 1. An alkali atomic beam (rubidium in our experiments) crosses first a circular state preparation zone (CB in Fig. 1). The atoms exit CB in a circular state $|e\rangle$, with a high principal quantum number (51 for instance), and maximum angular and magnetic quantum numbers. Though this state is a long lived one (lifetime $t_{\text{rad}} = 30$ ms), it is strongly coupled to millimeter-wave radiation. The dipole matrix element of the $51 \to 50$ ($|e\rangle \to |g\rangle$) transition, at $\omega_0/2\pi = 51.099$ GHz, is 1250 a.u. The atoms interact later with the superconducting cavity, resonant at a frequency close to the atomic one. The cavity field energy damping time, t_{cav}, may be also in the tens of ms range. A classical microwave source may be used to prepare a coherent field in C. The atomic state can be determined downstream by the selective and sensitive field ionization technique (detector D). Various atomic state manipulations can be made with the help of auxiliary microwave zones (R_1 and R_2). The atom can be, for instance, prepared in a superposition of states $|e\rangle$ and $|g\rangle$ before entering C. The zone R_2, in association with detector D, can be used to measure any atomic observable.

When the atom-cavity interaction is resonant, the atom may emit or absorb photons in the mode. This is the principle of the micromaser (6, 7). In this case, the atom plays a double role. At the same time, it "prepares" the field in the cavity, and it "probes" it. A simpler situation is encountered when the atom and the cavity are not exactly resonant. The interaction is then purely dispersive. No photon can be absorbed or emitted by the atom, provided the atom–cavity detuning δ is large enough. The effect of the coupling is then twofold. First, the atomic transition is shifted by the cavity field. Part of this shift exists even in an empty mode. It can be described as the Lamb shift of the upper level due to the interaction with the vacuum fluctuations. Another part of the shift is linear versus the field intensity, and is simply the light shift (8) (note that the Lamb shift effect is equal to the light shift produced by half a photon). These shifts are accessible experimentally with Rydberg atoms. The shift per photon is of the order of Ω^2/δ, where Ω is the atom-field coupling (half the Rabi precession frequency in the field of a single photon). With our experimental parameters ($\Omega/2\pi = 25$ kHz and $\delta/2\pi = 100$ kHz — much larger than the cavity and atomic linewidths), the shifts are a few kHz per photon. In parallel with the atomic levels modifications, the interaction shifts the cavity mode frequency. This level dependant shift can be attributed to the atomic index of refraction. As it is also in the kHz range, it means that a single Rydberg atom changes the index of the medium by about 1 part in 10^{+7}, quite a large value.

An important point in these experiments is that the atom and the field get correlated after the interaction. Their quantum state is an entangled one which cannot be cast, under any representation, as a tensor product of an atom and a field state. The two components of this correlated system have a lifetime much longer than the interaction time t_{int} (tens of ms compared to tens of μs). We obtain thus a nonlocal quantum state, very similar to the correlated pairs of the famous EPR paradox (9). Tests of quantum nonlocality based on Bell inequalities (10) could be undertaken with this system. It is also possible, as will be shown later, to prepare entangled states with more than two particles. Proposed for the first time by Greenberger et al. (11), these states provide tests of quantum nonlocality much more stringent than the "ordinary" pairs.

The atom–field correlation can be used also to gain some information on the field state. This is at the heart of a Quantum Non Demolition (QND) measurement method for the field intensity (12, 13). The measurement of a well chosen atomic observable allows one also to project the field onto an highly nonclassical state. "Schrödinger cats" of the field can be prepared in this way (13). They are quantum superpositions of fields differing through their macroscopic attributes. The study of the relaxation of the coherence between their components due to the coupling with the outside world can help us bridging the gap between the quantum world, where superposition is the rule, and the macroscopic world, where it is never observed (there is not such thing as a living *and* dead cat, following the provocative Schrödinger's wording (14)).

In the first part of this paper, we will describe an experimental evidence of the dispersive atom–cavity interaction. Using an interferometric technique (separated oscillatory fields Ramsey fringes (16)), we have measured the light shifts at the photon level and determined the Lamb shift (15). We will show then how this interferometric signal can be used for a QND measurement of the field intensity. In the same conditions, a "Schrödinger cat" can be produced and analyzed. Using the nonclassical properties of this state, one can prepare entangled triplets of atoms (à la GHZ). Finally, we will describe a striking application of quantum nonlocality: the quantum "teleportation" of an atomic state between two cavities.

FROM LAMB SHIFTS TO LIGHT SHIFTS

The energy shift of the upper level $|e\rangle$ at the center of a cavity containing n photons is $(n+1)\hbar\Omega^2/\delta$ (13). The n term accounts for the light shifts, the "1" term is the Lamb shift. Level $|g\rangle$ experiences, in the same conditions, a shift $-n\hbar\Omega^2/\delta$ (note that, within the approximations we use, $|g\rangle$ is not shifted in an empty cavity). The wavefunction of an atom crossing the cavity in level $|e\rangle$ experiences thus a phase shift $\varepsilon(n+1)$, with $\varepsilon = 2t_{\text{int}}\overline{\Omega}^2/\delta$. $\overline{\Omega}$ is the r.m.s. average of the atom field coupling along the atom's trajectory. An atom in state $|g\rangle$ would experience a dephasing $-n\varepsilon$. Note that the atom remains in the same state during the interaction, provided $\delta \gg t_{\text{int}}^{-1}$.

These dephasings correspond to the slowing down or acceleration of the atom in the cavity, due to the forces deriving from the position–dependant shifts (17, 18, 19). They are detected easily by an interferometric technique. We use the Ramsey method (16). The zone R_1 is set to prepare a superposition with equal weigths of $|e\rangle$ and $|g\rangle$. This superposition crosses the cavity, which dephases in different ways the wavepackets associated to $|e\rangle$ and $|g\rangle$. The atomic coherence phase modification is translated into a population information in zone R_2, which mixes again the two levels. Depending upon the relative phase of the atomic coherence and of the R_2 field, the effect of the second zone doubles or cancels the one of R_1. The atomic energy is then a sinusoidal function of the frequency $\nu_r = \omega_r/2\pi$ applied in R_1 and R_2, with a period equal to the reciprocal of the transit time across the apparatus, modulating the line profile associated to a single zone. The phase of these fringes reflects directly the shifts experienced in the cavity. This fringe signal reveals a quantum interference between two undistinguishable paths: a transition from $|e\rangle$ to $|g\rangle$, for instance, can occur either in R_1 or in R_2. Nothing in the final outcome can be used to tell which path the atom has followed.

The whole experimental set–up is contained in an Helium 4 croystat, operating at 1.4 K. The low temperature allows one to use superconducting cavities, and to get rid of most of the thermal radiation. The rubidium atoms

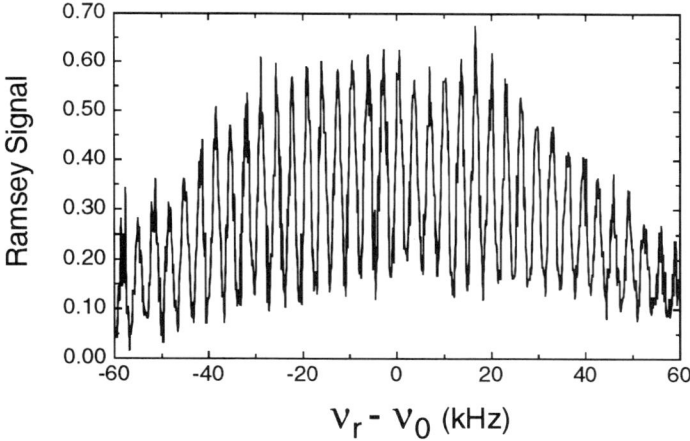

FIGURE 2: Ramsey fringes signal: $|e\rangle$ to $|g\rangle$ transfer rate versus the frequency difference $\nu_r - \nu_0$ between the fields in R_1 and R_2 and the atomic line. 45 fringes, 3.2 kHz apart, modulate the single zone interaction profile

are first promoted, by a laser–diode excitation in CB, in an "ordinary" low angular momentum Rydberg state ($51F, m = 2$). A "circularization" process then transfers most of the population to the circular state. The method makes use of an adiabatic rapid passage, during which the atom absorbs 48 radiofrequency (250 MHz) photons, on transitions between Stark levels in a static electric field of 2.5 V/cm. A small magnetic field along the quantization axis lifts the degeneracy between the angular–momentum–increasing transitions (σ^+ polarization), and the angular–momentum–decreasing ones. This very efficient technique has already been described in details elsewhere (20).

Each circularization sequence produces about 400 circular atoms. The sequence is repeated at a 1.6 kHz rate, well adapted to the 600 μs transit time across the apparatus. Circular atoms are stable only in a small electric field (21). In free space, they are degenerated with many other high angular momentum levels. Stray fields can induce transitions between these levels. A very homogeneous, carefully controlled, 0.3 V/cm field is thus applied along the atom's trajectory with the help of 40 adjustable electrodes. At the exit of the interferometer, the atoms are detected by the field ionization technique. The produced electrons are focused by electrostatic lenses on a dynode multiplier. The association of the time resolved preparation and detection allows us to perform a passive velocity selection (relative velocity dispersion 1.5 %). As optical interferometric signals, the Ramsey fringes are better observed with a "monochromatic" atom source.

Two low-finesse cavities are used for the Ramsey zones. They are 9 cm apart. This corresponds, for 300 m/s atomic velocity, to a 3.2 kHz fringe spacing. In order to observe a large fringe number, we broaden the single zone line by applying the microwave in R_1 and R_2 for a short time, while the atoms are close to the center of the zones. The observed transfer rate as a function of $\nu_r - \nu_0$ is displayed in Figure 2. The contrast (from 10 to 70 % transfer rate) is limited by the Ramsey field homogeneity on the beam size (1 mm diameter) and by the source width (200 Hz). Field inhomogeneities and stray magnetic field fluctuations contribute also to the contrast reduction

The superconducting cavity C is an open Fabry–Pérot. This is the only cavity compatible with the static electric field needed for circular states preservation. It is made of two spherical massive niobium mirrors (50 mm dia., 40 mm radius of curvature), placed 27 mm apart. They sustain two orthogonally polarized gaussian TEM_{900} modes, with 9 antinodes. The degeneracy of these modes is slightly lifted by geometric imperfections: the two frequencies are 146 kHz apart. The two modes can be tuned simultaneously by slightly translating one mirror with the help of a mechanical drive deforming elastically the mirror mount (10 MHz range). For fine tuning, the other mirror is held by a PZT stack (200 kHz range).

Both modes have a quality factor $Q = 8.10^5$. It is determined by monitoring the cavity transmission through two 0.2 mm diameter coupling holes drilled in the mirrors centers. With such a quality factor, the field damping time $t_{cav} = 2$ μs is much shorter that the atom–cavity interaction time $t_{int} = 25$ μs. The field is renewed many times by the sources during the interaction. The light shifts are then proportional to the average intensity, which is not quantized. Such a cavity is certainly not suited for exhibiting long–lived atom-field correlations. The main limitations of this first cavity were due to the field diffusion on the mirrors imperfections. Since then, we have developped optical polishing techniques, which enabled us to reach $Q = 2.10^8$, corresponding to $t_{cav} = 700$ μs, a value large enough for most of the experiments we will describe later.

The average atom–field coupling at cavity center, $\Omega/2\pi$, is a little lower than the 25 kHz expected value. This is due to a slight misalignment of the beam with respect to the cavity antinodes, and to the atomic beam size. The value deduced from a careful measurement of the beam position in the cold apparatus is 17(3) kHz.

Figure 3 presents the central fringes for an empty cavity and a cavity containing one photon (the field calibration will be discussed later). The noise on the signal is mainly statistical: only a few atoms are detected for each frequency channel. The thick line is a fit on a sine function. The precision on the fringes position, obtained through this fit, is about 25 Hz, corresponding to a 5.10^{-10} resolution. The daily stability of the fringe pattern is also 25 Hz.

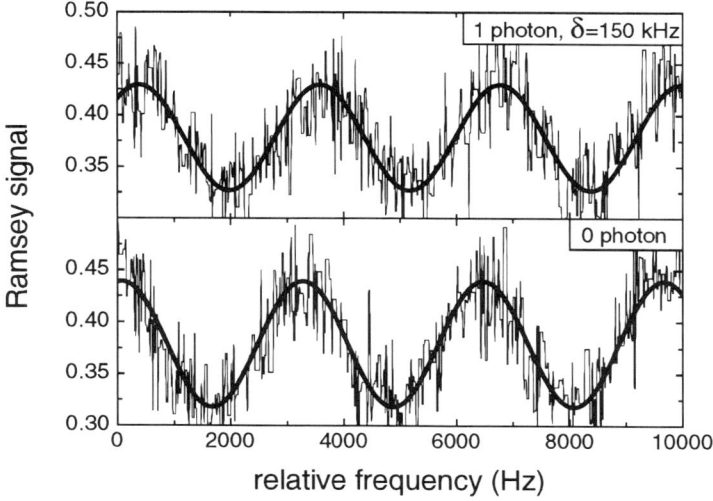

FIGURE 3: Ramsey fringes for a cavity containing zero or one photon. The fringes dephasing due to the light shift in the cavity is clearly apparent

A very careful screening of the magnetic field fluctuations, down to the 10 μG level, has been necessary to reach such a stability.

The fringes shift for a one photon field is 315 Hz, corresponding to $2\varepsilon = 0.6$ rd, for $\delta/2\pi = 150$ kHz. This corresponds to a few pm displacement of the atomic wave packets in the cavities. We checked the linearity versus field intensity and the dispersive nature of the shift (proportional to $1/\delta$). The precision of the fringes position determination allows us to detect 0.1 photon fields, corresponding to a 125 fm retardation of the atom's wave packet!

With such a sensitivity, the Lamb shift is easily measured. Figure 4 shows, as open circles, the position of the fringes versus δ for an empty cavity. In order to interpret these data, two effects should be taken into account. First, the two modes of the cavity contribute to the shift (for the light shift experiment, the narrow bandwith source is coupled to a single mode). One should also take into account the residual thermal field. The number of blackbody photons per mode has been found in an auxiliary experiment to be 0.32, corresponding to a 1.7 K radiation temperature. The contribution of the light shifts induced by this field have been removed from the raw data to obtain the pure Lamb shift effect, depicted by solid circles on Fig. 4. The solid line presents a fit on the theoretical values. The only adjustable parameter is the atom–field coupling. The obtained value, $\Omega = 16$ (0.5) kHz is in good agreement with, and more precise than, the one deduced from the beam position. It gives an excellent and independant calibration of the field in the light shift experiment.

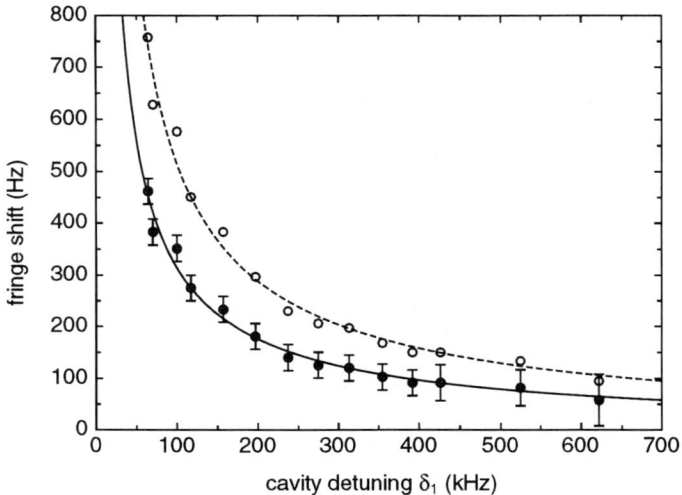

FIGURE 4: Shifts of the fringe system in an empty cavity versus atom–cavity detuning. Open circles: raw data. Full circles: data corrected for the residual thermal radiation. Solid line: theoretical fit.

Lamb shifts are of course well known, and have been, in some cases, measured with excellent precision. Ordinarily, however, they are due to the coupling of the atomic levels with a whole continuum of field modes. This experiment is, to our knowledge, the first one which singles out the effect of only one field mode and studies the continuous transition between Lamb shifts and light shifts.

QND MEASUREMENTS AND SCHRÖDINGER CATS

Used with a very high finesse cavity, our atomic interferometer opens the way to a wide variety of experiments. The simplest one, in progress now, uses the atom–field correlations to measure, without absorption, the intensity of a field containing 0 or 1 photon (13). The interferometer is tuned so that $\varepsilon = \pi/2$. The fringes phase shift per photon is therefore π. Note that this corresponds to realistic experimental parameters ($\delta/2\pi = 100$ kHz, $\Omega/2\pi = 25$ kHz and a velocity of 150 m/s). By a proper tuning of ω_r, the atom (incoming in $|e\rangle$) is certainly found in $|g\rangle$ if the cavity is empty (at least if we assume a perfect contrast for the fringes). It is also found in $|g\rangle$ if the cavity contains an even photon number. At variance, the atom is found in $|e\rangle$ if the cavity contains odd photon numbers. Assuming that the initial field has a low probability of containing 2 or more photons (small thermal or coherent field), the detection of a single atom is enough to pin down the field intensity and to count the photon number (a small bunch of atoms would be needed in an inperfect experiment).

At variance with most photodetections, this measurement does not change the photon number: the nonresonant atom cannot emit or absorb in the cavity. It is therefore a Quantum Non Demolition (QND) measurement (22). If one neglects the cavity relaxation, the photon left in the cavity after detection of an atom in $|e\rangle$ will remain forever, and all following atoms will exit the cavity in $|e\rangle$. This correlation between repeated measurements is one of the main features of such processes and can be used to check the QND character. Taking into account cavity relaxation, the photon has a small chance to be absorbed or diffused out of the cavity between two measurements. Assuming that the sources are switched off, one should observe a sequence of atoms in $|e\rangle$ (photon present) followed by an infinite sequence of atoms in $|g\rangle$ (cavity empty). This quantum jump, monitoring in real time the photon death, is very reminiscent of the quantum jumps observed on the fluorescence of a single trapped ion (23). The ordinary quantum mechanics prediction (exponential decay of the energy) is recovered only as a statistical average over many individual realizations of the experiment. Such features illustrate the properties of quantum mechanics applied to single systems, and not to large ensembles.

Let us assume now that the cavity initially contains a classical field with amplitude α, represented by a Glauber's coherent state (24) $|\alpha\rangle = \sum_n c_n |n\rangle$ with $c_n = \exp(-|\alpha|^2/2)\alpha^n/\sqrt{n!}$. The atomic index of refraction shifts transiently the cavity frequency and changes the field phase. With $\varepsilon = \pi/2$, an atom in state $|e\rangle$ dephases the field by $\pi/2$, changing the initial field $|\alpha\rangle$ into $|\beta\rangle = |i\alpha\rangle$. An atom in state $|g\rangle$ has an opposite effect, changing $|\alpha\rangle$ into $|-\beta\rangle$. These phase shifts are at the heart of the phase scrambling mechanism which explains how phase information is lost in a QND intensity measurement (13). The Ramsey zones R_1 and R_2 are set to achieve the transformations: $|e\rangle \rightarrow (|e\rangle - |g\rangle)/\sqrt{2}$ and $|g\rangle \rightarrow (|g\rangle + |e\rangle)/\sqrt{2}$. The atom enters then C in state $(|e\rangle - |g\rangle)/\sqrt{2}$ and the atom–field system is, after the interaction:

$$\frac{1}{\sqrt{2}}\Big(|e,\beta\rangle - |g,-\beta\rangle\Big), \tag{1}$$

where the first symbol inside the kets refer to the atom, the second to the cavity. This is clearly an entangled atom–field state. Performing at this stage a measurement of the atomic energy amounts in projecting the field state on $|\beta\rangle$ or $|-\beta\rangle$. In such a situation, the state of the atom in the cavity is known, and the interaction only produces a classical phase shift.

A more interesting situation is obtained by scrambling the information on the atomic state. This can be achieved by sending the atom through R_2, which mixes both levels. The final atom–field state is then:

$$\frac{1}{2}\big(|e\rangle[|\beta\rangle - |-\beta\rangle] - |g\rangle[|\beta\rangle + |-\beta\rangle]\big). \tag{2}$$

Measuring the atomic energy projects the field on one of the states:

$$|\Psi_\pm\rangle = \frac{1}{\mathcal{N}}(|\beta\rangle \pm |-\beta\rangle), \tag{3}$$

where $\mathcal{N}$ is a normalization factor, close to $1/\sqrt{2}$ when $|\beta|$ is large enough (this will be assumed in the following).

The $|\Psi_\pm\rangle$ states exhibit remarkable quantum features (13, 25). They are quantum superpositions of two classical fields with opposite phases. The name "Schrödinger cat" is often coined out for such superpositions of macroscopically distinguishable states, in reference to the famous "paradox" stated by Schrödinger (14). $|\Psi_+\rangle$, for instance, corresponding to an atom detected in $|g\rangle$, has no classical counterpart: the sum of two classical fields with opposite phases is zero. Moreover, it contains only even numbers of photons: the odd photon numbers probability amplitudes associated to the two components cancel out. This can be understood easily in terms of the QND measurement. The atom exits the interferometer in $|g\rangle$ only if the cavity contains an even photon number. Detecting $|g\rangle$ makes us sure that no odd photon number can exist in the cavity. In a similar way, $|\Psi_-\rangle$ contains only odd photon numbers. $|\Psi_+\rangle$ can be called an "even cat", and $|\Psi_-\rangle$ an "odd cat".

This peculiar photon number distribution explains why the "cats" are short lived. After a time of the order of t_{cav}/N (N is the average photon number in the cat), relaxation fills these gaps, and the cat turns into a mere statistical superposition of two fields. The larger the energy, the larger the "distance" between the two "cat components", the faster the decoherence. For macroscopic objects, the same decoherence occurs in times so short that quantum superpositions are never observed. This explains why the needle of a detector, measuring a quantum system in a coherent superposition of states, is never found in a quantum superposition of the corresponding positions (26). With the "cavity cats", the size of the system can vary continuously from microscopic to mesoscopic. A study of the "relaxing cat" is thus an exploration of the subtle border between the quantum and classical worlds.

The coherent nature of the "cat" can be revealed by the quantum interference effects it provides (27). Let us send a second atom, in the same conditions, through the cavity, a time T after the first (which has been detected in $|g\rangle$). If $T \ll t_{\text{cav}}/N$, this atom interacts with an intact cat, containing only even photon numbers. It exits therefore the apparatus in $|g\rangle$. When $T \gg t_{\text{cav}}$, this atom crosses an empty cavity, and exits also in $|g\rangle$. At variance, for $t_{\text{cav}}/N \ll T \ll t_{\text{cav}}$, the atom encounters a statistical superposition of coherent fields, containing odd and even photon numbers. The atom is then found in $|e\rangle$ or $|g\rangle$ with equal probabilities. The conditional probability of detecting two successive atoms in the same state thus exhibits a fast variation around $T = 0$, which reveals the decoherence. Let us stress that the orders of magnitude for the experimental realization are quite comparable to the ones for the QND measurement, as long as the "cat" is not too big.

TESTS OF QUANTUM NONLOCALITY

The states of atoms interacting with a Schrödinger cat are strongly correlated. This feature can be used to prepare triplets of correlated atoms (11), exhibiting in a striking way the nonlocal properties of quantum mechanics.

Following Mermin (28, 29), let us consider a set of three spins 1/2 (eigenstates $|+\rangle$ and $|-\rangle$ along a quantization axis Oz). The system is prepared, via a common interaction, in the entangled state:

$$(|+,+,+\rangle - |-,-,-\rangle)/\sqrt{2}. \tag{4}$$

Once the spins have been spatially separated, a measurement of their components along an axis Ox, orthogonal to Oz, is performed. Quantum mechanics predicts that the product of the three outcomes, $m_x^1 m_x^2 m_x^3$, should be equal to -1. On the other hand, any local theory, based upon the concept of "elements of reality" (hidden variables theory), predicts an outcome of $+1$. There is a striking difference between the two predictions, and, at least with an ideal set–up, a single realization of the experiment would be enough to confirm or contradict quantum mechanics. This is a much more stringent test than the ones based on particle pairs and Bell inequalities (10), which involve statistics performed on a large number of experiments.

The question of the preparation of the state described by Eq. (4) remains to be solved. Our two level atom can be substituted for the spin 1/2, the m_x measurement being replaced by an energy measurement after a proper $\pi/2$ microwave pulse. Let the first atom cross the cavity containing a coherent state $|\alpha\rangle$. The final atom–field state is given by Eq. 2. It is an entangled state where $|e\rangle$ is correlated to an odd cat, $|g\rangle$ to an even one. Let then a second atom cross the cavity before the first one is detected (and before the quantum coherences are washed out by relaxation). The final atoms–field state can be guessed by simple arguments. If the first atom was in $|e\rangle$, it leaves in the cavity an odd cat. The second atom will be also in $|e\rangle$, and leave the same cat (up to a global phase rotation of $\pi/2$). Similarly, if the first atom is in $|g\rangle$, the second will be in the same state, and the field will remain in a phase shifted "even cat" (phase rotation $-\pi/2$). The same arguments can then be applied to a third atom crossing the cavity (and even generalized to a sequence of N atoms). The final "three atoms+field state" is thus:

$$\frac{1}{2}\Big[|e,e,e\rangle(|-\beta\rangle - |\beta\rangle) - |g,g,g\rangle(|\beta\rangle + |-\beta\rangle)\Big]$$

$$= \frac{1}{\sqrt{2}}\Big[-|\beta\rangle \frac{1}{\sqrt{2}}(|g,g,g\rangle + |e,e,e\rangle)\Big] + |-\beta\rangle\Big[\frac{1}{\sqrt{2}}(|e,e,e\rangle - |g,g,g\rangle)\Big] \quad (5)$$

In the second line, we recognize two coherent states with opposite phases entangled with correlated atomic states à la Mermin (28). A simple homodyne measurement of the field phase projects the atomic system on one of these states. After the three atoms have crossed the cavity, a classical source adds an amplitude $-\beta$ to the field already present in the cavity. $|\beta\rangle$ becomes then $|0\rangle$ (vacuum) and $|-\beta\rangle$ becomes $|-2\beta\rangle$. A fourth atom is then sent in the cavity. It is prepared in state $|g\rangle$ and does not interact with R_1 and R_2. Inside the cavity, a controlled Stark effect tunes it into resonance with the mode. If the cavity is empty, it remains in state $|g\rangle$. If the cavity contains state $|-2\beta\rangle$, there is a probability it will absorb a photon an exit in state $|e\rangle$. A detection in state $|e\rangle$ will therefore project the other three atoms on state:

$$\frac{1}{\sqrt{2}}\big(|e,e,e\rangle - |g,g,g\rangle\big), \quad (6)$$

formally identical to the one given in Eq.(4).

The three atoms are then submitted to a $\pi/2$ pulse and detected. It is easily shown that quantum mechanics predicts an odd number of atoms in state $|g\rangle$, while classical arguments predict an even number. For an ideal experiment at least, a single measurement is enough to decide between the two theories.

The experimental parameters required to perform this experiment are quite similar to the ones for the QND measurement. A coherent field with a few photons only is enough. The "cat coherences" live therefore a time of the order of t_{cav}, and it is easy to have three atoms crossing the cavity in a much shorter time. The main problem is to make sure that exactly three atoms are used (unread atoms mess up the parity measurement). A brute force solution could be to prepare three bunches of atoms, each containing much less than one atom (say 0.01) on the average. Only the events yielding exactly three detections are to be kept. The probability for an unread atom is thus low enough (3%) to observe the quantum correlations. The price to pay would be very long acquisition times (a few events per day).

QUANTUM TELEPORTATION

Quantum nonlocality is one of the most striking predictions of modern physics. It has been suggested to apply it to quantum cryptography (30), or to quantum computing (31). More recently, Bennett et al. proposed to use it for a "teleportation" experiment (32). The rule of the game is to transmit from one "cabin" C_1 to a second "cabin" C_2, far away, the unknown quantum state of a spin 1/2 particle, a, entering C_1. In Bennett's proposal, the two cabins are

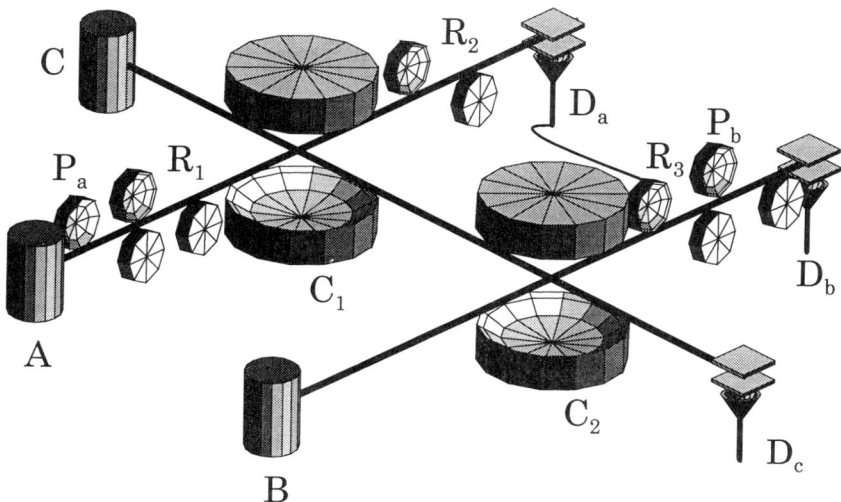

FIGURE 5: Scheme of a quantum teleportation experiment.

a correlated pair of EPR particles. Two measurements, performed on $a + C_1$, give two bits of information, and project a spin b (C_2 itself in this case) on a state differing from the initial one by a unitary transformation selected among a set of four. The two bits, sent to C_2 through a classical information channel, allow an observer to perform the right transformation, and to reconstruct the state of a, completing the teleportation scheme. Bennett's paper did not discuss the experimental realization. We have shown recently that CQED allows the practical realization of a teleportation scheme (33).

The principle of the experiment is sketched in Figure 5. C_1 and C_2 are two high–Q cavities. Three atomic beams A, B and C are used. An atom c of beam C, crossing the two cavities, correlate their quantum states. Atom a to be teleported belongs to beam A and crosses C_1 only. The atom b (replica of a) belongs to beam B and crosses C_2. Various microwave zones are used to manipulate the atomic states. The atoms are detected in D_a, D_b and D_c. Stark fields can be used to tune the atoms in resonance, close to resonance (dispersive interaction) or completely out of resonance (no interaction) during controlled time intervals.

The first step establishes a nonlocal quantum correlation between C_1 and C_2. Atom c is prepared in state $|e\rangle$, and is tuned into resonance with the cavities for such durations that it experiences a $\pi/2$ pulse in C_1 and a π pulse in C_2. The atom c + cavities system thus undergoes the transformations:

$$|e\rangle|0\rangle_1|0\rangle_2 \longrightarrow \frac{1}{\sqrt{2}}[|e\rangle|0\rangle_1|0\rangle_2 + |g\rangle|1\rangle_1|0\rangle_2]$$

$$\longrightarrow \frac{1}{\sqrt{2}}[|g\rangle|0\rangle_1|1\rangle_2 + |g\rangle|1\rangle_1|0\rangle_2], \qquad (7)$$

where the indices on field kets refer to the cavity number. Atom c leaves the cavities in $|g\rangle$. Its detection, indicating that the teleportation machine is ready to operate, leaves the fields in state:

$$|\Psi_c\rangle = \frac{1}{\sqrt{2}}[|0\rangle_1|1\rangle_2 + |1\rangle_1|0\rangle_2]. \qquad (8)$$

This is a new type of "Schrödinger cat", with one photon in the first or in the second cavity, the alternative being of quantum nature (34, 35). Note that, since this "cat" contains a single photon, its lifetime is t_{cav}.

The atom a is prepared, in zone P_a, in an arbitray combination of its states, unknown to the observers: $|\phi_a\rangle = c_e|e_a\rangle + c_g|g_a\rangle$. The initial state of the system "$a + C_1 + C_2$" is the tensor product of $|\Psi_c\rangle$ and $|\phi_a\rangle$, which can be written:

$$|\Psi\rangle = \frac{1}{2}\Big[|\Psi^{(+)}\rangle(c_e|1\rangle_2 + c_g|0\rangle_2) + |\Psi^{(-)}\rangle(c_e|1\rangle_2 - c_g|0\rangle_2)$$
$$+ |\Phi^{(+)}\rangle(c_e|0\rangle_2 + c_g|1\rangle_2) + |\Phi^{(-)}\rangle(c_e|0\rangle_2 - c_g|1\rangle_2)\Big], \qquad (9)$$

in terms of the Bell basis (36) of the $a + C_1$ system:

$$|\Psi^{(\pm)}\rangle = \frac{1}{\sqrt{2}}(|e_a\rangle|0\rangle_1 \pm |g_a\rangle|1\rangle_1), \qquad (10)$$

$$|\Phi^{(\pm)}\rangle = \frac{1}{\sqrt{2}}(|e_a\rangle|1\rangle_1 \pm |g_a\rangle|0\rangle_1). \qquad (11)$$

Each Bell state appearing in Eq.(9) is correlated to a superposition of 0 and 1 photon states in C_2. Besides order and sign changes, the coefficients of this superposition are the ones of $|\phi_a\rangle$. Measuring in which Bell state $a + C_1$ is, duplicating the state of C_2 on atom b, and finally restoring the order and signs of the coefficients are enough to complete the teleportation scheme.

To single out one Bell state among four, two atomic detections are necessary. First, a crosses R_1, C and R_2. It interacts dispersively with C and all parameters are set as for the QND measurement: the transfer rate from one level to the other is maximum for an empty cavity, zero for a cavity containing one photon. $a + C_1$ experiences therefore, during these interactions, the transformation: $|e_a\rangle|0\rangle_1 \rightarrow -|g_a\rangle|0\rangle_1$; $|e_a\rangle|1\rangle_1 \rightarrow -|e_a\rangle|1\rangle_1$; $|g_a\rangle|0\rangle_1 \rightarrow |e_a\rangle|0\rangle_1$;

$|g_a\rangle|1\rangle_1 \to |g_a\rangle|1\rangle_1$. Applying this transformation to the Bell states of Eq.(11), one obtains:

$$|\Psi^{(\pm)}\rangle \to -|g_a\rangle\frac{1}{\sqrt{2}}(|0\rangle_1 \mp |1\rangle_1), \qquad (12)$$

$$|\Phi^{(\pm)}\rangle \to -|e_a\rangle\frac{1}{\sqrt{2}}(|1\rangle_1 \mp |0\rangle_1). \qquad (13)$$

Detecting the state of a thus tells us wether $a + C_1$ was in a Ψ or Φ-type Bell state.

In order to distinguish the two remaining possibilities, a second atom a' from beam A is sent. It is prepared in $|g\rangle$, does not interact with R_1. The interaction with C_1 is resonant, and amounts to a π pulse when C_1 contains a photon. The atom interacts then with R_2, set to perform a $\pi/2$ pulse. The $a' + C_1$ system undergoes in C_1 and R_2 successively the transformations:

$$|g_{a'}\rangle\frac{1}{\sqrt{2}}(|0\rangle_1 \pm |1\rangle_1) \to \frac{1}{\sqrt{2}}(|g_{a'}\rangle \pm |e_{a'}\rangle)|0\rangle_1 \to \begin{cases} |e_{a'}\rangle|0\rangle_1 & \text{if } + \\ |g_{a'}\rangle|0\rangle_1 & \text{if } - \end{cases}. \qquad (14)$$

The final state of a' is thus correlated unambiguously to the sign in the Bell states. Finally, the detections of a and a' provide us with two bits of information, indicating in which Bell state the $a + C_1$ system is, with the correspondances: $g_a, g_{a'} \to |\Psi^{(+)}\rangle$, $g_a, e_{a'} \to |\Psi^{(-)}\rangle$, $e_a, g_{a'} \to |\Phi^{(+)}\rangle$, $e_a, e_{a'} \to |\Phi^{(-)}\rangle$.

The state of the second cavity contains then all the information needed to reconstruct $|\Phi_a\rangle$. This state is replicated on an atom b from beam B. Atom b, prepared in $|g\rangle$, interacts resonantly with C_2, experiencing a π pulse in the field of one photon. The system $b + C_2$ undergoes the transformation:

$$(\alpha|1\rangle_2 + \beta|0\rangle_2) \otimes |g_b\rangle \to (\alpha|e_b\rangle + \beta|g_b\rangle) \otimes |0\rangle_2, \qquad (15)$$

where α and β are both equal to $\pm c_e$ and $\pm c_g$. The state of b differs from the one of a only by coefficient signs and permutations. This can be overcome by applying in R_3 an unitary transformation on b, selected among a set of four possible transformations. Which one is telled by the two bits resulting from the detection of a and a', transmitted to R_3 by a classical information channel ("wire" on Figure 5).

We described here the whole process as if the detectors were perfect. This would not be the case in an actual experiment. We have shown however (33) that, when measuring the correlations between the state of a and the one of b (determined by "polarizer" P_b and detector D_b), one obtains values higher than the ones predicted for non quantum correlations between C_1 and C_2, provided the quantum efficiency is high enough ($\geq 70\%$). Like the GHZ/Mermin experiments, this teleportation scheme is a high order correlation experiment displaying in a striking way the nonlocal properties of quantum mechanics.

CONCLUSION

Long lived circular Rydberg atoms coupled to superconducting millimeter wave cavities offer unique possibilities to realize fundamental tests of quantum theory. We have shown experimentally that it is possible to realize and study these systems. The preservation of a two–level atomic coherence over a long path through the apparatus, recent progresses on the cavities quality factors, show that it is possible to preserve this system from relaxation and perturbations for a time long enough to perform many fascinating experiments. The realization of a QND detection of a single photon, the observation of quantum jumps due to the photon death are foreseeable in a near future, as well as the first experiments with Schrödinger cats. With a moderate increase in the quality factors, and in experimental complexity, tests of quantum nonlocality will also become possible.

REFERENCES

1. Haroche, S., "Cavity Quantum Electrodynamics", in *Fundamental Systems in Quantum Optics, Les Houches Summer School, Session LIII*, Dalibard, J., Raimond, J.M., and Zinn-Justin, J., eds., North Holland, Amsterdam (1992).
2. Haroche, S., and Raimond, J.M., "Manipulation of non classical field states in a cavity by atom interferometry", in *Cavity Quantum Electrodynamics, Special Issue of Advances in Atomic and Molecular Physics*, Berman, P., ed., Academic Press, New York (1994).
3. Haroche, S., and Raimond, J.M., *Scientific American*, April 1993.
4. Haroche, S., and Kleppner, D., *Physics Today*, January 1989.
5. Hulet, R.G., and Kleppner, D., *Phys. Rev. Lett.* **51**, 1430 (1983).
6. Walther, H., *Physica Scripta*, **T23**, 165 (1988); Raithel, G., Wagner, C., Walther, H., Narducci, L.M., and Scully, M.O., "The micromaser: a proving ground for quantum physics" in *Cavity Quantum Electrodynamics, Special Issue of Advances in Atomic and Molecular Physics*, Berman, P., ed., Academic Press, New York (1994).
7. Brune, M., Raimond, J.M., Goy, P., Davidovich, L., and Haroche, S., *Phys. Rev. Lett.* **59**, 1899 (1987).
8. Cohen-Tannoudji, C., Dupont-Roc, J., and Grymberg, G., *Atom-Photon Interaction*, Wiley, New York (1992).
9. Einstein, A., Podolski, B., and Rosen, N., *Phys. Rev.* **47**, 777 (1935).
10. Bell, J.S., *Physics* (Long Island City, N. Y.) **1**, 195 (1964).
11. Greenberger, D.M., Horne, M.A., Shimony, A., and Zeilinger, A., *Am. J. of Phys.* **58**, 1131 (1990).
12. Brune, M., Haroche, S., Lefèvre, V., Raimond, J.M., and Zagury, N., *Phys. Rev. Lett.* **65**, 976 (1990).
13. Brune, M., Haroche, S., Raimond, J.M., Davidovich, L., and Zagury, N., *Phys. Rev.* **A45**, 5193 (1992).
14. Schrödinger, E., *Naturwissenschaften* **23** 807, 823, 844 (1935); English translation by Trimmer, J.D., *Proc. Am. Phys. Soc.* **124**, 3235 (1980).
15. Brune, M., Nussenzveig, P., Schmidt-Kaler, F., Bernardot, F., Maali, A., Raimond, J.M., and Haroche, S., *Phys.Rev. Lett*, **72**, 3339 (1994).

16. Ramsey, N.F., *Molecular Beams*, Oxford University Press, New York (1985).
17. Haroche, S., Brune, M., and Raimond, J.M., *Euro. Phys. Lett.* **14**,19 (1991).
18. Englert, B.G., Schwinger, J., Barut, A.O., and Scully, M.O., *Europhys. Lett.* **14**, 25 (1991).
19. Ivanov, D., and Kennedy, T.A.B., *Phys. Rev.* **A47**, 566 (1993).
20. Nussenzveig, P., Bernardot, F., Brune, M., Hare, J., Raimond, J.M., Haroche, S., and Gawlik, W., *Phys. Rev.* **A48**, 3991 (1993).
21. Gross, M. and Liang, J., *Phys. Rev. Lett.* **57**, 3160 (1986).
22. Braginsky, V.B., and Khalili, F.Y., *Zh. Eksp. Theor. Fiz.* **78**, 1712 (1977) [*Sov. Phys. JETP* **46**, 705 (1977)].
23. Nagourney, W., Sandberg, J., and Dehmelt, H., *Phys. Rev.Lett.* **56**, 2797 (1986).
24. Glauber, R.J., "Optical coherence and photon statistics", in *Quantum Optics and Electronics, Les Houches Summer School*, de Witt, C., Blandin, A., and Cohen-Tannoudji, C., eds., Gordon and Breach, London (1965). *Phys. Rev.* **130**, 2529 (1963); **131**, 2766 (1963).
25. Yurke, B. and Stoler, D., *Phys. Rev. Lett.* **57**, 13 (1986); Yurke, B., Schleich, W., and Walls, D.F., *Phys. Rev. A* **42**, 1703 (1990); Milburn, G., *Phys. Rev. A* **33**, 674 (1986).
26. Zurek, W., *Physics Today*, **44**, 36 (1991).
27. Haroche, S., Brune, M., Raimond, J.M., and Davidovich, L., in *Fundamentals of Quantum Optics III*, ed. Ehlotzky, F., Springer-Verlag, Berlin, (1993).
28. Mermin, N.D., *Phys. Today*, **43**, 9 (1990)
29. Mermin, N.D., *Phys. Rev. Lett.* **65**, 1838 (1990)
30. Wiesner, S., *Sigact News* **15**, 78 (1983); Ekert, A.K., *Phys. Rev. Lett.* **67**, 661 (1991); Bennett, C.H., Brassard, G., and Mermin, N.D., *Phys. Rev. Lett.* **68**, 557 (1992); Bennett, C.H., *Phys. Rev. Lett.* **68**, 3121 (1992); Ekert, A.K., Rarity, J.G., Tapster, P.R., and Palma ,G.M., *Phys. Rev. Lett.* **69**, 1293 (1992).
31. Deutsch, D., *Proc. R. Soc. London A* **400**, 97 (1985); Deutsch , D., and Jozsa, R., *Proc. R. Soc. London A* **439**, 553 (1992); Berthiaume, A. and Brassard, G., in *Proceedings of the Seventh Annual IEEE Conference on Structure in Complexity Theory, Boston, June 1992*, IEEE, New York, 1992, p. 132.
32. Bennett, C.H., Brassard, G., Crépeau, C., Jozsa, R., Peres, A., and Wootters, W., *Phys. Rev. Lett.* **70**, 1895 (1993).
33. Davidovich, L., Zagury, N., Brune, M., Raimond, J.M. and Haroche, S., *Phys. Rev.* **50**, R1 (1994).
34. Meystre, P., in *Progress in Optics XXX*, edited by Wolf, E., Elsevier Science, New York, 1992.
35. Davidovich, L., Maali, A., Brune, M., Raimond, J.M., and Haroche, S., *Phys. Rev. Lett.* **71**, 2360 (1993).
36. Braunstein, S.L., Mann, A., and Revzen, M., *Phys. Rev. Lett.* **68**, 3259 (1992).

Quantum Optics With Strong Coupling

H. J. Kimble, O. Carnal*, N. Georgiades,
H. Mabuchi, E. S. Polzik**, R. J. Thompson***,
and Q. A. Turchette

*Norman Bridge Laboratory of Physics,
California Institute of Technology, Pasadena, CA 91125*

I. INTRODUCTION

In general terms the dynamics of open quantum systems can be characterized by two rates (g, Γ) which specify, on the one hand, the time scale for the irreversible interaction of system and environment and, on the other hand, the time scale for reversible evolution of the system itself.[1,2] From the perspective of the master equation, the ratio $n_o = \Gamma^2/g^2$ of these two rates is a natural scaling parameter which allows a demarcation between situations in which dissipation is dominant ($n_o \gg 1$ and hence weak coupling) and for which manifestly quantum dynamics come to the fore ($n_o \ll 1$ and hence strong coupling). In somewhat more specific terms, for optical systems dissipation is as simple as the escape of fields into an external environment at rate Γ, while reversible internal evolution is associated with the mutual interaction of constituents such as an atom and a high-Q mode of the electromagnetic field with coupling constant g. Within this context, optical physics is carried out almost exclusively in a domain of weak coupling (g$\ll \Gamma$) for which critical photon or atom numbers are much greater than unity. For example, a typical laser at threshold has critical photon number $\sqrt{n_o} \sim 10^3 - 10^4$, while an optical parametric oscillator has critical photon numbers for signal and idler of order $10^4 - 10^5$.

A noteable exception to this state of affairs can be found in the area of cavity quantum electrodynamics (CQED), where in recent years conditions for strong coupling have been achieved in both the optical and microwave domains. As illustrated in Figure 1, a two-state atom interacts with a single mode of a resonator as described by the interaction Hamiltonian [3-7]

$$\hat{H}_I = \hbar g [\hat{\sigma}_+ \hat{a} + \hat{a}^\dagger \hat{\sigma}_-], \tag{1}$$

where $(\hat{a}, \hat{a}^\dagger)$ are the annihilation and creation operators for the field mode, $\hat{\sigma}_\pm$ are Pauli operators for the atomic degrees of freedom, and $g \equiv (\mu^2 \omega_o / 2\hbar \varepsilon_0 V)^{1/2}$ is the assumed dipole coupling constant of atom and field. In addition to the coherent interaction specified by $\hat{H}_I$, we must also include coupling to external degrees of freedom to describe the decay of the cavity field at rate κ and spontaneous decay of the atomic polarization and inversion at rates $\gamma \equiv (\gamma_\perp, \gamma_\parallel)$ to field modes other than the one priviledged cavity mode. The condition for strong coupling in this

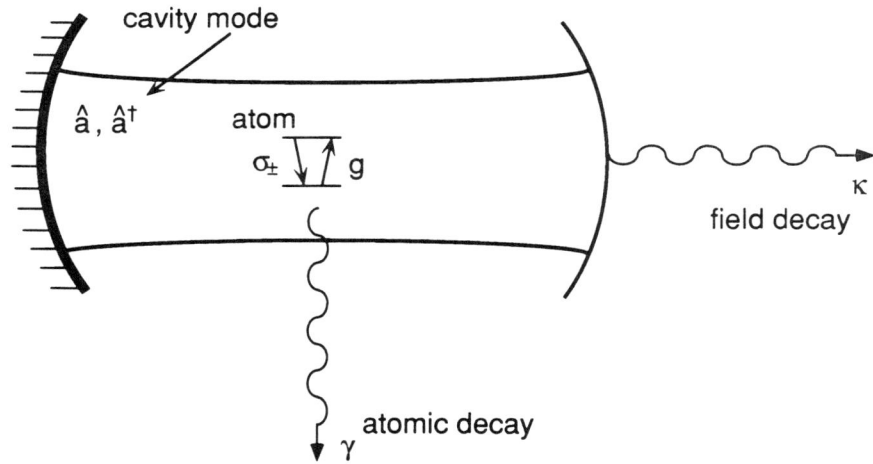

FIGURE 1. Illustration of model system. The coherent interaction of a two-state atom with a single mode of a resonator is characterized by the dipole coupling coefficient g. Dissipative interactions with the external environment proceed via decay of the cavity field at rate κ and of the atom at rate γ.

case is then stated as
$$g \gg (\kappa, \gamma). \qquad (2)$$
In terms of dimensionless parameters, we define $N_o \equiv \frac{2\gamma_\perp \kappa}{g_0^2}$ as a critical atom number (with $1/N_o$ as the cooperativity parameter C_1 per atom) and $m_o \equiv \left(\frac{\gamma_\perp \gamma_\parallel}{4g_o^2}\right) b$ as a critical or saturation photon number, where b depends upon the cavity geometry (b = 8/3 for a Gaussian standing-wave mode). Here (N_o, m_o) determine the roles of individual atoms and quanta in the system's dynamics. Strong coupling requires $(N_o, m_o) \ll 1$ and means in qualitative terms that a single atom can profoundly effect the cavity characteristics and the field associated with even a fraction of a quantum can saturate the atomic response. In fact in a domain of strong coupling, it is not appropriate to view the atom and cavity field as individual entities, but rather one must consider the dynamics of the composite system.

To date, strong coupling has been demonstrated in three experiments, with the parameter sets (g, γ, κ) that have been achieved summarized in Table 1. Since all of these experiments have employed atomic beams, the interaction time is determined by the transit time T_o, with the value of gT_o also of great importance. For example, for resonant phenomena as in the one-atom maser,[7,8] the condition for repeated Rabi nutations with $m = 1$ intracavity excitation is $gT_o \gg 2\pi$. Likewise, the scheme for quantum-state synthesis as described in Ref.[12] employs adiabatic passage with the vacuum and requires that $gT_o \gg 1$. Beyond the case of resonant

TABLE 1.

Experiments with strong coupling						
Group	$\omega/2\pi$	$g/2\pi$	$\kappa/2\pi$	$\gamma/2\pi$	gT_0	ϕ
Walther et al., Ref. [7,8]	21.5 GHz	7 kHz	0.4 Hz	500 Hz	$0.7\,\pi$	$0.07\,\pi$
Haroche et al., Ref. [9,10]	51.1 GHz	17 kHz	40 kHz	5 Hz	$0.9\,\pi$	$0.09\,\pi$
Kimble et al., Ref. [5,11]	353 THz	7.2 MHz	0.6 MHz	5 MHz	$4.3\,\pi$	$0.43\,\pi$

interaction, the dispersive regime (atom-cavity detuning $\Delta \equiv (\omega_c - \omega_A) \gg g$) has also been considered and should give rise to a wealth of new phenomena, such as atom-cavity phase shifts for quantum measurement via atom interferometry.[9,10] Here, transit of a ground-state atom through a cavity mode with one quantum of excitation is accompanied by a phase shift $\phi = g^2 T_0/\Delta$, where now ϕ is required to be of order 2π for appreciable effects. Similarly, quantum measurements of field statistics via atomic deflections in a dispersive regime have also been discussed.[13-17] Values for gT_o as well as for $\phi = g^2 T_o/\Delta$ (with $\Delta = 10g$) are given in Table 1 for the various experiments.

With reference to the table, note that the work in the microwave domain as in the groups of Professor Walther in Garching and Professor Haroche in Paris utilizes the large dipole moments of Rydberg atoms in high-Q superconducting cavities and has the distinct advantage of relatively small damping rates (κ, γ) relative to g. On the other hand, the time to reach steady state is usually much longer than the typical transit time T_o, so that steady states are reached as the result of many individual atomic transits $[g > T_o^{-1} > (\kappa, \gamma)]$. As well the energy scale $\hbar g$ is itself small when compared to atomic kinetic energies $(\hbar g/k_B \sim 1\mu K)$. By contrast since about 1980, our experiments in the optical domain have exploited dipole transitions of alkali atoms in very high finesse cavities, with the progress of this work over the intervening years documented in Figure 2. Although we have not yet obtained comparably small values for (N_o, m_o) as in the microwave domain, we have nonetheless achieved a situation of strong coupling for steady-states reached in the transits of individual atoms $[g > (\kappa, \gamma) > T_o^{-1}]$. Also note that the actual energy scale $\hbar g$ in our work can be much larger than that associated with the kinetic temperature of laser cooled atoms $(\hbar g/k_B \gtrsim 300\mu K)$ and that the intracavity field is directly accessible as an emitted Gaussian beam for photon counting or heterodyne detection. Beyond the brief overview provided in Table 1, more extended discussions by each of the three groups that have achieved strong coupling can be found in a recent review.[5,7,9]

From a somewhat broader perspective, although conditions appropriate for strong coupling have been demonstrated in the realm of cavity QED, research on related fronts is being actively pursued and is represented by other contributions to the 1994 ICAP. For example, while cavity QED centers on the coupling of the

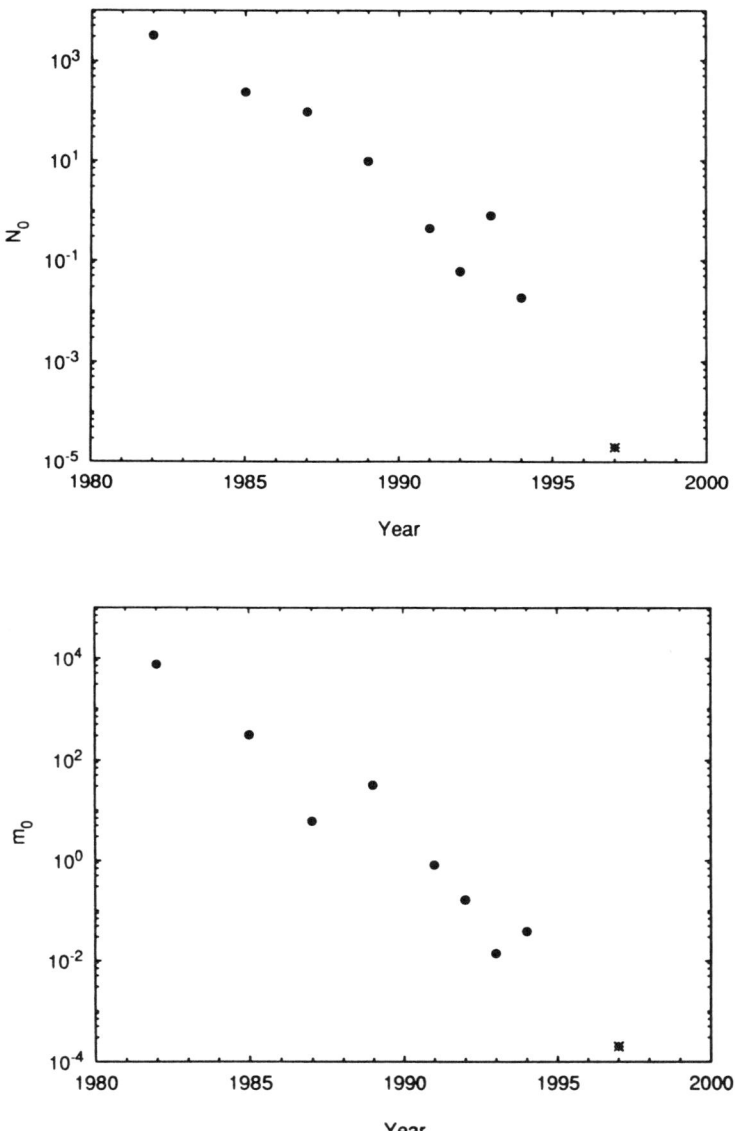

FIGURE 2. Progress of our group in the pursuit of strong coupling for which the critical atom and saturation photon numbers $(N_o, m_o) \ll 1$. The relevant literature citations are discussed in Ref. [5]. The point marked with an asterisk is a projection based upon cavity QED with optical whispering gallery modes as discussed in Section III.

electromagnetic field (photons) to the internal degress of freedom of an electronic transition in an atom, it should be possible to explore related phenomena for the coupling of the center of mass motion of a bound ion (phonons) to either a spin (by employing a magnetic field gradient)[18,19] or to an electronic transition (by utliliz- ing an electric field gradient).[20] Likewise, although parametric processes such as sub/second harmonic conversion via $\chi^{(2)}$ have been encountered experimentally ex- clusively in a domain of weak coupling, Agarwal[21] has recently considered a regime of strong coupling between two field modes of frequencies $(\omega, 2\omega)$ and has predicted a coupling-induced mode splitting analogous to that in the atom-cavity system.

Given this brief introduction to optical physics with strong coupling, our intent in subsequent sections is to describe in somewhat more detail the research activities of the Quantum Optics Group at Caltech in this area. In particular, we begin in Section II with a discussion of our direct spectral measurements of the normal- mode structure of the atom-cavity system (the so-called "vacuum − Rabi" splitting) and of the nonlinear modification of this structure with only $m \sim 0.1$ intracavity photons. In Section III we next turn to consider our work with quartz microspheres for which we have demonstrated cavity $Q \simeq 2.5 \times 10^9$ and where the prospects thus seem quite promising for achieving large values of g relative to (κ, γ). We as well suggest exploiting the spheres for binding atoms in stable orbits around an optical resonator and for QND measurement of photon number. Here the goal is to find a meeting ground between cavity QED and laser trapping and cooling (that is, between the quantization of internal and external degrees of freedom of the atom-cavity system). Against this backdrop of current and projected experimental capability, we then describe in Section IV a theoretical proposal for the synthesis of arbitrary quantum states of the electromagnetic field based upon transfer of Zeeman coherence from an atom to the cavity via adiabatic passage. Section V presents a brief discussion of the relationship of the physics of strong coupling to quantum computation. Finally, in Section VI we step back from the nonperturbative to the perturbative regime in cavity QED to discuss our attempts to make a "one- dimensional" atom which is then strongly coupled to squeezed light from an external source. The objective here is to investigate the fundamental alteration of atomic radiative processes in the presence of fields of manifestly quantum or nonclassical character.

II. NONLINEAR SPECTROSCOPY WITH 0.1 PHOTONS

The interaction of an atom with a single mode of the field leads to modifications of both the atomic radiative processes and of the cavity characteristics. For weak coupling with $g \ll (\kappa, \gamma)$, a perturbative description suffices and the atom and cavity maintain individual identities. However, for strong coupling with $g \gg (\kappa, \gamma)$ the system must be described in terms of the structure and dynamics of the composite atom-cavity entity. Indeed, one of the most striking characteristics in the domain of strong coupling is a normal-mode splitting variously refered to as the "vacuum-Rabi" or "Jaynes-Cummings" splitting.[22−25] The origin of this structure is familiar from the physics of simple coupled oscillators[26] and follows immediately either from the eigenvalues $\lambda_\pm$ of the interaction Hamiltonian of Eq. (1) (suitably generalized to include damping) or from the Maxwell-Bloch equations in the limit of weak excitation.[27-29] For the case of coincident atomic and cavity resonance frequencies ($\Delta = 0$), explicitly we have

$$\lambda_\pm = -\left(\frac{\kappa + \gamma_\perp}{2}\right) \pm \left[\left(\frac{\kappa - \gamma_\perp}{2}\right)^2 - g^2 \bar{N}\right]^{1/2}, \qquad (3)$$

where we now consider a collection of N_s two-state atoms distributed at various sites j within the cavity mode function $\psi(\vec{r})$ with coupling $g(\vec{r}_j) = g\psi(\vec{r}_j)$. In the strong coupling regime, Eq.(3) leads to a normal-mode splitting given approximately by $\text{Im}\lambda_\pm = g\sqrt{\bar{N}}$, with $\bar{N}$ defined as the effective intracavity atomic number; $\bar{N} \equiv \sum_{j=1}^{N_s} |\psi_j|^2$. In analogy with radiative frequency shifts associated with the perturbative coupling of an atom to many modes of the electromagnetic field, note that this coupling-induced structure in the nonperturbative regime might be termed a "Lamb splitting". Likewise, an interpretation associated with a Rabi splitting induced by the rms vacuum field could be offered. However, we have repeatedly emphasized that this phenomenon is more simply understood in terms of the normal-mode structure of coupled oscillators. Furthermore, this interpretation of the eigenvalues of coupled oscillators is equally valid for nonzero detuning with $\Delta \neq 0$, where alternative descriptions of the atom-cavity structure can be couched in more sophisticated terms (such as of cavity-induced Lamb shifts).[10] This perspective is elaborated both theoretically and experimentally in our previous work in which we documented the detuning dependence of the eigenvalue structure for the atom-cavity system.[5,27-29]

In somewhat more operational terms, the arrangement that we have employed for investigations of the structure of the atom-cavity system is as depicted in Fig. 3. The basic measurement strategy is that of heterodyne spectroscopy, where the transmission of a weak probe beam is recorded as a function of the frequency Ω of the probe.[23] An example of such a transmission spectrum in the weak-field limit is given in Fig. 4 for $\bar{N} = 1.1$ atom, with $\Omega = 0$ being the position of the common atom-cavity resonance in the absence of coupling. Note that the ordinate is normalized in terms of the intracavity photon number m_p associated with the probe beam, with the requirement of weak-field being that $m_p = 0.02 << m_o = 0.15$ photons. The two-peaked structure evidenced in Fig. 4 occurs at $\Omega \approx \pm g$ and is due to the dominance of coherent coupling g over dissipation (κ, γ). The measurement in Fig. 4 together with our previously reported work with somewhat smaller coupling represent the only direct observations of the "vacuum-Rabi" splitting for $N = 1$ atom and $m < 1$ photon. From the perspective of the energy spectrum, the measurements are the first to resolve directly the lowest lying excited states of the coupled atom-cavity system. We emphasize that the result of Fig. 4 is taken in the weak-field limit and is thus independent of the strength of the probe beam; increasing or decreasing the probe power results only in an overall scaling of the ordinate. The spectrum thus reflects the underlying structure of the composite atom-cavity system, independent of whether this structure is attributed to a vacuum-field induced effect[22,23] or to a normal-mode splitting[26-29] or to linear dispersion theory with the atom serving as a microscopic dielectric.[3,30,31]

Whatever one's favorite interpretation, the frequency scale in Figure 4 is associated with a position-dependent energy (and hence a force) via the dependence

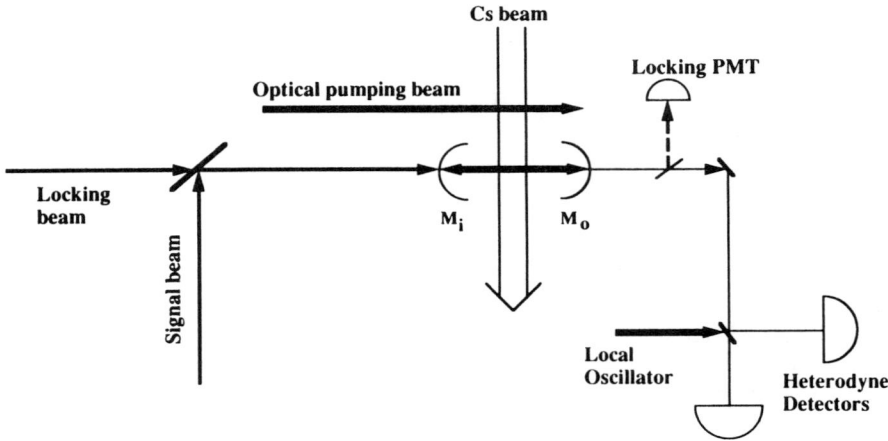

FIGURE 3. Schematic of the principal elements in our cavity QED experiments. The central component is the high finesse cavity formed by the mirrors (M_i, M_o) through which a well-collimated and optically pumped beam of atomic Cesium passes. Various signal inputs excite and probe the atom-cavity system. The transmission of these fields is recorded by heterodyne detectors operated in a balanced mode. Locking of the "empty-cavity" resonance (no atoms) is accomplished by an auxilliary beam and detector in a chopped sequence.

of the coupling coefficient $g(\vec{r})$ on position in the cavity mode. The energy scale in terms of temperature for this dependence is indicated on the spectrum by a bar corresponding to 100 μK (recall that the Doppler cooling limit for Cs is 125 μK). The larger magnitude of the normal-mode splitting relative to the temperature bar hopefully serves to emphasize the point of the potential for coupling the internal degrees of freedom of the atom-plus-cavity field to the external degrees of freedom of the atomic center-of-mass motion. Variations in $g(\vec{r})$ can lead to significant changes in potential energy for the center-of-mass motion, with the interplay of external and internal degrees of freedom affected at the scale of the vacuum field itself.

With this quantitative footing in the realm of linear spectroscopy, we next move to explore nonlinear spectroscopy of the atom-cavity system. The procedure is now that of pump-probe spectroscopy with the addition of an "intense" pump beam of fixed frequency. An example of our measurements in this case is presented in Fig. 5, which displays two transmisison spectra. Curve (i) is taken with the pump beam turned on with the frequency Ω_{pump} set to match the location of the upper peak and with an intensity that produces an intracavity photon number $m_{\text{pump}} \simeq 0.1$ photons. The modification of the weak-field spectrum for the probe beam due to the pump beam is quantified by the difference spectrum Curve(ii), which is the difference between Curve (i) (pump on) and the trace in Fig. (4) (pump off). Since

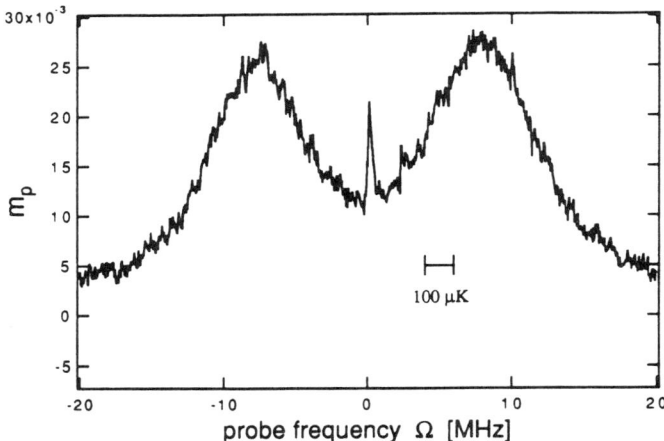

FIGURE 4. Transmission spectrum with $\bar{N} = 1.1$ intracavity atoms as a function of probe frequency Ω, with the transmission normalized in units of intracavity photon number m_p for the probe beam. The two broad peaks represent a normal-mode (or vacuum-Rabi) splitting for the atom-cavity system. The position of the common atom-cavity resonance ($\Delta = 0$) in the absence of coupling is $\Omega = 0$, as calibrated by the sharp peak at this position (which is otherwise of no significance). The spectrum is taken with weak excitation (m_p = 0.02 photons) so that the structure evidenced is independent of probe power. The bar indicates a frequency change equivalant to a change in energy of $100 \mu K$.

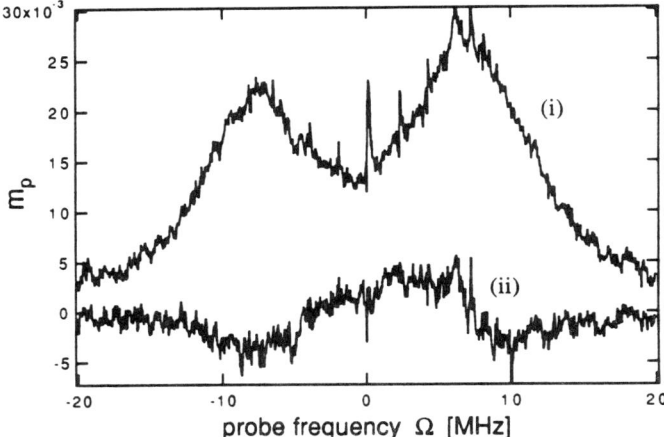

FIGURE 5. Transmission spectrum as in Figure 4 but now with the addition of a pump field (curve (i)). The intensity of the pump is such that there are only $m_{\text{pump}} \simeq 0.1$ intracavity photons, yet there is an appreciable change in the probe transmission (curve(ii)). The frequency of the pump field is Ω_{pump}= 7.5 MHz corresponding to the center of the upper peak in Fig. 4.

for a linear system the probe response would be independent of the pump (except of course at $\Omega_{\text{probe}} = \Omega_{\text{pump}}$), Curve (ii) represents a nonlinear "susceptibility" for the atom-cavity system with a characteristic scale for the nonlinearity of only about 0.1 intracavity photons.

Beyond the data presented in Figures 4 and 5, we have also made extensive measurements of nonlinear spectra over a range of intracavity pump photon numbers at fixed intracavity atomic number $\bar{N}$ to investigate the intensity dependence of the nonlinear repsonse. We have as well varied $\bar{N}$ to explore the transition from a quantum ($\bar{N} \sim 1$ atom) to a semiclassical domain ($\bar{N} \gg 1$ atoms) for the structural characteristics of the atom-cavity system.[24,32] Much of our effort has been directed toward the identification of a "quantum anharmonicity" in the nonlinear response due to the nature of the anharmonic level spacing of the Jaynes-Cummings ladder of states (i.e. the eigenstates associated with $\hat{H}_I$ in Eq.[1]). This program of research is being carried out in collaboration with Professor H. J. Carmichael, whose group is generating theoretical spectra by way of numerical simulations involving quantum trajectories.

Apart from "structural" issues as in our spectroscopic studies, we have also pursued a program of research aimed at exploring dynamical aspects of the atom-cavity system and in particular of manifestly quantum features in the strong-coupling regime. Towards this end, we have reported observations of photon statistics which display the nonclassical effects of photon antibunching and sub-Poissonian photon statistics.[33] Most importantly, we have observed that the magnitude of these effects is largely independent of intracavity atomic number as $\bar{N}$ increases, in agreement with theory in the strong-coupling domain[34]. By contrast, the usual quantum statistical theories of this effect are carried out in a domain of weak coupling and rely on system-size expansions for which nonclassical effects such as we have observed diminish as $1/N$.[3] Our measurements have thus identified one modest avenue by which a "small" quantum system can grow into a "large" system without the loss of quantum character that is usually dictated by the scaling laws for weakly coupled systems. It is interesting to note that, while on the one hand the quantum characteristics in structural terms are evidently lost with increasing N, on the other hand, certain manifestly quantum properties associated with the system's dynamics persist and can be traced precisely to the fact that the critical atom number $N_o < 1$. For a strongly coupled system, the dynamical processes associated with individual atoms and quanta cannot be scaled away, but can have profound consequence for the system's evolution even for $(N, m) \gg 1$. Our work in cavity QED thus offers strong support for a belief that there is much to be learned about the quantum-classical interface through the study of open quantum systems in a regime of strong coupling.

III. ATOM GALLERIES FOR WHISPERING ATOMS

In terms of future prospects for cavity QED in the optical domain, there are a variety of measurements that would benefit from a longer storage time for the cavity field. The general feature of such measurements is atomic transit on a relatively short time scale with $gT_o \gg 1$ and $\kappa T_o \ll 1$, with the interaction of atom and cavity having produced a mutual (and perhaps entangled) change of state. Of course information encoded in atomic hyperfine levels can have a very long lifetime and is ammenable to interrogation by various techniques. On the other hand, the combination of lifetime and transit time for our current cavities is hard pressed to satisfy the condition $\kappa T_o \ll 1$. Our next generation cavities for which we have demonstrated finesse 1.9×10^6 offers good prospects[35], but it seems wise to search for new technical avenues.

One very attractive possibility for increasing both the coherent coupling g and the cavity lifetime is an optical resonator formed from a quartz microsphere, as has been pioneered by Braginsky and co-workers.[36] The "whispering gallery" modes of these resonators can have very high quality factors Q. In collaboration with Professor Braginsky's group, we have demonstrated $Q \simeq 2.5 \times 10^9$,[37] with similarly high values recorded in Moscow [38] and Paris.[38] With regard to cavity QED, the general idea is to couple an atom to the external "evanescent" field of a whispering gallery mode for a quartz sphere of radius $10 < a < 500 \mu m$. Although the field external to the sphere is much smaller than the field circulating inside the quartz dielectric (as illustrated in Figure 6), the small mode volume means that large values of g can nonetheless be obtained. Quantitative values for the coupling coefficient g for an atom interacting with the external field of a whispering gallery mode are given in Ref.[37] as a function of size parameter $x = 2\pi a/\lambda$, with a as the sphere's radius and λ as the free-space wavelength. These calculations indicate that it may be possible to have $g(a)/\gamma_\perp \sim 50$ near the surface of very small spheres ($a \sim 10\mu m$), with similarly large values $g(a)/\kappa \sim 50$ for $Q = 2 \times 10^9$. Furthermore, if $Q = 10^{11}$ could be achieved (as projected in Ref.[36]), one would have $g/\kappa \gtrsim 10^3$. In such a domain of strong coupling, an atom interacting with the external evanescent field could repeatedly absorb and reemit a circulating photon, which is a situation that we have dubbed a "whispering atom".[37]

Beyond the perspective of cavity QED per se, there are also exciting possibilities for exploiting the external fields of optical whispering gallery modes for confining atoms in stable orbits around the microsphere.[37] We have previously presented an explicit scheme utilizing dipole-forces for an atom with a three-level "Vee" configuration to generate a toroidal atom trap (which we term an "atom gallery"), and there appears to be a variety of other avenues for binding atoms in orbit. The confinement of atoms in this fashion also suggests the possibility of an atomic (matter-wave) resonator with atoms confined with position uncertainty Δx around the equator of the sphere which is comparable to the circumference of the sphere ($\Delta x \simeq 2\pi a$) and with the associated prospect of de Broglie resonance phenomena.

Thus, quartz microspheres appear to have sufficient potential for diverse investigations in optical physics to warrant a serious experimental effort. In general terms, the prospects can be divided into those associated with (i) cavity QED with strong coupling ("whispering atoms"), (ii) confinement of the atomic center-of-mass motion ("atom galleries"), and (iii) phenomena at the intersection of these two areas which involve the mechanical consequences of strong coupling. Some sense of the nature of opportunities in the latter category is provided by the fact that the dipole force for a single photon circulating in a 20 μm sphere (with $\Delta = 10g$) can provide the centripetal acceleration for a Cs atom with an orbital kinetic energy of 20 mK! In terms of scattering instead of binding, we [16] (and independently Treussart et.

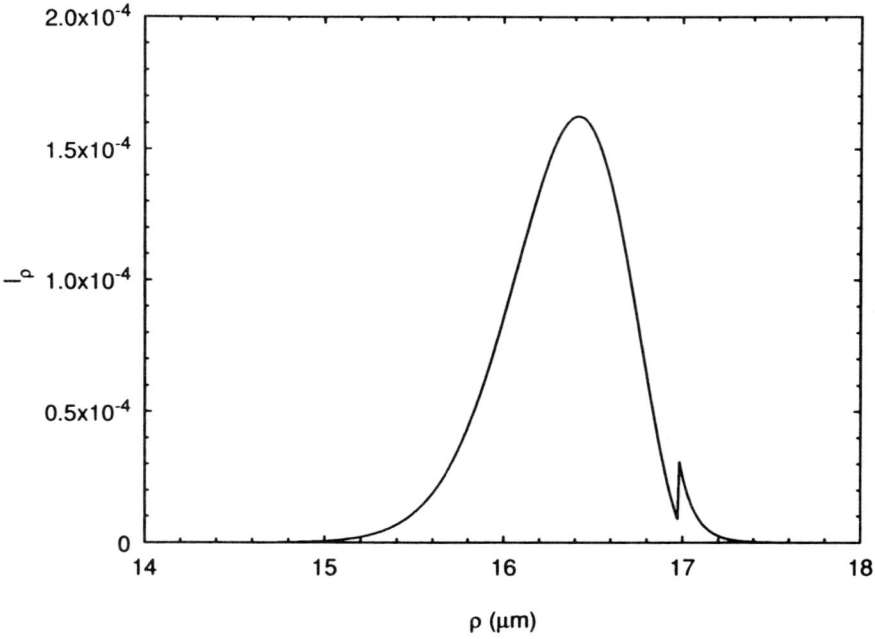

FIGURE 6. Intensity I_ρ of the radial electric field for a whispering gallery mode [TM145, l=m in the equatorial plane] as a function of radius ρ for a microsphere of radius $17\mu m$ ($\lambda = 1\mu m$). We propose cavity QED experiments by coupling atoms to the evanescent component external to the sphere.

al.)[17] have recently analyzed the possibility for quantum nondemolition detection of single photons by deflection from the fields of a microsphere or other open dielectric resonators. For these examples, it is important to include the nonresonant modifications of line widths and positions due to the close proximity of the atom to the dielectric sphere, as has been recently considered by Jhe and Kim.[40] Finally, for comparison with Table 1, note that the parameter $gT_o = 15\pi$ for a sphere with $a = 18\mu$m and for an atom with velocity $v = 30$m/s, while $\hbar g/k_B = 3$ mK.

IV. QUANTUM STATE SYNTHESIS

Over the past year and a half, Professor P. Zoller and his colleagues at JILA and our group at Caltech have developed and are working to implement an exciting new idea for the synthesis of quantum field states of the form $|\phi\rangle = \sum_m c_m |m\rangle_F$, where $|m\rangle_F$ are Fock states for a single-mode field and the c_m can be chosen experimentally with flexibility and broad lattitude.[12] To understand the basic idea, consider the setup shown in Fig.7 where an atom passes through the fields of a

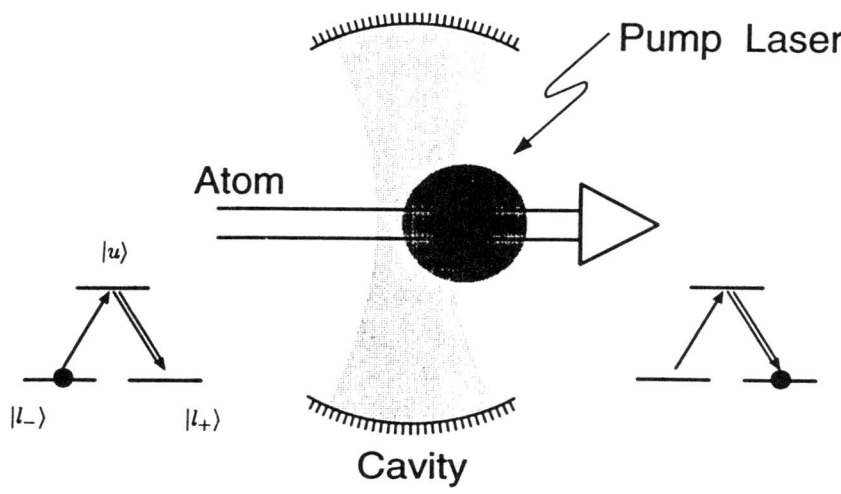

FIGURE 7. Proposed scheme for Fock-state generation via adiabatic passage as discussed in the text. The direction of propagation of the pump laser is perpendicular to the page. Note that the scheme can be extended for the transfer of Zeeman coherence from multiple ground states to the cavity field, and hence provides an avenue for generating general superpositions of Fock states.

resonant (quantized) cavity and of a classical (coherent-state) pump laser. The cavity is initially in the vacuum state $|0\rangle_F$ and the atom is initially in the ground state $|l_-\rangle_A$ as indicated. In its transit through the cavity, the atom sees two pulses, translated in time somewhat but nonetheless overlapping. We assume that the classical field is coupled to the $|l_-\rangle_A \leftrightarrow |u\rangle_A$ transition, while the quantized cavity field is coupled with coefficient g_+ to the $|l_+\rangle_A \leftrightarrow |u\rangle_A$ transition. By diagonalizing the Jaynes-Cummings Hamiltonian for this system (similar to Eq. (1)), it is straightforward to show that the initial state as chosen is the so-called "dark" dressed state and that this state evolves continuously into the final state shown, which is a state with 1 photon in the cavity mode and with the atom in the ground state $|l_+\rangle_A$. Note that throughout the transit, the system remains in the dark-state (which is a superposition of $|l_-\rangle_A$ and $|l_+\rangle_A$) and that the amplitude for the excited state is zero. Hence one atom in transit produces the change $|0\rangle_F \to |1\rangle_F$ for the intracavity field. More generally, the initial state $|l_-\rangle_A \langle l_-| \otimes \rho_F$ is transformed into a final state for which the atom leaves the cavity in state $|l_+\rangle_A$ with probability one and for which the cavity photon distribution is shifted by exactly one photon. Note that this process involves adiabatic passage "via the vacuum state" and that passing exactly N atoms through the cavity gives an N-photon Fock state for an initial vacuum state $|0\rangle_F$.

These ideas can be generalized to the situation of a transition of the form $J_{\text{lower}} \to J_{\text{upper}} = J_{\text{lower}} - 1$. In this case we assume for example σ_+ transitions driven by a classical field and π transitions coupled to a (quantized) cavity mode. The family of dark states for this circumstance can be computed by suitable diagonalization. For an atom initially in the state $|m_l = -J_l + k\rangle_A$ ($k = 0, 1, ...2, J_l - 1$) and with the cavity field initially in a vacuum state, we have a final state for the atom $|m_l = J_l - 1\rangle_A$ (independent of k) and for the field the state $|2J_l - 1 - k\rangle_F$. Thus the choice k for the initial Zeeman substate determines the photon number for the final state of the cavity field. An immediate extension is to inject atoms not in a single Zeeman state but in a coherent superposition of the form $|\psi\rangle_A = \sum_{m_l} c_{m_l} |m_l\rangle_A$. In this case, for a cavity initially in a vacuum state, the total final state will be $|J_l - 1\rangle_A \otimes \sum_{n=0}^{2J_l-1} c_{J_l-1-n} |n\rangle_F$, where now the coefficients c specify the (complex) weights of the Fock states $|n\rangle_F = |J_l - 1 - m_l\rangle_F$. A trace over the atomic state $|J_l - 1\rangle_A$ then leaves a general field state $\sum_{n=0}^{2J_l-1} c_{J_l-1-n} |n\rangle_F$ that has been synthesized from the initial choice of Zeeman coherences. We can thus generate "arbitrary" quantum states for the intracavity field, with the caveat that the maximum photon number is set by $2J_l - 1$. Note that the scheme is reversible, so that it offers a powerful new possibility for quantum measurement of optical fields.

Of course this discussion must be extended to include both atomic and cavity damping. Atomic decay plays no significant role since, for a broad range of conditions, the atom remains in the dark state throughout its transit with negligible excited state population and hence no spontaneous emission. The principal conditions for successful field-state generation are in the form of the following inequalities (note that no precise pulse areas are required):

$$gT \gg 1 \qquad (gT \simeq 12) \qquad \{4 \leq gT \leq 22\}$$

$$\kappa T \ll 1 \qquad (\kappa T \simeq 1) \qquad \{0.08 \leq \kappa T \leq 0.42\}. \qquad (4)$$

Parenthetically () we have indicated the various products for our Cesium system listed in Table 1 and described in Section II. Some indication of the possibilities for improvement are indicated in the final column in brackets {} where we assume a cavity of length $l = 100\mu m$, finesse $\mathfrak{F} = 1 \times 10^{6[35]}$ and mirrors of radii $R = 1m$, leading to the parameter set $[g, \gamma_\perp, \kappa]/2\pi = [18, 2.5, 0.33]$ MHz. We further assume that the transit time $T = 2w_0/\bar{v}$ can be controlled over the range $3.4 \times 10^{-8}s \leq T \leq 1.7 \times 10^{-7}s$ by seeding the Cs beam into a Helium carrier gas in a supersonic nozzle, which has the added advantage of reducing the dispersion in transit times. Exact solutions of the quantum master equation for various $J_l \to J_u$ transitions have been computed by Professor Zoller and colleagues for parameter sets as in (4). The evolution of the various ground state populations, of the intracavity photon number $\langle a^+ a \rangle$, and of the Mandel Q parameter indicate that the adiabatic passage scheme is quite effective in generating individual Fock states and superpositions thereof.

In the end, our goal is to achieve "dial-a-state" capabilities for the electromagnetic field as is summarized by the cartoon of Figure 8, where we depict Model 1 of our quantum state synthesizer (QSS-1). By appropriate settings of the dials for

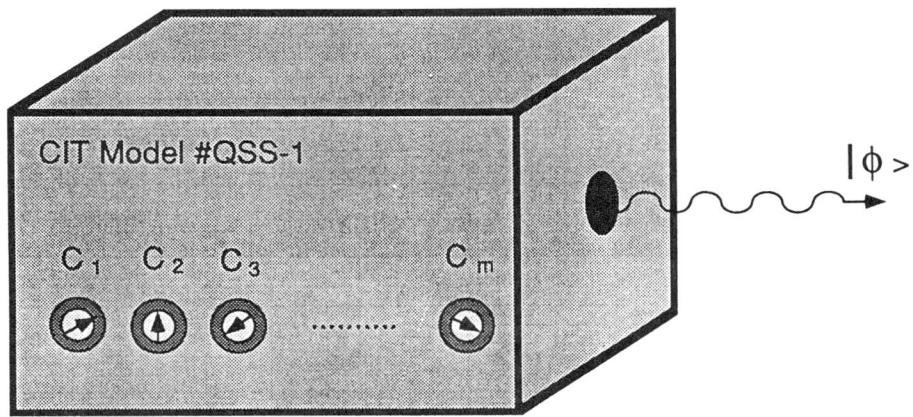

FIGURE 8. Model 1 of Caltech Quantum State Synthesizer for generating "arbitrary" quantum states of the electromagnetic field. Note that the dials for $\{c_j\}$ have inner and outer knobs for control of modulus and phase.

$\{c_j\}$, we would be able to specify the field state $|\phi>$ in the form

$$|\phi> = \sum_{j=0}^{n_{max}} c_j |j>, \qquad (5)$$

with $|j>$ as Fock states. The user thus tunes to her favorite state by appropriate choices of the complex coefficients c_j, where of course an individual dial's settings implies both a modulus and a phase. Although the number of states that can be generated is limited both by the maximum value n_{max} for the Fock state expansion ($n_{max} = 2J_l - 1$ in the preceeding discussion) and by the finite resolution by which the coefficients can be determined in practice, the number of states that can be accessed in the Hilbert space for the single-mode of the cavity field is nonetheless quite large.[41] To give an estimate, imagine that each value c_j is specified to k bits. Then the number of distinguishable states N that could be accessed for a given value of n_{max} is

$$N \simeq (2^k)^{(n_{max}+1)}. \qquad (6)$$

For even modest values of $n_{max} = 7$ photons (as would be appropriate for $J_l = 4 \rightarrow J_u = 3$ in Cs) and $k = 7$ bits (i.e., about 1% accuracy), the number of distinguishable states in the Hilbert space (and hence dial settings) is $N \simeq 10^{17}$ states, which should be a large enough number to satisfy even a Texan! Also note that as with long distance telephone companies, the user will have diverse choices for "dial-a-state" services, as in the work of Professor Schleich's group.[42]

The moral of this discussion is that Hilbert space is indeed a big place. Although a rich phenomenology of manifestly quantum states has emerged in the past 20 or so years (including antibunched light in 1977 and squeezed light in 1985), we believe that quantum optics has only begun to explore the complexity of possibilities associated with states in the Hilbert space and their dynamics.

V. QUANTUM COMPUTATION AND STRONG COUPLING

A spectacular recent result by P. Shor [43] in the field of quantum computation clearly makes the point of the tremendous potential associated with exploring and perhaps ultimately harnessing the complexity of quantum dynamics (for an overview, see the contribution by A. Ekert in this volume). Following the pioneering work by Feynman[44] and by Deutsch,[45] Shor has shown how to exploit quantum mechanics to yield an efficient (quantum) algorithm for factoring large numbers. The essential point is that for a given system (here, the quantum computer), the number of quantum states is exponentially larger than the number of associated classical states, yet this large number of states can be efficiently accessed by the unitary transformations of quantum evolution.

In the current flurry of excitement over quantum computation, a number of important questions remain unanswered (and indeed even difficult to pose). For example, in principle Shor's quantum algorithm must have access to the whole of the Hilbert space with its exponentially large set of possibilities. However, in practice there will undoubtedly be "localized" regions of dense trajectories in the Hilbert space (in the sense of a path integral), and one should thus seek to determine an effective strategy for "adaptive meshes" which best utilize finite computational resources. Likewise, a villain in this story is sure to be dissipation, and one might seek to optimize quantum algorithms to employ states with a robustness to the particular dissipative channels encountered and to explore possibilities of "stabilization" with external stimuli. One should go beyond the simplified (and often misleading) viewpoint of dissipation based upon solutions for ensemble averages, and consider descriptions based upon quantum trajectories for individual realizations of quantum (stochastic) dynamical processes. Lastly, it is important to realize that there is a "nonuniqueness" associated with dissipation; the dynamics of open quantum systems with strong coupling are conditioned upon the particular measurement strategy employed in the external environment, so that there is no unique dynamical process to be associated with the system's evolution.

While this discussion would seem to take us far afield of our topic of strong coupling, in fact the implementation of quantum computation requires an exploration of many issues for which our quantum-state synthesizer is well-suited, including the generation and survivability (against imperfections of various sorts as well as against dissipation) of states globally over a "large" Hilbert space. Furthermore, the elemental gates by which quantum computations might eventually proceed will inevitably involve the physics of strong coupling since outputs in the form of quantum bits (or "qubits") must be conditioned upon the state of input qubits. Hence the basic carriers of information (be they spins or photons) must be able to affect one another through interactions at the level of individual quanta in times short compared to any dissipative time scale, which is nothing more than our statement of strong coupling expressed generally in Section I and specifically for cavity QED. Following this theme, several groups (including our own) have suggested schemes for implementing elemental quantum logic gates by way of strong coupling in cavity QED.

VI. "ONE-DIMENSIONAL ATOMS" COUPLED TO SQUEEZED LIGHT

In this penultimate section, we wish to step back from the nonperturbative to the perturbative regime in cavity QED to discuss our attempts to make a "one-dimensional" atom which can then be strongly coupled to squeezed light from an external source. Following the seminal work of Gardiner,[46] our objective here is to investigate the fundamental alteration of atomic radiative processes in the presence of fields of manifestly quantum or nonclassical character. There is by now an extensive theoretical literature in this area; roughly speaking, any optical process (ranging from resonance fluorescence to photon echoes to optical gain and lasing) is modified in an important manner if the "regular" vacuum field is replaced by a "squeezed" vacuum.[47] Of course, from an experimental perspective, this replacement is easier said than done; we have presented various strategies towards this end in Ref. [48].

Here we will describe one particular avenue in the realm of cavity QED which relies on the dominance of coupling over spontaneous decay for an atom interacting with a cavity ($g \gg \gamma$), but for which cavity damping dominates coherent coupling ($\kappa \gg g$).[49,50] Stated more precisely, we move to the so-called "bad−cavity" limit for which

$$\kappa \gg g^2/\kappa \gg \gamma, \qquad (7)$$

with g^2/κ specifying the rate of emission into the cavity mode as follows from Eq. (3). That is, the pair of eigenvalues $\lambda_\pm$ for the atom-cavity can now be associated with an atom-like (cavity-enhanced atomic emission) and a cavity-like (atom-inhibited cavity decay) pair, where explicitly we have for the atom-like eigenvalue

$$\beta_o = \gamma_\perp(1 + 2C_1), \qquad (8)$$

where C_1 is the single-atom cooperativity parameter; $C_1 = 1/N_o = g^2/2\kappa\gamma_\perp$. Hence for $C_1 \gg 1$, the rate of emission into the cavity mode (g^2/κ) dominates that into free-space ($\gamma_\perp$). Since the input-output channel of the cavity is readily available to us as a collimated Gaussian beam, we can efficiently couple fields from the external world to the atom and thus overcome the difficulty of "mode matching" incident fields to the large angular content of the free-space atomic dipole emission.

The strategy then is to arrange for one intracavity atom in the domain specificied by Eq.(7) and to illuminate the cavity with squeezed light. In this way, we effectively replace the vacuum field of the cavity with a squeezed vacuum and can investigate the consequent alteration of (cavity-enhanced) atomic radiative processes. For the case of broad bandwidth squeezed light, Eq. (8) is replaced by

$$\beta_\pm = (1 + 2C_1 \Delta x_\pm^2), \qquad (9)$$

where $\Delta x_\pm^2$ specifies the variances of the quadrature-phase amplitudes for the squeezed field, with $\Delta x_\pm = 1$ for the vacuum state. Hence in the limit of large squeezing with $\Delta x_\mp \to \{0, \infty\}$, we find that $\beta_-/\beta_o \to 1/1+2C_1$, and $\beta_+/\beta_o \to \infty$, so that decay in one quadrature is suppressed and that in another is enhanced by the nonclassical fluctuations brought by the squeezed field. Here the dimensions of suppressed and enhanced fluctuations correspond to the $\{\hat{\sigma}_x, \hat{\sigma}_y\}$ components of the atomic polarization on the Bloch sphere. Note that this linewidth narrowing appears to be a manifestly quantum result in that stochastic fields of classical character can apparently only increase relaxation above the value β_o.

Our experimental technique for investigating cavity QED with squeezed light relies on the same basic arrangement depicted in Figure (3). Now, however, the cavity parameters are altered to place us in the domain specified by Eq.(7), namely $\{g, \kappa, \gamma_\perp\}/2\pi = \{20, 91, 2.5\}$ MHz, so that $2C_1 \simeq 1.8$ and $m_o \simeq 0.02$ photons. As well, the cavity is configured in a "single-port" arrangement with the transmission of the input coupling mirror (M_o in Fig. 3) dominating all other passive cavity losses. It is via this channel that we introduce squeezed light generated by parametric down conversion.[51] The behavior of the atom-cavity system is interrogated in the same fashion as in Figures (4,5) by way of a weak probe beam (injected via M_i in Fig. 3) whose frequency is tuned across the common atom-cavity resonance to produce a transmission spectrum. Figure (9) gives our theoretical prediction for the probe spectrum in the case of infinite bandwidth for the squeezed light and for the strict bad-cavity limit. The various curves in the figure are basically of the form of a "hole" in the otherwise smoothly varying broad transmission feature for the cavity. Note that as the number of intracavity photons brought by the squeezed-vacuum field increases, the initial dip associated with usual vacuum ($n = 0$) diminishes in size and narrows. In fact, the dip actually has two components; the first is a complex Lorentzian that narrows with increasing n (corresponding to coupling to Δx_- and reduced fluctuations of a squeezed vacuum relative to the normal vacuum), while the second broadens as n increases (coupling to Δx_+ and increased fluctuations). The size of the dip diminishes with increasing n because a squeezed vacuum brings real quanta which saturate the atomic response. Note that the phenomenology displayed in Fig. (9) is precisely as for the original problem of an atom in free-space considered by Gardiner,[46] but here is referenced to the dynamical processes of the cavity mode. Modes outside the cavity to which the atom is coupled are unaffected by the squeezed light (as evidenced by the factor "1" in the expression $(1 + 2C_1)$ in Eq. (9)).

In practice, the bandwidth of our squeezing is neither infinite nor are we strictly in the bad-cavity limit. Hence, a realistic calculation for our experiment must deal with the intertwining time scales of the cavity, the squeezing, and the atom coupled to the cavity. Dr. A. S. Parkins is collaborating with our group on this problem and has carried out numerical integrations of the relevant master equation. Probe spectra derived in this fashion confirm the qualitative trends evidenced in Fig.(9), but allow for a quantatitive comparison with experiment.

Although coordinating the two laboratories involved in this experiment (one for squeezed-state generation[51] and one for cavity QED) is not completely trivial, we have nonetheless obtained initial results for illumination with squeezed light. Figure (10) gives an example of probe spectra obtained with and without squeezing. Note that since $m_o \simeq 0.02$ photons, the probe power must be quite small if the probe

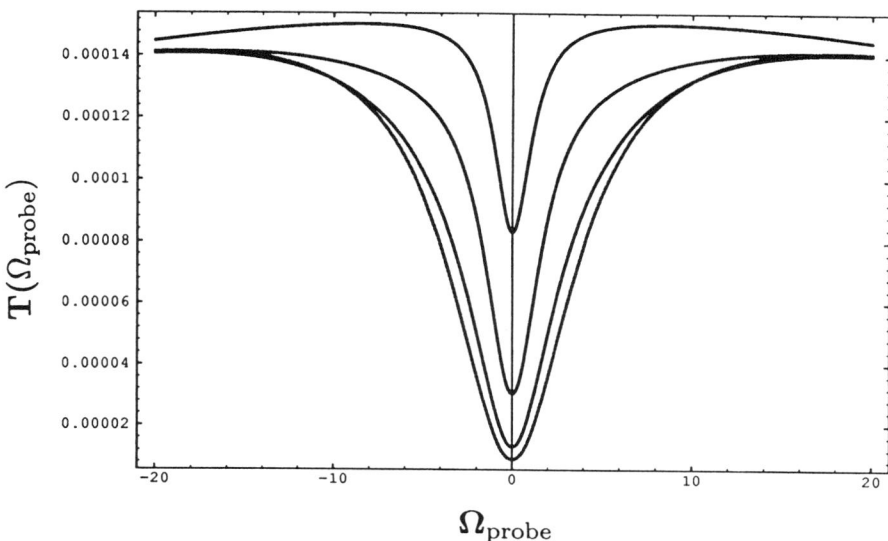

FIGURE 9. Transmission spectrum $T(\Omega_{\text{probe}})$ for a weak probe beam in the presence of squeezed light. The progression starting with the curve with the deepest dip at $\Omega_{\text{probe}} = 0$ is $n = \{0, 0.1, 0.5, 2.5\}$, where n is intracavity photon number associated with the squeezed-vacuum field. The curves are calculated following the formalism of Ref. [49] {i.e., strict bad-cavity limit}, but are evaluated with the parameter set $\{g, \kappa, \gamma_\perp\} = \{40, 200, 2.5\}$, with Ω_{probe} normalized in units of $\gamma_\perp$.

is to function as a nonintrusive measure of the atomic response. In Figure (10), the intracavity photon number attributed to the probe at the extreme edges of the scan is only $m_{\text{probe}} \simeq 1.6 \times 10^{-3}$ photons and diminishes near line center. On the other hand, the intracavity photon number attributed to the squeezed field is approximately $n \sim 0.05$ photons. While the squeezed field brings an alteration of the probe spectrum following the qualitative expectation of Fig.(9), we are seeking to quantify the changes by way of measurements involving excitation with "thermal" as well as with squeezed excitation. Here the thermal field is generated by the same parametric process but now in a nondegenerate mode. Hence the spectral distributions for squeezed and thermal fields should be identical, with the quintessential difference being the presence of a nonclassical character for the squeezed field. Although we have made a number of measurements as in Fig.(10) for various operating conditions, we have not yet seen reproducible evidence for linewidth narrowing. A principal suspect is the efficiency with which the squeezed light is transported to and coupled into the atom-cavity system. Nonetheless, data as in Fig.(10) are quite remarkable in their own right due to the small photon number associated with

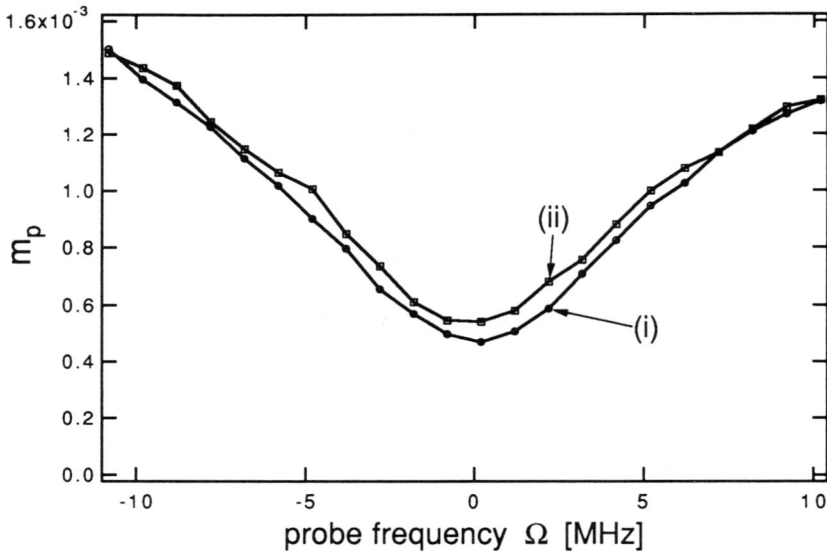

FIGURE 10. Excitation of the atom-cavity system with squeezed light. Displayed are measured probe spectra m_p versus probe frequency Ω without (curve (i)) and with (curve (ii)) squeezed excitation. For these data the intracavity atomic number $\bar{N} \simeq 1$ atom. For curve (ii) the squeezed vacuum is injected from the output of a subthreshold optical parametric oscillator and represents an intracavity field of only about 0.05 photons driving the atom.

nonlinear behavior ($n \sim 0.05$ photons) and to the sensitivity of the probe technique (spectra can be acquired with reasonable signal-to-noise ratio in several minutes with $m_{\text{probe}} \sim 10^{-4}$ photons).

VII. SUMMARY

Our objective here has been to provide a broad overview of the physics of strong coupling and its realization in the area of cavity QED. The particular progress of the Quantum Optics Group at Caltech is summarized in Figure 2, which illustrates a steady downward trend for the critical atom N_o and saturation photon m_o numbers; in fact, reductions of 5 - 6 orders of magnitude have been achieved over the fifteen years that we have been pursuing research in this area, so that now we are well into the domain of strong coupling $[(N_o, m_o) \ll 1]$. Although the work originally began in the context of optical bistability, the goal from the outset has been an investigation of the quantum dynamical processes for the atom-cavity system, for which a rich theoretical and experimental literature developed from the mid

1970's into the 1980's. A related field of research over the same period has been "cavity QED", which in its infancy was concerned principally with cavity-modified atomic emission, the essential features of which follow from classical antenna theory. Only in recent years have these two tracks converged, driven mostly by advances in technical capability. Unfortunately, relevant and seminal work in the area of optical bistability has been almost universally ignored in modern-day discussions and reviews of cavity QED, which we view as a most unfortunate situation.

Such concerns about the sociology of science aside, we do wish to emphasize that optical physics with strong coupling represents a conceptual frontier for the investigation of quantum dynamics of open (dissapative) systems. Historically, the emphasis has been on weakly coupled systems for which traditional methods such as system size expansions are adequate. In this case the route from the microscopic quantum to the macroscopic classical domain is well charted. Quantum dynamics become essentially classical in character (albeit with quantum coefficients), with small bits of quantum fluctuations persisting. By contrast, we suggest that there are startling new discoveries to be made in the realm of strong coupling, with this perspective grounded in the belief that the complexity of quantum dynamical processes has yet to be adequately explored.

This work was supported in part by the National Science Foundation (Grant PHY-9014547) and by the Office of Naval Research (Grant N00014-90-J-1058). We gratefully acknowledge our ongoing collaborations with the research groups of Professor H. J. Carmichael and Professor P. Zoller and with Dr. A. S. Parkins.

*University of Fribourg, Perolles, CH-1700 Fribourg, Switzerland
**University of Aarhus, Institute of Physics and Astronomy, Aarhus, Denmark
***National Institute of Standards and Technology, Gaithersburg, Maryland

REFERENCES

1. Kimble, H. J., in Fundamental Systems in Quantum Optics, Les Houches Session 53, Dalibard, J., Raimond, J. M., and Zinn-Justin, J., eds. (Elsevier Science Publishers, Amsterdam, 1992), pp. 549-674.
2. Carmichael, H. J., in An Open Systems Approach to Quantum Optics, Lecture Notes in Physics, m18 (Springer Verlag, Berlin, 1993).
3. Lugiato, L. A., in Progress in Optics, Vol. 21, Wolf, E., ed. (Elsevier Science Publishers B. V., Amsterdam, 1984), p. 71.
4. Meystre, P., in Progress in Optics, Vol. 30, Wolf, E., ed. (Elsevier Science Publishers, B. V., Amsterdam, 1992), p. 261.
5. Kimble, H. J., in Cavity Q. E. D., Advances in Atomic, Molecular and Optical Physics Suppl. 2, Berman, P., ed. (Academic Press, New York, 1994), p. 203.
6. Haroche, S., and Raimond, J. M., in Advances in Atomic Molecular, and Optical Physics, Vol. 20, Bates, D., and Bederson, B., eds. (Academic Press, New York, 1985), p.347.

7. Raithel, G., Wagner, C., Walther, H., Narducci, L. M., and Scully, M. O., in Advances in Atomic, Molecular, and Optical Physics, Suppl. 2, Berman, P., ed. (Academic Press, New York, 1991), p. 57.

8. Rempe, G., Schmidt-Kaler, F., and Walther, H., Phys. Rev. Lett. 64, 2783(1990).

9. Haroche, S., and Raimond, J. M., in Cavity Q. E. D., Advances in Atomic, Molecular and Optical Physics, Suppl. 2, Berman, P. ed. (Academic Press, New York, 1994), p. 123.

10. Brune, M., Nussenzveig, P., Schmidt-Kaler, F., Bernardot, F., Maali, A., Raimond, J. M., and Haroche, S., Phys. Rev. Lett. 72, 3339(1994).

11. Thompson, R. J., Rempe, G., and Kimble, H. J., Phys. Rev. Lett. 68, 1132(1992).

12. Parkins, A. S., Marte, P., Zoller, P., and Kimble, H. J., Phys. Rev. Lett. 71, 3095(1993), and Phys. Rev. A (submitted April, 1994).

13. Meystre, P., Schumacher, E., and Stenholm, S., Opt. Comm. 73, 443(1989).

14. Holland, M. J., Walls, D. F., and Zoller, P., Phys. Rev. Lett. 67, 1716(1991).

15. Herkommer, M., Akulin, V. M., and Schleich, W. P., Phys. Rev. Lett. 69, 3298(1992).

16. Matsko, A. B., Vyatchanin, S. P., Mabuchi, H., and Kimble, H. J., Phys. Lett. A 192, 175(1994).

17. Treussart, F., Hare, J., Collot, L., Lefevre, V., Weiss, D., Sangdoghdar, V., Raimond, J. M., and Haroche, S., Opt. Lett. (1994).

18. Dehmelt, H., Science 247, 4942(1990).

19. Wineland, D. J., Bollinger, J. J., Itano, W. M., and Heinzen, D. J., Phys. Rev. A 50, 67(1994).

20. Cirac, J. L., Blatt, R., Parkins, A. S., and Zoller, P., Phys. Rev. Lett. 70, 762(1993).

21. Agarwal, G. S., Phys. Rev. Lett. 73, 522(1994).

22. Sanchez-Mondragon, J. J., Narozhny, N. B., Eberly, J. H., Phys. Rev. Lett. 51, 550(1983).

23. Agarwal, G. S., Phys. Rev. Lett. 53, 1732(1984).

24. Jaynes, E. T., and Cummings, F. W., Proc. IEEE 51, 89(1963).

25. Kleppner, D., Physics Today (October, 1990), p. 9.

26. Agarwal, G. S., J. Opt. Soc. Am. B 2, 480(1985).

27. Carmichael, H. J., Brecha, R. J., Raizen, M. G., and Kimble, H. J., Phys. Rev. A 40, 5516(1989).

28. Raizen, M. G., Thompson, R. J., Brecha, R. J., Kimble, H. J., and Carmichael, H. J., Phys. Rev. Lett. 63, 240(1989).

29. Thompson, R. J., Rempe, G., and Kimble, H. J., Phys. Rev. Lett. 68, 1132(1992).

30. Zhu, Y., Gauthier, D. J., Morin, S. E., Wu, O., Carmichael, H. J., and Mossberg, T. W., Phys. Rev. Lett. 64, 2499(1990).

31. Raizen, M. G., Doctoral Thesis, University of Texas at Austin (1989).

32. Tavis, M., and Cummings, F. W., Phys. Rev. 170, 379(1968).

33. Rempe, G., Thompson, R. J., Brecha, R. J., Lee, W. D., and Kimble, H. J., Phys. Rev. Lett. 67, 1727(1991).

34. Carmichael, H. J., Brecha, R. J., and Rice, P. R., Opt. Comm. 82, 73(1991).

35. Rempe, G., Thompson, R. J., Kimble, H. J., and Lalezari, R., Opt. Lett. 17, 363(1992).

36. Braginsky, V. B., Gorodetsky, M. L., and Ilchenko, V. S., Phys. Lett. A 137, 393(1989).

37. Mabuchi, H., and Kimble, H. J., Opt. Lett. 19, 749(1994).

38. Braginsky, V. B., Gorodetsky, M. L., and Ilchenko, V. S., in Laser Optics '93– Proceedings of the S.P.I.E. (1994).

39. Collot, L., Lefevre-Sequin, V., Brune, M., Raimond, J. M., and Haroche, S., Europhys. Lett. 23, 327(1993).

40. Jhe, W., and Kim, L. W., Phys. Rev. A (submitted Sept. 6, 1994).

41. Professor C. M. Caves has brought these points to our attention.

42. Vogel, K., Akulin, V. M., Schleich, W. P., Phys. Rev. Lett. 71, 1816(1993).

43. Shor, P., in Proc. 35th Symposium on Frontiers of Computer Science (IEEE Press, Nov., 1994).

44. Feynman, R., Intl. J. of Theoretical Phys. 21, 467(1982).

45. Deutsch, D., Proc. Royal Soc. Lond., A400, 96(1985).

46. Gardiner, C., Phys. Rev. Lett. 56, 1917(1986).

47. For a review, see Parkins, A. S., in Modern Nonlinear Optics (Wiley, New York, 1993).

48. Kimble, H. J., Polzik, E. S., Georgiades, N., and Mabuchi, H., in Laser Spectroscopy – XIth Int'l. Conference, eds. Bloomfield, L., Gallagher, T., and Larson, D., (AIP Press, New York, 1994), p. 340.

49. Rice, P., and Pedrotti, L., JOSA B9, 2008(1992).

50. Cirac, J. I., Phys. Rev. A 46, 4354(1992).

51. Polzik, E. S., Carri, J., and Kimble, H. J., Phys. Rev. Lett. 68, 3020(1992) and Appl. Phys. B 55, 279(1992).

Coherence and Control of Atomic Electrons

C. R. Stroud, Jr.

The Institute of Optics and Department of Physics and Astronomy
University of Rochester, Rochester, New York 14627-0186

Abstract. Recent theory and experiments studying the interaction of short intense optical pulses with atoms are reviewed. It is shown that the conventional energy eigenstate expansions of atomic physics are not an efficient way to describe the response of atoms to these field pulses. Instead a description in terms of ensembles of classical paths of the electrons is proposed and illustrated with examples from the area of Rydberg electron wave packets.

INTRODUCTION

The beginnings of atomic physics can be traced to the discovery that atoms and molecules absorb and emit light at certain discrete frequencies. Almost all of the knowledge that we have gained since that time about the *structure* of atoms has come from increasingly precise spectroscopic measurements of the same basic type as these initial ones. The tool used for these precise measurements is generally an extremely stable, narrowband, tunable laser. Indeed, at this conference are reports of spectroscopy at millihertz resolution.

It is often said that these measurements along with associated measurements of structural properties such as oscillator strengths or cross sections gives one everything that there is to know about atoms. In other words, *everything is contained in the structure*. Very short and very intense laser pulses that are available today are not very good at measuring structure. They are blunt instruments when a scalpel is needed. But, of course, these pulses are appropriate to measure *dynamics* instead of structure. Now, what can dynamics tell us that is different from the structure? One might plausibly argue that nothing new can be gained from such measurements; afterall, if we know the eigenfunctions $\psi_n(r)$, and the eigenfrequencies ω_n, then any time-dependent state can be written in the form

$$\Psi(r,t) = \sum_n c_n \psi_n(r) e^{-i\omega_n t}, \tag{1}$$

C. R. Stroud, Jr.   337

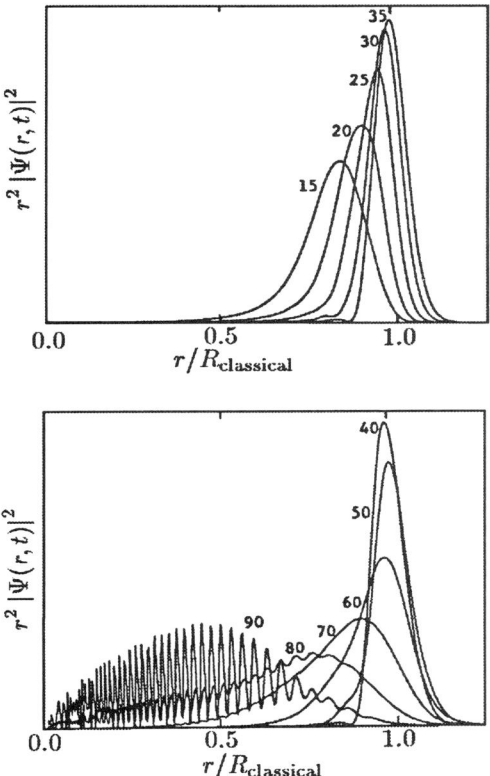

FIGURE 1. Evolution of Radial Rydberg Wave Packet During First Orbit. A 10 ps laser pulse is used to excite a hydrogen atom from the ground state to a superposition of Rydberg states centered about n=85. The probability distribution for finding the electron at various radial distances from the nucleus is plotted at various times in picoseconds after the center of the laser pulse. The radial coordinate is scaled by the distance from the nucleus to the outer turning point of a classical electron with energy equivalent to the energy of the quantum state. In upper figure we see that shortly after excitation the electron state is a well defined wave packet moving out toward the classical turning point, and slowing as it approaches it. In the lower figure we see that at later times the packet begins to return to the nucleus. As it approaches the nucleus it speeds up, and it also develops a high frequency modulation as the portion of the wave packet that is approaching the nucleus interferes with the portion that has already passed by the nucleus and is beginning to go back out toward the outer turning point. After Parker and Stroud (2).

where the c_n are just constant coefficients determined by the initial conditions.

The response to this comment might be that by this argument, if we know the vocabulary and grammar of any language we know all there is to know about how to write a novel, a sonnet, or a physics paper in that language. From a simplistic point of view this is true, but not in the real world. Similarly, the dynamics of an atom can be described in principle by simple knowledge of its structure, but it is not true in practice. For example, consider the following question: what is the dynamic response of an atom to a picosecond pulse that excites the atom from the ground state up to the manifold of Rydberg states? A one picosecond laser pulse has a transform bandwidth of about 30 cm^{-1} so that it can simultaneously and coherently excite all of the Rydberg levels from approximately $n = 67$ to the continuum limit. This complex superposition in fact forms a spatially localized wave packet that oscillates periodically from near the nucleus out to a turning point. The period of the orbit is just that of a classical Kepler orbit with the same energy, and the turning point is that of the corresponding classical orbit.

The motion of the Rydberg wave packet can be described accurately by an energy eigenstate expansion, but it simply is not a very appropriate representation. A description in terms of classical orbits is much more attractive. In the very early days of quantum theory Heisenberg stated very strongly that classical orbits have no place in the modern quantum mechanics.(1) We will see that this dictum is not true in the case of atomic or molecular wave packets so long as we are careful to take into account the wave properties of the electrons.

Let us begin by reviewing the situation with some simple examples. Radially localized wave packets have been studied both theoretically(2-4) and experimentally. (5-7)

In Fig. 1 we see a series of graphs showing the evolution of the probability distribution as a function of time for a wave packet excited by a 25 ps pulse tuned to resonantly excite the level with principal quantum number 85 from the ground state. We see that it is indeed radially localized, and oscillates at the classical period. But, of course, if the ground state is an s-state, and we excited the atom with a single photon transition, then by the dipole selection rules all of the levels in the excited state superposition will be p-states. The corresponding probability distribution will be cosine squared as a function of polar angle.

ENSEMBLES OF CLASSICAL ORBITS

The connection with classical orbits is not obvious in this case, but there is one. If one considers, not a single classical particle, but an ensemble, then most of the behavior of the wave packet for this trip around the orbit can be reproduced. The trick is to simply make up an ensemble of elliptical orbits whose principal axes are distributed in space according to the cosine-squared distribution, but start each

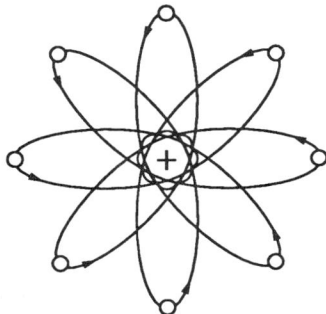

FIGURE 2. Ensemble of classical orbits equivalent to a radial wave packet. Each member of the ensemble travels along an elliptical orbit according to Kepler's laws. The angular orientations of the orbits are random. The short optical pulse prepares the ensemble so all move in phase around their respective orbits.

of the classical particles in the ensemble with exactly the same phase in its orbit. In that case, the ensemble is almost indistinguishable from the quantum probability distribution for the first several orbits of its evolution, except when the packet is near the nucleus. When the wave packet is near the nucleus, as in the case of the 90 ps graph in Fig. 1, there is interference between the head and tail of the wave packet. This wave phenomenon cannot be reproduced in a classical ensemble.

Atomic electron wave packets are not limited to these radial wave packets. One can also localize the wave packets in the angular variables. In order to do this, one must produce a superposition of various angular momentum states as well as principal quantum numbers. This cannot be done by simply exciting the atom with a single laser field, one must either use an rf field to supply the extra angular momentum, or possibly some combination of dc electric and magnetic fields to break the symmetry. In the very first Rydberg wave packet experiment Yeazell and Stroud used an rf field to dress the atom and then a short laser pulse to populate the dressed atomic states. When the rf field was adiabatically turned off the electron was left in a superposition of high angular momentum states $l = m \approx 30$. This superposition formed a wave packet in the form of a stationary pie-shaped wedge about 20° wide, sitting on one side of the nucleus.

This wave packet can also be well described in terms of a classical ensemble. In this case the classical elliptical orbits all have their major axes oriented approximately along a line, and the planes of the orbits are also approximately aligned, but the locations of the particles along the orbits is random. The probability distribution is stationary in time so that the probability of finding a particle at a particular location is inversely proportional to velocity of a particle at that point in the orbit. The particles are most likely to be found at the outer turning point, just like comets in the Oort cloud. This ensemble reproduces the quantum mechanical probability distribution quite well.

In this case the classical ensemble turned out to be a particularly useful way of analyzing the system. Experimental detection of these wave packets was by use of a pulsed electric field. The ionization probability depended on the orientation of the axis of the wave packet with respect to the electric field. When the field was along the orbit the atom was easily and rapidly ionized, while the ionization cross section was much smaller for wave packets oriented perpendicular to the field. Detailed theoretical modeling of this ionization process proved to be difficult because an enormous number of states was involved as the high angular momentum states essentially diffused out through the higher Rydberg states. In fact, a full quantum treatment of this problem has not yet been carried out. It is very similar to the problem of microwave ionization of hydrogen which has also been resistant to full quantum calculations. However, classical calculations are quite simple using the ensemble that we have just described, and they agree very well with the experimental results.

What about the case for fully three-dimensionally localized wave packets? To date such wave packets have not been produced in the laboratory. Various groups have studied the theory of such wave packets. (8-12) Gaeta, Noel and Stroud have just proposed a technique for producing these states in the laboratory.(13) The technique relies strongly on the intuition obtained by modeling the system with a classical ensemble. One begins with a circular orbit eigenstate. Several techniques have been used to make such states, and they have been produced in the laboratory with principal quantum numbers as large as 100. The eigenstate is modeled classically by a uniform distribution of particles moving around in a circular Kepler orbit with classical energy corresponding to that of the quantum eigenenergy. Then a half-cycle electric field pulse is applied, with the electric field vector in the plane of the orbit, and the pulse duration shorter than the orbital period.

As is illustrated in Fig. 3, the electrons that are moving parallel to the electric field will see a force opposing their motion and thus slowing them down, while the electrons that are moving anti-parallel to the field will be speeded up by the field. Those with motions perpendicular to the field will not have their velocities modified much. Initially, the spatial distribution of the classical ensemble will not be modified by the pulse, but the velocity distribution will be. After a few orbital periods the velocity distribution will cause the faster electrons to overtake the slower electrons and the spatial distribution will become bunched – a wave packet will be formed. In fact, the wave packet will be essentially identical to that predicted by the full quantum theory. The classical model predicts nothing that regular quantum theory cannot, but it does it in a an intuitive fashion that allowed us to arrive at this method in the first place.

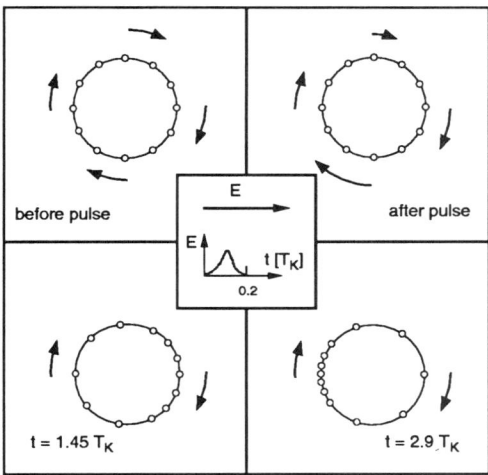

FIGURE 3. Creation of three dimensionally localized wave packet using a half-cycle field pulse. The atom first excited to a circular state which corresponds to a classical ensemble, all with the same angular velocity, uniformly spread around the circular orbit. A short field pulse in the plane of the orbit then modifies the velocity distribution so that the members of the ensemble pile up into a localized wave packet. From Gaeta *et al.* (13)

EHRENFEST TIME

Of course, ensembles of classical particles cannot mimic all features of the quantum theory of wave packets. The wave packets do orbit the nucleus just like classical particles for a while, but in order to produce a radial localization we must include states of different energy in our ensemble. That means that the ensemble includes particles of different velocities, so inevitably, after a few orbits the wave packet begins to disperse. Even this is properly described by the classical ensemble. A classical ensemble that has a gaussian distribution in space, and a gaussian distribution in momentum, with the width in momentum space that given by Heisenberg's relation, will spread exactly like a quantum mechanical wave packet. In free space this is the whole story, but in an atom the wave packet will eventually spread all the way around the orbit and the head and tail will overlap. At that point the head and tail can interfere. Since the classical ensemble has no wave properties, it cannot produce interference. The time for onset of this nonclassical overlapping has been characterized as the Ehrenfest time by Tomsovic and Heller.

This interference eventually produces a complex system of decays, revivals, and fractional revivals that are the subject of a great deal of current research. It was first noted by Parker and Stroud (2) and has been studied a great deal since, both

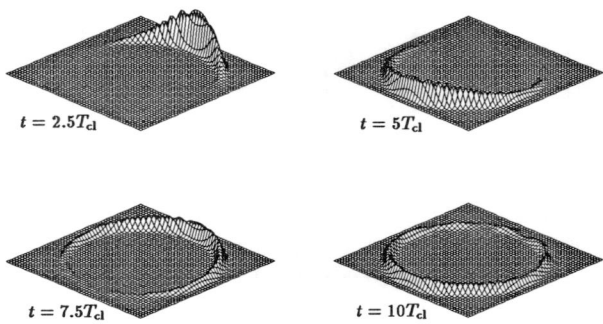

FIGURE 4. Evolution of circular wave packet during first few orbits. The wave packet spreads just as would a classical ensemble with the same range of energies until the head and tail of the wave packet begin overlap at the Ehrenfest time.

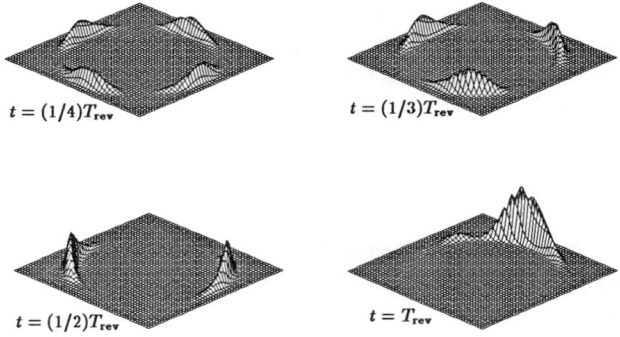

FIGURE 5. Evolution of circular wave packet for times longer than the Ehrenfest time.

theoretically (9,10) and experimentally.(7, 14-16) Actually, with the addition of a small ansatz, even most of these features can be explained in terms of a classical ensemble. Assume that our classical ensemble was made up not with a continuous range of velocities, but with a discrete set of velocities, just those satisfying Bohr's quantization conditions, *ie.*

$$v_n = \alpha c n^{-2}. \qquad (2)$$

An ensemble of these classically evolving particles will undergo spreading, revival, and even fractional revivals – but there is one thing wrong with this model. If the central energy of the wave packet exactly matches with one of the Bohr levels, then the classical ensemble will revive at exactly the right time, but on the wrong side of the orbit. In general the fractional revivals also occur at the right times, but the sub-wave packets in this case are also at different locations in the orbits than are the quantum wave packets. What is going on? To understand we must delve deeper into the justification of the classical ensembles.

SEMICLASSICAL THEORY

What exactly is the role of classical paths in quantum theory? Schrödinger said (in rough English translation) that *classical physics is to quantum mechanics as ray optics is to wave optics*. Both optical fields and quantum mechanical wave functions satisfy wave equations. It is well known that in the limit in which the optical wave length is much shorter than the distance over which the index of refraction varies significantly the eikonal approximation can be applied to reduce Maxwell's wave equation to the equations of geometrical optics.(17) In the same way, if the quantum mechanical potential does not vary appreciably over one deBroglie wave length, then Schrödinger's equation can be approximated by classical mechanics.

This is a very old idea in quantum theory that was first investigated by Van Vleck in 1927, (18) and developed much more extensively in recent times by Gutzwiller,(19) and by Tomsovic and Heller.(20) The approach is closely related to the Feynman path integral formulation of quantum mechanics (21) in which it is shown that the Schrödinger wave function can be propagated exactly from some initial point x and time t to another point x' at a later time t' by summing the exponential of the action over *all* paths that go from the initial point at the initial time to the final point at time t'. Of course, there are in general an infinite number of such paths, and along most of the paths the exponential is rapidly oscillating, so carrying out the sum is very difficult. However, one can approximate the sum by a stationary phase approximation. To do this one finds paths that minimize the action. Of course, this is just the condition that defines the classical paths. In general the nonclassical paths give rapidly oscillating contributions that cancel out in the sum over paths.

There are two difficulties that arise in carrying out these sums even when we limit consideration to classical paths. First, in general the paths may form caustics. When we calculate the action along a classical path that passes through a caustic we must add a phase shift. This effect is exactly the same as the phase shift that occurs in optics when a wave front passes through a focus. Gutzwiller has shown that this is easily accounted for by simply adding a phase shift of $\pi/2$ to the action integral of any path that passes through a caustic.(19) The second difficulty is that there are in general an infinite number of classical paths that that pass from the initial to final point in the given time interval so that the sum is quite difficult to carry out. Tomsovic and Heller showed that it is not necessary to sum even over all classical paths. It is sufficient to sum over a set of *reference orbits* and then to include the other nearby classical orbits by a simple perturbation theory.(20)

We have applied this technique to the problem of circular orbit wave packets.(12) In the case of atomic Rydberg wave packets one generally carries out an experiment using a pump-probe technique. A first pulse creates the wave packet and then a second pulse later probes the state of the atom. Such experiments are described by a simple correlation function of the wave packet at the initial time with the wave packet at a later time. The wave packet localization greatly limits the number of classical paths that contribute. In the case of the circular orbit wave packets the reference orbits are simply the classical circular orbits that have passed from the initial to the final point after one orbit, two orbits, *etc.* This simplication allows one to actually determine an analytic expression for the correlation function in the case in which the initial wave packet is a gaussian in all three dimensions. The analytic expression is

$$C_{\rm sc}(t) = \sum_k \frac{\sigma_\phi}{\alpha_k} \exp\left(-\frac{(n_k - \bar{n})^2}{\beta_k^2} + iR_k^0\right). \tag{3}$$

Here we have used the full classical action for the kth reference trajectory

$$R_k^0 = 2\pi k n_k + \frac{t}{2n_k^2}, \tag{4}$$

and

$$n_k = \left(\frac{t}{2\pi k}\right)^{1/3}, \tag{5a}$$

$$\alpha_k^2 = \sigma_\phi^2 - i\chi_k^2, \tag{5b}$$

$$\beta_k^2 = \frac{1}{\sigma_\phi^2} + i\frac{1}{\chi_k^2}, \tag{5c}$$

$$\chi_k^2 = \frac{\Gamma_k^s}{2}t. \tag{5d}$$

The summation index k runs over the positive integers counting the number of full orbits in the reference trajectories, the other parameters are defined in terms of the width of the original wave packet in σ_ϕ, and rate of spreading or "shearing" of the classical ensemble. A detailed derivation of this expression is given in (12) . Here we will content ourselves by seeing that it is a quite simple expression, and

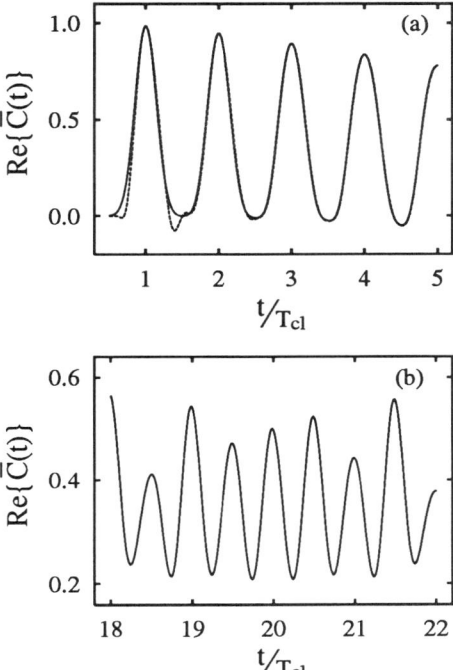

FIGURE 6. Comparison of semiclassical and quantum autocorrelation functions. The dashed line is the quantum expression while the solid line is the semiclassical approximation. The upper figure gives the comparison during the first few orbits while the lower figure makes the comparison near the first one-half fractional revival. From (12).

by comparing its predictions with those made by a conventional quantum energy eigenstate evaluation of the correlation function.

As can be seen in Fig. 6, the semiclassical expression is extremely accurate. Only in the first orbit do the semiclassical and quantum results disagree by more than the linewidth in the graph. Even in the regions where quantum effects are most striking, for example at the one-half fractional revival, the two theories agree extremely well. The agreement gets better with time as the reference orbits become more closely spaced. The particular parameters in the case illustrated here are the average quantum number excited is given by $\bar{n} = 120$, and the width of the initial wave packet is $\sigma_\phi = 2\pi/10$.

SUMMARY AND CONCLUSIONS

We have seen that the conventional energy eigenstate basis is inconvenient and illsuited to describe the dynamic response of an atom after excitation by a very short intense pulse. The dynamics are nearly those of an ensemble of classical particles each travelling along a classical Kepler orbit. The use of such classical models is useful in developing an intuitive understanding of the dynamics. Eventually however, this classical model will become an inaccurate representation of the quantum dynamics. The time required for this breakdown of the accuracy of the classical model has been characterized the *Ehrenfest time*. It is just the time required for the classical ensemble to spread all the way around the orbit so that the head and tail of the packet overlap and begin to interfere. Even in this nonclassical regime a convenient alternative exists for the energy eigenstate representation: the *semiclassical path integral representation*. The range of validity of this formalism is not yet clear, but for the particular case of circular-orbit wave packets it is extremely accurate and actually gives a simple analytic expression for the autocorrelation function of the wave function. Furthermore, it greatly aids the intuition so that one can use it to develop excitation schemes to engineer and control the atomic electron wave function to produce produce any desired geometric distribution.

ACKNOWLEDGEMENTS

I would like to acknowledge my present and former students who carried out much of the work that I have reviewed. In particular, Dr. Jonathan Parker and Dr. John Yeazell who carried out the early work, and Dr. Mark Mallalieu who developed much of the semiclassical theory. At present Jake Bromage, John Corless, Zagorka D. Gaeta, Michael W. Noel, and James West are making large contributions. I would also like to acknowledge the financial support of the Army Research Office.

REFERENCES

1. Pais, A., *Niels Bohr's Times, in Physics, Philosophy, and Polity,* Oxford: Clarendon Press, 1991, ch. 13.
2. Parker, J. A., and C. R. Stroud, Jr., *Phys. Rev. Lett.* **56**, 716 (1986).
3. Alber, G, H. Ritsch, and P. Zoller, *Phys. Rev. A* **34**, 1058 (1986).
4. Parker, J., and C. R. Stroud, Jr., *Phys. Scr.* **T12**, 70 (1986).
5. ten Wolde, A., L. D. Noordam, H. G. Muller, Lagendijk, A., and H. B. van Linden van den Heuvell, *Phys. Rev. Lett.* **61**, 2099 (1988).
6. Yeazell, J. A., M. Mallalieu, J. Parker, and C. R. Stroud, Jr., *Phys. Rev. A* **40**, 5040 (1989).
7. Meacher, D. R., P. E. Meyler, I. G. Hughes, and P. Ewart, *J. Phys. B* **24**, L63 (1991).

8. Brown, L. S., *Am. J. Phys.* **41**, 525 (1973).
9. Gaeta, Z. D. and C. R. Stroud, Jr., *Phys. Rev. A* **42**, 6308 (1990).
10. Averbukh, I. S., and N. F. Perelman, *Phys. Lett. A* **139**, 449 (1989).
11. Nauenberg, M., *J. Phys. B* **23**, L385 (1990).
12. Mallalieu, M. and C. R. Stroud, Jr., *Phys. Rev. A* **49**, 2329 (1994).
13. Gaeta, Z. D., M. W. Noel, and C. R. Stroud, Jr., *Phys. Rev. Lett.* **73** 128 (1994).
14. Yeazell, J. A., M. Mallalieu and C. R. Stroud, Jr., *Phys. Rev. Lett.* **64**, 2007 (1990).
15. Yeazell, J. A., and C. R. Stroud, Jr., *Phys. Rev. A* **43**, 5153 (1991).
16. Wals, J., H. Fielding, J. Christian, L. Snoek, W. van der Zande, and H. B. van Linden van den Heuvell, *Phys. Rev. Lett.* **72** 3783 (1994).
17. Born, M., and E. Wolf, *Principles of Optics* New York: Pergamon Press, 1959, ch. 3.
18. Van Vleck, J. H., *Proc. Nat. Acad. Sci. USA* **14**, 178 (1928).
19. Gutzwiller, M., *J. Math. Phys.* **8**, 1979 (1967).
20. Tomsovic, S., and E. Heller, *Phys. Rev. Lett.* **67**, 664 (1991).
21. Feynman, R. P., and A. R. Hibbs, *Quantum Mechanics and Path Integrals,* New York: McGraw-Hill, 1965.

COLLISIONAL AND OPTICAL PROPERTIES OF ULTRACOLD ATOMS

Cold Collision Phenomena

B.J. Verhaar

Department of Physics, Eindhoven University of Technology,
P.O. Box 513, 5600 MB Eindhoven, The Netherlands

Abstract. The recent developments in laser cooling, together with a preceding decade of work on cryogenically cooled atomic hydrogen, have opened possibilities for a large variety of experiments in which collisions between very slow atoms play an important role. Examples of such experiments are reviewed with an emphasis on theoretical interpretation. We focus primarily on open problems in this field and on subjects not yet dealt with in the existing reviews. Most of the examples deal with collisions between ground state atoms. Although a future development is foreseen in the direction of longitudinally cooled and bright beams, all present examples deal with experiments on gas samples. In that connection some attention is devoted to the step required to go from a microscopic to a macroscopic description.

INTRODUCTION

In recent years the experimental possibilities to cool atomic gas samples by means of laser beams and to store and manipulate them have undergone a tremendous development. This development was preceded by experiments going on since the end of the seventies that dealt with cryogenically cooled spin-polarized atomic hydrogen. Cold atomic samples offer challenging opportunities like the realization of Bose-Einstein condensation or the construction of atomic clocks with unprecedented accuracy. The interest is also stimulated by the prospect that collisions between atoms will display unusual features at the low temperatures that are being realized.

The temperature region where new collision phenomena are to be expected and in which both the recent work on laser cooled alkali atoms and the work on cryogenically cooled atomic hydrogen is taking place, is the so-called *quantum regime*, characterized by the condition

$$\lambda_{th} > r_0, \tag{1}$$

i.e. the thermal de Broglie wavelength is large compared to the range of the interatomic interaction.

In the regime given by (1) atoms approach each other during collisions more closely than their wavelength. This "quantum overlap" provides special phenomena, examples of which will be discussed like the phenomenon of spin waves in dilute gases and the collisional frequency shift in an atomic fountain, in both cases with a "quantum cross section" going to infinity for $T \rightarrow 0$. We will go into the consequences for the construction of an improved cesium clock. Collisions in this temperature regime are referred to as *ultracold collisions*. The superlative "ultra" might sound like an adjective which is subject to inflation. However, there exists a well-defined meaning: an ultracold collision is one which is dominated by s-wave scattering. This is equivalent to condition (1).

A further temperature regime is that in which the thermal wavelength is not only larger than the range of the interatomic interaction but exceeds even the mean distance of nearest neighbors in the gas:

$$\lambda_{th} > n^{-1/3}, \tag{2}$$

in which n is the gas density. In this *degenerate quantum regime*, as yet unrealized in a dilute gas, atoms are permanently within a wavelength from other atoms. The ideal degenerate quantum gas is a fundamental paradigm of quantum statistical mechanics. The same is true for the weakly-interacting Bose gas for which a theory was already formulated in 1957 [1]. The realization of a weakly interacting quantum gas in the degenerate regime will provide the opportunity to observe the spectacular phenomenon of Bose-Einstein condensation (BEC) in a well-characterized system, calculable from first principles. As we will see, collisions between atoms also play a crucial role in experiments aiming at the realization of circumstances where this phenomenon occurs.

Much effort is presently also going into the subject of collisions of alkali atoms in a near-resonant laser field exciting one of the collision partners during the collision (sometimes called "optical collisions"). This subject is of importance for understanding loss processes in optical traps. It is also of fundamental interest, since it is one of the few situations where the collision time is comparable to the spontaneous emission time. In view of the existence of a comprehensive review [2] only some recent developments in this subject will be dealt with in this talk.

As pointed out previously, the cold atom experiments to be described deal with gas samples. This makes it necessary to make a step from the microscopic level of a single collision to the macroscopic level of a collective phenomenon. In our work this step was made by means of the quantum Boltzmann equation [3],[4],[5]. This equation has a form similar to the classical Boltzmann equation and describes the time evolution of the one-body density matrix $\rho(\vec{r}, \vec{p}, t)$ in terms of a single-particle term and a (two-body) collision term, the latter being a complicated expression in terms of S-matrix elements, for which we refer to Ref. [5]:

$$\left[\frac{\partial}{\partial t} + \frac{\vec{p}}{m} \cdot \nabla\right] \rho(\vec{r}, \vec{p}, t) = \left(\frac{\partial \rho}{\partial t}\right)_{\text{s.p.}} + \left(\frac{\partial \rho}{\partial t}\right)_{\text{coll.}} \quad (3)$$

The equation is based on a quantal treatment of the internal dynamics and a classical treatment of the translational motion *between* collisions. It is emphasized that the translational motion during collisions, as summarized in the substituted values of the S-matrix (or equivalently T-matrix) elements, can only be treated quantum mechanically due to condition (1). ρ acts in the internal state space and can be obtained from the exact quantum density operator by means of a Wigner transform. Its diagonal elements characterize the occupation of the internal states and its non-diagonal elements describe their coherence.

SPIN WAVES

Although the subject of spin waves in dilute cold gases is already rather old on the typical time-scale of the rapidly developing cold-atom field, we briefly review it here because I consider it the most striking example of a quantum phenomenon for cold-atom collisions in the regime (1). Long before, spin waves were a well-known phenomenon in dense systems like the solid state and the ^{3}He liquid. There they occur due to the strong Heisenberg correlation between neighboring spins arising from the permanent overlap of adjacent atoms. A spin wave is essentially a wavelike propagation of a local tilt of spins from a preferential orientation (see Fig. 1).

Lhuillier and Laloë, and Bashkin (see, e.g., the review [6]) made the remarkable observation that such effects can also occur in the gas phase without a permanent overlap, for instance in spin-polarized atomic hydrogen at a density of 10^{16} at/cm^3. In these circumstances atoms spend most of the time far from their neighbors. Only during a very small fraction of the time they overlap as in the solid state.

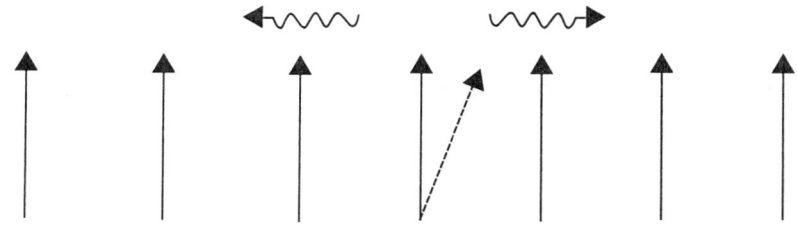

FIGURE 1. Simple picture of spin wave in solid state and ^{3}He: propagation of spin locally tilted from preferential direction.

In 1984 it turned out experimentally that the rare quantum overlap during collisions is sufficient for spin waves to occur [7],[8],[6]. In a number of papers (see again the review [6]), microscopic theories for spin waves in dilute gases have been formulated. We restrict ourselves to linear versions of spin-wave theory, i.e. small tilt angles. Important parameters predicted by these theories are D_0, the spin-diffusion constant in the unpolarized gas, and Q, the quality factor of the spin-wave modes. In the quantum temperature regime where the experiments take place Q is given by the simple expression

$$Q = \frac{3\sqrt{2}}{16} \frac{\lambda_{th}}{|a|} \qquad (4)$$

with a the scattering length.

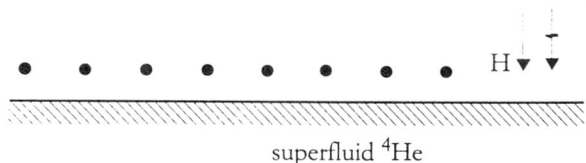

FIGURE 2. Quasi-2D gas of hydrogen atoms adsorbed at surface of superfluid ^{4}He.

An open problem is the observability of two-dimensional spin waves in the quasi-2D gas of hydrogen atoms adsorbed at a surface of superfluid ^{4}He (see Fig. 2), which have been predicted in Ref. [9]. Both the quality factor (4) and the corresponding expression for 2D spin-waves,

$$Q = \frac{2}{\pi}\left[-\gamma + \ln\left(\frac{4\sqrt{2}}{3\pi}\frac{\lambda\text{th}}{|a_{2D}|}\right)\right] \tag{5}$$

illustrate the limitation of the spin-wave phenomenon to the quantum regime, the absolute magnitudes of the 2D and 3D scattering lengths being a quantitative measure for the "range" in eq. (1). Only under the condition (1) Q is sufficiently large to make spin waves observable. In the 2D case both the predicted Q (about 2) and the residence time of atoms at the surface, which is larger than the damping time of the most weakly damped spin wave mode, seem to make observation possible at temperatures of 0.1 K and below. This would have both fundamental interest and interest from the point of view of measuring for the first time the 2D triplet H + H scattering length, which contains information on the role of the superfluid ^{4}He dynamics in the 2D cold collision process. Note that the most advanced existing calculation [10] of a_{2D} neglects the dynamics of the superfluid surface.

On the microscopic scale the quantum nature of the collective spin-wave phenomenon in dilute gases is due to the so-called *identical-spin rotation effect* (ISR, see review [6]), a mutual rotation of the spins of two indistinguishable particles about their vector sum due to interference effects when their wave packets of dimension λ_{th} overlap. It has been illustrated schematically in Fig. 3. By the ISR effect each atom feels an effective magnetic field due to the average spin orientation of atoms in its environment and spin-precesses around it.

As a second striking aspect, in addition to the occurrence of spin waves in a dilute gas, the interatomic potential is independent of the orientation of the spins participating in the rotation effect: only the nuclear spins are tilted, the atoms simply interact according to the triplet potential. The ISR effect does not need a spin-dependent potential: it is only due to the indistinguishability of the atoms, which plays a role in collisions when condition (1) is fulfilled. This indistinguishability gives rise to interference of the undistorted ongoing wave and a 180° scattered wave.

As a third surprise it was found in 1990 that spin waves can even occur in a Knudsen H-gas [11], in which the hydrogen atoms collide predominantly with the walls. At first sight it is amazing that spin waves can occur for such low densities: the mean free path l as it is usually defined is larger than the cell dimensions L, so that it is difficult to understand how an atom can feel the average spin orientation in its environment. It is emphasized, however, that l is only small relative to L for the usual type of elastic collision with a (low-temperature) cross section of order a^2. The special type of collisions involved in spin waves (see above), occurs with the "quantum cross section" λa, which gives rise to a much shorter mean free path at low T. A fourth surprise is that Knudsen spin waves have only been studied in one experiment [11], despite their fundamental significance.

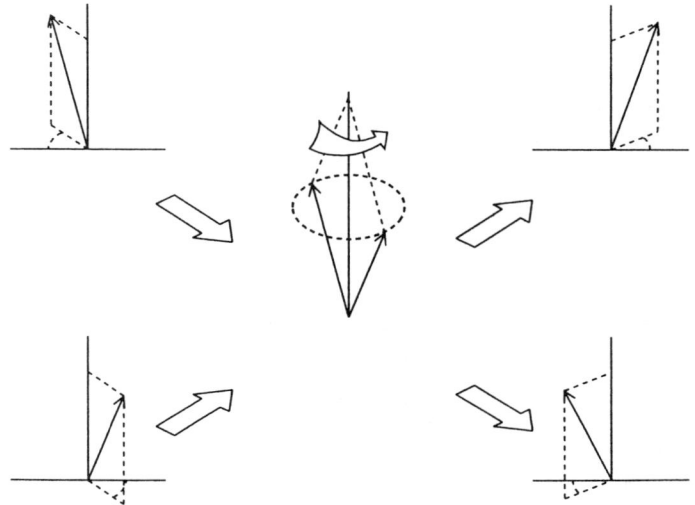

FIGURE 3. Identical Spin Rotation effect in collision: rotation of proton spins about their vector sum during collision. In zero-temperature limit ISR effect survives.

CESIUM FOUNTAIN CLOCK

In this case we deal with another example of an elastic collision process for which the cross section goes as $E^{-1/2}$. As a consequence, the collisional frequency shift of the fountain clock goes to a finite value for $T \to 0$.

The foregoing example dealt with cryogenically cooled atomic gas samples. The recent atomic fountain experiments (see for instance [12],[13]) make use of laser cooling techniques. As is well-known the conventional cesium clock is based on the comparison of the phase change of a microwave field with that of an atomic magnetic moment oscillating at the Bohr frequency of the pair of $F, M_F = 3, 0$ and $4, 0$ ground state hyperfine states (see Fig. 4) during the time interval between two passages of the atoms through a microwave cavity. An increase of this time interval Δt increases the change of the phase difference and thus the accuracy. This is the basic idea of using an atomic fountain with ultracold atoms to improve the conventional cesium clock: Δt can be increased to almost a second, implying a decrease of the linewidth compared with conventional set-ups by two orders of magnitude.

The trade-off involved in this set-up is the increase of the signal-to-noise ratio achievable by increasing the number of atoms in the atomic cloud undergoing the fountain orbit, compared with the increased perturbation due to interatomic collisions. The overlap of the electron clouds during a collision causes the valence electron spins to precess around their vector sum, which perturbs the hyperfine precession in each of the individual atoms and thus shifts the Bohr frequency. For a derivation of this shift on the basis of the quantum Boltzmann equation I refer to [14]. Instead, it is illustrative to give the (non-vanishing!) shift in the zero-temperature limit, which turns out to be valid approximately in the extreme circumstances realized experimentally [12], in terms of (complex) scattering lengths:

$$i\Delta\omega_{\text{at}} - \Gamma_{\text{at}} = i\frac{4\pi\hbar^2}{m}\sum_i (1 + \delta_{i1} + \delta_{i2})n_i \arg\left[a^*_{2+i\to 2+i} - a_{1+i\to 1+i}\right], \quad (6)$$

with n_i the partial density of atoms in each of the hyperfine states i with which an atom in the superposition of clock states $|1> \equiv |3,0>$ and $|2> \equiv |4,0>$ collides. This expression can be interpreted in terms of a Hartree-Fock mean field:

$$H_{\text{eff}} = \frac{4\pi\hbar^2}{m}\sum_i n_i \left[(1 + \delta_{i1})a_{1+i\to 1+i}P_1 + (1 + \delta_{i2})a_{2+i\to 2+i}P_2\right], \quad (7)$$

with P_1 and P_2 projection operators on the two clock transition states. Note that the P_2 term also contains an absorptive contribution due to the negative imaginary part of the corresponding scattering length describing the loss of flux to other channels. In this sense, there is an analogy to the "optical model" for scattering of projectiles by atomic nuclei in nuclear physics. Clearly, the loss of flux determines the collisional damping of the coherence of the clock transition states described by Γ_{at}. The 1 and 2 terms in Eq. (7) shift the two clock states in energy by a different amount, which leads to the change of the Bohr frequency given in Eq. (6).

In 1991 it became clear [15] that collisions between atoms in the atomic cloud during the fountain orbit shift the Bohr frequency by an amount of order 1 mHz for typical densities of 10^9 cm^{-3}. This is a sizable shift compared with the accuracy of the best conventional Cs clocks ($\Delta\nu = 1$ to 0.1 mHz, corresponding with a relative accuracy $\Delta\nu/\nu = 10^{-13}$ to 10^{-14}) and the predicted accuracy of a cesium fountain clock ($\Delta\nu = 1$ μHz, implying a relative accuracy 10^{-16}).

The experimental determination of the collisional frequency shift thereby became an important priority. Beginning 1993 the predicted order of magnitude was confirmed by observations by Gibble and Chu at Stanford University [12]. Since then these and more recent data were analyzed in a more detailed way [16], making it possible to estimate the scattering length for elastic triplet scattering, which is relevant, e.g., for the possibility of realizing BEC in a cold gas of Cs atoms (see following section).

It is of interest to point out that the measured fountain frequency shifts have made it possible to study collisions at the coldest temperatures ($T \simeq 1.5\mu K$) at which atomic collisions have ever been studied. In particular, as pointed out above the temperature is so low that the $T = 0$ limit is rather well satisfied. It is emphasized also that the same collision calculations predict the collisional line broadening Γ_{at}. It turns out that at presently achievable densities this broadening is not observed.

COLLISIONAL CONSTRAINTS FOR BEC

The experiments considered in the previous sections pertain to experimental circumstances in which the condition (1) is fulfilled. Since the end of the seventies attempts are going on to approach the temperature-density region where also condition (2) is obeyed. Initially, the attempts aimed at the realization of BEC in atomic hydrogen, cooled and stored in a cryogenic set-up, the atoms occupying the doubly-polarized spin-down state relative to a strong external magnetic field. These attempts were frustrated by the occurrence of a loss mechanism due to three-body collisions [6]. One of the ways out for both atomic hydrogen (see for instance [17],[18]) and the atomic alkalis (for instance [19]) is to (evaporatively) cool a gas sample in a static magnetic trap, thus allowing the realization of the condition (2) at much lower density-temperature combinations, where three-body collisions are unimportant.

In the achievement of BEC along this line, to which we restrict ourselves in the following, collisions play a key role [20], [21],[19]. In the first place elastic collisions are needed for an efficient (collision-induced or laser-induced) evaporative cooling scheme, giving rise to replenishment of the high-energy atoms escaping from the trap. They are also needed for the establishment of a BEC kinetic equilibrium state. In contrast, inelastic two-body collisions and three-body collisions are unfavorable. Three-body collisions lead to formation of bound triplet and singlet states and inelastic hyperfine-changing collisions to the formation of untrappable high-field seeking states. In both cases the consequences are heating and loss of density. For atomic hydrogen this was investigated theoretically in Ref. [22].

The challenge is therefore to adjust the experimental circumstances (magnetic field, density, temperature) so that the elastic two-body collision rate is large relative to both inelastic and three-body rates. On the elastic two-body time scale, no bound levels are formed and a theoretical description of the properties of the gas can be given in terms of a single quantity, the appropriate elastic scattering length. In particular, all gases with the same positive or negative value of a have properties identical to that of a gas of hypothetical atoms without (two-body) bound states but with the same a value.

Since some decades it is known that a positive sign of a is crucial. $a > 0$ represents an effectively repulsive potential and collisions can be modelled as those between hard spheres of radius a. The thermodynamic properties of such a gas can be obtained as a power series expansion in the parameter $(na^3)^{\frac{1}{2}}$ [1]. For negative values of the scattering length the interaction is effectively attractive and the Bose-condensed state is predicted to be unstable [23]. Also other features are lost which are characteristic of BEC [24]. A theoretical treatment of the inhomogeneous time-dependent Bose system with $a < 0$ does not at present exist in the literature, nor has it been studied experimentally. The subject of the $a < 0$ Bose gas presents an interesting challenge both theoretically and experimentally. It is already clear now that in the inhomogeneous situation of a trap experiment the self-consistent solution of the Hartree-Fock equation for the condensate wave function, known as the Gross-Pitaevskii equation [25], collapses as a function of time. This is easily understood. The mean field (Hartree-Fock potential) describing the average interaction of a condensate particle with the remaining condensate atoms, is proportional to the local condensate density n_0 multiplied by a [23] (see Eq. 7):

$$U(\vec{r}) = n_0(\vec{r})a\frac{4\pi\hbar^2}{m}. \tag{8}$$

For negative a an unstable situation arises, i.e. a collapse of the linear dimensions l of the condensate, due to the fact that the growing attractive potential energy ($\sim n_0 \sim l^{-3}$) beats the increase of the kinetic energy following from the uncertainty relation ($\sim (\Delta p)^2 \sim l^{-2}$).

Clearly, it is of great importance to determine the value of a and in particular its sign. Despite the evident importance the value of a was until recently only known for atomic hydrogen, in which case the interaction potential has been rigorously calculated [26]. For doubly-polarized H (i.e. the pure triplet H + H case) the value of a is positive [27]. For the various alkali gases for which BEC is being attempted experimentally, no reliable theoretical potentials are available. Taking advantage of the special features of cold collisions, however, it has been possible to extract the needed information for several alkali atoms reliably from experimental data. Table 1 summarizes the results.

TABLE 1. Summary of known signs of scattering lengths

Atom	Sign $a_{F_>}$	Sign $a_{F_<}(B)$	Method	Ref.
H	>0	Non-ex.	Rig. pot.	27
^{7}Li	<0	>0	Bound di-at. sp.	28,29
^{23}Na	>0	>0	Bound di-at. sp.	29
^{85}Rb	?	?		
^{87}Rb	?	?		
^{133}Cs	<0	<0 (res.>0)	Cs fountain	16

The sign of the scattering length for the doubly-polarized $F = F_> \equiv I + \frac{1}{2}$ Cs ground state, as determined recently from the first measurements of the Cs fountain frequency shifts (see above) is negative. Contrary to atomic hydrogen, there exists a second promising trappable hyperfine state in the alkalis: the upper state of the lower hyperfine manifold $F_<$. This low-field seeking state exists only for nuclear spin I larger than 1/2 (see Fig. 4).

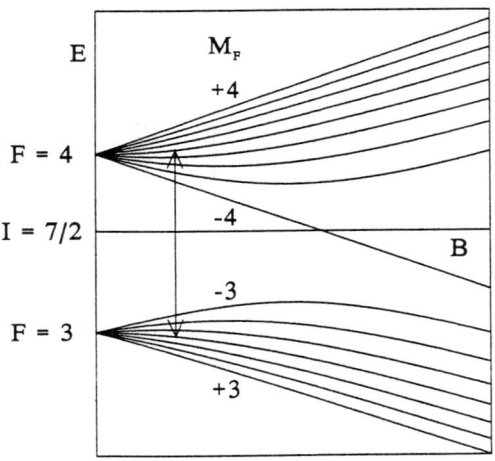

FIGURE 4. Presence of an ultrastable weak-field seeking state in Cs in the lowest manifold, absent in H

It is a very attractive candidate for achieving BEC, since the inelastic decay of the density is suppressed at low B by a strong centrifugal barrier, only weakly-exothermal $l = 2$ final channels being available [19],[20], [21]. In the case of Cs this 3,-3 state is also predicted to have a negative scattering length [16]. However, an interesting resonance structure in the theoretical field dependence is predicted to exist (see Fig. 5), which can be exploited experimentally: it is expected that for suitable field strengths in the center of the trap scattering Feshbach resonances occur with which a positive sign can be realized[21].

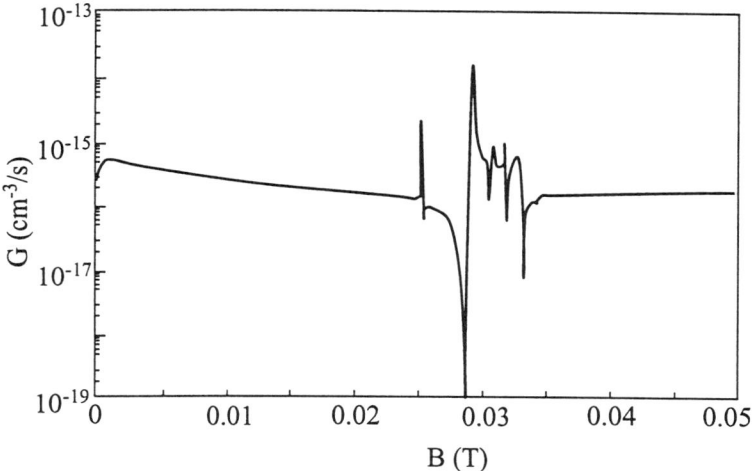

FIGURE 5. Resonance structure in decay rate of trapped hyperfine state $F, m_F = 3, -3$ of Cs (schematic)

In the environment of such a resonance the elastic S-matrix element has the approximate form of a background and a resonance contribution:

$$S_{l=0}(k, B) = e^{-2ika_0} + e^{-2ika_0} \frac{i\Gamma(k, B)}{E_0(B) - E - \frac{1}{2}i\Gamma(k, B)}, \quad (9)$$

with $E_0(B)$ the field-dependent energy of the resonance relative to threshold. Taking into account that the resonance width is proportional to k, we find also for the scattering length a sum of a background and a resonance contribution:

$$a(B) = a_0 - \frac{C}{E_0(B)}, \quad (10)$$

with C a positive constant. Since the quasi-bound Feshbach state arises from the promotion of at least one atom to the upper manifold, for which the

362   Cold Collision Phenomena

Zeeman splitting is opposite, the resonance energy is expected to cross the threshold from above, so that we get the schematical behavior presented in Fig. 6: at resonance the scattering length *decreases* through $\pm\infty$ for increasing B. This applies when B is in the low-field seeking interval for the $F_<, m_F = -F_<$ state. Outside this interval the situation can be different [28]

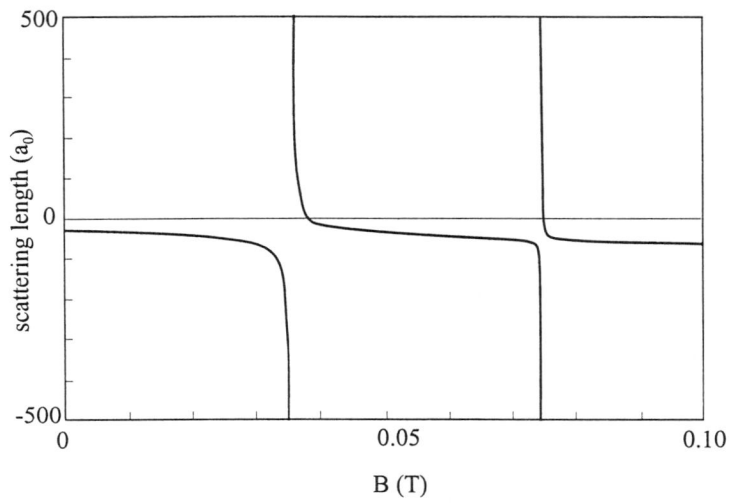

FIGURE 6. Expected background + resonance behavior of scattering length. Change of sign near resonance energy.

This behavior is in agreement with the general expectation for a resonance state crossing threshold into the bound spectrum. An interesting possibility thus presenting itself is a transition from the stable to the unstable regime by a small change of the field in the center of the trap. The width of the stronger resonances is sufficient to select a single sign, or alternatively to change the sign over part of the sample. Also, a time-varying magnetic field in the vicinity of a resonance would be possible, giving rise to a condensate which is only unstable during a limited time, thus enabling one to study the time-dependence of the collapse for negative a. The present experimental data do not allow to predict the positions of such resonances reliably [16]. Note that for a doubly-polarized state B-dependent resonances do not exist.

^{7}Li and ^{23}Na are other examples for which the presently available experimental data allow a determination of the sign of the triplet scattering length and that for the alternative $F_<$ hyperfine state. In both cases we applied a new method to find the scattering length by extrapolating spectroscopic data on the highest bound di-atom states to the continuum. We used the simple fact that the difference between bound and continuum radial wave functions in a small range of energies around $E = 0$ is very small up to a rather large radius R_0. Within R_0 the detailed shape of the potential is irrelevant. The only relevant information is the cumulated information contained in the phase of the rapidly oscillating wave function at R_0. We determined this from the energy *differences* of the highest bound states. Combining this with the (dispersion) potential outside R_0, which is generally known more reliably from ab-initio calculations, (signs of) scattering lengths could be obtained, [29] [28] yielding for the first time definite predictions for alkali atoms on which some of the BEC experiments are concentrating. For Na, for instance, we found in this way [28] $+76 < a_T < +185 a_0$ and $+22 < a_S < +24 a_0$, taking in both cases the dispersion parameters from [30].

OPTICAL COLLISIONS

All examples discussed in the foregoing dealt with collisions between atoms which remain in the electronic ground state during the complete collision. Strong interest exists presently for collisions (sometimes called "optical collisions") between atoms in a near-resonant red-detuned laser field, which causes one of the colliding atoms to be excited to an $-C_3/r^3$ attractive potential curve near the Condon point r_C during the collision. Owing to the low collision velocity the probability of spontaneous emission during the same collision is comparable to 1, analogous to the situation in thermal neutron scattering by atomic nuclei studied in the fifties. This complicates the theoretical description considerably.

The first treatment of optical collisions, containing the essential physics, was proposed by Gallagher and Pritchard [31] (see also Ref. [32]). It treated the heavy-particle motion classically and the laser excitation by a local equilibrium model. Much of the subsequent theoretical work has been done by Julienne and co-workers (see the review [2]). They improved the electronic part of the problem by means of a quantummechanical Optical Bloch Equation description of the internal states, based on a 2×2 density matrix

$$\rho(r(t)) = \begin{pmatrix} \rho_{ee}(r(t)) & \rho_{eg}(r(t)) \\ \rho_{ge}(r(t)) & \rho_{gg}(r(t)) \end{pmatrix}, \quad (11)$$

thus treating the nuclear motion classically.

Measured trap loss coefficients have been shown to be in agreement with these models within a factor of two to three for Cs, Na and Li, and with a larger discrepancy for Rb (see summary in Ref. [33]).

The first fully quantum mechanical treatment was presented by Boesten et al. [34], calculating the losses to first order in the laser intensity. It is based on a time-independent Schrödinger equation, taking the spontaneous-emission loss of coherent flux in the excited channel into account by an imaginary absorptive term. The linear-intensity limit is not an essential restriction compared with the OBE method: the latter would have to include a number of coupled hyperfine states of the ground-state and excited channel and a number of coupled excited electronic states for a calculation at higher intensities to be meaningful. Moreover, an extension of the single-manifold coupled-channel description with one or more additional incoherent lower manifolds is feasible. It is emphasized that the single-manifold description of Ref. [34] is rigorous to first order in the laser intensity at arbitrarily low energies, in contrast to a statement by Suominen et al. [35]: there is no perturbation type of argument involved in its justification [36].

The central result of Ref. [34] was a "quantum suppression" of atom losses for sub-mK energies. It is related to the rapidly growing difference in local wavelength of the radial wavefunctions in the ground-state and excited channels with decreasing R left of the Condon point, due to the fact that the nuclear kinetic energy is no longer large compared to the difference in the channel potentials. This precludes the application of the classical limit to the coupled problem [38]. Wave packets in the two channels thereby separate too rapidly to justify the assumption of a common classical position, which reduces effectively the coupling between the two channels.

In several more recent publications a more complete picture of optical collisions was obtained [37] [35] [39] [33]. In the first of these references the suppression was shown to decrease rapidly with increasing detuning. Furthermore, a simple picture for optical collisions was presented which agrees closely with the quantummechanical calculations for arbitrary single partial waves and thus for the sum over partial waves. The probability for the atomic system to remain in the excited state was written as a product of a Landau-Zener factor for excitation and a WKB survival factor. Finally, this paper contained a comparison of this simple picture with fully quantum mechanical results and with the Gallagher-Pritchard and OBE descriptions.

In Ref. [35] similar conclusions with respect to the Landau-Zener treatment were reached independently for s-wave scattering by a different fully quantum mechanical method: the Quantum Monte Carlo approach. Essentially, it solves the equation for the full quantum density matrix including the translation by a Monte Carlo state vector method analogous to the highly successful single-atom treatment including driving fields and dissipation. If it turns out to be practically possible to extend the calculations to higher intensities, a complete comparison between theory and experimental collisional loss rates would be achievable, thus shedding light on the above discrepancies with experiment which is after all the main stimulus for improvement of the theoretical descriptions. It is not clear presently whether the needed computational effort would be less than that required for the above-mentioned extension of the single-manifold calculation of Ref. [34].

An important recent development is the technique of photoassociation spectroscopy, first proposed by Thorsheim et al. [40] and in the meantime observed by several groups [41] [42] [43] [44]. It is a high resolution probe of molecular rovibrational states formed by photoexciting two colliding cold atoms. A landmark in the theoretical analysis of such spectra is a recent publication by Napolitano et al. [45]. It deals with the asymmetric lineshapes observed as a function of the laser frequency and with their Breit-Wigner resonance description. For these and related recent developments I refer to D.J. Heinzen's lecture at this conference.

ACKNOWLEDGEMENTS

The work of our group described in this talk is part of the research program of the Stichting Fundamenteel Onderzoek der Materie (FOM), which is financially supported by the Nederlandse Organisatie voor Wetenschappelijk Onderzoek (NWO).

REFERENCES

1. Lee, T.D., Huang, K., and Yang, C.N., *Phys. Rev.* **106**,1135-1145 (1957).

2. Julienne, P.S., Smith, A.M., and Burnett, K., "Theory of collisions between laser cooled atoms", in *Advances in Atomic, Molecular and Optical Physics*, edited by D.R. Bates and B. Bederson (Academic Press, San Diego, 1993), p.141-195.

3. Hess, S., *Z. Naturforsch.* **22a**, 1871-1889 (1967).

4. Lhuillier, C., and Laloë, F., *J. Phys. (Paris)* **43**, 225-241 (1982).

5. Koelman, J.M.V.A., Thesis Eindhoven University, 1988 (unpublished).

6. Silvera, I.F., and Walraven, J.T.M., "Spin-polarized atomic hydrogen", in *Progress in Low Temperature Physics*, 1986, Vol. 10, p. 139.

7. Johnson, B.R., Denker, J.S., Bigelow, N., Lévy, L.P., Freed, J.H., and Lee D.M., *Phys. Rev. Lett.* **52**, 1512-1515 (1984).

8. Nacher, P.J., Tastevin, G., Leduc, M., Crampton, S.B., and Laloë, F., *J. Phys. (Paris)* **45**, L441-448 (1984).

9. Koelman, J.M.V.A., Noteborn, H.J.M.F., de Goey, L.P.H., Verhaar, B.J., and Walraven, J.T.M., *Phys. Rev.* B **32**, 7195-7198 (1985).

10. van den Eijnde, J.P.H.W., Reuver, C.J., and Verhaar, B.J., *Phys. Rev.* B **28**, 6309-6315 (1983).

11. Bigelow, N.P., Freed, J.H., and Lee, D.M., *Phys. Rev. Lett.* **63**, 1609-1612 (1990); Thesis Bigelow, N.P., Cornell University, 1990.

12. Gibble, K., and Chu, S., *Phys. Rev. Lett.* **70**, 1771-1774 (1993).

13. Thompson, R., *Nature* **362** (1993) 789-790.

14. Verhaar, B.J., Koelman, J.M.V.A., Stoof, H.T.C., Luiten, O.J., and Crampton, S.B., *Phys. Rev.* A **35**, 3825-3831 (1987); Koelman, J.M.V.A., Crampton, S.B., Stoof, H.T.C., Luiten, O.J., and Verhaar, B.J., *Phys. Rev.* A **38**, 3535-3547 (1988).

15. Tiesinga, E., Verhaar, B.J., Stoof, H.T.C., and van Bragt, D., *Phys. Rev.* A **45**, 2671-2673 (1992).

16. Verhaar, B.J., Gibble, K., and Chu, S., *Phys. Rev.* A**48**, R3429-3432 (1993).

17. Doyle, J.M., Sandberg, J.C., Yu, I.A., Cesar, C.L., Kleppner, D., and Greytak, T.J., *Phys. Rev. Lett.* **67**, 603-606 (1991).

18. Setija, I.D., Werij, H.G.C., Luiten, O.J., Reynolds, M.W., Hijmans, T.W., and Walraven, J.T.M., *Phys. Rev. Lett.* **70**, 2257-2260 (1993).

19. Monroe, C.R., Cornell, E.A., Sacket, C.A., Myatt, C.J., and Wieman, C.E., *Phys. Rev. Lett.* **70**, 414-417 (1993).

20. Tiesinga, E., Moerdijk, A.J., Verhaar, B.J., and Stoof, *Phys. Rev.* A **46**, 1167-1170 (1992).

21. Tiesinga, E., Verhaar, B.J., and Stoof, H.T.C., *Phys. Rev.* A **47**, 4114-4122 (1992).

22. Lagendijk, A., Silvera, I.F., and Verhaar, B.J., *Phys. Rev.* A **33**, 626-628 (1986).

23. Fetter, A.L., and Walecka, J.D., *Quantum Theory of Many-Particle Systems* New York: McGraw-Hill, 1971, p. 222.

24. Tiesinga, E., Moerdijk, A.J., Verhaar, B.J., and Stoof, H.T.C., "Bose-Einstein condensation in ultra-cold cesium: Collisional constraints", presented at the International Workshop on Bose-Einstein Condensation, Trento, Italy, May-June, 1993.

25. Gross, E.P., *Nuovo Cim.* **20**, 454-461 (1961); Pitaevskii, L.P., *Sov. Phys. JETP* **13**, 451-454 (1961).

26. Kolos, W., Szalewicz, K., and Monkhorst, H.J., *J. Chem. Phys.* **84**, 3278-3283 (1986).

27. Friend, D.G., and Etters, R.D., *J. Low Temp. Phys.* **39**, 409-415 (1980).

28. Moerdijk, A.J., and Verhaar, B.J., *Phys. Rev. Lett.*, July 25 1994 issue.

29. Moerdijk, A.J., Stwalley, W.C., Hulet, R.G., and Verhaar, B.J., *Phys. Rev. Lett.* **72**, 40-43 (1994).

30. Marinescu, M., Sadeghpour, H.R., and Dalgarno, A., *Phys. Rev. A* **49**, 982-987 (1994).

31. Gallagher, A., and Pritchard, D.E., *Phys. Rev. Lett.* **63**, 957-960 (1989).

32. Julienne, P.S., *Phys. Rev. Lett.* **61**, 698-701 (1988).

33. Julienne, P.S., Suominen, K-A, Band, Y., *Phys. Rev. A* **49**, 3890-3896 (1994).

34. Boesten, H.M.J.M., Verhaar, B.J., and Tiesinga, E., *Phys. Rev. A* **48**, 1428-1433 (1993).

35. Suominen, K.-A., Holland, M.J., Burnett, K., and Julienne, P.S., *Phys. Rev. A* **49**, 3897-3902 (1994).

36. Boesten, H.M.J.M., and Verhaar, B.J., Internal Report Eindhoven University 1992 (to be published).

37. Boesten, H.M.J.M., and Verhaar, B.J., *Phys. Rev. A* **49**, 4240-4242 (1994).

38. Nikitin, E.E., and Ymanskii, S.Ya., *Theory of Slow Atomic Collisions*, New York: Springer, 1984, p. 254.

39. Holland, M.J., Suominen, K.-A., and Burnett, K., *Phys. Rev. Lett.* **72**, 2367-2370 (1994).

40. Thorsheim, H.R., Weiner, J., and Julienne, P.S., *Phys. Rev. Lett.* **58**, 2420-2423 (1987).

41. Lett, P.D., Helmerson, K., Phillips, W.D., Ratliff, L.P., Rolston, S.L., and Wagshul, M.E., *Phys. Rev. Lett.* **71**, 2200-2203 (1993).

42. Abraham, E.R.I., Ritchie, N.W.M., McAlexander, W.I., and Hulet, R.G., unpublished (1994).

43. Bagnato, V., Marcassa, L., Tsao, C., Wang, Y., and Weiner, J., *Phys. Rev. Lett.* **70**, 3225-3228 (1993).

44. Miller, J.D., Cline, R.A., and Heinzen, D.J., *Phys. Rev. Lett.* **71**, 2204-2207 (1993).

45. Napolitano, R., Weiner, J., Williams, C.J., and Julienne, P.S., *Phys. Rev. Lett.*, submitted.

Collisions of Ultracold Atoms in Optical Fields

D. J. Heinzen

Dept. of Physics, The University of Texas, Austin TX, 78712

Abstract. Two recent developments in ultracold atomic collision physics are discussed. The first is the "optical shielding" of ultracold atomic collisions by optical radiation that is tuned to the blue of an atomic resonance. This radiation couples a pair of colliding atoms to repulsive excited states and prevents their close approach. This effect may substantially suppress inelastic collision rates, and is an interesting example of the control of a cold collision process by a weak perturbing field. The second topic is ultracold atom photoassociation spectroscopy. In this work, high resolution free-bound absorption spectra are obtained which provide detailed information on the long-range states of the two interacting atoms. The physics of these states has an appealing simplicity, and in addition is of importance for a complete understanding of the effects of atomic interactions in cold atom experiments. The photoassociation spectra also provide a unique probe of cold collisions.

INTRODUCTION

Recent advances in laser-cooling and trapping have opened up the new field of ultra-cold (T < 1 mK) atomic collision physics (1,2). This field has attracted considerable interest, largely because ultracold collisions exhibit novel features related to their low relative velocity and long duration. For example, these collisions are very sensitive to weak interactions, and to the long-range part of the interatomic potential. Also, few partial waves may contribute to the collisions, and the deBroglie wavelength of the colliding atoms may be much greater than the range of the interatomic potential. Other unusual features arise because the duration of the collision may be much longer than the spontaneous emission lifetime. In this case the atom may be repeatedly optically excited and de-excited during the collision. Understanding the physics of such an "open" collision system presents a significant challenge.

Aside from their fundamental interest, ultracold collisions can play a dominant role in applications of ultracold atoms. For example, collisional frequency shifts may limit the accuracy of cold atomic clocks (3), or of searches for permanent electric dipole moments of laser cooled atoms (4). The density and lifetime of trapped atoms can be limited by inelastic collisions which transfer sufficient energy to the atoms to eject them from the trap (1,2). Inelastic collisions which do not lead to trap loss can heat trapped atoms. Finally, cold collisions are crucial to efforts to achieve Bose-Einstein condensation in an ultracold atomic gas; for a successful experiment, the ratio of elastic to inelastic collisions must be very favorable (5).

Another reason for the interest in ultracold collisions is that data from these experiments provides information on the long-range interactions between atoms, and also on purely atomic properties. This information can be difficult to obtain with conventional methods. In particular, the newly developed technique of cold atom photoassociation spectroscopy is proving to be a powerful probe of long-range interactions between atoms (6-12). With this technique, the free-bound absorption rate of pairs of colliding atoms is monitored as a function of the frequency of an optical field. This directly yields the spectrum of excited, weakly bound states, and the long-range interaction potentials. These long-range states are of interest in their own right, and exhibit a variety of interesting behaviors related to the recoupling of atomic angular momenta, to retardation effects, and to nonadiabatic effects (13-20). The photoassociation spectra also provide a unique probe of cold collisions. For example, the lineshapes and peak heights of these spectra directly give the distribution of energy and angular momentum states participating in the collisions.

Another important recent development has been the experimental realization of the "optical shielding" of ultracold collisions (21,22). In these experiments, a blue-detuned laser couples the ground state of the collision to a repulsive excited potential. This prevents the close approach of the atoms and may suppress inelastic collisions as a result. It also provides an interesting example of the control of a cold collision process by a weak perturbing field.

In this article, we discuss the new developments of optical shielding and of ultracold atom photoassociation spectroscopy. Several important related topics are not included but are well-treated in recent review articles (1,2). In particular, purely ground state ultracold collisions have been studied; these are reviewed in references 1 and 2 and in the contribution of B. J. Verhaar to these proceedings. Also, trap loss collisions have been extensively studied, and are reviewed in references 1 and 2. We briefly note here one important development that has taken place since the publication of these reviews: fully quantum mechanical calculations of the optical excitation of colliding, ultracold atoms have been carried out, using either a complex potential (23-25) or a quantum Monte-Carlo wavefunction approach (26,27). These calculations show that earlier semiclassical calculations can seriously overestimate trap loss collision rates, particularly at low temperature (23). This shows that to some extent earlier qualitative agreement between semiclassical calculations and experiments (1,2) was fortuitous. However, these calculations have also substantiated the validity of other semiclassical approaches, including one based on a simple Landau-Zener approximation. The quantum Monte-Carlo wavefunction calculations also give the spreading of the atomic momentum distribution due to long-range ultracold optical collisions (27).

OPTICAL SHIELDING OF ULTRA-COLD COLLISIONS

One of the most interesting aspects of ultracold atomic collisions is the possibility that these collisions might be controlled with weak perturbing fields. For instance, the scattering lengths and collision cross sections of ground state atoms can be modified by a weak magnetic field (28). Another idea, which has now been realized experimentally (21,22), is that of "optical shielding". By illuminating the atoms with a laser field that is tuned to the blue of the atomic resonance, it is possible to form a repulsive barrier at long range. This "shields" the collision in the sense that the atoms are prevented from reaching small separation where most inelastic mechanisms occur. This shielding can in principle substantially reduce the rate of ground state inelastic collisions such as hyperfine changing collisions of alkali atoms (1,2). In fact, a substantial reduction in the rate of Penning ionization collisions of trapped metastable Xe atoms by optical shielding has been observed (22), and is discussed in the contribution of W. D. Phillips to these proceedings.

The concept of optical shielding is illustrated in Fig. 1, in a simplified two level approximation. Two atoms approach each other on a ground state potential curve. The atoms are illuminated by a laser field of frequency ω_s that is tuned to the blue of the $|S\rangle \leftrightarrow |P\rangle$ transition, where $|S\rangle$ and $|P\rangle$ denote the ground and excited atomic states. The excited atoms experience a resonant dipole interaction of the form $V(R) = \pm C_3/R^3$, where the sign and magnitude of C_3 depend on the electronic wavefunction symmetry (18). By comparison the ground state potential is relatively flat. The laser field couples the ground state to the repulsive excited potential in the vicinity of the Condon point R_c. As shown in Fig. 1(a), at low intensity the shielding can be thought of as due to the diversion of flux from the ground state channel to the excited state, after which the atoms roll down the repulsive potential. Of course, this is itself an inelastic collision and may lead to trap loss if the kinetic energy acquired is greater than the trap depth (29).

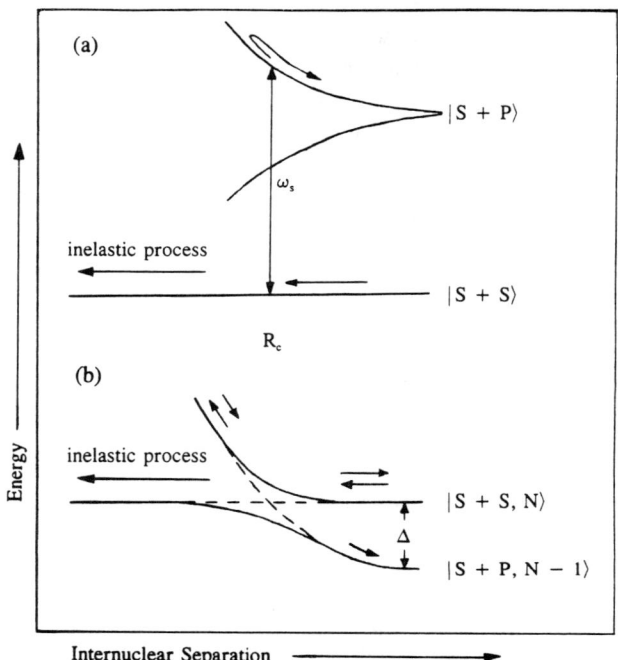

FIGURE 1. Optical shielding of ultracold collisions. (a) Flux incident on the ground state potential is diverted onto a repulsive excited potential. (b) Dressed-state view of the same process, making clear the possibility of elastic reflection at R_c.

At higher laser intensities a dressed state model is more appropriate, as is shown in Fig. 1(b). An avoided crossing between the dressed atom-field states occurs at separation R_c. The atoms are incident from the right on the $|S+S, N\rangle$ channel, where N denotes the number of photons in the laser mode. It is apparent that the atoms may either be transmitted to short distance on the $|S+S\rangle$ ground state potential, be transferred to the $|S+P\rangle$ potential as discussed above, or adiabatically follow their potential and be elastically reflected. In the latter case a repulsive wall is effectively formed at R_c on the ground state potential, and there is the intriguing possibility that the collision might become entirely elastic at high laser intensity. It is also interesting that this provides a mechanism to control the ground state scattering length.

The first experimental evidence for optical shielding was obtained by Marcassa *et al.* (21). In their work, laser-cooled Na atoms trapped in a magneto-optical trap (MOT) collide in the presence of a tunable probe laser field of frequency ω_p and

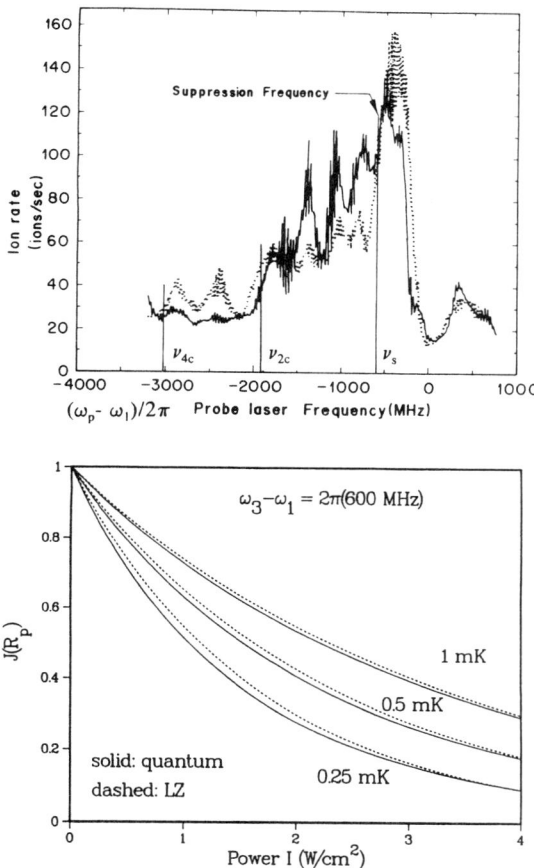

FIGURE 2. (a) Ion production rate as a function of probe laser frequency for the experiment of Marcassa et al., with (dotted line) and without (full line) the suppression laser frequency ω_s present. The atomic resonance frequency is near ω_1. Optical shielding is evident for probe laser detunings between −600 and −1500 MHz. (b) Theoretical shielding factor as a function of laser power for three different temperatures. (From Ref. 21.)

a second laser field at fixed frequency ω_s. The second field functions as the shielding laser and has a detuning $\Delta/2\pi = 600$ MHz and intensity 0.8 W/cm². The combination of the probe laser and an additional laser frequency induces photoassociative ionization (PAI) at a distance $R_p < R_c$. This consists of the double excitation process $|S+S\rangle \rightarrow |S+P\rangle \rightarrow |P+P\rangle$ followed by autoionization of the doubly excited state. Thus the rate of ion production is proportional to the number of atoms reaching internuclear separation R_c. PAI is the collisional process which is to be suppressed by the shielding laser.

Fig. 2(a) shows the observed ionization signal vs. probe laser detuning, once with the shielding laser frequency ω_s turned on and once with it turned off. When the probe laser is tuned more than 600 MHz to the red of the atomic resonance a shielding of the collision (reduced PAI) is observed at the level of about 30%. (For smaller probe laser detunings, the Condon point R_p for probe absorption lies outside R_c, and the PAI is not shielded but actually enhanced.) Calculations of the flux that penetrates to R_p on the ground state, normalized to the incident flux, are shown in Fig. 2(b) for several different temperatures as a function of laser power, for a detuning of -600 MHz. Both a semiclassical Landau-Zener and a quantum mechanical calculation are shown, and agree fairly well with each other. Also, the predicted shielding of 27 to 45% at 1 W/cm^2 is in agreement with the observed shielding. An extension of this theoretical work has recently been carried out (30).

A related experiment has been reported by Bali *et al.* (29). They measure the loss of ^{85}Rb atoms from a magneto-optical trap induced by a blue-detuned probe laser beam of frequency ω_s. This laser couples the ground state of colliding ^{85}Rb atoms to repulsive excited states as shown in Fig. 1. The detuning of the probe laser from atomic resonance is 10 GHz, which is large enough that the kinetic energy gained by a pair of atoms rolling down the repulsive excited potential is sufficient to eject them from the trap. The probe-induced loss rate of atoms from the trap is $<\beta n>$, where n is the density and the brackets denote a spatial average over the trap. Their measured loss rate coefficient β is shown in Fig. 3 as a function of probe laser intensity. At low intensities the loss increases linearly with intensity. This loss results from the excitation to the repulsive state. However, at higher intensities, the loss rate decreases with increasing intensity. This is due to the onset of elastic reflection of the atoms along the ground state potential. The solid line shows the result of a semiclassical calculation based on a Landau-Zener model. The model takes into account crossings with all four repulsive electronic states that are optically coupled to the ground state. Reasonable agreement between the model and experiment is obtained, further confirming the validity of the Landau-Zener theory.

The practical limits to optical shielding have not yet been established. Detailed quantum mechanical calculations of the optical shielding process have been carried out that include the effects of spontaneous emission (30). These calculations show that, within a two-level approximation, it is possible in principle to obtain nearly complete shielding and very low diffusive heating at the same time. However, the validity of the two-level approximation needs to be examined. In particular, the laser field can couple angular momentum state J

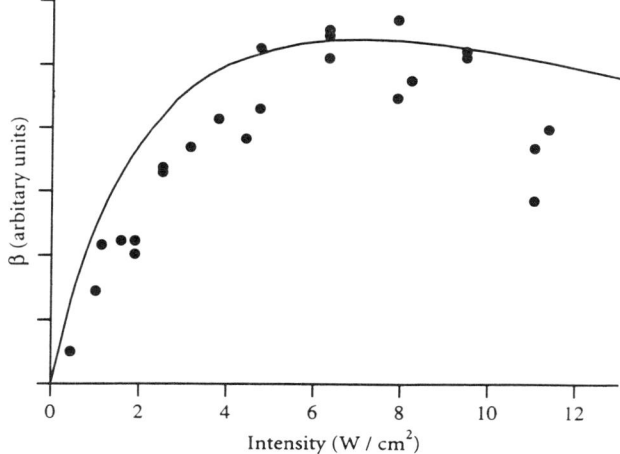

FIGURE 3. Measured rate coefficient for the loss of ^{85}Rb atoms from a magneto-optical trap, induced by a probe laser tuned 10 GHz to the blue of atomic resonance. The solid curve is the result of a calculation based on a Landau-Zener model of the excitation. (From Ref. 29.)

according to the selection rule J ↔ J ± 1, so that in reality successively higher angular momentum states are coupled together (30). A number of issues, including this angular momentum "ladder climbing", require further investigation.

ULTRA-COLD PHOTOASSOCIATION SPECTROSCOPY.

A dramatic possibility, first discussed in detail by Thorsheim, Weiner, and Julienne (6), is that colliding, ultracold atoms can display a resolved photoassociation spectrum. As illustrated in Fig. 4, photoassociation is the process A + B + $\hbar\omega_L$ → AB*, in which a colliding pair of atoms A and B absorbs a photon of frequency ω_L to produce a bound excited molecule AB*. As ω_L is tuned, resonant absorption peaks occur when $\hbar\omega_L$ matches the energy difference between bound excited molecular states and the initially free state. A crucial point is that the energy spread of the initial collisional state is so small at low temperature (e.g. $k_B T/h$ = 21 MHz at T = 1 mK) that the free-bound absorption lines can have a sharpness comparable to those normally associated with spectroscopy between bound states.

Photoassociation spectroscopy has now been realized with laser-cooled, trapped Rb (7,8), Na (9, 10, 11), and Li (12) atoms. The photoassociation spectra provide

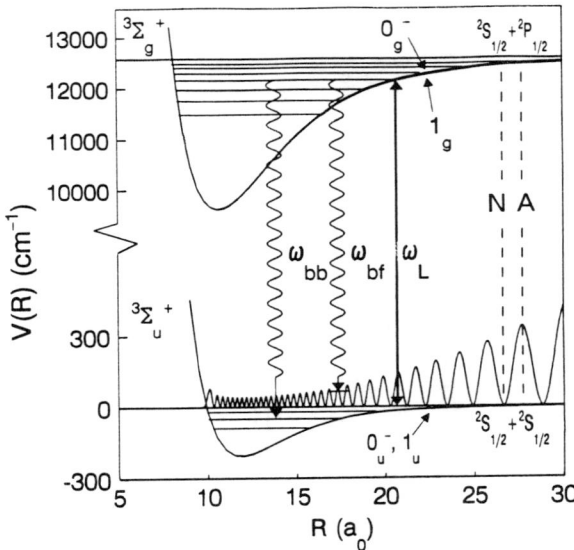

FIGURE 4. Cold atom photoassociation. Colliding atoms incident on the ground state potential are excited by a laser of frequency ω_L to bound excited states. At low temperature, the photon absorption rate exhibits a highly resolved peak when ω_L is tuned across a particular free-bound transition. The triplet states of Rb_2 are shown in this example. The solid oscillating curve shows the square of an approximate radial wave function $u(R)$ of the colliding atoms. (From Ref. 7).

a wealth of information on the long-range interactions between the cold atoms. This information is very timely, since it may help to determine the role these interactions play in collisional trap loss, cold atom frequency standards, efforts to observe Bose-Einstein condensation, and other cold atom experiments. The technique is also complementary to conventional methods. Because the initial state is formed in a collision, both singlet and triplet states may be probed. Also, it is well suited to probing highly excited, long range molecular states which are difficult to reach with conventional bound-bound molecular spectroscopy. (Multiphoton techniques to do this have been developed in some cases (31).)

All of the experiments carried out thus far probe long-range alkali excited molecular states that correlate to the $^2S_{1/2} + {}^2P_{1/2}$ and $^2S_{1/2} + {}^2P_{3/2}$ separated atom limits. The physics of these long range states is of interest in its own right, and is appealing in its simplicity (13-18). At short distance the states are well described by the usual Hund's case "a" labels. But at a range of about 20 a_0, the atomic electronic angular momenta begin to recouple. For separations larger than 20 a_0 the Hamiltonian is fairly well approximated by the sum of the separate

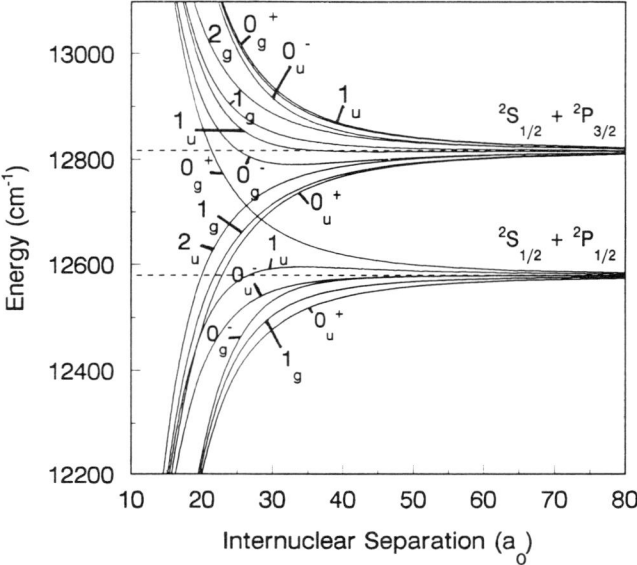

FIGURE 5. Long-range states of Rb_2 asymptotic to the $5^2S_{1/2} + 5^2P_{1/2}$ and $5^2S_{1/2} + 5^2P_{3/2}$ limits. These curves are of the form given by Movre and Pichler (14).

atomic spin-orbit interactions and the resonant dipole interaction between the atoms. The eigenvalues and eigenstates of this Hamiltonian take a simple form (14) and are shown in Fig. 5 for the case of Rb_2. All the alkalis show curves of a similar structure. The states are labelled by Hund's case "c" quantum numbers $\Omega_{g/u}^{+/-}$, where Ω gives the projection of total electronic angular momentum onto the internuclear axis. For still larger separations of several hundred a_0, the nuclear and electronic angular momenta within the atoms recouple, the Hund's case "c" quantum numbers lose their validity, and the potential curves exhibit a complex hyperfine structure (20) (sometimes referred to as "hyperfine spaghetti"). Nonadiabatic effects can be important in the hyperfine recoupling region, and retardation effects are important at very long range. Another interesting feature is the potential wells that exist for the upper 0_g^- and middle 1_u states. These arise from avoided crossings with the lower 0_g^- and 1_u states. These "pure long-range states" (15) lie entirely outside the range of an ordinary chemical bond, and their bound levels had not previously been seen experimentally.

Several methods to detect the photoassociation resonances have been used. One method which is possible in principle is to monitor directly the fluorescence from the AB* excited state (6), but to our knowledge this has not been tried yet. A method which has been applied in Na, Rb, and Li, is to monitor the loss of atoms from the trap. This loss occurs because the excited AB* molecule decays predominantly to free states with a kinetic energy that is too large to remain trapped. Some fluorescence may also occur to bound molecular ground states; this also effectively results in the loss of atoms from the trap. This increased loss rate results in fewer trapped atoms, which is detected as a reduced laser induced fluorescence signal. The advantages of the trap loss method are its relative simplicity and generality. Its primary disadvantage is that it is not a zero-background technique. The dynamic range is limited, and the photoassociation rate must be comparable to the other loss rates in the trap or the signals will be difficult to detect.

Another detection technique which has been used in the Na experiments is to photo-ionize the AB* excited molecule and to detect the resulting AB$^+$ molecular ion. The ionization may either occur by direct excitation to the ion, AB* → AB$^+$ + e$^-$, or by excitation to an autoionizing doubly excited state, AB* → AB** → AB$^+$ + e$^-$. This technique has the advantage of high sensitivity, low background, and wide dynamic range. With suitable lasers it is presumably also a fairly general technique.

A typical ^{85}Rb$_2$ photoassociation spectrum that was recorded in our laboratory is shown in Fig. 6. In this experiment (7), laser-cooled ^{85}Rb atoms are trapped in a far-off resonance optical dipole force trap (FORT) (32). The FORT consists of a single, linearly polarized, TEM$_{00}$-mode Gaussian laser beam focussed to a waist $w_0 = 10.2 \pm 1.2$ μm. Its power is between 1.3 and 1.6 W, its linewidth 0.4 cm^{-1}, and its wavelength λ_L between 798 nm and 860 nm. The FORT contains about 3,000 atoms at a density of about 1×10^{12} cm^{-3}, and a temperature of 0.4 to 0.7 mK. In order to record this spectrum, the FORT laser is tuned; that is, the same laser both traps the atoms and induces the photoassociation transitions. When the laser is tuned to a strong photoassociation resonance the atomic density decays with a volume-averaged initial decay rate of $<\beta n> = 20$ s^{-1}, which corresponds to a photoassociative rate coefficient β between 1.4×10^{-11} and 1.7×10^{-10} cm^3/s. The data are recorded with the trap-loss technique. In Fig. 6, the signal is proportional to the number of trapped atoms, with increasing signal plotted downward from origin at the top of the graph.

Remarkably, a spectrum consisting of more than 150 well-resolved lines is observed, spanning the range from 80 to 940 cm^{-1} below the $5^2S_{1/2} + 5^2P_{1/2}$ dissociation limit. The density of lines in the spectrum increases as the dissociation limit is approached, as expected from the roughly 1/R^3 form of the potentials (33). Also, the intensity of the lines increases as the dissociation limit is approached, which is expected because the free-bound Franck-Condon factors become larger and because the atomic density in the trap increases. The most

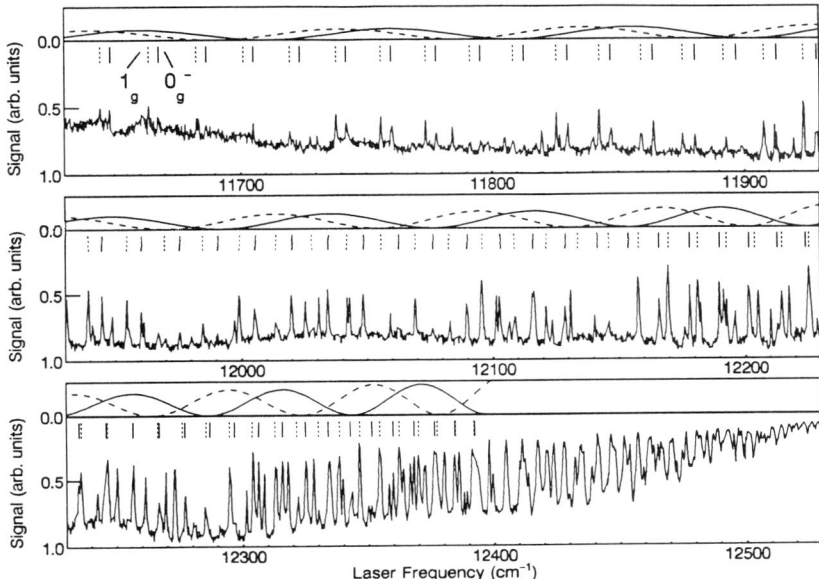

FIGURE 6. Photoassociation spectrum of ^{85}Rb$_2$ for very large detunings below the $5^2S_{1/2}$ + $5^2P_{1/2}$ limit. The inset dashed and solid curves show the approximate variations in the Franck-Condon factors associated with the 1_g and 0_g^- series, respectively. The $5^2S_{1/2} \leftrightarrow 5^2P_{1/2}$ atomic transition occurs at 12578.9 cm^{-1}. (From Ref. 7.)

striking features of the spectrum are the two regular vibrational series labelled by the solid and dashed vertical lines in Fig. 6. These series have been assigned to the (lower) 0_g^- and 1_g potentials shown in Figs. 4 and 5, both of which connect to the $^3\Sigma_g^+$ potential at short range. These two series account for about 65% of the identifiable lines. Some of the remaining lines have been assigned to 0_u^+ levels on the basis of high resolution spectra, while others remain unassigned.

A further very noticeable feature of the spectrum is the oscillation in the intensity of the lines, as illustrated by the inset curves of Fig. 6 for the 0_g^- and 1_g series. These oscillations arise from the variation in Franck-Condon factors, and their periodicity is governed by the ground state wavefunction. Roughly speaking, if the outer turning point of the excited vibrational level lies above an antinode in the initial ground state wavefunction u(R), the Franck-Condon factor will be large (dashed line "A" of Fig. 4). At a slightly lower energy the outer turning point will lie above a node in u(R) and the Franck-Condon factor will be small (dashed line "N" of Fig. 4). In this way a further lowering of the energy will produce a series of maxima and minima that effectively maps out the oscillations in the ground state wavefunction.

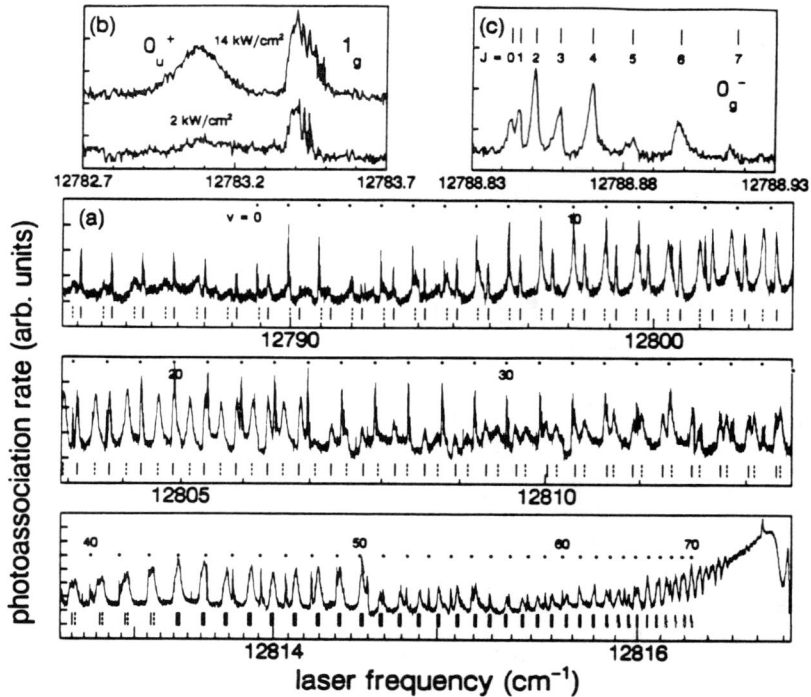

FIGURE 7. High-resolution photoassociation spectrum of ^{85}Rb$_2$. (a) Extended scan over the range within 35 cm^{-1} of the the $5^2S_{1/2} + 5^2P_{3/2}$ dissociation limit. The solid lines give the positions of the 1_g levels, the dashed lines give the positions of the 0_u^+ levels, and the dots give the positions of the first 70 levels belonging to the 0_g^- "pure long range" state. (b) Expanded view of a single 0_u^+ and 1_g vibrational level. (c) Expanded view of the v = 0 vibrational level of the 0_g^- state. (From Ref. 8.)

With the single laser technique described above, we were unable to obtain high resolution spectra or to probe states much closer than 50 cm^{-1} to the Rb$_2^*$ dissociation limit. This is because the single laser must be strong enough to trap the atoms at high density, which tends to power broaden and saturate the resonances. More recently we have developed a new technique which solves this problem. We use one fixed frequency FORT laser beam to trap the atoms, and superimpose on the trap a separate, tunable, 1 MHz linewidth probe laser beam to drive the photoassociation resonances. This allows us to tune the probe laser to any desired wavelength and power, while at the same time maintaining a strong trap. The FORT laser is tuned to 12289 cm^{-1}, which is between two well-resolved photoassociation resonances, and is rapidly alternated in time with the probe laser to eliminate power broadenings and shifts.

Using this method, we have recorded the ^{85}Rb photoassociation spectrum within 35 cm^{-1} of the $5^2S_{1/2} + 5^2P_{3/2}$ dissociation limit, as shown in Fig. 7. Except for the addition of the probe laser, the experimental conditions are similar to those of Fig. 6. The probe laser intensity is changed in several discrete steps from 2 kW/cm^2 at the low frequency end of the plot to 1 W/cm^2 at the high frequency end. Three vibrational series are observed, which are associated with the three attractive potentials of Fig. 5 just below $5^2S_{1/2} + {}^2P_{3/2}$ that are optically coupled to the ground state. Resolved vibrational levels of the 0_u^+ and 1_g potentials are observed from the low end of the scan to within 0.3 cm^{-1} of the dissociation limit. In addition, we observe the first 70 vibrational levels of the 0_g^- pure long-range state. These series also exhibit Franck-Condon factor oscillations. The last few resolved levels have outer turning points of 200 - 250 a_0.

Each of the three states shows a distinctive substructure. In Fig. 7(b) we show a high resolution scan over a 0_u^+ and a 1_g vibrational level. The 0_u^+ level is featureless and broad, due to fine structure predissociation to $^2S_{1/2} + {}^2P_{1/2}$. The width of this state gives a direct measure of a Landau-Zener curve crossing probability that is important to the understanding of fine structure changing collisions (18,34). On the other hand, the 1_g state exhibits hyperfine structure. Its overall width is consistent with a calculated 1.4 GHz hyperfine splitting in the 1_g potential curves, (20,35) together with some additional rotational structure. The 21 most strongly bound 0_g^- vibrational levels exhibit a simple rotational spectrum, as shown for the v = 0 level in Fig. 7(c). For v > 20 a more complex structure is observed, which is not understood at present. The rotational spectrum is cut off at J = 7, because at these energies photoassociation occurs inside the ground state centrifugal barrier at 100 - 150 a_0 which excludes higher angular momentum states. Because J = 7 is the highest observed rotational level, $\ell = 5$ is the highest partial wave that contributes to the collision.

The Franck-Condon factor oscillations and the rotational line intensities and lineshapes contain detailed information on the ground state potential (7,36). We have obtained photoassociation spectra of the lower 0_g^- state with doubly spin polarized ^{85}Rb atoms, which results in certain simplifications of interpretation. An analysis of this data yields the ^{85}Rb ground state triplet scattering length and C_6 coefficient, as well as other information on the cold collisional process (37).

We have compared the binding energies and rotational constants of the 0_g^- pure long-range state with those of a very simple model, as shown in Fig. 8. The potential is taken to be $V(R) = V_{MP}(R) - C_6/R^6$, where V_{MP} is the form of the potential given by Movre and Pichler (14). Its only parameters are the square of the atomic dipole matrix element, which we take to be $d^2 = 8.90$ (ea$_0$)^2 from the weighted average of measured $5^2P_{3/2}$ atomic lifetimes (38), and the measured fine structure splitting. The term $-C_6/R^6$ models in an approximate way the

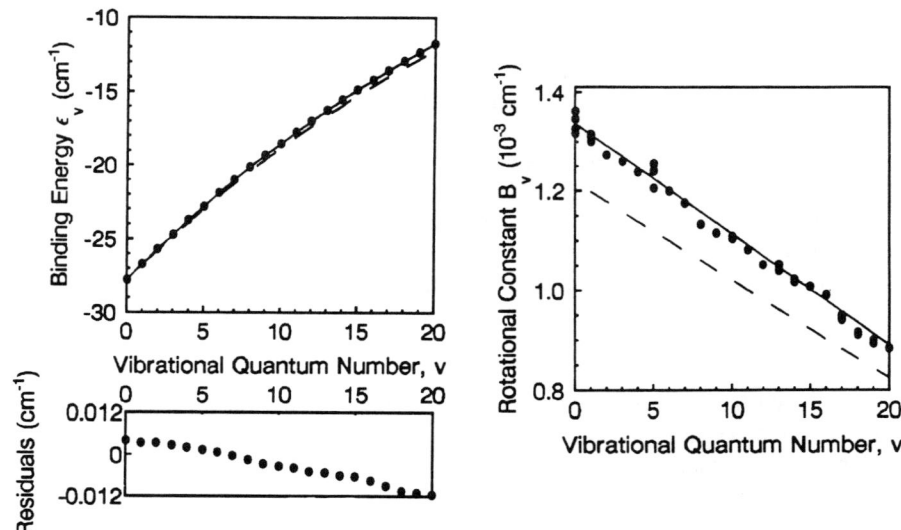

FIGURE 8. Binding energies and rotational constants for the first 21 0_g^- levels shown in Fig. 7. The dashed line shows the theoretical calculation of Ref. 17, and the solid line shows the prediction of a simple modification of the Movre and Pichler potential (14). The residuals show the deviation of the solid line from the data.

contributions of the dispersion interactions (16), and we take $C_6 = 10,825$ a.u, which provides a good fit and is within the range of predicted values of C_6 (16). Excellent agreement is obtained from such a simple model - the binding energies are correct to within 0.012 cm^{-1} out of more than 10 cm^{-1}. A more precise analysis is in progress from which it should be possible to extract an accurate value for d^2 and for the Rb atomic lifetime. The advantage of the 0_g^- state is that the analysis does not depend on poorly known short distance potentials. The most serious limitation appears to be uncertainties in the dispersion interaction coefficients. A new calculation of these coefficients, in progress (39), may help to alleviate this difficulty.

Very similar results have been obtained in Na by Lett and collaborators (10, 11). Na atoms trapped in a MOT at a temperature of ≤ 500 μK and a density of 10^{10} cm^{-3} were illuminated by a tunable probe laser beam. The probe laser drives the two step process $Na(3^2S_{1/2}) + Na(3^2S_{1/2}) \rightarrow Na_2^*(\Omega_{g/u}^{+/-}) \rightarrow Na_2^+ + e^-$. The second step is without significant structure, so the ion production rate displays photoassociation resonances that are associated with the first step. Early work provided well-resolved spectra of the middle 1_g state over the range within 3 cm^{-1} of the $^2S_{1/2} + ^2P_{3/2}$ dissociation limit (10).

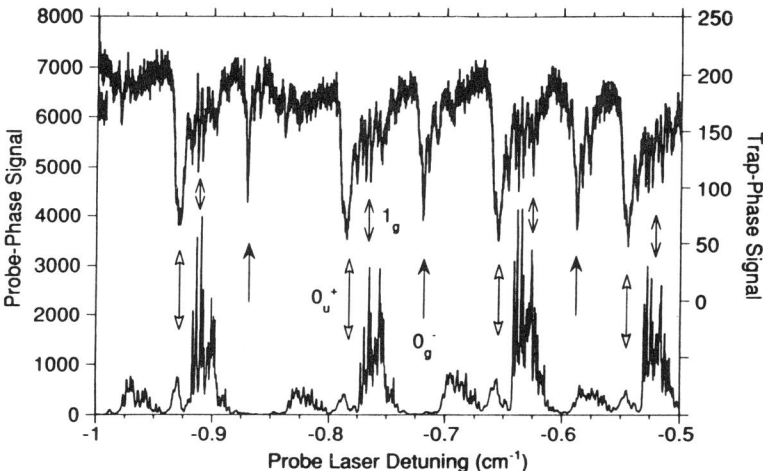

FIGURE 9. Photoassociation spectrum of Na_2. Trap loss signals similar to those shown in Figs. 6 and 7 for Rb are shown on the top curves. The lower curve shows the ionization rate, which is proportional to the photoassociation rate. (From Ref. 11).

In more recent work, a "dark spot" MOT (40) was applied to these measurements (11). This trap provides a significantly higher density and greater signal levels. Spectra were obtained over a much greater range of detuning of 170 cm^{-1}. A typical spectrum is shown in Fig. 9. The top curve show a trap loss spectrum. The lower curve shows the ionization signal. As for the Rb spectrum, the three allowed states 1_g, 0_g^-, and 0_u^+ are clearly identified. For small detunings from atomic resonance, the 1_g state exhibits hyperfine structure. However this state connects adiabatically to the $^1\Pi_g$ state at short range, and therefore as the detuning is increased the hyperfine structure diminishes, and the vibrational levels show clean rotational series. The rotational series span the range from J = 1 to J = 4. From this, one can conclude that the d-wave is the highest partial wave that contributes to these collisions.

The measured rotational constants are shown in Fig. 10. B_{ij} is the rotational constant derived from the splitting between peaks with rotational quantum numbers i and j, ignoring centrifugal and lineshape distortions (36). The solid line of Fig. 10 shows the result of adiabatic potential calculations by Williams and Julienne (20). Excellent agreement with the theory is obtained.

With very high resolution the photoassociation resonances exhibit asymmetric lineshapes (8,11). An analysis of the photoassociation lineshapes obtained in Ref. 11 has been carried out by Napolitano and collaborators (36). The results are

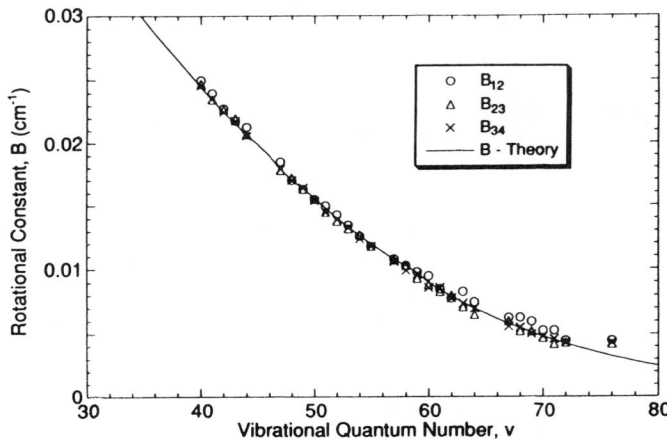

FIGURE 10. Rotational constants derived from Na_2 photoassociation spectra for the $^1\Pi_g$ state. The solid line shows a theoretical prediction. (From Ref. 11.)

shown in Fig. 11. The data shows high-resolution scans of the J = 1, 2, 3, and 4 peaks of the $^1\Pi_g$(v = 48) vibrational level, plotted on a logarithmic scale. The solid curves show the lineshapes calculated using a resonant scattering formalism (36). Residual hyperfine structure is evident in the rotational peaks, and is reasonably well reproduced by the theory. The peaks all show an exponential tail for red laser detuning. Similar tails have been observed in Rb_2 (8). This corresponds to the excitation of higher energy atoms, in which the atoms' kinetic energy brings the free-bound transition into resonance with the red shifted laser. This exponential tail essentially follows the Boltzmann energy distribution of the laser-cooled atoms. The temperature may be derived from these lineshapes, but the whole lineshape must be fit rather than just the tail if accurate temperatures are to be obtained (36). The temperature assumed for the solid curves in the figure is 600 μK, which is in reasonable accord with the expected temperature.

The onsets to the blue side of the peak illustrate Wigner threshold law behavior for low energy collisions (36). That is, for sufficiently small collision energy ϵ, we expect an optical excitation rate that for a given ℓ scales as $\epsilon^{(2\ell+1)/2}$. For s-waves, a very sharp onset is expected, whereas for higher partial waves the onset is increasingly gradual. Evidence of this behavior may be seen in the data, since the sharp edge on the right of the J = 2 peak is known from the calculation to arise from an s-wave feature, whereas the more gradual onset on the right side of the J = 4 peak arises from a d-wave feature. These lineshape effects significantly shift the peaks from the rotational line position corresponding to zero kinetic energy, and must be taken into account in an accurate rotational

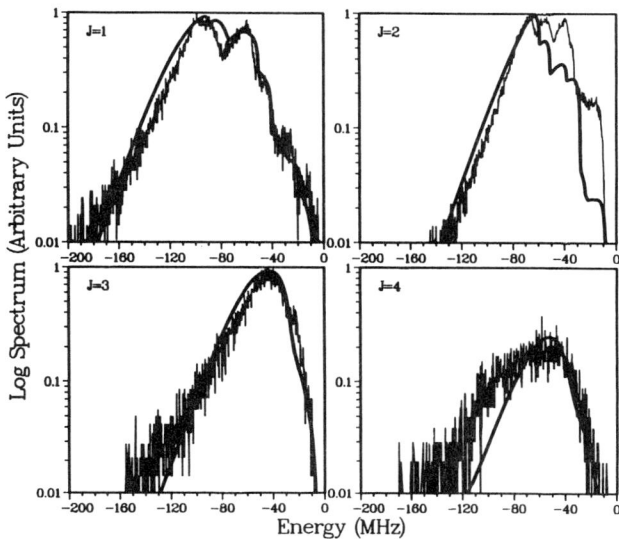

FIGURE 11. Lineshapes of single rotational lines for the v = 48 vibrational level of the Na_2 $^1\Pi_g$ state. The data shows the ion signal vs. laser tuning recorded in the experiment of Ref. 11. The solid curves show the lineshapes calculated from a resonance scattering formalism. (From Ref. 36.)

analysis. It is also of interest that the lineshapes are very sensitive to the ground state potential (36). The rotational peak height and lineshape analysis provide a beautiful example of the detailed information that photoassociation spectra provide both on long-range atomic interactions and on the cold collisions themselves (36,37).

A third example of photoassociation spectroscopy is the work of McAlexander and collaborators (12). They measured the photoassociation spectrum of Li_2 using the trap loss technique in a MOT. The results of the experiment are shown in Fig. 12. Owing to the small fine structure of Li, the Hund's case "a" labels are valid even at relatively large internuclear separation. Two clear vibrational series are observed, the strongest of which is associated with the $^3\Sigma_g^+$ state, and which spans a range of binding energies up to 90 cm^{-1}.

Li_2 has several features that simplify the analysis of these spectra relative to the heavier alkalis. Accurate RKR potentials and ab-initio potentials exist which span the short range part of the curve. Also, dispersion, spin orbit, and hyperfine interactions play a reduced role in the Li spectra. The arrows below the spectra show the energy levels calculated from a model potential, the most important

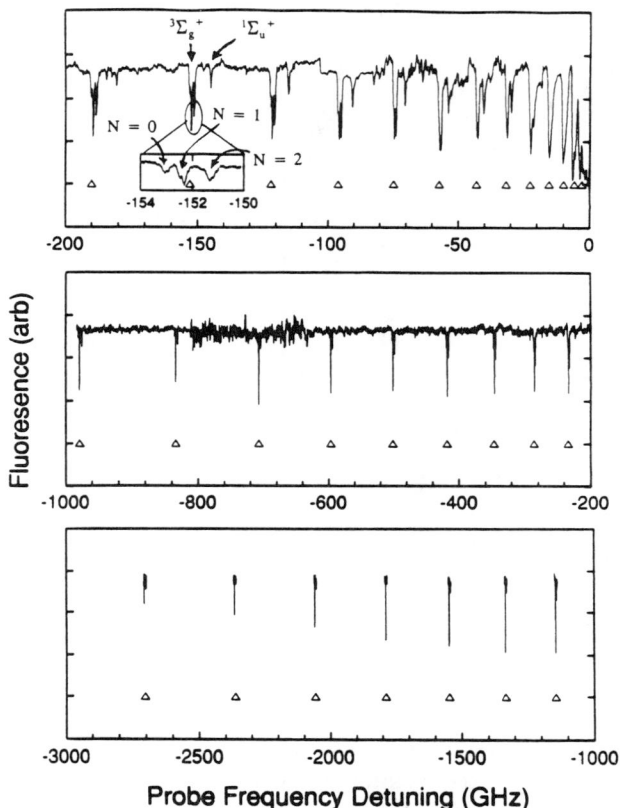

FIGURE 12. Photoassociation spectrum of Li_2. Vibrational series of the $^3\Sigma_g^+$ and $^1\Sigma_u^+$ potentials are visible. The inset shows a high-resolution scan over a single vibrational level (From Ref. 12.)

parameter of which is the atomic dipole matrix element, or equivalently the Li atomic lifetime. An analysis shows that this data can be used to determine the Li lifetime to an accuracy of at least 0.4% (12). It should be possible to derive accurate lifetimes for all of the alkalis from photoassociation spectra. Although it is possible to obtain approximate matrix elements or lifetimes by considering only the long-range part of the potential (33, 8, 10), for a very accurate lifetime determination the whole potential must be accounted for, as in the analysis as of Ref. 12.

CONCLUSION

There has been a great deal of progress in the field of ultracold collisions over the past several years. In addition to new studies of trap loss and ground state collisions not addressed in this article, the first steps towards the control of cold collisions have been taken in the form of the optical shielding experiments, and the new technique of photoassociation spectroscopy has been developed. Similarly many new developments should occur over the next several years. Certainly the one-laser photoassociation spectroscopy of the alkalis and the corresponding analysis of their interactions should advance substantially. It also seems likely that double resonance, free-bound-bound photoassociation spectroscopy will be developed, which should provide access to higher excited states or direct access to the ground states. In fact, some work on higher excited states and preliminary experiments on ground states has already been carried out (41). Further, it is clear that photoassociation produces some ground state ultracold molecules, and that these molecules could be held in a dipole or a magnetic trap (6,7). If such cold, trapped molecules could be produced in sufficient numbers and detected, it could lead the way towards studies of ultracold molecular interactions. Other directions that are of interest include studies of optical shielding in the high intensity limit, or the control of cold collisions by a magnetic field (28). Lower temperature experiments might more clearly reveal quantum threshold behaviors (2). Because of these and numerous other interesting problems and directions that have been suggested (1,2), ultracold atomic collision physics should continue to be a challenging and dynamic area of investigation.

ACKNOWLEDGEMENTS

The Rb_2 spectroscopy presented here was carried out in collaboration with R. A. Cline and J. D. Miller. The support of the A.P. Sloan Foundation, the National Science Foundation, and the R. A. Welch Foundation for this work is gratefully acknowledged. Finally, we thank Paul Julienne, Carl Williams, Boudewijn Verhaar, John Weiner, Paul Lett, Bill Phillips, Randy Hulet, Thad Walker, Bill Stwalley, and Alex Dalgarno for helpful and stimulating discussions.

REFERENCES

1. T. Walker and P. Feng, to be published in *Advances in Atomic, Molecular, and Optical Physics*, vol. 34, B. Bederson and H. Walther, eds.
2. P. S. Julienne, A. M. Smith, and K. Burnett, in *Advances in Atomic Molecular, and Optical Physics,* vol. 30, D. R. Bates and B. Bederson, eds. (Academic Press, San Diego, 1993) pp. 141-198.

3. K. Gibble and S. Chu, *Phys. Rev. Lett.* **70**, 1771 (1993); E. Tiesinga, B. J. Verhaar, H. T. C. Stoof, and D. van Bragt, *Phys. Rev. A* **45**, R2671 (1992).
4. M. Bijlsma, B. J. Verhaar, and D. J. Heinzen, *Phys. Rev. A* **49**, R4285 (1994).
5. E. Tiesinga, A. J. Moerdijk, B. J. Verhaar, and H. T. C. Stoof, *Phys. Rev. A* **46**, R1167 (1992).
6. H. R. Thorsheim, J. Weiner, and P.S. Julienne, *Phys. Rev. Lett.* **58**, 2420 (1987).
7. J. D. Miller, R. A. Cline, and D. J. Heinzen, *Phys. Rev. Lett.* **71**, 2204 (1993).
8. R. A. Cline, J. D. Miller, and D. J. Heinzen, *Phys. Rev. Lett.* **73**, 632 (1994).
9. V. Bagnato *et al.*, *Phys. Rev. Lett.* **70**, 3225 (1993).
10. P. D. Lett, K. Helmerson, W. D. Phillips, L. P. Ratliff, S. L. Rolston, and M. E. Wagshul, *Phys. Rev. Lett.* **71**, 2200 (1993).
11. L. P. Ratliff, M. E. Wagshul, P. D. Lett, S. L. Rolston, and W. D. Phillips, *J. Chem. Phys.* **101**, 2638 (1994).
12. W. I. McAlexander, E. R. I. Abraham, N. W. M. Ritchie, C. J. Williams, H. T. C. Stoof, and R. G. Hulet, to be published.
13. W. C. Stwalley, *Contemp. Phys.* **19**, 65 (1978).
14. M. Movre and G. Pichler, *J. Phys. B* **10**, 2631 (1977).
15. W. C. Stwalley, Y.-H. Uang, and G. Pichler, *Phys. Rev. Lett.* **41**, 1164 (1978).
16. B. Bussery and M. Aubert-Frécon, *J. Chem. Phys.* **82**, 3224 (1985).
17. B. Bussery and M. Aubert-Frécon, *J. Mol. Spect.* **113**, 21 (1985).
18. P. S. Julienne and J. Vigué, *Phys. Rev. A* **44**, 4464 (1991).
19. M. Marinescu, H. R. Sadeghpour, and A. Dalgarno, *Phys. Rev. A* **49**, 982 (1994).
20. C. J. Williams and P. S. Julienne, *J. Chem. Phys.* **101**, 2634 (1994).
21. L. Marcassa *et al.*, *Phys. Rev. Lett.* **73**, 1191 (1994).
22. S. L. Rolston *et al.*, to be published.
23. H. M. J. M. Boesten, B. J. Verhaar, and E. Tiesinga, *Phys. Rev. A* **48**, 1428 (1993).
24. H. M. J. M. Boesten and B. J. Verhaar, *Phys. Rev. A* **49**, 4240 (1994).
25. P. S. Julienne, K.-A. Suominen, and Y. Band, *Phys. Rev. A* **49**, 3890 (1994).
26. K.-A. Suominen, M. J. Holland, K. Burnett, and P. S. Julienne, *Phys. Rev. A* **49**, 3897 (1994).
27. M. J. Holland, K.-A. Suominen, and K. Burnett, *Phys. Rev. Lett.* **72**, 2367 (1994).
28. E. Tiesinga, B. J. Verhaar, and H. T. C. Stoof, *Phys. Rev. A* **47**, 4114 (1993).
29. S. Bali, D. Hoffmann, and T. Walker, *Europhys. Lett.* **27**, 273 (1994).
30. K.-A. Suominen, M. Holland, K. Burnett, and P. Julienne, to be published.
31. H. Knöckel *et al.*, *Chem. Phys.* **152**, 399 (1991); E. F. McCormack and E. E. Eyler, *Phys. Rev. Lett.* **66**, 1042 (1991); A. M. Lyyra *et al.*, *Phys. Rev. Lett.* **66**, 2724 (1991).
32. J. D. Miller, R. A. Cline, and D. J. Heinzen, *Phys. Rev. A* **47**, R4567 (1993).
33. R. J. LeRoy and R. B. Bernstein, *J. Chem. Phys.* **52**, 3869 (1970); W. C. Stwalley, *Chem. Phys. Lett.* **6**, 241 (1970).
34. O. Dulieu, P. Julienne, and J. Weiner, *Phys. Rev. A* **49**, 607 (1994).
35. C. J. Williams and P. S. Julienne, private communication.
36. R. Napolitano, J. Weiner, C. J. Williams, and P. S. Julienne, *Phys. Rev. Lett.* **73**, 1352 (1994).
37. J. R. Gardner, R. A. Cline, J. D. Miller, D. J. Heinzen, H. M. J. M. Boesten, and B. J. Verhaar, to be published.
38. J.K. Link, *J. Opt. Soc. Am.* **56**, 1195 (1966); R.W. Schmieder *et al.*, *Phys. Rev. A* **2**, 1216 (1970); S. Svanberg, *Phys. Scr.* **4**, 269 (1971).
39. A. Dalgarno, private communication.
40. W. Kettrle *et al.*, *Phys. Rev. Lett.* **70**, 2253 (1993).
41. Paul Lett, private communication.

Experiments with Atomic Hydrogen in Magnetostatic Traps

I.D. Setija, O.J. Luiten, M.W. Reynolds, H.G.C. Werij, T.W. Hijmans and J.T.M. Walraven

Van der Waals – Zeeman Institute, University of Amsterdam, Valckenierstraat 65/67, 1018 XE Amsterdam, The Netherlands

ABSTRACT

Abstract Atomic hydrogen is an important model system for the investigation of the properties of ultra-cold gases. We summarize the current experimental situation for magnetically confined hydrogen, putting emphasis on recent optical experiments with Lyman-α radiation. As the research is aimed at gas-phase properties, the samples are studied at densities where interatomic collisions are frequent and internal thermal equilibrium may be assumed. Interestingly, although optically thick, the samples still may be cooled with light. We discuss both Doppler cooling and light induced evaporative cooling.

INTRODUCTION

'Nearly ideal gases' serve as important model systems in statistical physics. They enable an accurate microscopic description of complicated many-body behavior. For classical systems this has been established in detail during the development of the kinetic theory of gases. For quantum systems the situation is different. Although the canonical treatments of degenerate quantum behavior in fluids start with a discussion of the nearly ideal gases, in this case even a single example of such behavior in gases is absent [1]. Not surprisingly, interesting properties of nearly ideal quantum gases have remained untested. This holds in particular for optical properties.

In this lecture we discuss the case of spin-polarized atomic hydrogen, which is the best-documented example of a nearly ideal Bose gas [2, 3, 4]. Aside from the statistics, quantum gases are characterized by the condition

$$\Lambda \gg R_0 \tag{1}$$

where $\Lambda = [2\pi\hbar^2/mk_BT]^{1/2}$ is the thermal de Broglie wavelength (with m the atomic mass and T the temperature) and R_0 is the characteristic radius of interaction between the atoms. For hydrogen $\Lambda \approx 55$ nm at $T = 1$ mK.

Condition (1) defines the s-wave scattering regime, where binary collisions may be described asymptotically with a relative wavefunction of the type $\Psi(r) \simeq (1 - a/r)$. The parameter a is the s-wave scattering length; for spin-polarized hydrogen $a \approx 0.72$ Å. Bose gases are 'nearly ideal' provided that

$$na^3 \ll 1, \qquad (2)$$

where n is the gas density. Here, a is assumed to be positive; for $a < 0$ the gas is unstable against collapse. In the thermodynamic properties quantum corrections show up for $n\Lambda^3 \gtrsim 0.1$. Thus far, quantum gases have only been investigated in the regime of quasi-classical statistics

$$n\Lambda^3 \ll 1. \qquad (3)$$

To observe perhaps the most intriguing phenomena in ultra-cold gases, degenerate quantum behavior - in particular Bose-Einstein condensation (BEC), higher densities or lower temperatures are required. The critical density for BEC in a dilute Bose gas is given by the expression $n_c\Lambda^3 \approx 2.612$. Hence for H at $T = 30$ μK one should reach $n_c \approx 10^{14}$ cm^{-3}. Note that with these numbers the conditions (1) and (2) are extremely well satisfied.

H is created by dissociation of H_2. In this process ground state atoms are produced in all four hyperfine states, labeled a, b, c and d in order of increasing energy. Only atoms in the c and d states (spin-up polarized hydrogen, H↑) are low-field-seeking and can be confined in magnetostatic traps. Such traps are formed by a local minimum in the modulus of a static magnetic field. Atoms in the a and b states (spin-down-polarized hydrogen, H↓) are driven toward high fields where they can be confined with liquid helium covered surfaces. The spin-polarization impedes the formation of molecules and results in metastable samples that can be studied for periods of seconds to hours, depending on the density. H has been studied intensively since 1979 [5]. Over the last decade the advent of laser cooling has resulted in several new quantum gases, in particular alkali systems. Many more are to follow. All these systems are metastable. Hydrogen stands out among these gases as the system which is best suited for theoretical analysis.

TRAPPING ATOMIC HYDROGEN

Surface-free confinement of H↑ in a magnetostatic trap was proposed by Hess [6] and first demonstrated at MIT [7] and in Amsterdam [8]. The approach of Hess exploits the anomalously small adsorption energy of H to the surface of liquid helium ($\epsilon_a/k_B \approx 1$ K). This allows a cryogenic filling method in which heat exchange with the surrounding walls serves to carry away the heat of trapping. The experimental arrangement used in Amsterdam [9, 10] is

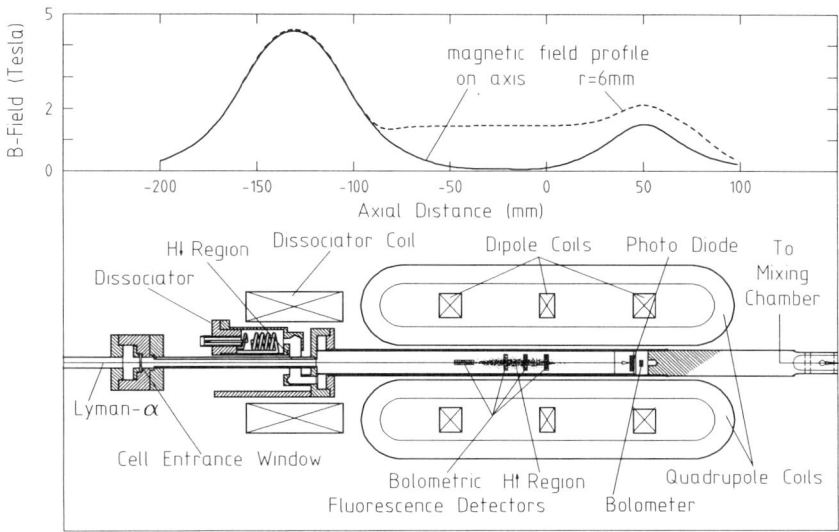

Figure 1: Schematic drawing of the trap configuration used in Amsterdam. Shown are the experimental cell and the confinement coils with corresponding field profile. Solid line: field on the z axis; dashed line: field 6 mm off-axis.

shown in Fig.1. A cold puff of H gas emerging from an rf dissociator operated at 0.6 K distributes itself over the experimental cell according to its spin polarization. The H↑ collects in the trapping region surrounded by He covered walls. Once the density is sufficiently high to result in frequent interatomic collisions, atoms start to be trapped. Those atoms emerging from a collision with low kinetic energy are trapped whereas the collision partners carry away the excess energy. Optical access is provided by a MgF_2 window sealing off the experimental cell. The cryogenic filling method enables large filling fluxes as compared to the optical methods used to trap alkali atoms and is therefore especially suited to produce thermal gas samples.

In thermal equilibrium under condition (3) the density distribution of H↑ may be written as

$$n(\mathbf{r},T) = n_0 \exp[-U_p(\mathbf{r})/k_B T] \;, \tag{4}$$

where n_0 is the density in the center of the trap and $U_p(\mathbf{r})$ is the trapping potential. The effective volume occupied by the gas is $V_{1e} = \int [n(\mathbf{r})/n_0]d\mathbf{r} = N/n_0$, where N is the total number of trapped atoms. V_{1e} decreases with decreasing T, reflecting the compression of the gas towards the trap center upon cooling. Similarly to V_{1e} one may define the effective sample radius R_{1e} and length L_{1e}. For a typical configuration used in Amsterdam and $T = 100$ mK, we have $V_{1e} \approx 0.25$ cm^3, $L_{1e} \approx 4$ cm and $R_{1e} \approx 0.1$ cm. Hence the samples have a pencil-like shape.

The trapped samples always remain dilute from a gas collisional point of view. With an elastic (s-wave) cross section $\sigma_{el} = 8\pi a^2 = 1.3 \times 10^{-15}$ cm^2, even for the highest densities achieved with trapped H ($n_0 \approx 10^{14}$ cm^{-3}) the collisional mean free path $\ell_c = (n_0 \sigma_{el})^{-1} \approx 7.5$ cm and we have

$$\ell_c \gg R_{1e}. \tag{5}$$

Although the gas is dilute, the elastic collision time of the atoms in the trap is short in comparison to the experimental time scale τ_{lab},

$$\tau_c \ll \tau_{lab}. \tag{6}$$

This is an essential condition for thermalization of the gas. The thermalization rate is of the order of the collision rate per atom $\tau_c^{-1} \sim n_0 \bar{v} \sigma_{el}$, where $\bar{v} = [8 k_B T/\pi m]^{1/2}$ is the average thermal speed. For $n_0 = 10^{12}$ cm^{-3} and for $T = 100$ mK we have $\tau_c \approx 1$ s.

STABILITY OF H↑ GAS

The processes that limit the stability of gaseous H↑ were analyzed and calculated in detail by Kagan et al. [11], Lagendijk et al. [12] and Stoof et al. [13]. The main causes for decay are spin exchange and magnetic dipolar interactions between the atoms. Spin exchange is most efficient (provided the hyperfine mixing angle is not too small) but only leads to relaxation in collisions between two c state atoms. For c-d collisions spin-exchange only proceeds via odd partial waves which are not populated under condition (1). Hence, in the case of d-d and c-d collisions, relaxation proceeds more slowly because it is induced by the relatively weak dipolar interaction. This implies a preferential depletion of the c-state component and ultimately leads to a pure d-state gas [12]. For a pure d-state sample the decay rate, $\tau_{dip}^{-1} \sim n_0 \langle G_{dd} \rangle$, where $\langle G_{dd} \rangle$ is the loss rate constant for dipolar relaxation averaged over the trap. For the trap used by Setija et al.[10] with $n_0 = 10^{12}$ cm^{-3} and $T = 10$ mK the average yields $\langle G_{dd} \rangle \approx 2 \times 10^{-15}$ cm^3/s and $\tau_{dip} \approx 1000$ s. For spin-exchange in a pure c-state gas under the same conditions $\langle G_{cc} \rangle \approx 10^{-13}$ cm^3/s and $\tau_{ex} \approx 20$ s. The dipolar relaxation rate has been established experimentally by Van Roijen et al. [8] and Masuhara et al. [14]. The spin exchange rates have not been measured in magnetic traps, thus far.

Comparing the dipolar decay rate with the elastic collision rate it is clear that

$$\tau_c \ll \tau_{dip}. \tag{7}$$

Hence, thermal equilibrium can be achieved without a severe loss of sample. This condition is expected to remain satisfied even down to 1 μK.

EVAPORATIVE COOLING

Hess et al. [7] allowed atoms to escape from a magnetostatic trap across a magnetic potential energy barrier of height $\epsilon_{tr} \equiv \eta k_B T$. It was established, by measuring the quantity of gas remaining in the trap after a certain holding time, that this procedure causes the gas to cool to a temperature well below the temperature of the surrounding walls. This method, in which the most energetic atoms are preferentially removed from the gas, is known as *evaporative cooling*, and was also proposed in ref.[6]. To ensure that the atoms are removed at high enough energy, η should be chosen sufficiently large. As the evaporation rate is given by $\tau_e^{-1} \sim \eta \tau_c^{-1} e^{-\eta}$, it is possible to remove the atoms under quasi-equilibrium conditions, slow in comparison to the elastic collision rate (thermalization) τ_c^{-1} but fast as compared to the dipolar decay rate

$$\tau_c \ll \tau_e \ll \tau_{dip}. \tag{8}$$

Magnetic relaxation leads to heating because the atoms preferentially relax near the center of the trap where the density is highest and are hence removed at lower than average potential energy. For a fixed ϵ_{tr} value η increases with decreasing temperature and the evaporation rate is suppressed exponentially until the cooling stops at a temperature where the cooling due to evaporation is balanced by the relaxation heating. In the experiments of Hess et al. [7] this point was reached at $T = 40$ mK. The dynamics of this type of evaporative cooling was studied optically by Luiten et al. [9].

To assure evaporation at a constant rate while the temperature is decreasing, ϵ_{tr} has to be ramped down in order to keep η constant. This procedure was introduced by Masuhara et al. [14] and is known as *forced evaporative cooling*. The method was further investigated by Doyle et al. [15] who also reported the highest density achieved with this method $n_0 = 8 \times 10^{13}$ cm^{-3} at $T \approx 0.1$ mK [16]. This corresponds to $n\Lambda^3 \approx 0.4$, within one order of magnitude from the degenerate quantum regime.

OPTICAL EXPERIMENTS

Optical spectroscopy of H↑ in magnetostatic traps was introduced by Luiten et al.[9] in Amsterdam, thus enabling the *in situ* determination of n_0 and T of the trapped gas. Optical excitation of H↑ involves Lyman-α (L$_\alpha$) transitions from the $1^2S_{1/2}$ electronic ground state to the 2^2P excited states, with an energy separation of 10.2 eV. The corresponding wavelength $\lambda_\alpha = 121.56$ nm is located in the vacuum ultraviolet (VUV) part of the spectrum and puts severe practical limitations on the experimental possibilities [17, 18]. The light source used in Amsterdam [9, 10] is based on non-resonant third-harmonic generation of L$_\alpha$ using frequency-doubled pulse-amplified light from a tunable cw dye-laser operated at 729.4 nm. The amplifiers are pumped by a XeCl excimer laser. A detailed description is given elsewhere [19]. The source yields typically 2×10^9 L_α photons per 10 ns pulse in a 100 MHz bandwidth at a repetition rate of 50 Hz ($\sim 10^{-6}$ duty cycle). The light polarization is adjustable. About 3×10^7 photons/pulse reach the sample. For a beam of 1 mm in diameter this corresponds to a pulse intensity of 0.25 W cm^{-2}, well below the saturation level of 3.6 W cm^{-2} for the L_α transitions. Spectra are recorded by sweeping the frequency of the dye-laser and observing the transmission of the L_α light with a GaAsP photodiode in the experimental cell (at $T < 0.2$ K).

In a magnetic trap the Zeeman effect lifts the degeneracies of the $1S$ and $2P$ levels and causes the transition frequency to become position dependent. For a magnetically trapped gas this results in inhomogeneous broadening of the spectral lines. Further broadening is caused by the Doppler effect as a result of the thermal motion of the trapped atoms. Starting from the hyperfine d state there are five allowed electric dipole transitions, labeled σ_1, π_1, σ_2, π_2 and σ_3 in the notation of Hijmans et al.[18]. The broadening mechanisms in combination with the total absorption enable the determination of both n_0 and T by absorption spectroscopy at least down to a few millikelvin. As the gas is spin-polarized, the transition probabilities depend on the angle between the wavevector $\mathbf{k}_\alpha$ and the direction of the $\mathbf{B}(\mathbf{r})$ field as well as on the polarization vector of the incident light. Both the amplitude and the polarization of the light change while propagating through the sample. In this paper we emphasize optical cooling in hydrogen rather than the hydrogen spectroscopy. We discuss both Doppler cooling and light-induced evaporative cooling.

DOPPLER COOLING

Doppler cooling of H↑ was demonstrated by Setija et al.[10] using a *single* beam of Lyman-α radiation from the pulsed narrow band VUV light source, described above, for σ_1 excitation. The σ_1 transition ($1^2S_{1/2}, m_j = \frac{1}{2} \to 2^2P_{3/2}, m_j = \frac{3}{2}$) provides a closed optical pumping cycle (for d-state atoms). All other L_α transitions have branching ratios that lead to H↓ and escape from the trap. To minimize the optical pumping losses the σ_1 transition is spectrally isolated from the other transitions by applying a small offset field $B_0 = 0.1$ T at the center of the trap. Moreover, circularly polarized light is used. We estimate the optical pumping loss for d-state atoms to be ~ 0.3 % per scattered photon.

As is already clear from the description of the light source, Doppler cooling of H↑ differs in several ways from the usual implementation of Doppler cooling in ultra-cold gases. Doppler cooling of H↑ is only done with trapped gas. The loading is done by cryogenic methods. The cryogenic environment provides the the best conceivable background vacuum and enables long trapping times. As the gas is trapped anyhow, unusually long Doppler cooling times τ_D (up to 15 minutes) are available. This allows the use of *pulsed* light with a duty factor $\sim 10^6$ times smaller than typical in experiments with alkali atoms. The cryogenic filling method yields high densities of trapped gas, $n_0 = 10^{11} - 10^{14}\,\mathrm{cm^{-3}}$. At such densities condition (6) is satisfied, $\tau_c \ll \tau_D$, which means that, during Doppler cooling, we have a quasithermal gas. A nice feature of the slow cooling is that T and n_0 can be followed while the gas cools. The trapped situation and the typical densities also allow the use of only a *single* beam of L_α radiation, relying on thermalizing collisions to assure three dimensional cooling. Interestingly, as the duration of the L_α pulses is long in comparison to natural lifetime $\tau_\alpha = \Gamma^{-1}$, the cooling process may still be described as if we were dealing with cw radiation, accounting for the pulsed character of the light by multiplying the pulse intensity with the experimental duty factor.

To describe laser cooling of H↑ we use the approach of Wineland and Itano [20], who described Doppler cooling of a thermal gas of neutral atoms with a single beam of cw radiation at an intensity far below the saturation limit. Many aspects of their approach apply to our experimental situation. However, for our case we have to deal with atoms in an inhomogeneous magnetic field and with optically thick samples.

To find the Doppler cooling rate τ_D^{-1}, one has to determine the change in internal energy of the gas per scattered photon. If the density of the gas is sufficiently low,

$$n_0 \lambdabar_\alpha^3 \ll 1, \qquad (9)$$

polariton effects may be neglected [21] and the change in internal energy can be obtained with the expressions for conservation of energy and momentum in the scattering of a photon from an isolated atom. As $\lambdabar_\alpha \equiv \lambda_\alpha/2\pi$ is very small, in the case of hydrogen the isolated atom approximation only breaks down for $n_0 \gg 10^{14}$ cm^{-3}, well above the densities employed in our cooling experiments. The energy change of an isolated atom, averaged over all scattering angles is then $\Delta E_{kin} = \hbar(\omega - \omega') = 2E_r + \hbar \mathbf{k} \cdot \mathbf{v}$, where $\mathbf{v}$, $\mathbf{k}$ and $\omega = c|\mathbf{k}|$ are respectively the velocity of the atom, the photon wavevector and the resonant frequency in the laboratory reference frame, before scattering; ω' is the photon frequency after scattering. $E_r = \hbar^2|\mathbf{k}|^2/2m$ is the photon recoil energy (for hydrogen, $E_r/k_B = 0.64\,\mathrm{mK}$). Integrating over the total volume of the gas the total rate of change of internal energy U of the gas due to Doppler cooling is given by the following expression,

$$\dot{U}(\omega,T) = 2E_r \int_V \frac{I(\mathbf{r},\omega,T)}{\hbar\omega} \left(1 - \frac{k_B T}{\hbar}\frac{\partial}{\partial \omega}\right) \sigma_V(\mathbf{r},\omega,T) n(\mathbf{r},T)\, d^3r \qquad (10)$$

where $\sigma_V(\mathbf{r},\omega,T)$ is the Doppler broadened absorption cross section (Voigt lineshape - see [22]) of an atom at position $\mathbf{r}$ and the function $I(\mathbf{r},\omega,T)$ describes the intensity profile of the light beam. Note that the intensity at site $\mathbf{r}$ depends on the extinction and therefore on the frequency of the incident radiation ω

and the sample temperature T. In general $I(\mathbf{r},\omega,T)$ has to be obtained from a numerical solution of the equations describing the propagation of the electromagnetic field through the sample. The first term in Eq.(10) describes the recoil heating, i.e., the total absorption rate times $2E_r$. The second term describes the Doppler cooling rate of a thermal sample, which is seen to be proportional to the derivative of the lineshape $\partial\sigma_V/\partial\omega$. To have cooling at site $\mathbf{r}$ one should have $\partial\sigma_V/\partial\omega > 0$, i.e., the light should be red detuned with respect the local resonance frequency. For a given frequency ω the minimum temperature is reached when the Doppler cooling equals the recoil heating and $\dot{U}(\omega,T) = 0$ (not considering other heating mechanisms). The lowest temperature for which the condition $\dot{U}(\omega,T) = 0$ can be satisfied is known as the Doppler limit T_D.

Optically Thin Samples

Let us first turn to the case of very dilute gases where the incident light beam is not appreciably attenuated when passing through the sample, $I(\mathbf{r},\omega,T) \approx I_0(\mathbf{r})$. This is the regime in which magnetically trapped Na was investigated by Helmerson et al.[23]. For very dilute gases and eqn.(10) may be rewritten as

$$\dot{U}(\omega,T) = \left(1 - \frac{k_B T}{\hbar} \frac{\partial}{\partial \omega}\right) P(\omega,T), \qquad (11)$$

where $P(\omega,T) = 2E_r \int_V [I_0(\mathbf{r})/\hbar\omega]\sigma_V(\mathbf{r},\omega,T)n(\mathbf{r},T)\,d^3r$ is the total power absorbed from the incident light beam at frequency ω, i.e., the absorption spectrum observed experimentally at temperature T. The frequency for which the Doppler limit is reached follows from $d\ln P(\omega,T_D)/d\omega = \hbar/k_B T_D$. For most atomic systems the root mean square Doppler spread $\delta = |\mathbf{k}|\sqrt{k_B T_D/m}$ is negligibly small ($\delta \ll \Gamma$ at the Doppler limit) and the Voigt line may be replaced by the corresponding Lorentz line, $\sigma_V(\mathbf{r},\omega,T) \approx \sigma_L(\mathbf{r},\omega)$. For a homogeneous sample sample $\sigma_L(\mathbf{r},\omega) = \sigma_L(\omega)$ and the well-known expression $k_B T_D = \frac{1}{2}\hbar\Gamma$ is obtained. For hydrogen T_D/m is relatively large ($\delta \approx \frac{1}{2}\Gamma$) and T_D is slightly higher, $T_D \approx 0.62\,\hbar\Gamma/k_B \approx 3$ mK.

For inhomogeneous samples such as trapped H↑ the resonant frequency shifts with position and it is not a priori clear at what frequency optimal cooling is obtained. Tuning to a frequency $\omega < \omega_0$, where ω_0 corresponds to the σ_1 resonant frequency in the center of the trap, the light is red shifted with respect to the local resonant frequency anywhere in the trap. For $\omega > \omega_0$ the light is 'blue shifted' for the gas enclosed within a σ_1 resonant shell coinciding with an equipotential surface of the trap, and 'red shifted' for the gas outside this shell. As it turns out the optimal frequency depends strongly on the density n_0 of the trapped gas. For the optically transparent regime eqn.(11) applies and since $P(\omega,T)$ is inhomogeneously broadened with respect

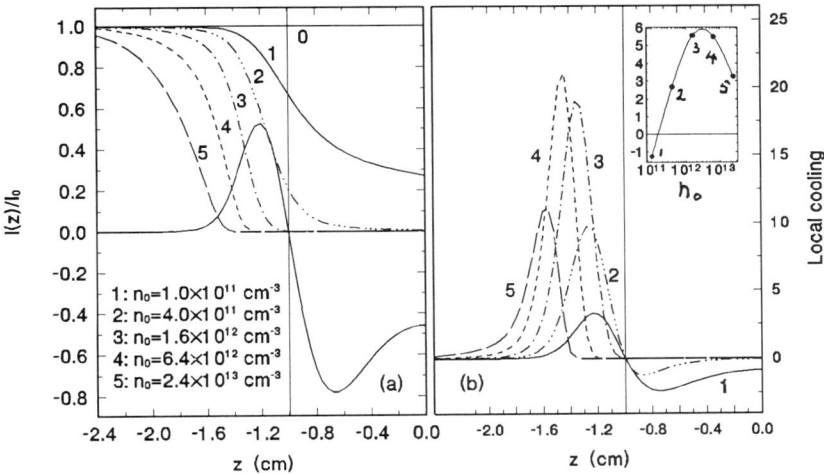

Figure 2: (a) Intensity profile along the z axis for densities ranging from the optically thin to the optically thick regime. Also plotted is the frequency derivative of the normalized absorption coefficient. (b) Local Doppler cooling per incident photon in units of $2E_r$ for the same densities as in (a). The inset shows the total cooling rate for these densities.

to $\sigma_V(\omega, T)$ the Doppler limit will be raised with respect to the homogeneous situation. For H↑ in the trap used by Setija et al.[10] $T_D \approx 0.69\,\hbar\Gamma/k_B \approx 3.3$ mK; this value is obtained for $\omega \approx \omega_0$.

Optically Thick Samples

The essential features of Doppler cooling in thermal gases with substantial extinction of the incident light but neglecting multiple scattering are illustrated in fig.2. Fig.2a shows the L_α intensity profile $I(\mathbf{r}, \omega, T)$ on the z axis of the trap of fig.1 for gas densities ranging from fully transparent (curve 0) to optically thick at Lyman-α (curve 5). Also shown is $\partial \sigma_V / \partial \omega$. The frequency ω of the incident light gives rise to a σ_1 resonant shell intersecting the z axis at $z = \pm 1$ cm. The gas temperature T is fixed at 40 mK. The local cooling rate is shown in fig.2b. The inset shows the total cooling rate obtained by numerical integration of eqn.(10) along the z axis. In accordance with the discussion in the previous section we find cooling outside the resonant shell ('red detuning') and heating on the inside ('blue detuning'); because $\omega > \omega_0$ the net effect is heating for the lowest densities (see inset). With increasing density more and more light is scattered before the resonant shell is reached; and cooling becomes dominant. The Doppler cooling rate is optimal if most

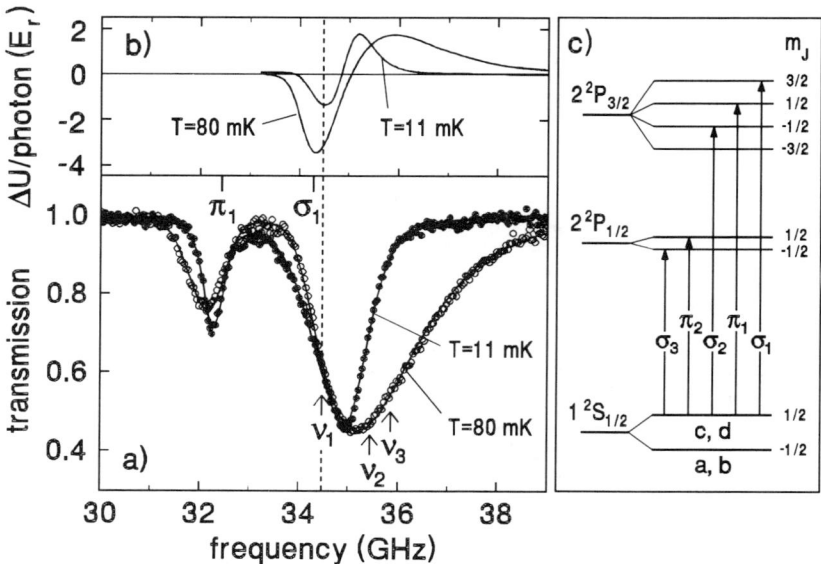

Figure 3: (a) Absorption spectrum before (open circles) and after (closed circles) Doppler cooling. The solid lines are calculated spectra. The frequency is relative to $(3/4)R_\infty(1 + m_e/m_p)$, where R_∞ is the Rydberg constant and m_e/m_p is the electron/proton mass ratio. The resonant frequencies for the σ_1 and π_1 transitions at the trap center are indicated by a small bar. (b) Change in internal energy U per incident photon in units of the recoil energy. (c) Energy level diagram showing the allowed L_α transitions in H↑.

of the light is absorbed just outside the resonant shell, where $\partial \sigma_V / \partial \omega$ is large. For the highest density all light will be scattered in the wing of the density distribution, far outside the resonant shell. Here $\partial \sigma_V / \partial \omega$ is small and cooling is not effective.

This example shows that during cooling the optimal cooling frequency ω_{opt} will change. Starting at high temperature under optically thin conditions, the optimal frequency $\omega_{opt} \approx \omega_0$. During cooling the sample will contract, causing n_0 to increase and the sample to become optically thick; ω_{opt} will increase and with it the corresponding resonant shell. As the sample continues to contract with further cooling, at a given stage also the resonant shell (and ω_{opt}) has to become smaller again until it reaches a minimum value at the Doppler limit. Numerical analysis shows that the Doppler limit itself hardly depends on n_0.

Clearly, when the extinction length becomes very short multiple scattering cannot be avoided, but, as we will show using the picture of resonant shells, its influence may be kept small by properly shaping the samples. The resonant shells will not be vanishingly thin but may be attributed a thickness $d_{ax} =$

$d_{ax}(\partial B/\partial z)^{-1}$ in the axial direction and $d_{rad} = \Delta_V(\partial B/\partial \rho)^{-1}$ in the radial direction, where Δ_V is the width of the Voigt line in magnetic field units. For the trap shown in fig.1, typically $d_{ax} \approx 50\, d_{rad}$, which means that the sample can be optically thick in the axial direction while remaining essentially transparent in the radial direction. As Doppler cooling is optimal when the extinction length of the incident light at the shell equals the shell thickness d_{ax} double scattering events will be rare, except in cases where the photons are initially scattered in forward direction. By numerical simulation it was established that for these elongated samples multiple scattering only starts to have a substantial effect on the Doppler cooling rate for densities $n_0 \gtrsim 10^{12}$ cm^{-3}.

Doppler cooling is demonstrated experimentally in fig.3a. We show two spectra, one taken before and one after irradiating the sample for 15 minutes at a fixed frequency indicated as ν_1. The observed spectral lines correspond to the σ_1 and π_1 transitions. The conditions are such that the diameter of the L$_\alpha$ beam is larger than the radial size of the sample so that the sample casts a shadow on the photodiode. The energy level diagram and allowed L$_\alpha$ transitions in H↑ are shown in fig.3c. The initial and final values of n_0 and T are obtained by a fitting procedure with calculated spectra: $T = 80(10)$ mK, $n_0 = 8(2) \times 10^{10}$ cm^{-3} and $T = 11(3)$ mK, $n_0 = 1.3(4) \times 10^{12}$ cm^{-3}. The cooling reflects itself in the narrowing of the absorption lines. The maximum extinction does not change very much as with decreasing temperature the shadow due to the sample on the photodiode becomes smaller and approximately compensates for the increase in optical thickness. The frequency ν_1 is chosen close to the maximum slope in the red wing of the σ_1 line. This frequency more or less optimizes the cooling rate along the full cooling trajectory as is illustrated in fig.3b where we show the change in internal energy of the gas per incident photon in units of the recoil energy.

LIGHT-INDUCED EVAPORATIVE COOLING

Aside from Doppler cooling, Setija et al.[10] used a new cooling method: Light-Induced Evaporation (LIE). This is an optical version of forced evaporative cooling where the removal of high energy atoms is accomplished by optical pumping to high-field seeking states. The method works only for optically thick samples and can be understood with the above described picture of resonant shells. In LIE, rather than using the closed cycle σ_1 transition, the sample is excited on the optical pumping transition σ_2 ($1S_{\frac{1}{2}}, m_j = \frac{1}{2} \rightarrow 2P_{\frac{3}{2}}, m_j = -\frac{1}{2}$) which preferentially decays back to high field seeking hyperfine states (the branching ratio is 2/3 for $B_0 = 0.1$ T in the trap center) and is very well spectrally separated from the other transitions. For the method to work efficiently the density should be sufficiently high ($n_0 \gtrsim 3 \times 10^{12}$ cm^{-3} for our trap) to fully absorb the incident light outside a resonant shell located in the tail of the density distribution at a potential energy far above $k_B T$. To achieve the large starting densities and low initial temperatures we used regular forced evaporative cooling. By filling a 320 mK deep trap and subsequently lowering the trap barrier to 57 mK, we were able to achieve initial densities in the range 10^{12}–10^{13} cm^{-3} and initial temperatures between 9 and 13 mK. With LIE in H↑ we reached temperatures $\sim 3(1)$ mK, starting from $T = 11(2)$ mK and $n_0 = 4.8(2.1) \times 10^{12}$ cm^{12}. To reliably handle samples at lower temperatures we should get better control over the thermometry at these very low temperatures. In principle it should be possible to use this method down to temperatures well below the recoil limit.

ACKNOWLEDGMENTS

One of the authors (JTMW) wishes to thank the Joint Institute for Laboratory Astrophysics (JILA) in Boulder for hospitality and assistance during the preparation of the manuscript. The hydrogen research in Amsterdam is supported by the Stichting voor Fundamenteel Onderzoek der Materie (FOM), by the Nederlandse Organisatie voor Wetenschappelijk Onderzoek (NWO-PIONIER) and by the EEC-SCIENCE program.

References

[1] See for example L.D. Landau and E.M. Lifshitz, *Statistical Mechanics*, (third edition) Pergamon, New York 1980.

[2] T.J. Greytak and D. Kleppner, in: *New Trends in Atomic Physics*, Vol.II G. Grynberg and R. Stora (Eds.), Elsevier Sci. Publ., Amsterdam (1984) 1125.

[3] I.F. Silvera and J.T.M. Walraven, Prog. Low Temp. Phys., Vol. X, D.F. Brewer (Ed.), Elsevier Sci. Publ. (1986) 139.

[4] I.F. Silvera and M.W. Reynolds, J. Low Temp. Phys. **87** (1992) 343.

[5] I.F. Silvera and J.T.M. Walraven, Phys. Rev. Lett. **44** (1980) 164.

[6] H.F. Hess, Phys. Rev. B **34** (1986) 3476.

[7] H.F. Hess, G.P. Kochanski, J.M. Doyle, N. Masuhara, D. Kleppner and T.J. Greytak, Phys. Rev. Lett. **59** (1987) 672.

[8] R. van Roijen, J. Berkhout, S. Jaakkola and J.T.M. Walraven, Phys. Rev. Lett. **61** (1988) 931.

[9] O.J. Luiten, H.G.C. Werij, I.D. Setija, M.W. Reynolds, T.W. Hijmans and J.T.M. Walraven, Phys. Rev. Lett. **70** (1993) 544.

[10] I.D. Setija, H.G.C. Werij, O.J. Luiten, M.W. Reynolds, T.W. Hijmans and J.T.M. Walraven, Phys. Rev. Lett. **70** (1993) 2257.

[11] Yu. Kagan, I.A. Vartanyantz and G.V. Shlyapnikov, Sov. Phys. JETP **54** (1981) 590.

[12] A. Lagendijk, I.F. Silvera en B.J. Verhaar, Phys. Rev. B **33** (1986) 626.

[13] H.T.C. Stoof, J.M.V.A. Koelman and B.J. Verhaar, Phys. Rev. B **38** (1988) 4688.

[14] N. Masuhara, J.M. Doyle, J.C. Sandberg, D. Kleppner and T.J. Greytak, Phys. Rev. Lett. **61** (1988) 935.

[15] J.M. Doyle, J.C. Sandberg, N. Masuhara, I.A. Yu , D. Kleppner and T.J. Greytak, J. Opt. Soc. Am. B **6** (1989) 2244.

[16] J.M. Doyle, J.C. Sandberg, I.A. Yu, C.L. Cesar, D. Kleppner and T.J. Greytak, Phys. Rev. Lett. **67** (1991) 603.

[17] P.D. Lett, P.L. Gould and W.D. Phillips, Hyp. Int. **44** (1988) 335.

[18] T.W. Hijmans, O.J. Luiten, I.D. Setija and J.T.M. Walraven, J. Opt. Soc. Am. B **6** (1989) 2235.

[19] O.J. Luiten, H.G.C. Werij, M.W. Reynolds, I.D. Setija, T.W. Hijmans and J.T.M. Walraven, Feature issue on high resolution laser spectroscopy, Applied Physics B, to be published in 1994.

[20] D.J. Wineland and W.M. Itano, Phys. Rev. A **20** (1979) 1521.

[21] B.V. Svistunov and G.V. Shlyapnikov, Sov. Phys. JETP **70** (1990) 460.

[22] See for example: L. Allen and J.H. Eberly, *Optical Resonance and Two-Level Atoms*, John Wiley, New York 1975.

[23] K. Helmerson, A. Martin and D.E. Pritchard, J. Opt. Soc. Am. **9** (1992) 1988.

ATOMS IN INTENSE FIELDS

Atoms in Intense Laser Fields

P.B. Corkum, N.H. Burnett, P. Dietrich and M. Y. Ivanov

Steacie Institute for Molecular Sciences
National Research Council of Canada
Ottawa, Ont. K1A 0R6
Canada

Abstract: We describe theory and experiment of ultrashort laser pulses ionizing atoms. We relate three phenomena: harmonic emission, correlated double ionization and high kinetic energy photo-electron production. The experimental ellipticity dependence is consistent with double ionization and high harmonic generation both having a common origin.

Introduction:

Just as lasers lead to the rapid development of perturbative nonlinear optics, and indeed to development of nonperturbative theories, intense ultrashort pulse sources have provided an impetus to the experimental development of nonperturbative nonlinear optical regimes and to the renaissance of associated theories. In the high photon number limit, where nonperturbative theories are required, there are many quantum states that superimpose and the system often evolves classically.

In this paper we will make use of the insight that classical mechanics allows, to explain a number of strong field processes. These are: the electron kinetic energy in a multiphoton ionization processes (above-threshold-ionization), the harmonics emitted during the ionization process, and the process of double ionization.

This understanding also allows us to identify new applications[1]. One general approach to finding applications is to identify means of controlling high field processes. Any nonlinear process will be sensitive to the structure of the optical field and the optical field structure can be changed by superimposing two or more optical pulses with related frequencies and, or phases. The modification of nonlinear processes by varying the relative phase of two or more beams is often known as "coherent control". It will be the subject of another paper in this volume.

A Free Electron in a laser field:

An electron at rest before being illuminated by a pulse of plain wave linearly polarized ($\alpha=0$) radiation that is turned on adiabatically

$$E(t) = E_0\cos(\omega t)e_x + \alpha E_0\sin(\omega t)e_y \qquad (1)$$

will oscillate in the field with a velocity $v(t)$ and an amplitude $x(t)$

$$v(t) = \frac{qE_0}{m\omega}\sin(\omega t) \quad (2)$$

$$x(t) = \frac{qE_0}{m\omega^2}\cos(\omega t) \quad (3)$$

It will also experience motion in the direction of propagation. In a weak field:

$$v_z(t) = \frac{q^2 E_0^2}{2m^2\omega^2 c}\sin^2(\omega t) \quad (4)$$

$$z(t) = \frac{q^2 E_0^2}{2m^2\omega^3 c}[\frac{1}{2}\omega t - \frac{1}{4}\sin(2\omega t)] \quad (5)$$

It is useful to note the magnitude of this motion. For 10^{15}Wcm^{-2} of 800 nm light $x=32\cos(\omega t)$ Angstroms and $z=0.4[\frac{1}{2}\omega t - \frac{1}{4}\sin(2\omega t)]$ Angstroms.

Of course after the laser pulse has passed, the electron returns to rest since energy and momentum conservation requires that a free electron can not absorb energy in a single photon event. It can, however be displaced from its original position.

A Semi-Classical Perspective on Strong Field Atomic Processes:

The above discussion is only valid for a free electron. However, we can use the basic classical approach to help understand atomic ionization processes. Consider the following procedure[2]. First one determines the probability of ionization as a function of the laser phase using tunnel ionization models. For all calculations in this paper, we will assume that the ionization rate is given by[3]:

$$W_{dc} = \omega_s |C_{n^*l^*}|^2 G_{lm}(4\omega_s/\omega_t)^{2n^*-m-1}\exp(-4\omega_s/3\omega_t) \quad (6)$$

where $\omega_s = E_s/\hbar$, $\omega_t = eE(2m_e E_s^0)^{-1/2}$, $n^* = (E_s^h/E_s^0)^{1/2}$, $G_{lm} = (2l+1)(l+|m|)!/(2^{|m|}|m|!(l-|m|)!)$, $|C_{n^*l^*}|^2 = 2^{2n^*}[n^*\Gamma(n^*+l^*+1)\Gamma(n^*-l^*)]^{-1}$. In Eq. (6), E_s^0 is the ionization potential of the atom of interest, E_s^h is the ionization potential of hydrogen, l and m are the azimuthal and magnetic quantum numbers, E is the electric field amplitude. The effective quantum number l^* is given by $l^*=0$ for $l<<n$ or $l^*=n^*-1$ otherwise. The tunnelling model describes the formation of a sequence of wave packets, one at each peak of the laser electric field.

As a second step classical mechanics can be used to approximately describe the dynamical evolution of an electron wave packet. For simplicity, we consider only the electric field of the laser, ignoring, for example, the electric field of the nucleus. The initial conditions of velocity and position equal 0 (the position of the nucleus) at the time of ionization have been justified previously in the long wavelength limit by the comparisons with above-threshold ionization experiments. The electron motion in the field for all times subsequent to tunnelling is given by:

$$v_x = v_0 \sin(\omega t) + v_{0x}, v_y = -\alpha v_0 \cos(\omega t) + v_{0y} \qquad (7)$$

$$x = x_0[\cos(\omega t) + t] + x_{0x}, y = \alpha x_0[\sin(\omega t) + \alpha t] + y_{0y} \qquad (8)$$

where $v_0 = qE_0/m_e\omega$, $x_0 = qE_0/m_e\omega^2$, α describes the polarization of the light with $\alpha=1$ describing circularly polarized light and v_{0x}, v_{0y}, x_{0x} and y_{0y} are determined by the initial conditions at the moment of birth. Setting $\alpha=0$, equations 7 and 8 differ from Eqs 2 and 3 through the constants of integration, v_{0x}, v_{0y}, x_{0x} and y_{0y}. The motion described by these constants of integration remain even after a plane wave short pulse has passed. The energy associated with the velocities v_{0x} and v_{0y} constitute the energy of the photo-electrons that are produced. That is, it is the quantity known as the "above threshold ionization" energy in ultrashort pulse experiments. Equation 8 indicates that, for circularly polarized light, the electron trajectory never returns to the vicinity of the nucleus.

It should be noted that there is a second purely quantum aspect to the dynamical evolution of the electron wavepacket produced by tunnelling electrons which is not described by these classical trajectory considerations. That is to say, the free electron wavepacket will spread by diffraction at a rate determined by quantum mechanics. In principle this rate of spread can be estimated by applying the quantum uncertainty principle in time-energy or position-momentum form to the initial tunnelling event[2]. We will return to this point which although not crucial to final electron energy distributions is very important in determining the strength of re-interaction of electrons with their parent subsequent to tunnelling.

When introduced[2], it was thought that the quasi-static approach was valid only in the long wavelength limit. However, the experiments of Mohideen et al[4] indicate that Eq. (6) accurately describes electron energy distributions produced when light of 800 μm wavelength is used to ionize Helium. Figure 1 shows the calculated above-threshold ionization spectrum obtained using the above equations. With no free parameters, the difference between experiment and model is very small. Multiplying the intensity axis for the model by 0.8 results in excellent agreement. With the experiment of Mohideen et al. as justification we can safely use the above procedure in what follows.

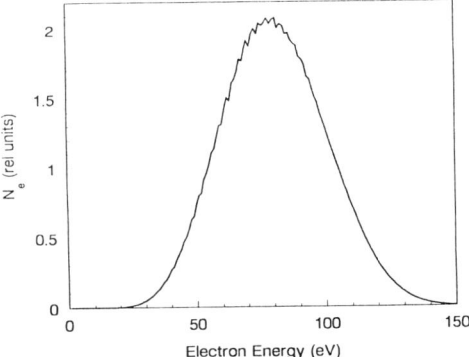

Figure 1: Above threshold ionization spectrum calculated for ionization of helium with circularly polarized 800 nm light.

For linearly polarized light we can show[5] that half of the electrons will pass close to the nucleus once during the first laser period following ionization. The remaining electrons never return. Using the semiclassical model outlined above, we can determine the energy distribution of electrons (wavepackets) that do re-encounter the nucleus (Fig. 2).

Figure 2 shows that the most likely velocity of an electron passing the nucleus corresponds to an instantaneous kinetic energy of 3.17 times the ponderomotive potential (3.17U_p). An electron ionized by tunnelling at $\omega t \approx 17, 197$, etc. degrees will arrive at the nucleus with this velocity.

One consequence of this interaction can be immediately understood. If the 3.17U_p electron energy exceeds the e-2e scattering energy, ejection of a second electron should be observed. Figure 3 shows the calculated ion yields obtained as a function of the laser intensity for 0.6μm light interaction with helium. To obtain this curve, the known collision cross-section of He$^+$ was used. The only free parameter in the model is the transverse spread of the electron wave function, or equivalently, the range of possible impact parameters. To obtain Fig. 3 wave function radius of 2 Å was used. Figure 3 agrees very well with recent experiment[6], however this value of the wavefunction radius is considerably (by about a factor of four) smaller than that estimated from simple quantum principles. This implies that either additional sources contribute to double ionization or that the collision cross section differs greatly from the field free value. Spin correlation and ionization of excited atoms are at least partially responsible for the difference.

The electron can also scatter elastically. Any electron that scatters, either elastically or inelastically, absorbs photons leading to very hot photo-electrons[5].

There are other consequences of the electron-nuclear interaction[5]. Classically, we know that as the electron passes the nucleus, it will be accelerated,

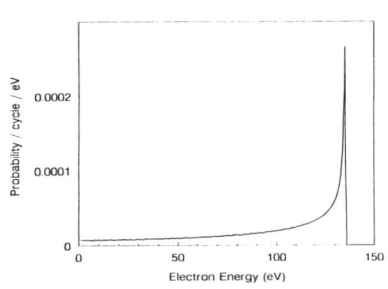

Figure 2: Velocity distribution for electrons passing the ion. Laser intensity 7×10^{14} W/cm^2 and wavelength 800 nm.

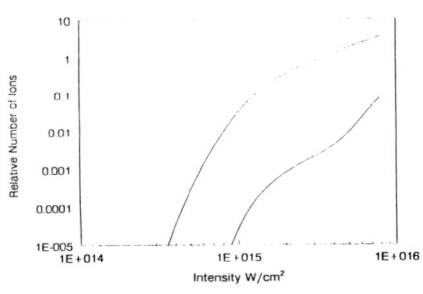

Figure 3: Ion yield for helium plotted as a function of the peak laser intensity for 625 nm, 100fs pulses.

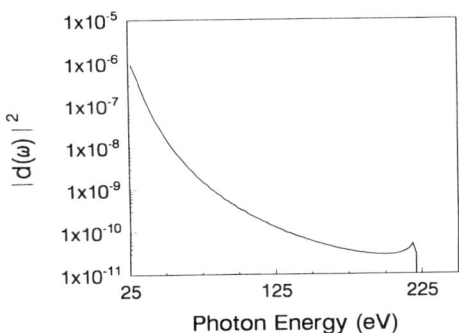

Figure 4: Dipole moment squared in atomic units calculated for 1.06µm light, helium and an intensity of 6×10^{14} W/cm^2.

leading to emission of light. If the ground state is not significantly depleted, the wave packet will pass the nucleus in the same way during each cycle of the laser light. Thus, any light that is emitted will be at a harmonic of the laser frequency. The emission can be calculated from the expectation value of the dipole operator $<\psi|er|\psi>$. If we assume that $\psi=\psi_g+\psi_c$ where ψ_g is the ground state wave function and ψ_c is the continuum wave function then the dipole moment can be rewritten as $<\psi|er|\psi>=<\psi_g|er|\psi_c> + <\psi_c|er|\psi_c>$ + cc. Figure 4 shows the calculated harmonic spectrum of the absolute value of the dipole moment squared. Neglecting continuum-continuum contributions, $<\psi_c|er|\psi_c>$. (We shall return to this assumption below.) The calculation was performed for 1 µm light interacting with helium just as in a recent Schrodinger equation simulation of high harmonic generation[7]. The only "free" parameter in this calculation is the wave function spread and it was assumed to be the same as that used for the double ionization calculation above.

Figure 4 bears a remarkable similarity with the results of the Schrodinger equation simulation. The plateau region has the same structure, beginning at 10^{-6} and decaying in an extended plateau until about a photon energy of 225 eV. Even the magnitude of the high harmonics agrees within less than an order of magnitude with the Schrodinger equation simulations. Clearly the quasi-static model catches the essence of high-harmonic generation.

Ellipticity Dependence of Strong Field Atomic Processes:

The semiclassical picture gives a very physical way of looking at the ellipticity dependence of strong field processes. Consider an elliptically polarized laser field with the major axis along the x-axis, $E_x(t) = E_{ox} \cos \omega t$ and a weak perpendicular component $E_y(t) = E_{oy} \sin \omega t$ with ellipticity $\varepsilon \equiv E_{oy}/E_{ox} \ll 1$. Since the tunnelling probability depends exponentially on the width of the potential barrier, the electron tunnels mostly near the peak of the field E_x when $|\cos \omega t| \approx 1$. After ionization the electron drifts in y-direction. By the time the electron can return to the position of the ion its displacement in the y-direction from the origin is $\Delta y = [-\sin \omega t + \sin \omega t' + (\omega t - \omega t')\cos \omega t']\varepsilon eE_{ox}/m\omega^2$ where t' is the moment the electron is born in the continuum and t the time the electron returns to the ion. Electrons that return to the ion with energies close to 3 U_p and produce high harmonics are born around $\omega t' \sim 0.3$ and come back at $\omega t - \omega t' \sim 4$, which gives $\Delta y \approx 5\varepsilon eE_{ox}/m\omega^2$. When the returning electron wavepacket is displaced by more than its transverse dimension harmonic production will be quenched. For example, at laser intensity I~10^{15}W/cm^2, ellipticity=0.1 and ω in the optical frequency range, $\Delta y \approx 10$Å.

Under conditions where the wave function spread is much in excess of atomic dimensions, the yield of harmonic radiation or double ionization will simply be a map of the returning electron density in the nuclear vicinity, for electrons with appropriation energy to produce these effects. For example, the highest harmonics will have a yield $Y(\varepsilon)$ which can be directly equated with $\psi_c^2(r)$ where $r \approx 5\varepsilon e E_{ox} / m\omega^2$. From an experimental perspective ellipticity measurements provide a direct test of models used to predict the nonlinear polarizability of ionizing gases, a test which is insensitive to phase matching (since small changes in the ellipticity do not change the focusing properties of the laser beam nor the ionization rate).

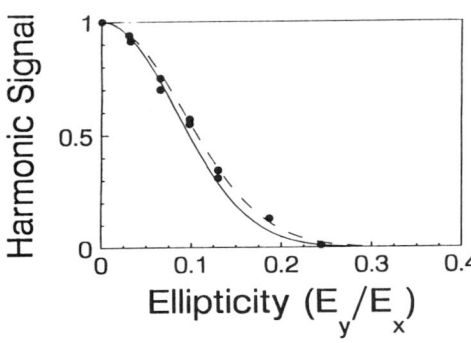

Figure 5: Harmonic yield of the 41st harmonic of neon plotted as a function of the ellipticity of the fundamental 775 nm radiation.

We now describe the experimental studies of the ellipticity dependence of high harmonic emission performed with a compact all Ti:Sapphire laser system. We refer the reader to a similar experiment has recently been published [8]. The output of the cw mode-locked oscillator is amplified using chirped pulse amplification in a regenerative amplifier followed by a two-pass final amplifier to produce 10 mJ, 200 fs pulses at a repetition rate of two Hertz and wavelength of 775 nm. After compression in a two-grating double-pass compressor the laser pulses are linearly polarized to better than 500:1 in intensity.

Ellipticity is introduced by rotating a zero order λ/4 waveplate placed just before the focusing lens and target chamber window. The pulses are focused by an f/30 thin lens onto a 3.1mm diameter stainless steel tube with 150 μm thick walls which has been squeezed to an internal thickness of 750 μm. A 150 μm diameter hole is drilled by slightly defocusing the laser and irradiating the target for a few hundred shots. The laser is then focused to 75 μm diameter for harmonic production. The laser intensity averaged over its 90% energy circle is estimated to be 8x10^{14} Wcm^{-2}. The target tube is transiently filled with approximately one hundred mbar of Ne with a fast pulsed valve.

The harmonic radiation is dispersed with a 1200 gr/mm Hitachi variable space grating with a slit placed in its focal plane to isolate one harmonic. Harmonic yield is monitored on an electron multiplier placed behind the slit. Rotating the λ/4 waveplate rotates the major axis of the polarization ellipse with respect to the plane of incidence of the diffraction grating. At the experimental angle of incidence (87°) the change in the diffraction efficiency due to this effect is negligible for the present purposes. Zero order (specular reflection) radiation from the grating is collected and transmitted through a window to monitor the third harmonic yield (by means of a grating spectrograph and UV photomultiplier).

Using this apparatus, harmonic radiation extending out to $N > 41$ in neon.

Harmonic yield as a function of the ellipticity ε is shown as data points in Fig. 5 for the 41st harmonic in neon. Data were also collected from other harmonics in the plateau region neon. It did not significantly differ from $N=41$. The curves plot theoretical results which will be discussed below.

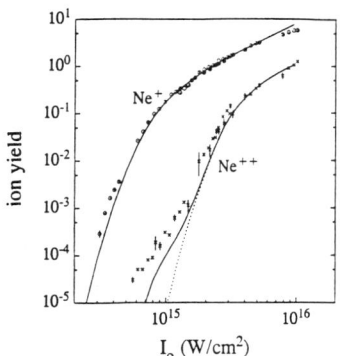

Figure 6: Yield of singly and doubly charged neon ions plotted as a function of the peak laser intensity of a 625 nm ionization beam.

We now describe our study of two-electron multiphoton ionization. The experiment uses the amplified output of a colliding-pulse mode-locked laser operated at 625 nm. The 70 fs pulses are amplified in a dye laser amplifier chain pumped by a frequency-doubled Nd:YAG laser and then compressed in a two-grating, single-pass pulse compressor. A vacuum spatial filter combined with an aperture to select the central part of the Airy pattern ensures a good spatial profile. A small portion of the beam is sampled to measure the pulse energy and to produce second harmonic radiation in a KDP crystal whose energy is also measured. The pulses have an energy of 150-200 μJ and a duration of about 100 fs measured with a single shot autocorrelator. Just before entering the vacuum chamber, the pulses pass through a linear polarizer and a λ/4 wave plate which can be rotated to change the ellipticity of the pulse. The residual ellipticity in the case of linear polarization is $ε ≤ 0.02$.

The pulses are focussed inside the time-of-flight mass spectrometer using an on-axis paraboloidal mirror with 50 mm focal length. We use f/5 focussing geometry resulting in a focal diameter of ~ 3 μm. Neon is leaked into the vacuum chamber with a background pressure of $6 × 10^{-9}$ mbar. The operating pressure was adjusted between 10^{-7} mbar and 10^{-5} mbar depending on the pulse intensity. The ions are extracted by an electric field of 500 Vcm^{-1} and detected by a microchannel plate. The amplified signals are integrated using boxcar integrators. The signals from both ion channels were measured without neon in the chamber and subtract from the actual neon signals for background suppression. Furthermore we measured the boxcar output between laser pulses to correct for drifting offsets of the boxcar integrators.

The experimental data from all pulses were binned according to their energy, $E_{1ω}$, and discriminated on the basis of the energy of the second harmonic, $E_{2ω}$. Only pulses with $α = E_{1ω}^2/E_{2ω}$ within ± 10% of a preset value were accepted. For pulses whose temporal shape can be described by one parameter τ, α is proportional to the pulse duration τ. This selection ensures the reproducibility of the data. Each data point shown in Fig. 6 is an average of typically 1000 shots. The error is the statistical error of the mean.

The intensity dependence of the ionization yields of Ne^{n+} (n = 1,2) in the range from 10^{14} - 10^{16} Wcm^{-2} for linear polarization is shown in Fig. 6. The intensity scale was obtained by fitting the experimental Ne$^+$ yields with the calculated yields from the tunnel ionization model by scaling both intensity and yields. The experimental data for Ne^{++} show a clear bump at low yields. This increase of the ion yields compared to a simple tunnel ionization model is similar to published results and it has been proposed that it is the result of two-electron multiphoton ionization[6]. The curves show theoretical results which will be discussed below.

We now study the ellipticity dependence of the correlated two-electron multiphoton ionization signal by measuring the Ne$^+$ and Ne^{++} yields at a peak intensity of 9×10^{14} Wcm^{-2} as a function of the ellipticity. The Ne$^+$ yields (not shown) do not vary significantly over the measured range. The Ne^{++} yields decrease rapidly (data points in Fig. 7). The curves plot theoretical results that will be discussed below. The qualitative agreement between the measurements in Figs. 5 and 7 lends strong support to the hypothesis that the underlying physics investigated by both diagnostics is the same. However, any model of high harmonic generation or strong field multiphoton ionization must be *quantitatively* consistent with experimental results. The remainder of the paper will deal with this quantitative comparison.

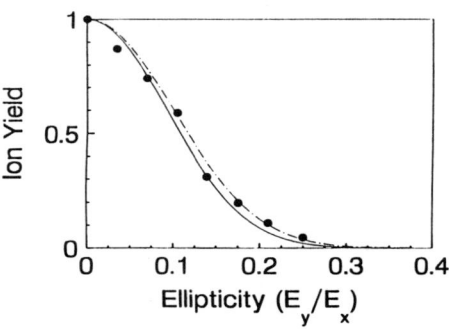

Figure 7: Yield of doubly charged neon ions obtained at a peak intensity of 9×10^{14} Wcm^{-2} plotted as a function of the ellipticity.

The signal measured in any harmonic generation experiment is the result of the interplay between the nonlinear response of the atom and the coherent interaction between the propagating wave produced at different positions in the focus (phase matching). Phase matching issues have been extensively studied [9] and there are two main sources of dephasing: (1) the natural phase advance that occurs in any beam as it passes through its geometric focus and (2) dispersion in the nonlinear medium. Both of these should be unchanged by the very small ellipticity characteristic of this experiment. Consequently, our measurements must reflect changes in the atomic nonlinear response.

Quantum Theory of High Harmonic Emission:

Quantum mechanically the initial kinetic energy of the electron in the continuum is not exactly zero and can compensate for the drift in y-direction. In other words, the quantum mechanical wave packet spreads in all directions. As noted above, the spread can be estimated by the position/momentum or the time/energy uncertainty relation[2]. For a quantitative fit we use a quantum mechanical generalization, based on a Keldysh-Faisal-Reiss-like approximation which is fully described in Ref [10]. We summarize the main features below and note the important equations.

The model quantum system includes one bound (ground) state $|g>$ and a 3D continuum of states labelled with their momentum $|\mathbf{p}>$. The model applicable in the region $U_p > I_p$ assumes that the electron motion in the continuum is dominated by the laser field, and that the depletion of the ground state is negligible. Under these assumptions the field induced dipole moment of the atom $d(t) = <\Psi(t)|\mathbf{d}|\Psi(t)>$ is determined by the overlap between the continuum part of the wave function and the ground state. Solution of the Schrodinger leads to the following expression (in atomic units) for the field-induced dipole moment of the atom.

$$d(t) = i \int_0^t dt' \int d^3p \ (E(t').d_g(p)) \ d_g^+(p+A(t')-A(t)) \ \exp[-i(S(p,t,t')+I_p(t-t'))] + c.c. \tag{9}$$

Here $\mathbf{E}(t) = (E_x, E_y)$ is the total electric field, $\mathbf{d}_g(\mathbf{p}) = <\mathbf{p}|\mathbf{d}|g>$ is the matrix element of bound-free transition, $\mathbf{A}(t)$ is the field vector-potential, and $S(\mathbf{p}, t, t')$ is the classical action at the moment t for the electron born in the continuum at the moment t' with initial momentum $\mathbf{p}$. The harmonic intensities are given by the squared modules of the Fourier transform of $\mathbf{d}(t)$.

For linearly polarized light the mathematical analysis of Eq. (9) using the stationary phase and saddle point methods recovers all three of the basic assumptions of the semi-classical picture outlined above. (1) The main contribution to the integral over $d^3\mathbf{p}$ is given by the values of $\mathbf{p} \approx \mathbf{p}_o(t, t')$, where $\mathbf{p}_o(t, t')$ is the initial momentum for which the electron born in the continuum at the moment t' will return back to its initial position at the moment t. (2) The main contribution to the integral over dt' (t' is the moment of birth) is given by $t' \approx t_0$, where t_o is the moment of time for which the corresponding $\mathbf{p}_o(t, t')$ has zero x-component: $\mathbf{p}_{ox}(t, t' = t_0) = 0$. (3) The stationary point t_N of the Fourier integral at frequency $N\omega$ is given by the condition $N\hbar\omega = I_p + \mathcal{E}(t_N)$. However, Eq. 1 can be applied for any ellipticity.

We use Eq. 9 to calculate the ellipticity dependence of N-th harmonic intensity assuming $\varepsilon^2 \ll 1$. The result is:

$$R_N(\varepsilon) = \frac{I_N(\varepsilon)}{I_N(\varepsilon = 0)} = Ai^2\left[\frac{[I_p + p^2_{0y}(t_N, t_{0N})/2m]/\hbar\omega}{[2U_p/\hbar\omega]^{1/3}}\right] Ai^{-2}\left[\frac{I_p/\hbar\omega}{[2U_p/\hbar\omega]^{1/3}}\right] \tag{10}$$

Here $Ai(z)$ is the Airy function. The ellipticity dependence appears through $p_{oy}(t_N, t_{0N})$ - the initial momentum in y-direction which the classical electron born at t_{0N} must have in order to return to the same place at the moment t_N. It is related to the laser field by

$$p_{0y}(t_N, t_{0N}) = \frac{i}{t_{0N} - t_N} \int_{t_{0N}}^{t_N} \frac{q\varepsilon E_{ox}}{\omega}[\cos(\omega t) - \cos(\omega t_{0N})]dt \tag{11}$$

Finally, the times t_N and t_{0N} of collision with the parent ion should satisfy the energy conservation law $\mathcal{E}(t_N) \approx p^2_x(t_N, t_{0N})/2m = N\hbar\omega - I_p$, with the x-component of the initial momentum equal to zero: $p_{ox}(t_N, t_{0N}) = 0$.

The experimental intensity distribution is inhomogeneous in both space and time. However, high harmonic emission occurs over a narrow range of intensities. In the above model it is directly related to the bound-free transition rate. In Fig.

5 we present two curves for the 41st harmonic of neon, the solid curve is for the peak intensity in the experiment (8×10^{14} Wcm^{-2}) which is roughly the saturation intensity for ionization. The dashed curve is for 5×10^{14} Wcm^{-2} which corresponds to a factor of 10 decrease in the ionization rate. The model has no free parameters.

The Keldysh-like model discussed above only deals with harmonic generation. However, the experimental results on the ellipticity dependence of multiphoton ionization can be *quantitatively* analyzed if we assume that the semi-classical model [5] correctly identifies correlated two-electron multiphoton ionization as an inelastic scattering event (A$^+$ + e → A^{++} + 2e). The ellipticity dependence of this process will be given by the ratio

$$R_{ion}(\varepsilon) = \int_{\mathcal{E}_1}^{\mathcal{E}_2} N(\mathcal{E},\varepsilon)\sigma(\mathcal{E})d\mathcal{E} / \int_{\mathcal{E}_1}^{\mathcal{E}_2} N_0(\mathcal{E})\sigma(\mathcal{E})d\mathcal{E} \qquad (12)$$

where $N(\mathcal{E},\varepsilon)$ is the number of electrons returning to the nucleus with energy $\mathcal{E}$ and $N_0(\mathcal{E}) = N(\mathcal{E}, \varepsilon = 0)$, $\mathcal{E}_1$ is the minimum energy required to remove the second electron, and $\mathcal{E}_2 = 3.17 U_p$ is the maximum instantaneous energy the electron can have near the ion. Finally, $\sigma(\mathcal{E})$ is the cross-section for collisional ionization in the presence of an intense laser field. The value of $\sigma(\mathcal{E})$ is not known, although its zero-field value is known. However, a very good estimate for the ratio Eq. (11) can be obtained without knowing $\sigma(\mathcal{E})$. The ratio $R_N(\varepsilon) \equiv R(\mathcal{E},\varepsilon)$ in Eq. (10) measures the relative number of electrons returning to the nucleus with energy $\mathcal{E} = N\hbar\omega - I_p$. We can use Eq. (10) to calculate the ratio $N(\mathcal{E},\varepsilon)/N_0(\mathcal{E})$. Then, according to the middle point theorem, the value R_{ion} in Eq. (3) is equal to $R(\mathcal{E}^*,\varepsilon)$ where $\mathcal{E}^*$ is some point from the interval [$\mathcal{E}_1, \mathcal{E}_2$]. As $R(\mathcal{E}^*,\varepsilon)$ is a monotonically decreasing function of $\mathcal{E}$ (i.e., of harmonic number N) at fixed ε, we get the following result for the e-2e scattering probability:

$$R(\mathcal{E}_1,\varepsilon) \geq R_{ion}(\varepsilon) \geq R(\mathcal{E}_2,\varepsilon) \qquad (13)$$

The result using Eq. (12) is shown in Fig. 7 together with experimental data on Ne^{++} ion yield as a function of ellipticity measured at the intensity I = 9×10^{14} Wcm^{-2}.

The semiclassical model [5] has only one free parameter, the transverse spread of the electron wave function. The data presented in Fig. 7 remove this freedom. We estimate the absolute yield of Ne^{++} vs laser intensity by approximating the e-2e cross-section in a strong laser field by its field-free value $\sigma_0(\mathcal{E})$. $\sigma_0(\mathcal{E})$ underestimates the correct value because in the laser field both virtual and real excitation will can lead to ionization. Using published cross-section data [11], ADK tunnelling rates to calculate ionization probabilities and classical trajectory calculations to determine the velocity distribution of electrons passing the ion, we obtain the solid curves shown in Fig. (6).

In conclusion, semi-classical physics gives insight into many strong field atomic physics problems. The semi-classical model correctly suggests that strong field effects will be very sensitive to the ellipticity. We observe that the ellipticity variation of high-harmonics, correlated double ionization and Keldysh-like theory are in excellent agreement. Even the strength of the double ionization signal is remarkably well fit by including contributions from e-2e scattering.

In this paper we have concentrated on the plateau harmonics which are largely due to free bound transitions. However, it is clear that for harmonics with photon energy lower than the ionization potential must have another origin. The reader is referred to Ref. [12] for a account of the origins of the low harmonics and to Ref[1] for a discussion of the implications of the model for such practical considerations as attosecond pulse formation.

References:

[1] M. Ivanov, P. B. Corkum, T. Zuo and A. Bandrauk, submitted to Phys Rev Lett.
[2] P.B. Corkum, N.H. Burnett, and F. Brunel, Phys. Rev. Lett. 62, 1259 (1989).
[3] M.V. Ammosov, N.B. Delone, and V.P. Krainov, Sov. Phys. JEPT. 64, 1191 (1986).
[4] 4. U. Mohideen et al., Phys. Rev. Lett. 71, 509 (1993).
[5] P.B. Corkum, Phys. Rev. Lett. 71, 1994 (1993).
[6] D.N. Fittinghoff, et al., Phys. Rev. Lett., 69, 2642 (1992).
[7] J.L. Krause, K.J. Schafer, and K.C. Kulander, Phys. Rev. Lett. 68, 3535 (1992)
[8] K. S. Budil, et al. Phys. Rev. A48, R3437 (1993).
[9] A. L'Huillier et al. in **Atoms in Intense Fields** M. Gavrila, ed, Academic Press 1992, 139-206.
[10] M. Lewenstein et al., Phys Rev A49, 2117 (1994).
[11] H. Tawara and T. Kato, Atom Data and Nuc. Data Tables, 36, 167 (1987).
[12] N. H. Burnett, C. Kan and P.B. Corkum, submitted to Phys. Rev. Lett.

BEYOND ATI: STRONG FIELD QUANTUM CONTROL

P. H. Bucksbaum

Department of Physics, University of Michigan, Ann Arbor, MI 48109-1120

Abstract. Above-threshold ionization (ATI) is the ionization of atoms by more than the minimum number photons necessary to overcome the Coulomb binding energy. Although ATI is contrary to the predictions of minimum-order perturbation theory, it is the dominant mode of ionization whenever the light field becomes comparable to the static fields in the atom. We have learned a great deal by studying ATI over the past several years. We now know that by controlling these high intensity optical fields, we can control atomic and molecular processes such as ionization and dissociation. We can even engineer wave-functions.

1. INTRODUCTION

This paper reviews the current state and future directions of super-intense laser-atom physics (SILAP). The investigations of above-threshold ionization (ATI) and high-harmonic generation (HHG) in the past few years have led to a view of the SILAP interaction in which wavepacket dynamics play a dominant role (1,2). In a companion article in this volume, P. Corkum describes a successful model which explains many of the general features of ATI and HHG as the result of the scattering of an electron wavepacket by the atomic core, and the subsequent evolution of the wavepacket in the presence of the oscillating laser field (3).

Wavepacket scattering models such has those put forward and developed by Corkum, Schafer, Kulander, and others, have existed in one form or another since the first ATI experiments were performed (4,5) and even before (6-8). These are not the only way to view SILAP interactions; other models and techniques, such as dressed state analyses, have been similarly valuable. But these wavepacket models still represents a real breakthrough in the field, because they enable us to go beyond phenomenological investigations. Using these theories as guides, we can employ strong laser fields to probe atomic structure, to alter quantum dynamics, and even to construct new quantum structures. These are new and important directions in the science.

2. THE ATI SPECTRUM

The ATI spectrum of xenon photoionized by 1.06 μm radiation is shown in figure 1. Superimposed on this spectrum is a diagram showing how wavepacket production and scattering may produce the main features of ATI. Our description will be brief; for details, see the companion article by Corkum (3), and references therein.

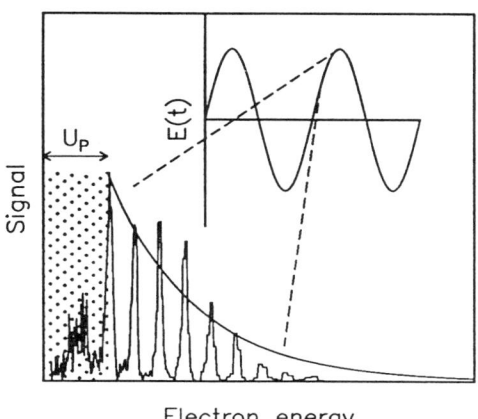

Figure 1. ATI spectrum of xenon ionized by 1064 nm 100 psec laser pulses. The spectrum is shifted to higher energy by the ponderomotive energy of the beam. Its general shape follows the predictions of tunneling during each optical cycle, followed by the evolution of the free photoelectron in the optical field.

The whole spectrum of decreasing peaks is offset to higher energy by 1-2 eV. This phenomenon, which has been called "channel closure," is due to the light-induced energy shift of the continuum relative to the ground state of the atom. Stated in terms of classical physics, the electrons produced in photoionization, even those with zero kinetic energy produced at threshold, are wiggling in the laser field. The wiggle energy, known as ponderomotive energy, is given by $U_P = e^2 F^2/(4 m_e \omega^2)$, where F is the peak amplitude of the laser field, m_e is the electron mass, and ω is the laser angular frequency. As photoelectrons leave the laser focus on the way to detection, their wiggle energy is converted to translational kinetic energy. The process is conservative if the laser intensity is constant during the ~3-5 psec traversal of the focus. If the wiggle energy is the principal source of AC Stark shift in the intense laser field, then this adiabatic transformation of wiggle energy to translational kinetic energy means that electrons are observed with energies that differ from the ground state energy only by the absorption of an integer number of photons. In other words, one observes multiple peaks in the spectrum which obey a generalization of Einstein's photoemission formula: $E = E_{ion} - E_{atom} + n\hbar\omega_{laser}$.

Beyond ponderomotive effects, the essential feature of strong field ionization is that *the ionization per optical cycle is no longer negligible*. Therefore, the overall shape of the spectrum is controlled by variations in the tunnel-ionization rate during each optical cycle. The instantaneous ionization rate $\Gamma(t)$ for an atom with binding energy E_0 is fairly well described by the "ADK" formula (9),

$$\Gamma(t) = 4 E_0^{5/2} \frac{1}{F(t)} \exp\left[-\frac{2}{3} E_0^{3/2} \frac{1}{F(t)}\right] \quad \text{(in atomic units)},$$

which is an extension of the WKB tunneling rate (10) to the case of monochromatic light (11). Once ionized, the energy of the electrons evolves under the influence of the periodic driving force of the light.

Most of the features of the ATI spectrum can be derived from classical mechanics. For example, the peak ionization rate should occur at the peak of the electromagnetic field cycle. Electrons emitted by tunneling at this time enter the field with very small initial kinetic energy, and begin to wiggle in the field. Over the course of one cycle, they accelerate away from the ion, decelerate as the electric force switches direction after 1/4 cycles, accelerate back towards the ion in the third 1/4 cycle, and finally return to rest at their starting point one full cycle after they were ionized. If the ion Coulomb field is neglected, it is clear that these electrons just wiggle in place, with little or no drift momentum. Ponderomotive gradients in the laser focus will eventually (3-10 psec) allow the electrons to drift out of the focus, with translational energy only equal to the ponderomotive potential energy they had when created.

Electrons that tunnel out of the atom slightly before the peak of an electric field cycle experience a similar evolution, except that they have a slightly longer initial acceleration. This they keep, as drift momentum. Therefore they emerge from the focus with larger kinetic energy. There are fewer of them, because the tunneling rate when they were created is smaller. In a monochromatic laser field, the exponential increase of the tunneling rate with field strength leads to an exponential decrease of the photoelectron spectrum. This general trend is modulated into an envelope of sharp ATI peaks because of the coherent interference of ionization on successive cycles of the field.

This simple classical picture can predict most of the features of ATI experiments (11,1). Furthermore, simple extensions of this picture have been used successfully to explain other high field phenomena such as high harmonic generation (12) and some aspects of multiple ionization (13,14). This is truly remarkable, since the picture suffers from some obvious deficiencies: it relies on tunneling, although most ATI phenomena are observed in the intensity regime where the ionization rate per optical cycle is much less than 1; it ignores internal structure in the atom; and it treats the electron as a classical point particle. Nevertheless, full quantum dynamics calculations confirm that many features of SILAP interactions are essentially classical (15).

2.1 Intermediate Resonances

Although the tunneling picture explains many of the phenomena of ATI, it cannot be correct on several points: the ionization rate is not given by the simple ADK model; in fact, the ionization is not even purely monotonic with light intensity, but is dominated by many sharp resonances (16). This was first observed in 1987, when ATI experiments were first obtained with very short pulses (17). If the pulses are shorter than about 3 psec, the ponderomotive acceleration of the free electrons is not complete. Basically, the light turns off before the electrons get out of the focus. For pulses shorter than 1 psec, new sharp peaks occur in the photoelectron spectrum. These are due to intensity-dependent multiphoton resonances between the ground state and excited states in the atom. The resonances occur in all atoms; they are caused by the huge AC stark shifts of the excited states relative to the ground state. To a good approximation, the ground state stark shift is negligible compared to the excited states, and the excited states shift by an amount equal to the ponderomotive potential shift of the continuum (18). At intensities where ATI occurs, the shift is generally on the order of one photon energy, which means that all excited states have shifted into resonance at some intensity or other during the pulse.

2.2 Dressed State Picture

It is easier to visualize the effect of these resonances in a dressed state picture, where they appear as avoided crossings between the ground state and excited states as the intensity passes through resonance. The probability of a transition during this passage depends on the size of the avoided crossing; the coupling to the continuum, which controls the ionization probability during the crossing; and the rate of crossing, which is proportional to the rate change of intensity during the pulse.

De Boer and Muller (19) pointed out that the avoided crossing leads to two different pathways to produce a photoelectron with the same energy. (a) The atom may ionize from the ground state at the intensity corresponding to a multiphoton resonance, which is basically the mechanism first proposed by Freeman et al. (17); or else (b) population may be transferred to the excited state during the crossing, and then ionization out of the excited state may occur later in the pulse, when the intensity is higher. Gibson et al. (20,21) have found that the probability of ionization through one or the other of these mechanisms can be controlled in various ways.

Recently, Story et al. (22) demonstrated this ability to control multiphoton excitation during ATI using either pulse length or the intensity where resonance occurs as control parameters. The experiment is summarized in figure 2, which shows the dressed state eigenvalues in the potassium atom as a function of time during the laser pulse. They excited potassium with laser pulses of 0.5 to 13 psec duration, and laser frequencies of 17,180 cm^{-1} or 17,300 cm^{-1}. With the lower frequency light, the atom could be intensity-shifted into resonance with the 14d state and higher, but the higher frequency light could only lead to resonances with 17d and higher. The residual population in these high Rydberg states was found to depend on the rate of passage through resonance, as expected.

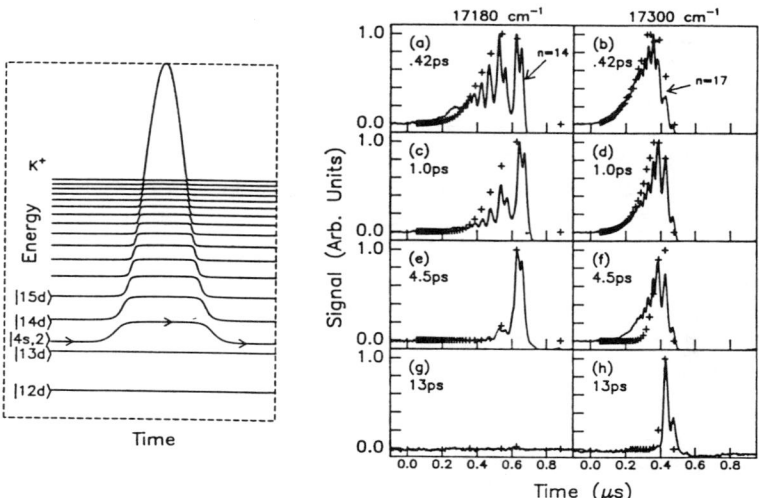

Figure 2. Dressed-state eigenvalues for 2-photon resonant, 3-photon ionization of potassium. Rydberg states are ac-Stark shifted into 2-photon resonance as the laser pulse turns on and off. The rate of intensity increase or decrease controls the probability for trapping population in the intermediate Rydberg states. (From reference 22; reprinted with permission from author.)

Passage through an avoided crossing is generally neither diabatic nor adiabatic, but rather leads to a superposition of the ground and excited states. Since the atom must pass through resonance twice, during the rising and falling edge of the laser pulse, the two passages may interfere either constructively or destructively, leading to coherent control over the left over population in the excited state after the pulse. In order to exploit this, however, it is necessary to isolate regions of the focused laser pulse where the intensity is fairly uniform; otherwise, averaging over regions with different intensity histories will wash out any coherent effects.

The experimental difficulties in observing the coherence between the two avoided crossings in the laser pulse have been recently overcome by Jones at the University of Virginia. In work reported at this ICAP meeting, Jones shows that population in sodium Rydberg states excited via multiphoton absorption can be enhanced or decreased, depending on the phase of the wave function at the two crossings.

In further work on this subject, Jones has exploited this phase control to enhance the ionization probability. The scheme works like this: if two pulses are incident on the atom with just the right time delay, the population transferred from the ground to the excited state on each crossing can coherently add. This can lead to an ionization enhancement for two pulses as compared to one pulse of nearly a factor of 20! This has been observed by Jones.

3. ATOMIC STABILIZATION

Strong field coherent control can also be employed to suppress the ionization rate. This "stabilization" of the atom in an intense field has been an area of active research over the past two years. Much work has concentrated on an idea proposed by Gavrila and Pont several years ago (23,24, and see ICAP '92 proceedings for details and further references). Briefly, they showed that in a very intense and very high frequency laser field, bound states become distorted along the laser polarization axis. Eventually, the electron wavefunction becomes delocalized from the atomic potential. This reduces the binding energy of the state, and also reduces the ionization cross section.

The high frequency approximation used by Gavrila and Pont states that $\hbar\omega \gg E_0$. This is not realized for ground state atoms in optical laser fields; however, since the distortion of the wave function eventually reduces the binding energy of all bound states, they conclude that even tightly bound states eventually reach the high frequency stabilization regime. The lifetime of the atom in a laser field can be plotted vs the laser intensity. There are three distinct regions. In the lowest intensity region, the one of ordinary ATI, the lifetime decreases steeply with intensity. As the intensity rises, eventually the atom reaches the tunneling regime where the lifetime is comparable to a single optical cycle. This is the second region, which has become known as "Death Valley." Beyond this is the third region, the regime of high frequency stabilization. The experimental challenge has been to devise a way for an atom to survive Death Valley as the laser intensity rises on the leading edge of the laser pulse.

Recently, de Boer and coworkers have found evidence for high frequency stabilization (25). They employed several experimental tricks to achieve the stabilization regime. First, they worked with atoms in highly excited states, so that the high frequency approximation is valid throughout the experiment. Their most important innovation was to prepare states with high l and m quantum numbers, so-called "circular states." In these states, Death Valley is considerably shallower, thus insuring that some population will survive the rising edge of a 100 fsec laser pulse.

The experiment was performed on 5g states in neon (25). Figure 3 is a schematic of the experimental arrangement. 5-photon excitation from the ground state with circularly polarized 1 psec 286 nm pulses prepares neon in the 5g m=4 state. Actually, the experiment utilized adiabatic transfer of population at an intensity-induced avoided crossing between the ground state and the 5g state, just as in the experiment of Story et al. described above. The crossing occurs at $I = 8.6 \times 10^{13}$ W/cm^2. The ionization of the 5g state saturates at 4×10^{14} W/cm^2, so it is possible to excite a significant fraction of the ground state without ionizing.

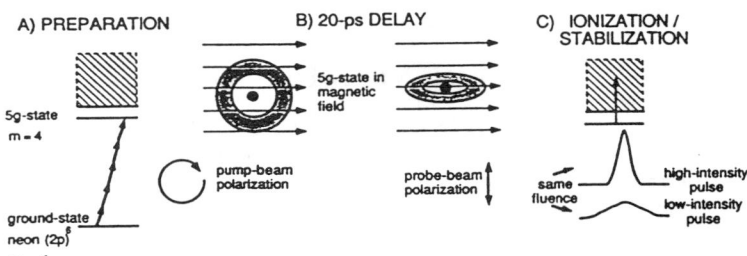

Figure 3. The experimental scheme employed in reference 25. Multiphoton excitation of the 5g circular state in Ne is followed by irradiation by an intense pulse. Leftover population is monitored to find evidence of stabilization. (Reprinted with permission from author.)

For circular states, Gavrila's stabilization theory requires a laser polarization along $\hat{z}$, that is, normal to the plane of the orbit. To accomplish this, the researchers allowed the atom to precess $\pi/2$ radians in a 0.9 T magnetic field. Then they illuminated it with a 620 nm probe pulse of various duration and peak intensity, and observed the ionization of the 5g state. Figure 4 shows the resulting spectra. The 5g peak appears to saturate if the pulses are intense enough. The results are undramatic, but significant. This is the first evidence that high frequency adiabatic stabilization occurs.

4. MOLECULAR STABILIZATION

There has also been considerable effort to control molecular processes with short lasers. In H_2^+, the the AC Stark effect creates gaps in the dressed state internuclear potentials for the ground state and first excited state (see figure 5). These are quite analogous to the dressed-state avoided crossings in atoms (see figure 2 above). Population may transfer from one state to the other at these gaps during an intense laser pulse, and control may be possible.

One of the more interesting aspects of these laser-induced avoided crossings is the presence of new light-induced vibrational potential wells formed above them. Recent work by Mies and Giusti suggest that population bound in the vibrational states of H_2^+ will be partly trapped in these new vibrational states if the laser pulse is sufficiently short (26). Zavriyev et al. (27) has obtained experimental evidence for this light-induced structure in recent experiments.

5. CONTROL STRATEGIES

The experiments described above all achieve some degree of control over the ionization process by adjusting the duration, wavelength, and intensity of a nearly

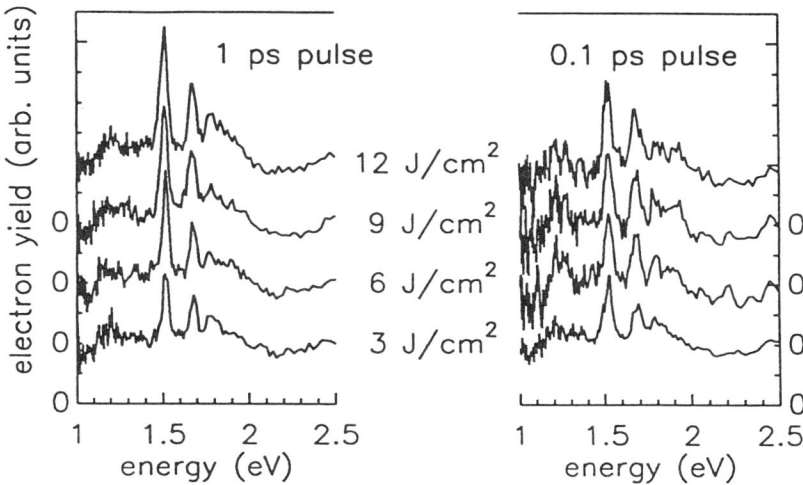

Figure 4. Photoelectron spectra showing evidence of adiabatic stabilization. For long pulses, higher fluence leads to less population left over in the 5g state in neon. For short pulses, the population decrease saturates. (From reference 25; reprinted with permission)

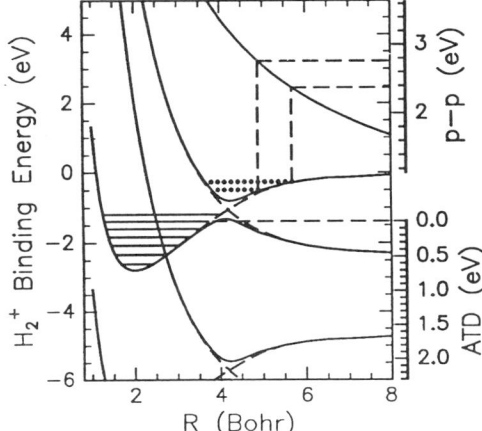

Figure 5. Laser-dressed energy levels of H_2^+ in a 3.5TW/cm^2, 532 nm laser field. Energies of proton fragments following dissociation are shown on the right side.

Gaussian laser pulse. However, current short pulse laser technology allows much more significant alteration of the form of short laser pulses. In the remainder of this paper, we will discuss several new methods for exerting more sophisticated control over atoms with strong laser fields.

5.1 Optical Frequency Multiplexing

One important method of increasing control over atomic photoionization and molecular dissociation is by combining two or more laser colors. In weak field experiments, this can produce multiple pathways which may interfere depending on the relative phase of the two laser fields. If the frequencies are commensurate, such a phase can always be defined(28,29). In a strong field experiment, we enter the tunneling regime described in figure 1: the instantaneous tunneling rate during each field cycle depends on the magnitude of the instantaneous electric field, which is just the superposition of the various laser fields. 2-color experiments are critical tests of these theories (30).

There is no well-defined field magnitude where perturbative multiphoton ionization gives way to tunneling; however, the transition is fairly rapid in high order multiphoton processes, since the ionization rate is changing as a high power of the intensity. The Keldysh parameter $\gamma = \sqrt{E_0/2U_P}$ (31) is a convenient dimensionless figure of merit for this. γ is roughly one over the product of the tunneling rate and one-quarter of an optical period, which is the time available for tunneling near each half-cycle maximum of the laser field. When $\gamma < 1$, ionization is expected to be dominated by tunneling. Under these conditions, the phase-dependence of ionization in a 2-color field is just due the phase-dependence of the field shape: highest rates occur when the fields add constructively, and the tunnel barrier is most suppressed.

Several experiments have begun to test these ideas (32-35). Schumacher (35) recently showed that it is possible to focus an intense laser beam and its second harmonic into an atomic gas so that the phase shift between the two colors is stable, measurable, and well-controlled. He used this control to test the two-step model of above-threshold ionization outlined above.

In Schumacher's experiment, there is a two-color pulsed laser field $\xi(t)$ consisting of a field $F_1(t)$ with fundamental frequency ω produced by an amplified mode-locked Nd:YAG laser ($\tau = 100$psec, $\lambda = 1064$nm), and its second harmonic, $F_2(t)$. $\xi(t)$ is the coherent sum field

$$\xi(t) = F_1(t)\cos\omega t + F_2(t)\cos(2\omega t + \phi)$$

The fields are equal in magnitude, and they are phase-locked to each other because $F_2(t)$ is produced via second harmonic generation in a KD*P crystal. Care is taken to focus the two beams so that their relative phase in the focus is constant and stable. Their polarizations are parallel. The amplitude and shape of $\xi(t)$ depends on the relative phase ϕ, which can be determined absolutely using nonlinear frequency mixing and optical rectification in a second KD*P crystal placed inside the vacuum chamber where the experiment takes place.

The experiment was performed in Krypton, where at least 12 (1064 nm) photons are required to ionize. Figure 6 shows a 2-color ATI spectrum for $\phi = 0$, and $F_1 = F_2 = 1.2$V/Å. (Total intensity of 4×10^{13}W/cm^2.) Electrons are detected

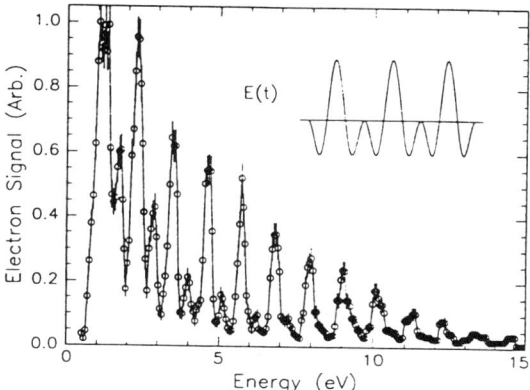

Figure 6. Krypton ATI photoelectron spectrum for equal amplitude 532 nm and 1064 nm radiation, with relative phase $\phi = 0$. (From reference 35, reprinted with permission.)

emerging along the polarization axis, but in only one direction. (This makes it possible to observed left/right asymmetries in the ionization spectrum.) There are two series of peaks (large and small) corresponding to two different final states in the ion. The ϕ-dependence for the third through sixth large peaks ($^2P_{3/2}$ final state) are shown in figure 7 together with the predictions of the 2-step model. It is quite clear that by controlling the phase, one can control the spectrum. For example, the phases with the highest ionization rates do not produce the highest energy electrons.

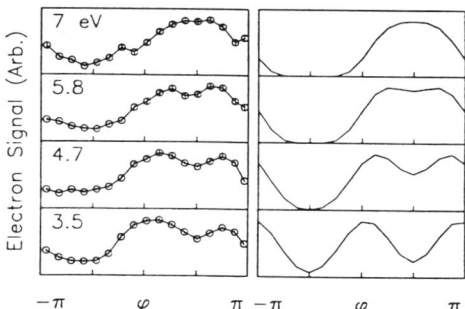

Figure 7. ϕ dependence of ATI peaks 3 through 6, for total intensity of 4×10^{13} W/cm^2. Tunneling model predictions shown at right. (From reference 35, reprinted with permission.)

As the intensity goes up, the data start to deviate from the predictions of the simple model. The two-step model describes the essential physics for low intensities and/or for low energy electrons; but high energy electrons are produced at high intensities by some additional process. Again the phase dependence is critical to this observation because the excess high energy electrons are produced predominantly in one direction. Excess high energy electrons have been seen in other experiments as well (2,36). The mechanism for producing the excess is still not clear, but one plausible idea proposed by several researchers is core rescattering following ionization (2). The electrons scattered by their parent ion are reaccelerated by the laser field, and on average they gain initial kinetic energy. This is essentially the same mechanism as radio-frequency heating in plasmas, except that here it is the parent ion that scatters the electron.

Schumacher also tested this hypothesis by comparing his linear polarization results with those obtained if he introduced a small amount of elliptical polarization. The idea is that the probability of core rescattering is greatly reduced in elliptically polarized light, since the electron is diverted from the parent ion by the out-of-phase component of the field. Indeed, the large phase asymmetry seen for high intensity 2-color ATI disappeared, lending support to the rescattering argument.

5.2 Mixing Colors in Dissociation Experiments

2-color strong-field laser control has also been discussed as a means of controlling molecular dissociation (37,38). Here the mixing of ω and 2ω breaks the left-right symmetry along the polarization axis. Recent calculations show how frequency multiplexing may be used to align molecules and to select the channel or even the directions of the fragments (39). The simplest case is again H_2^+, where the fragments are a proton and a hydrogen atom. Here, different above-threshold dissociation channels can open and close, and the proton can be preferentially directed in either sense along the polarization direction (39). In the case of HD, Mies and Giusti suggest that isotopic selectivity may also be possible, in the sense that the deuteron and proton may be ejected in different preferred directions (40). Experiments to test these ideas are underway at Brookhaven and Michigan.

5.3 Pulse Shaping

Frequency mixing is not the only way to modulate intense laser fields. A more general technique is pulse-shaping, where the laser electric field may by sculpted into a desired form by frequency and phase modulations. This is no longer a hypothetical possibility; liquid crystal modulators have now been employed to provide independent, programmable amplitude and phase control in amplified sub-picosecond pulses.

A group at the University of Michigan used amplified programmable shaped optical pulses recently to engineer Rydberg wavepackets (41). In the strong field regime, any arbitrary wavepacket superposition can be obtained by manipulating the light that creates the excited state. The demonstration experiment uses 110fs

785 nm pulses containing about 10 nm of bandwidth, obtained from a Ti:Sapphire Kerr-lens modelocked laser. The pulse shaper is built in a a zero-dispersion grating-pulse expander. This is essentially two diffraction gratings with an inverting telescope inbetween. This device can be configured to produce no net dispersion; however, in the middle of the telescope where where the light is focused, all the frequency components are dispersed in the horizontal plane. Two liquid crystal displays (LCD's) similar to those found in laptop computers intersect the beam at this point (42). These produced independent computer-controlled phase delays or polarization shifts, providing both phase and amplitude control over each color in the pulse. The pulses are amplified to 1 mJ in a standard chirped-pulse amplifier (43). The shape following amplification is characterized using a frequency-resolved optical gate (44).

The nonstationary states excited by the shaped pulses in the demonstration experiment are constructed from the np Rydberg states in Cs. Population in the Rydberg states is monitored using ramped-field ionization. In this common analysis technique, a slowly ramping dc electric field after the laser pulse ionizes the wavepacket as it reaches its field-ionization threshold ($F = 1/(2n^*)^4$). In this way, the np Rydberg states making up the wavepacket are dispersed in time in the detector.

The simplest way to see the effect of the pulse shaper is to observe spectral shaping in the Rydberg populations observed during ramped field ionization. Figure 8 shows two ramped-field spectra: one from a near-Gaussian pulse, and the second from a pulse in which frequencies corresponding to all but four np states have been selectively removed. Such a pulse is easy to produce in the pulse shaper.

A more sophisticated technique for observing shaped pulses is the "optical Ramsey method" as follows (45): The shaped light is split into two identical pulses with a Michelson interferometer. The time separation between the pulses can be varied over several picoseconds. Each pulse in the pair excites Cesium from a pre-excited 7s state to a sculpted wave function or "wavepacket", i.e. a coherent superposition of np states centered around n=27. The two wavepackets are phase-coherent with respect to each other so there is interference between them. This changes the total population excited into each Rydberg state as the time delay is moved. The population vs. time is an interferometric autocorrelation of the complex scalar wavefunction of the excited atom.

In the limit of little depletion of the 7s state, this autocorrelation is related to the fourier transform of the spectral amplitude function of the pulse, much as a Michelson autocorrelation of an optical field can be used for spectroscopy. In strong fields, The depletion of the 7s stated by the first pulse changes the wavepacket excited by the second pulse. This additional feature might be used to extract phase information about the wavepacket as well.

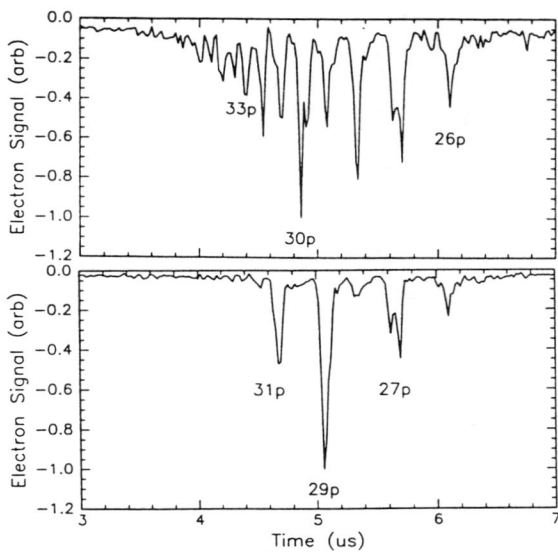

Figure 8. Top: ramped field spectrum from a Gaussian pulse. Bottom: ramped field spectrum from a shaped pulse.

Figure 9 shows the autocorrelation of the Cs wavepacket produced by the shaped pulse shown above. A calculation is also shown for comparison.

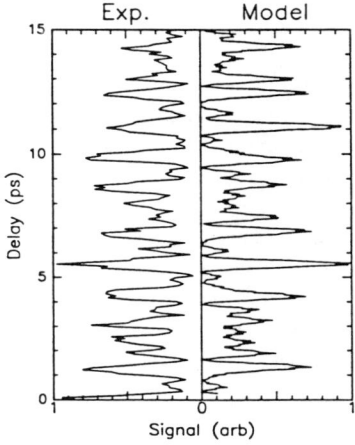

Figure 9. Autocorrelation of a 3-state cesium wavepacket produced by a shaped optical pulse. Also shown is a calculation of the expected autocorrelation function.

5.4 Half-Cycle Pulses

Ultimate quantum control requires the ability to sculpt any arbitrary wavefunction. This is certainly beyond reach with present technology. However, for a large number of problems in ionization, dissociation, or bond-selective chemistry, we can set a more modest goal of manipulating the wavefunction on a timescale comparable to the quantum dynamics. For Rydberg atoms, and for most molecules, present ultrafast laser technology ($\tau_{pulse} \geq 20$fsec) is nearly sufficient, provided we have total control of the light field in this timescale. Therefore, a great emphasis is placed on the ability to expand the usable coherent bandwidth in the pulse.

In this final section, we will describe recent advances in manipulating ultrashort far infrared pulses. The number of field cycles in these pulses can actually approach the dc limit where the field merely traces out the upper or lower envelope of the optical pulse train. We call these "half-cycle pulses" (HCP), since they contain less than one full optical cycle.

Large (~10 – 100kV/cm) HCP's are produced when a 100 fsec optical pulse is incident on a field-biased direct-gap semiconductor such as GaAs or InP (46, 47). The electric field in the pulse is produced by the rapidly changing photocurrent in the semiconductor following excitation, and propagates in the direction of the incident laser due to elementary phase-matching requirements. Since the HCP is freely propagating, it can be reflected from metal mirrors, focused, and directed into an atomic or molecular beam or cell. Typically, the pulse is about 400 fsec in duration, with a coherent bandwidth of more than 1 THz.

The "half-cycle" nature implies that the time integral of the field is not zero. This is possible because the current distribution in the semiconductor is not the same before and after the pulse, i.e. there is no current prior to excitation, but ~1psec afterwards a dc current is flowing through the semiconductor. There is a slow recharge period of several hundred μsec, when the dc current turns off. This produces a tiny but persistent negative tail in the field, so that, over milliseconds, the time integral really does vanish. However, for the purpose of manipulating wavefunctions with dynamical timescales of 100fsec to a few psec, the field is nearly unipolar, hence "half-cycle."

Rydberg states in alkali atoms show off the unique features of HCP's. Here we will show the results of two experiments in our laboratory. In the first, Jones et al (48) studied the stability of Rydberg ns and nd states in Na to HCP's. The states are bound by 100 cm^{-1} or more, so that single photons within the HCP coherent bandwidth cannot ionize the atom. Furthermore, dc field ionization is not expected, because the duration of the pulse is much shorter than the Kepler orbit time of the atom. Expressed in terms of wave mechanics, this means that most of the wavefunction does not sample the saddle-point in the potential during the pulse. The data are summarized in figure 10, and show some remarkable trends. First, the ionization vs field strength for each n state displays neither a power law nor a sharp threshold behavior; instead, there is a slow turn-on, followed by a rapid rise to nearly 100% ionization.

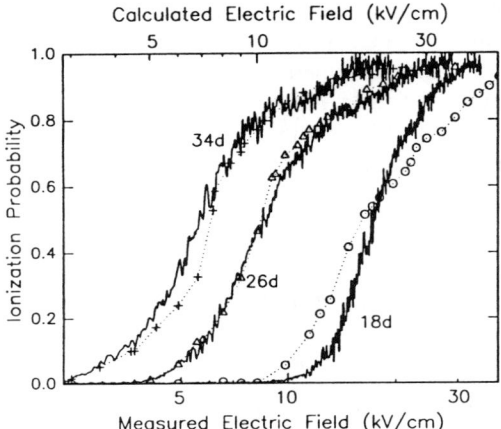

Figure 10. Ionization signal as a function of the peak electric field in the half-cycle pulse, for 18d, 26d, 34d initial states in Na. The smooth curves are classical simulations (48) for these states. Best agreement occurs for a rescale factor of 1.3 in the experimental electric field. (From ref. 47, reprinted with permission.)

A pulse of the same duration and bandwidth, but centered at optical frequencies, has a completely different behavior (49-51). For optical pulses, ionization can only take place near the ion core, where the electron can exchange momentum with the ion. Most of the state probability density is located at large distances from the core, so the ionization probability saturates at only a few percent ionization. This behavior has been studied extensively, and is now usually described as the formation of a trapped nonstationary state called a "dark wavepacket."

We can understand the different behavior of HCP ionization from simple classical dynamics arguments. The key is that an HCP electric field has a nonzero time integral, so that it can transfer much more momentum to an electron than the photon momentum hc/λ. The electron receives a kick from the field which changes its momentum by

$$\delta \mathbf{p} = \int_{-\infty}^{\infty} \mathbf{F}(t) dt,$$

where $\mathbf{F}(t)$ is the HCP field. Even electrons at rest can receive substantial energy from this. More generally, we can integrate the energy change:

$$\Delta E = -\int_{-\infty}^{\infty} \mathbf{F}(t) \cdot \mathbf{v}(t) dt,$$

where $\mathbf{v}(t)$ is the velocity of the electron. The results of the classical simulation are shown in figure 10.

Perhaps the most dramatic illustration of the effect of HCP's on atoms was in a recent experiment performed at Virginia and Michigan on "oriented" Rydberg states (52). Such states are produced when an electric field is used to split the l-degeneracy of Rydberg atoms. This permits the selective excitation of states with permanent electric dipole moments, where the electron wave function is predominantly on one side of the atom. Then the ionization probability depends not only on the magnitude of the HCP, but also on its orientation relative to the atom. Atoms ionize most easily when the HCP produces a force on the electron in the direction away from the ion core. These results can be qualitatively understood classical arguments. Figure 11 shows ionization curves for two states with the opposite orientation, along with classical simulations.

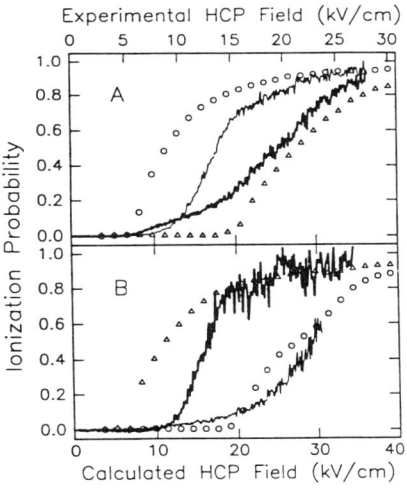

Figure 11. Ionization signal n=17 states in Na. with two different orientations: (a) dipole moment along $-\hat{z}$ (b) dipole moment along $+\hat{z}$. Light and dark lines are for HCP field directions $-\hat{z}$ and $+\hat{z}$, respectively. (o) and (Δ) trace classical simulation results for $-\hat{z}$ and $+\hat{z}$ HCP directions, respectively. From ref. 52, with permission.

6. ACKNOWLEDGEMENTS

We thank Robert R. Jones, Douglass W. Schumacher, Harm G. Muller, and M. P. de Boer for valuable discussions, and we thank several authors for permission to reprint their figures in this work. This work was supported by the National Science Foundation. Part of this work was supported by a Nato Collaborative Research Grant.

7. REFERENCES

1. Corkum, P.B., Physical Review Letters 71, 1994, (1993).
2. Schafer, K.J., Yang, B., DiMauro, L.F., Kulander, K.C., Physical Review Letters 70 1599-1602 (1993)
3. Corkum, P.B., in this volume.
4. P. Agostini, F. Fabre, G. Mainfray, G. Petite, and N. Rahman, Phys. Rev. Lett. 42, 1127 (1979).
5. P. Kruit, J. Kimman, H. G. Muller, and J. J. van der Wiel, Phys. Rev. A28, 248 (1983).
6. L. V. Keldysh, Sov. Phys. JETP 20, 1307 (1965).
7. F. H. M. Faisal, J. Phys. B 6, L89 (1973);
8. H. R. Reiss, Phys. Rev. A 22, 1786 (1980).
9. Ammosov, M. V., Delone, N. B., and Krainov, V. P. Sov. Phys. JETP 64, 1191 (1986).
10. See, for example, L.D. Landau and E.M. Lifshitz, *Quantum Mechanics*, Course of Theoretical Physics Vol. 3 (Pergamon, New York, 1965),
11. P. B. Corkum, N. H. Burnett, and F. Brunell, Phys. Rev. Lett. 62, 1259 (1989).
12. M. Lewenstein, P. Balcou, M.Y. Ivanov, A. L'Huillier, P.B. Corkum, Physical Review A 49, 2117 (1994).
13. K. Kondo, A. Sagisaka, T. Tamida, Y. Nabekawa, and S. Watanabe, Phys. Rev. A 48, R2531 (1993).
14. B. Walker, E. Mevel, Baorui Yang, P. Breger, J.P. Chambaret, A. Antonetti, L.F. DiMauro, P. Agostini, Physical Review A 48, R894 (1993).
15. K. C. Kulander, K. J. Schafer, J. L. Krause, Laser Physics 3, 359 (1993).
16. P. H. Bucksbaum, L. Van Woerkom, R. R. Freeman, and D. W. Schumacher, Phys. Rev. A 41, 4119 (1990).
17. R.R. Freeman, P.H. Bucksbaum, H. Milchberg, S. Darack, D. Schumacher, and M.E. Geusic, Phys. Rev. Letters 59, 1092 (1987).
18. L. Pan, L. Armstrong, Jr., and J. H. Eberly, J. Opt. Soc. Am. B 3, 1319, (1986).
19. M. P. de Boer and H. G. Muller, Phys. Rev. Lett. 68, 2747, (1992).
20. G.N. Gibson, R.R. Freeman, T.J. McIlrath, Physical Review Letters 69, 1904 (1992).
21. G.N. Gibson, R.R. Freeman, T.J. McIlrath, and H.G. Muller, Physical Review A 49, 3870 (1994).
22. J.G. Story, D.I. Duncan, T.F. Gallagher, Physical Review Letters 70, 3012 (1993).
23. M. Pont, N.R. Walet, M. Gavrila, and C.W. McCurdy, Phys. Rev. Lett. 61, 939 (1988).
24. M. Pont, Phys. Rev. A 40, 5659 (1989).
25. M. P. de Boer, J. H. Hoogenraad, R. B. Vrijen, L. D. Noordam, H. G. Muller, Physical Review Letters 71, 3263 (1993).
26. A. Giusti-Suzor, F.H. Mies, Physical Review Letters 68, 3869 (1992).
27. A. Zavriyev, P.H. Bucksbaum, J.Squier, and F. Saline, Physical Review Letters 70, 1077 (1993).
28. Yi-Yian Yin, Ce Chen, and D.S. Elliot, Phys. Rev. Lett., 69, 2353 (1992).
29. Ce Chen, Yi-Yian Yin, and D.S. Elliot, Phys. Rev. Lett., 64, 507 (1990).
30. K. J. Schafer and K. C. Kulander, Phys. Rev. A., 45, 8026 (1992).
31. L.V. Keldysh, Sov. Phys. JETP, 20, 1307 (1965).
32. H.G. Muller, P.H. Bucksbaum, D.W. Schumacher, and A. Zavriyev, J. Phys. B, 23, 2761, (1990)
33. D. Normand, L.A. Lompre, A. L'Huillier, J. Morellec, M. Ferray, J. Lanvancier, G. Mainfray, and C Manus, J. Opt. Soc. Am. B 6, 1513 (1989)
34. K. Kondo, Y. Nabekawa, and S. Watanabe, in abstracts of *High field interactions and short Wavelength Generation*, St. Malo, August 1994.
35. D.W. Schumacher, F. Weihe, H. G. Muller, and P. H. Bucksbaum, Phys. Rev. Lett. 73, 1344 (1994).
36. U. Mohideen, M. H. Sher, H. W. K. Tom, G. D. Aumiller, O. R. Wood, R. R. Freeman, J. Bokor, and P. H. Bucksbaum, Phys. Rev. Lett. 71, 509 (1993).

37. E. Charron, A. Giusti-Suzor, F. H. Mies, Physical Review Letters **71**, 692 (1993).
38. A. D. Bandrauk, J.-M. Gauthier, J. F. McCann, Journal of Chemical Physics **100**, 340 (1994).
39. E. Charron, A. Giusti-Suzor, F. H. Mies, Physical Review A **49**, R641 (1994).
40. E. Charron, A. Giusti-Suzor, F. H. Mies, submitted.
41. P. H. Bucksbaum, D. W. Schumacher, J. H. Hoogenraad, Jeffrey L. Krause, Kent R. Wilson, Proceedings of Ultrafast '94, ed. by P. Barbara and W. Knox, Springer Verlag, (1994).
42. A. M. Weiner, D. E. Leaird, J.S. Patel, and J. R. Wullert II, IEEE JQE **28**, 908 (1992).
43. J. Squire, F. Salin, G. Mourou, and D. Harter, Op. Lett. **16**, 324 (1991).
44. D. J. Kane, R. Trebino, Opt. Lett. **18** 823 (1993).
45. R. R. Jones, C. S. Raman, D. W. Schumacher, and P. H. Bucksbaum, Physical Review Letters **71**, 2575 (1993).
46. D. You, R. R. Jones, D.R. Dykaar, and P.H. Bucksbaum, Optics Letters **18**, 290 (1993).
47. R. R. Jones and P. H. Bucksbaum, Comments on Atomic and Molecular Physics, in press (1994).
48. R.R. Jones, D. You, and P.H. Bucksbaum, Physical Review Letters **70**, 1236 (1993).
49. R.R. Jones and P.H. Bucksbaum, Phys. Rev. Lett. **67**, 3215 (1991)
50. H. Stapelfeldt, D.G. Papaioannou, L.D. Noordam, and T.F. Gallagher, Phys. Rev. Lett. **67**, 3223 (1991).
51. R.R. Jones, D.W. Schumacher, and P.H. Bucksbaum, Phys. Rev. A **47**, R49 (1993).
52. R.R. Jones, N.E. Tielking, D. You, C. Raman, and P.H. Bucksbaum, submitted.

IMPACTS OF ATOMIC PHYSICS
IN OTHER DISCIPLINES

Collision Processes in Astrophysical Environments

Kate P. Kirby

Harvard-Smithsonian Center for Astrophysics
60 Garden Street
Cambridge, MA 02138

Abstract. Collisions of atoms, ions, electrons and molecules play an important role in the interpretation of astronomical observations and in our understanding of the physical characteristics and the history and evolution of astronomical objects. Two very different astrophysical environments are explored here. Collision processes leading to the formation of molecules in the early universe are discussed. Because of their unique cooling capabilities, molecules may have been significant in enabling the first objects to form from primordial clouds. Atomic collisions are central to the interpretation of emission lines in astrophysical plasmas. Recent work in predicting element abundances in planetary nebulae is highlighted.

INTRODUCTION

The collision events which will be described here have not received the same media attention that the recent series of collisions of Comet Shoemaker-Levy with the planet Jupiter have received. I will be discussing collision processes involving atoms, ions, and electrons in several different astrophysical environments: first, examining the role of atomic collisions in molecule formation in the early universe; and second, elucidating atomic collision processes in astrophysical plasmas in order to understand emission lines of ions in gaseous nebulae.

A detailed characterization of many astronomical objects—including chemical abundances, ionization balance, heating and cooling processes, formation and destruction of molecules—involves understanding the relevant atomic collision processes. Signatures of these processes can provide important diagnostics of these environments—clues as to the physical characteristics such as temperature, densities and radiation field in these regions. In addition, the atomic collision processes themselves, affect critically the evolution of the objects in question.

I would like to acknowledge the pioneering work of Sir David Bates in the area of atomic collision processes relevant to astrophysics and atmospheric physics. Bates' contributions were many and he established at the Queen's University, Belfast an internationally recognized research group in the area of atomic and molecular processes. Although he had officially retired several years ago, Bates continued his exceptionally productive research up until the day he died, on January 5, 1994. This paper is dedicated to him.

FORMATION OF MOLECULES IN THE EARLY UNIVERSE

Most civilizations, if they had an interest in astronomy or science at all, have pondered the origin of the universe. Modern-day physicists and astronomers are no exception. But as Steven Weinberg in his book *The First Three Minutes* (1) noted, there has always been an aura of the disreputable surrounding such research, because of its highly speculative nature. Observational evidence and a solid theoretical foundation on which to construct a credible scenario for the early universe has been lacking until recently. Thus it is intriguing to bring atomic collision processes into our consideration of the early universe, as I think atomic physics can lend a certain respectability to this discipline. In particular, the focus will be on atomic collisions central to the formation of primordial molecules.

Although cosmology is still very much "a data-starved science," (2) there has been considerable progress in the last several decades in our understanding of the early universe. There is now "the standard cosmological model" which more or less incorporates elements of "the big bang theory" and there are at least two somewhat different models of primordial nucleosynthesis. There is the discovery of and continuing research on the large-scale structure of the universe, in which the clustering of galaxies has been shown to exist in filamentary and/or sheet-like structures, together with large voids in which there is no visible matter. There have been observations of objects out to a red-shift of almost 4. Recent results from measurements by the COBE satellite have shown the existence of small temperature fluctuations in the cosmic microwave background which are thought to arise due to small density inhomogeneities in the universe at the time of decoupling of matter and radiation. And just this month there was reported the first observational evidence that the intergalactic medium consists of a highly ionized plasma of hydrogen and helium (3), consistent with Big Bang nucleosynthesis theory.

There are many questions still to be answered, particularly with respect to the formation of the first density inhomogeneities and the evolution of primordial clouds from which the first objects formed. Atomic collision processes may have played a significant role in this scenario. The chemistry of the early universe has been explored previously by Lepp and Shull (1984) (4), Lepp and Dalgarno (1987) (5), and summarized by Dalgarno and Fox (1994) (6) and much of the following discussion is taken from these references.

In the Beginning, according to standard cosmology, in the first 10^{-2} seconds the temperature was 10^{12} K and the universe consisted of a "soup" of elementary particles, including photons, leptons, quarks and neutrinos. In the first 100 seconds, as the temperature dropped to 10^9 K, primordial nucleosynthesis took place, producing from the protons and neutrons a few other nuclei in the following abundance ratios with respect to hydrogen: D^+ (D/H~5×10^{-5}), $^7Li^{+3}$ (Li/H~1×10^{-10} to 1×10^{-8}), $^4He^{+2}$(^{4}He/H~10^{-1}) and $^3He^{+2}$(^{3}He/H~10^{-5}). The universe was hot and dense, with radiation and matter in thermal equilibrium maintained by the scattering of photons by free electrons. The space occupied by the universe expanded adiabatically and the universe cooled.

At first the radiation field was energetic enough that all atoms remained fully stripped. With the expansion occurring, the temperature steadily decreased. The process of *radiative recombination*, in which nuclei capture electrons and emit photons, took place and the universe started to become less ionized as the photon energies become low enough that the reverse process of photoionization could no longer take place. The first species to recombine was Li^{+3}:

$$Li^{+3} + e \rightarrow Li^{+2} + h\upsilon ,$$

followed rapidly by

$$Li^{+2} + e \rightarrow Li^{+} + h\upsilon . \qquad \text{(I.P.~75.6 eV)}.$$

Then He^{+2} recombined

$$He^{+2} + e \rightarrow He^{+} + h\upsilon ,$$

followed by

$$He^{+} + e \rightarrow He + h\upsilon . \qquad \text{(I.P.~24.6 eV)}.$$

With the formation of neutral helium, the first molecules could begin to be made (6). Otherwise, the impact of these recombinations was relatively minor because the abundances of these species was small compared to hydrogen.

The first molecule formation took place through the process of *radiative association* in which two nuclei collide and are stabilized as a molecule by the emission of a photon:

$$He^{+} + He \rightarrow He_2^{+} + h\upsilon$$
$$Li^{+} + He \rightarrow LiHe^{+} + h\upsilon$$
$$H^{+} + He \rightarrow HeH^{+} + h\upsilon$$

For $LiHe^+$ and HeH^+, radiative association takes place entirely through the ground state ($^1\Sigma^+$) potential curve by spontaneous emission from the vibrational continuum into the bound vibrational levels. As this process is governed by linear and higher order terms in the dipole moment of the ground state, it is very improbable, with a very small rate coefficient. For He_2^+, however, the atoms can approach each other along an excited state potential curve ($^2\Sigma_g^+$) from which the system can radiate to the ground state ($^2\Sigma_u^+$). The rate coefficient is determined by the dipole transition moment ($^2\Sigma_g^+ - {}^2\Sigma_u^+$) in this case and is significantly enhanced (7).

Destruction mechanisms for these molecules include *photodissociation*, *dissociative recombination*, and collisions with neutral hydrogen. The radiation field is still hot enough that photodisssociation is rapid for He_2^+

$$\text{He}_2^+ + h\upsilon \to \text{He}^+ + \text{He} .$$

However, HeH$^+$ and LiHe$^+$, at high temperatures pertaining for large z, have such high effective thresholds for photodissociation, ~11eV and ~19eV, respectively, that they are not easily destroyed by photons.

Dissociative recombination rates are known only for HeH$^+$ (8):

$$\text{He}_2^+ + e \to \text{He} + \text{He}$$
$$\text{LiHe}^+ + e \to \text{Li} + \text{He}$$
$$\text{HeH}^+ + e \to \text{He} + \text{H} .$$

Collisions with atomic hydrogen can become significant as cooling continues

$$\text{He}_2^+ + \text{H} \to \text{He} + \text{He} + \text{H}^+$$

and

$$\text{HeH}^+ + \text{H} \to \text{H}_2^+ + \text{He} .$$

After approximately 100,000 years, at a redshift, z, of about 1000, the temperature decreased to ~3000K. Photoionization of atomic hydrogen was no longer effective in reversing the radiative recombination of protons and electrons:

$$\text{H}^+ + e \to \text{H} + h\upsilon .$$

During this period, known as the recombination epoch, the universe started to become neutral. The particle density at this time was approximately 10^3 cm^{-3}. At this point the radiation and matter began to lose thermal contact with each other, due to the disappearance of free electrons, and from then on the temperature of matter decreased more rapidly than the temperature of the radiation. With the advent of the recombination era, neutral molecules could be formed—H$_2$ being the most abundant.

Unlike the other molecules described so far, molecular hydrogen is not formed by radiative association of ground state atoms. Because H$_2$ lacks a dipole moment, formation directly through the ground state $X^1\Sigma^+$ potential curve is not possible. Formation along the excited $^3\Sigma_u^+$ potential curve would necessitate a highly forbidden triplet-singlet transition.

The two main mechanisms for H$_2$ formation involve protons and electrons as catalysts. The first mechanism to become operative while the universe was still partially ionized involves *radiative association* to form H$_2^+$:

$$\text{H}^+ + \text{H} \to \text{H}_2^+ + h\upsilon$$

followed by *charge transfer* with neutral hydrogen, regenerating the proton:

$$H + H_2^+ \rightarrow H_2 + H^+ .$$

The second mechanism involves as a first step *radiative attachment* to form H^-:

$$H + e \rightarrow H^- + h\upsilon$$

followed by *associative detachment* to form H_2:

$$H^- + H \rightarrow H_2 + e .$$

This mechanism can only become a significant source for H_2 when the radiation field has cooled sufficiently that the reverse process of photodetachment of H^- is not effective.

Two additional ways of contributing to the H_2 production by first producing H_2^+, which then undergoes charge transfer to produce H_2, include:

$$HeH^+ + H \rightarrow H_2^+ + He$$

and

$$H(n=2) + H \rightarrow H_2^+ + e . \quad (associative\ ionization)$$

Mutual neutralization of H^- by reactions with H_2^+, H^+ and He^+ diminishes slightly the formation of H_2 from H^-. Once formed, H_2 is not easily destroyed in the early universe due to the steadily decreasing temperature of the radiation field.

The formation of molecules from the trace constituents Li and D is of interest because such molecules have non-zero dipole moments and therefore, despite their low abundances, they may have played a significant role in cooling the primordial gas. HD is formed in ways exactly analogous to H_2, but in addition

$$D^+ + H_2 \rightarrow H^+ + HD$$

and *radiative association* is possible through the ground state of HD because of the small permanent dipole moment. The rate of this latter process is very small (4).

After most of the H^+ has recombined, the ^7Li still exists as Li^+ because the ambient radiation field has not cooled sufficiently to prevent photoionization of Li (I.P. 5.4 eV). It is not clear whether the Li^+ will ever be able to recombine, because by the time the radiation field has cooled sufficiently, almost all the free electrons have been recombined with the hydrogen, leaving a density n_e/n_H as low as (4) 1×10^{-4}, and the overall particle density may be so low that collisions have become exceedingly rare.

However, LiH^+ may be formed by radiative association of Li^+ and H:

$$Li^+ + H \rightarrow LiH^+ + h\upsilon .$$

Stancil, Kirby and Dalgarno (9) have calculated this rate coefficient to be ~ 10^{-23} cm^3 s^{-1}. The photodissociation threshold of LiH$^+$ is ~8eV and once the radiation field has cooled sufficiently the molecule will not be easily photodissociated. Additional destruction channels include dissociative recombination:

$$\text{LiH}^+ + e \rightarrow \text{Li} + \text{H}$$

and collisions with neutral hydrogen:

$$\text{LiH}^+ + \text{H} \rightarrow \text{Li}^+ + \text{H}_2 \ .$$

Dissociative recombination results in neutral lithium. The radiation association of Li with both H and H$^+$ has also been studied:

$$\text{Li} + \text{H}^+ \rightarrow \text{LiH}^+ + h\upsilon$$

and

$$\text{Li} + \text{H} \rightarrow \text{LiH} + h\upsilon \ .$$

Preliminary studies appear to indicate that the formation of LiH$^+$ will be favored significantly, even though the ratio $n(\text{H}^+)/n(\text{H})$ is ~1×10^{-4}. These results should be of considerable interest to astronomers carrying out searches for highly redshifted ro-vibrational lines of LiH thought to be emitted in collapsing proto-clouds (10), and to those who would like to use LiH and LiH$^+$ observations at high red-shifts to constrain primordial nucleosynthesis models (11).

To summarize the preceding discussion, Figure 1 shows the log of the fractional abundances of the various atoms, ions and most abundant molecules as a function of the redshift, z. The cosmological redshift is the result of the fact that space is expanding and the wavelength of the individual photons increases in proportion to the size of the universe. The red-shift is defined as: $z = \dfrac{\lambda_{obs} - \lambda_{lab}}{\lambda_{lab}}$. Thus $z=0$ is the present epoch, and as we go to higher redshifts we go back in time to earlier epochs. From Figure 1, obtained from J. H. Black (12), one can see the disappearance of the ions, He^{++}, He$^+$ and H$^+$ as recombination occurs, and the formation of molecules starting at redshifts around z~1000. The LiH$^+$ and LiH is not included in this figure because the quoted rate coefficients and formation processes have not yet been included in a time-dependent code for evolution of the early universe. As can be seen, the total fractional abundance of molecular material approaches ~8×10^{-6}, and is overwhelmingly H$_2$.

Despite the low fractional abundance, the role of primordial molecules may be quite significant. There have been recent papers addressing the effect of primordial

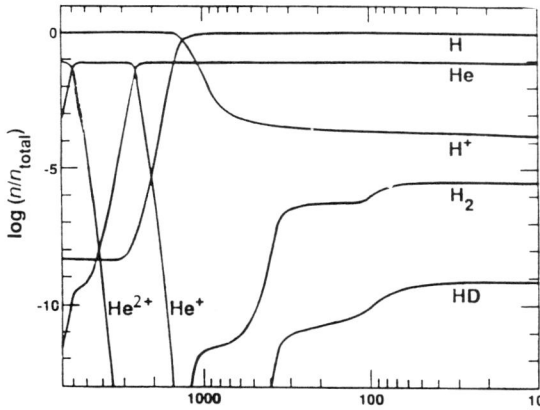

FIGURE 1. The fractional abundance of the various ions of hydrogen and helium as a function of red shift (see Reference 12).

molecules on the thermal evolution of the early universe (13) and the influence of such molecules on cosmic background radiation anisotropies at redshifts between 400 and 10 (14).

Molecules are more effective coolants than atoms like H and He, particularly for temperatures ranging from 100-10,000K. Cooling is accomplished when kinetic energy of the gas is transformed through collisional excitation into internal energy which is then radiated away. The numerous vibrational and rotational energy levels in molecules, which can be excited at much lower temperatures than electronic energy levels, allow for ample cooling. Molecules with large dipole moments can be significant even at extremely low abundance. Molecular hydrogen, with no dipole moment, must cool through electric quadrupole transitions.

The cooling function of molecules may have been critical to the formation of the first objects, enabling gravitational collapse to occur. After collapse is initiated there must be mechanisms to radiate away the energy. Otherwise, the temperature will increase and collapse will stop because gravity will not be able to overcome the gas pressure. Although there are attempts to observe the photons emitted by primordial molecules at large redshifts, the molecules themselves were probably destroyed by ultraviolet radiation from the first stars.

ATOMIC COLLISION PROCESSES IN ASTROPHYSICAL PLASMAS

Atomic physics plays a pre-eminent role in the interpretation of observations of emission lines in astrophysical plasmas—such as are found in stellar atmospheres, including the sun, and gaseous nebulae. There is not enough space to do justice to this subject, but it is an area that has experienced significant progress over the past decade both with respect to the observations and the atomic collision data. The idea for discussing this subject came from several extremely interesting recent papers of Kastner and Bhatia (15,16), in which they examine old assumptions regarding excitation mechanisms for line emission using better atomic data and improved observational facilities.

For the sake of simplicity, this discussion will focus on planetary nebulae. A typical planetary nebula consists of gas which has been ionized by an embedded central star with an effective temperature in the range of 35,000-100,000K. Heating of the surrounding gas occurs through photoionization by ultraviolet photons coming from the central star, and is mediated by collisions between electrons and between electrons and ions.

Optical and uv emission lines of C, N, O, Si, Fe, and other metals, in various stages of ionization are used to deduce the abundances of these species. Abundances relate directly to the degree of stellar processing of the material, and therefore the history of the object. Some of these emission lines are used as important diagnostics of electron density and temperature in these regions. If there are lines which arise due to charge transfer, they can provide a measure of the neutral component in the ionized plasma.

The subject of emission line ratios as diagnostics of electron density and electron temperature has been treated in some detail by Osterbrock (17) and others (18), and will not be discussed here. However, once the temperature and density in a nebula are known, abundances of various ions can be deduced from the observed line strengths if the mechanism exciting the line is known.

There are a number of atomic collision processes which create excited ions, $(A^{i+})^*$, that then radiate:

Electron impact excitation:

$$A^{i+} + e \rightarrow (A^{i+})^* + e$$

Radiative recombination:

$$A^{(i+1)+} + e \rightarrow (A^{i+})^* + h\upsilon$$

Dielectronic recombination

$$A^{(i+1)+} + e \rightarrow (A^{i+})^{**} \rightarrow (A^{i+})^* + h\upsilon$$

Charge transfer:

$$A^{(i+1)+} + H \rightarrow (A^{i+})^* + H^+ \ .$$

There are some simple rules linking the kinds of emission lines observed with the processes operative in exciting them. Thermal electrons, with $T \sim 5 \times 10^3 - 2 \times 10^4 K$, colliding with ions excite low-lying energy levels of the ions, giving rise to prominent forbidden-line spectra. Electron densities, in the range of $10^2 - 10^4$ cm^{-3}, are too low for collisional de-excitation to be important. Radiative recombination, the reverse process of photoionization, and dielectronic recombination, the reverse process of autoionization, populate many highly excited levels of the recombined ion that decay by radiative cascade to the ground state. The resulting spectra generally consist of many weak, allowed transitions. In dielectronic recombination, capture proceeds through a doubly-excited resonance state of the recombined ion, which stabilizes through the emission of a photon, leaving a Rydberg state of the recombined ion. Because it is a resonance process, the cross sections for dielectronic recombination are generally larger than for radiative recombination, but it is often most efficient to combine the cross sections to produce an effective recombination rate coefficient (19).

If there is a substantial component of neutral gas present, charge transfer can be important and an excited state of the recombined ion may be created. Usually only one or two characteristic lines result from the charge transfer process at low temperatures. Radiationless processes such as charge transfer are usually much faster than radiative processes in which photons are produced, by four to five orders of magnitude (20). If one writes the ratio of probabilities for neutralization by charge transfer versus recombination as $k(T)N(H)/(\alpha(T)N_e)$, with $k \sim 10^{-8}$ cm^3 s^{-1} and $\alpha \sim 10^{-12}$ cm^3 s^{-1} (N_e is electron density, N(H) is hydrogen atom density), it is clear that for $N(H)/N_e > 10^{-4}$, charge transfer can become the dominant recombination mechanism (21).

Clearly there are a number of different lines to be used in making abundance determinations. Until recently, however, only the collisionally excited forbidden lines were used for elements other than He and H. In an interesting paper appearing this year by Liu, Storey, Barlow and Clegg (22), which focuses on the rich OII recombination spectrum in NGC 7009, the abundances of C, N and O are derived based on recombination line measurements.

Liu et al. (22) identified over seventy emission lines from ~3800 to 4700 Å resulting from the recombination of OIII:

$$O^{2+} + e \rightarrow O^+ (nl) + h\upsilon .$$

The emission lines of O^+ (nl) are what is observed. They carried out new quantal calculations of the O^+ radiative recombination coefficients, by transforming coefficients calculated in L-S coupling to an intermediate coupling case. They show that the departure from L-S coupling is very important in obtaining consistent abundances of O^{2+} from different emission lines.

The most important point made by Liu et al (22) is that abundances obtained for C, N and O based on recombination lines are all greater by approximately a factor of 5 than abundances obtained using the collisionally excited forbidden lines. This confirms previous findings (23) for C^{2+}. Instead of blaming the atomic data for the discrepancies in the abundances deduced from recombination as opposed to collisionally-excited lines, the authors explore whether the same physical process may be responsible in all three cases. The factor of 5 consistency for all three elements appears to rule out an error in the atomic data. Temperature fluctuations are suggested as a source of the problem, as forbidden-line excitation is very sensitive to temperature. In addition, shocks are suggested as possibly enhancing the fluxes in certain forbidden lines which lead one to derive higher temperatures and therefore lower abundances of ions than are correct.

Over the last decade, there has been tremendous progress in the quantity and quality of the atomic data available. The theoretical work on the part of a number of people involved in the Opacity Project should particularly be noted here. There has been extraordinary experimental activity in the field of radiative and dielectronic recombination.

It is only within the last four years that there have been the first experimental measurements of radiative recombination—constituting the first direct test of the theory for this process. Examples of work in this area come from the group at Aarhus (Anderson et al.) (24) on fully stripped carbon:

$$C^{6+} + e \rightarrow C^{5+} (n) + h\upsilon$$

and from the group at Stockholm (Schuch et al.) (25) on recombination of deuterons with electrons:

$$D^+ + e \rightarrow D (n) + h\upsilon \ .$$

There have been substantial technological advances which have led to new experimental apparatus with which to measure recombination. Examples include the EBIT (Livermore), EBIS (Kansas State) and the heavy-ion storage rings at Aarhus, Darmstadt, Heidelberg and Stockholm. There has been extremely productive interplay between theorists and experimenters in exploring the dielectronic recombination process, e.g. Pindzola, Badnell and Griffin (1990) (26).

There is recognition by both theorists and experimenters of the significance of external electric fields in the recombination process and efforts are underway to measure such effects accurately, as demonstrated in a poster paper by Savin et al. (27) at this conference. The effects of both fields and collisional mixing may be important to our understanding of the recombination process in astrophysical plasmas.

Much of the current work in the field of recombination is well summarized in a proceedings of a NATO Workshop held in 1991 (28).

Astronomers have been making considerable progress on the observations of emission lines also. In the ultraviolet, the Goddard High Resolution Spectrograph (GHRS) aboard the Hubble Space Telescope is making a tremendous impact. Figure 2, taken from an article by Leckrone et al. (29), compares the resolution obtained with IUE (launched in 1978) with the resolution obtained by GHRS over a two angstrom region around 1938 Å. The spectrum of the star χ-Lupi clearly exhibits lines from many heavy element ions, features heretofore unseen. Leckrone et al. (29) issue a challenge to atomic physicists: "We face the bleak prospect of attempting to analyze spectra of 1% precision, obtained at great expense, with atomic parameters that can be inaccurate by factors of 2 or 10 if they exist at all. The GHRS will be relentless in highlighting this problem." The authors (29) further conclude that now and in the future atomic physicists are and will be playing an absolutely essential role in the analysis and interpretation of the astrophysical spectroscopic data.

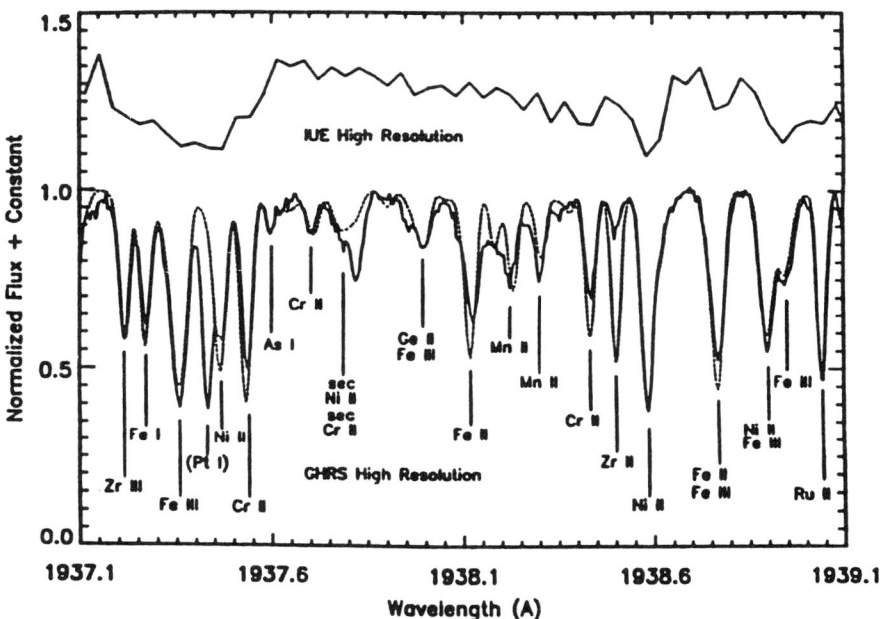

FIGURE 2. Comparison of high dispersion spectra of χ-Lupi obtained with the International Ultraviolet Explorer (IUE) satellite (top) and with GHRS (bottom). Courtesy of D. Leckrone (see Ref. 29).

Significant advances in the astronomical observations and in the atomic data ultimately challenge the underlying astrophysical assumptions upon which models are based and lead to new understanding.

ACKNOWLEDGEMENTS

I would like to thank A. Dalgarno, M. Geller and J. Raymond for helpful and stimulating discussions, and P. Stancil for carrying out the LiH$^+$ and LiH studies so thoroughly and rapidly. In addition, I thank J. Black and D. Leckrone for permission to use Figures 1 and 2.

REFERENCES

1. Weinberg, S., *The First Three Minutes*, New York: Basic Books, Inc., 1977.

2. Kolb, E. W. and Turner, M. S., *The Early Universe*, Redwood City, CA: Addison Wesley, 1990.

3. Jakobsen, P., Boksenberg, A., Deharveng, J., Greenfield, P., Jedrzejewski, R. and Paresce, F., *Nature* **370**, 35 (1994).

4. Lepp, S. and Shull, J. M., *Astroph. J.* **280**, 465 (1984).

5. Lepp, S. and Dalgarno, A., in *Astrochemistry* ed. Vardya and Tarafdar, 1987, pp. 109-120.

6. Dalgarno, A. and Fox, J. L., in *Unimolecular and Bimolecular Reaction Dynamics*, ed. Ng, Baer and Powis, New York: John Wiley & Sons, 1994, pp. 1-85.

7. Stancil, P. C., Babb, J. F. and Dalgarno, A., *Astrophys. J.* **414**, 672 (1993).

8. Yousif, F. B. and Mitchell, J.B.A., *Phys. Rev. A* **40**, 4318 (1989); Yousif, F.B., Mitchell, J.B.A., Rogelstad, M., LePaddelec, A., Canosa, A., and Chibisov, M.I., *Phys. Rev. A.*, **49**, 4610 (1994); Guberman, S.L., *Phys. Rev A* **49**, R4277 (1994); Sarpal, B.K., Tennyson, J. and Morgan, L.A. (unpublished).

9. Stancil, P., Kirby, K. and Dalgarno, A. (1994) to be published.

10. de Bernardis, P., Dubrovich, V., Encrenaz, P., Maoli, R., Masi, S., Mastrantonio, G., Melchiorri, B., Melchiorri, F., Signore, M. and Tanzilli, P. E., *Astron. Astrophy.* **269**, 1 (1993).

11. Signore, M., Verdrenne, G. de Bernardis, P., Dubrovich, V., Encrenaz, P., Maoli, R., Masi, S., Mastrantonio, G., Melchiorri, B., Melchiorri, F.and Tanzilli, P., *Ap. J. Suppl.* **92**, 535 (1994).

12. Black, J. H., in *Molecular Astrophysics*, ed by T. Hartquist, Cambridge: Cambridge University Press, 1990, pp. 473-480. Also, W. B. Latter, Thesis, University of Arizona (1989)

13. Puy, D., Alecian, G., Le Bourlot, J., Leorat, J. and Pineau des Forêts, G., *Astron. Astrophys.* **267**, 337 (1993).

14. Maoli, R., Melchiorri, F.and Tosti, D., *Astrophys .J.* **425**, 372 (1994).

15. Kastner, S. O. and Bhatia, A. K., *Astrophys. J.* **406**, 708 (1993).

16. Kastner, S. O. and Bhatia, A. K., *Astrophys. J.* **381**, L59 (1991).

17. Osterbrock, D. E., *Astrophysics of Gaseous Nebulae and Active Galactic Nuclei*, Mill Valley, CA: University Science Books, 1989.

18. e.g. Dupree, A.K. in *Advances in Atomic and Molecular Physics ,Vol 14.*, ed. by Bates and Bederson, New York: Academic Press, 1978, pp. 393-431.

19. Péquignot, D., Petitjean, P. and Boisson, C. *Astron. Astrophys.* **251**, 680 (1991); Storey, P.J., *Astron. Astrophys.* **282**, 999 (1994).

20. Bates, D. R., in *Molecular Astrophysics*, ed. T. Hartquist, Cambridge: Cambridge University Press, 1990, pp. 211-231.

21. Shields, G. A., in *Molecular Astrophysics*, ed. T. Hartquist, Cambridge: Cambridge University Press, 1990, pp. 461-472.

22. Liu, W.-X., Storey, P. J., Barlow, M. J. and Clegg, R. E. S., *Mon Not. R. Astron. Soc.*, to be published (1994).

23. Kaler, J.B., *Astroph. J.* **308**, 337 (1986).

24. Andersen, L.H., Bolko, J. and Kristegaard, P., *Phys. Rev. Lett.* **64**, 729 (1990).

25. Schuch, R., Quinteros, T., Pajek, M., Haruyama, Y., Danared, H., Hui, G., Andler, G., Schneider, D. and Starker, J., *Nucl. Instr. Methods Phys. Res.* **B79**, 59 (1993).

26. Pindzola, M. S., Badnell, N. R. and Griffin, D. C., *Phys. Rev. A* **42**, 282 (1990).

27. Savin, D. W., Reisenfeld, D. B., Gardner, L. D., Young, A. R. and Kohl, J. L. *14th International Conference on Atomic Physics, Abstracts of Contributed Papers*, 1L-4 (1994).

28. *Recombination of Atomic Ions*, ed. by W. G. Graham, W. Fritsch, Y. Hahn and J. A. Tanis, New York: Plenum Press 1992.

29. Leckrone, D. S., Johansson, S., Wahlgren, G. M. and Adelman S. J., *Physica Scripta* **T47**, 149 (1993).

Quantum Computation

Artur Ekert

Clarendon Laboratory, University of Oxford
Oxford OX1 3PU, U.K.

Abstract: As computers become faster they must become smaller because of the finiteness of the speed of light. The history of computer technology has involved a sequence of changes from one type of physical realisation to another - from gears to relays to valves to transistors to integrated circuits and so on. Quantum mechanics is already important in the design of microelectronic components. Soon it will be necessary to harness quantum mechanics rather than simply take it into account, and at that point it will be possible to give data processing devices new functionality.

COMPUTATION IS A PHYSICAL PROCESS!

Computers are physical objects, and computations are physical processes. This sentence, innocuous at first glance, leads to non-trivial consequences, some of which we wish to explore in this paper. In particular we want to convince the reader that the theory of computation is not a branch of pure mathematics. Fundamental questions regarding computability and computational complexity are questions about physical processes that reveal to us properties of abstract entities such as numbers or ideas. Those questions belong to physics rather than mathematics [1, 2].

For the purpose of this presentation we define a computation as a physical process that produces final states, *outputs*, that depend in some desired way on given initial states, *inputs*. Quantum computers are defined as physical devices whose unitary dynamics can be regarded as the performance of a computation. Unlike classical computers, quantum computers can operate on quantum superpositions of different numbers.

In recent years a new quantum theory of computation has been developed [3–8]. It is in many ways different from the classical, purely mathematical theory of computation. A single quantum computer can follow many distinct computation paths all at the same time and produce a final output depending on the interference of all of them. In particular, it has been shown recently that quantum computers can efficiently perform classes of computation, e.g. factorisation, which are believed to be intractable on any classical computer [9]. This makes it highly desirable to construct such devices. This paper is aimed at physicists rather than computer scientists. We hope that it will stimulate physicists, especially those working in the fields of quantum optics, atomic physics, and condensed matter physics, to search for physical processes that will eventually make quantum computation practical.

SLOW AND FAST ALGORITHMS

Before we go on and show that quantum computers are more powerful than their classical counterparts we have to pause and explain what *powerful* really means. In order to solve a particular problem computers follow a precise set of instructions that can be mechanically applied to yield the solution to any given instance of the problem. A specification of this set of instruction is called an algorithm. Examples of algorithms are the procedures taught in elementary schools for adding and multiplying whole numbers; when these procedures are mechanically applied, they always yield the correct result for any pair of whole numbers. Some algorithms are fast (e.g. multiplication) other are very slow (e.g. factorisation, playing chess). Consider, for example, the following factorisation problem

$$\Box \times \Box = 29083. \tag{1}$$

How long would it take you, using paper and pencil, to find the two whole numbers which should be written into the two boxes (the solution is unique). Probably about one hour. Solving the reverse problem

$$127 \times 229 = \Box, \tag{2}$$

again using paper and pencil technique, takes less than a minute. All because we know fast algorithms for multiplication but we do not know equally fast algorithms for factorisation.[1] It does not mean that the fast *classical* algorithm for factorisation does not exists, maybe it does and maybe it will be discovered one day, at the moment we simply do not know it.

---
[1] Currently, the best classical factoring algorithms are: the Multiple Polynomial Quadratic Sieve [10], for numbers less than 120 decimal digit long, and the Number Field Sieve [11], particularly good for numbers more than 110 decimal digits long. Still, even the fastest algorithms would need a couple of billions years to factorise a 200-digit number.

Skipping details of the computational complexity we only mention that computer scientists have a rigorous way of defining what makes an algorithm fast (and usable) or slow (and unusable) [12, 13]. For an algorithm to be fast, the time it takes to execute the algorithm must increase no faster than a polynomial function of the size of the input. Informally think about the input size as the total number of bits needed to specify the input to the problem, for example, the number of bits needed to encode the number we want to factorise. If the best algorithm we know for a particular problem has the execution time (viewed as a fuction of the size of the input) bounded by a polynomial then we say that the problem belongs to class P. Problems outside class P are known as hard problems. Thus we say, for example, that multiplication is in P whereas factorisation is not in P and that is why it is a hard problem.

NUMBERS

Traditionally we view computation as operations on numbers. However, a number, although an abstract entity, is represented in practice by a state of a physical object. If a physical object can be put into two different, distinguishable states then this object can represent two different numbers. We call any two state system a physical bit; when the system is quantum and the two states are two orthogonal *quantum* states we refer to it as a quantum bit or simply a qubit. Both a single classical bit and a qubit can represent at most two different numbers, however, qubits are different because apart from the two orthogonal basis states, which we label as $|0\rangle$ and $|1\rangle$, they can also be put into infinitely many other states of the form $|\Psi\rangle = c_0|0\rangle + c_1|1\rangle$.

Consider now a register composed of m physical bits. There are 2^m different states of this register therefore the register can represent 2^m different numbers, e.g. from 0 to $2^m - 1$. A quantum register of size m is described by a vector in the 2^m-dimensional Hilbert space. For example for $m = 3$ the most general state of the register can be written as

$$\begin{aligned}|\Psi\rangle &= c_{000}|0\rangle|0\rangle|0\rangle + c_{001}|0\rangle|0\rangle|1\rangle \\ &+ c_{010}|0\rangle|1\rangle|0\rangle + c_{011}|0\rangle|1\rangle|1\rangle \\ &+ c_{100}|1\rangle|0\rangle|0\rangle + c_{101}|1\rangle|0\rangle|1\rangle \\ &+ c_{101}|1\rangle|0\rangle|1\rangle + c_{111}|1\rangle|1\rangle|1\rangle,\end{aligned} \quad (3)$$

or,

$$|\Psi\rangle = \sum_{\bar{x}\in\{0,1\}^3} c_{\bar{x}}|\bar{x}\rangle, \quad (4)$$

where $\{0,1\}^3$ stands for all strings of zeros and ones of length 3 and the bar in $\bar{x}$ indicates that the number x is encoded using binary notation. From time to time we will find convenient to relabel the states and, instead of the binary notation $\bar{x}$, to use the decimal notation x,

$$|\Psi\rangle = c_0 |0\rangle + c_1 |1\rangle + \ldots + c_7 |7\rangle = \sum_{x=0}^{2^m-1} c_x |x\rangle. \tag{5}$$

In order to prepare a specific number in the register we have to perform m elementary operations - we set each of the m physical bits into one of the two states. Quantum registers accept more general states which represent not only numbers but also coherent superpositions of several numbers. Indeed, m elementary unitary transformations performed bit by bit can prepare the register in a coherent superposition of all 2^m numbers that can be stored in the register. Take the register initially in $|0\rangle |0\rangle \ldots |0\rangle$ state and apply the unitary operation

$$T = \frac{1}{\sqrt{2}} \begin{pmatrix} 1 & 1 \\ 1 & -1 \end{pmatrix} \tag{6}$$

to each qubit. This operation, introduced by Deutsch and Jozsa [4], is very useful in quantum computation so for our later convenience we will denote it by $\hat{A}$. The resulting state of the register is an equally weighted superposition of all 2^m numbers,

$$|\Psi\rangle = \hat{A} |\bar{0}\rangle = \overbrace{T|0\rangle T|0\rangle \ldots T|0\rangle}^{m \text{ times}} = \frac{1}{2^{m/2}} \sum_{\bar{x} \in \{0,1\}^m} |\bar{x}\rangle = \frac{1}{2^{m/2}} \sum_{x=0}^{2^m-1} |x\rangle. \tag{7}$$

It is quite remarkable that in quantum registers m elementary operations can generate a state containing all 2^m possible numerical values of the register. In contrast, in classical registers m elementary operations can only prepare one state of the register representing one specific number. This property of quantum registers can be used for a quantum parallel processing. If, after preparing the register in a coherent superposition of several numbers, all subsequent computational operations are unitary (i.e. preserve the superpositions of states) then with each computational step the computation is performed on all the numbers in the superpositions. This type of computation is particularly useful for problems which, in classical case, involve performing the same computation several times for different input data e.g. to calculate a period of a given function $f(x)$.

ALGORITHMS AND PHYSICS

Although algorithms are abstract specifications of instructions they depend on the physics of computation. Algorithms tell you - do this, do that, then do something else; what can be done to a number depends on the physical representation of the number. Classical algorithms cannot tell you, for example, "...and now prepare a coherent superposition of several numbers", but quantum algoritms can! Consequently we know a very fast method for factorisation, not with paper and pencil though but with quantum devices.

We already mentioned that the reason why quantum computers are faster is due to their ability to process quantum superpositions of many numbers in one computational step; each computational step being a unitary transformation of quantum registers. If quantum computers could speed up some computations, say, twice, or even several thousands times, they would be of some interest from the technological point of view but of no interest to the theory of computational complexity. However, quantum computers are of great interest to computer scientists because the speed-up is exponential. Some problems which are hard with respect to classical computation can end up in class P with respect to quantum computation. As an example let us describe the Fourier-Hadamard sampling problem [5].

Consider a string of 2^m real numbers a_x, $x = 0, 1, \ldots 2^m - 1$. We define its Fourier-Hadamard transform as a string b_y such that

$$b_y = \frac{1}{2^{m/2}} \sum_{\vec{x}} (-1)^{\vec{x}\cdot\vec{y}} a_x, \qquad (8)$$

where $\vec{x}$ and $\vec{y}$ are vectors formed out of the m-bit binary representation of $\bar{x}$ and $\bar{y}$ respectively, for example, for $x = 2$ we have $\vec{x} = [0, 0, \ldots, 1, 0]$. We define the Fourier-Hadamard sampling problem by providing as input an arbitrary string $\{a_x\}$ and requiring as output any y with probability proportional to $|b_y|^2$.

Classical computers, in order to sample with the probability proportional to the Fourier-Hadamard power spectrum, need exponential time as a function of the length of the string m. Quantum computers can solve the problem in time proportional to m.

The quantum algorithm is quite simple. The string a_x (properly renormalised) is encoded in the initial state of a quantum register of size m,

$$|\Psi\rangle = \sum_{\bar{x} \in \{0,1\}^m} a_{\bar{x}} |\bar{x}\rangle. \qquad (9)$$

We tacitly assumed that this state preparation can be performed efficiently. Then we apply $\hat{A}$ to the register i.e. apply the unitary transformation T to each of the m qubits in the register. The result of the transformation when applied to a number $|\bar{x}\rangle$ is

$$\hat{A}|\bar{x}\rangle = \overbrace{T \otimes T \otimes T \ldots \otimes T}^{m \text{ times}} |\bar{x}\rangle = \frac{1}{2^{m/2}} \sum_{\bar{y} \in \{0,1\}^m} (-1)^{\vec{x}\cdot\vec{y}} |\bar{y}\rangle, \qquad (10)$$

thus the state $|\Psi\rangle$ is transformed to a new state $|\tilde{\Psi}\rangle$ given by

$$|\tilde{\Psi}\rangle = \sum_{\bar{y}\in\{0,1\}^m} \left(\frac{1}{2^{m/2}} \sum_{\bar{x}\in\{0,1\}^m} (-1)^{\bar{x}\cdot\bar{y}} a_{\bar{x}} \right) |\bar{y}\rangle = \sum_{\bar{y}\in\{0,1\}^m} b_y |\bar{y}\rangle. \qquad (11)$$

The measurement of the state of the register will yield a number y with probability $|b_y|^2$, which solves the sampling problem.

The algorithm, proposed by Bernstein and Vazirani [5], makes use of quantum superpositions and non-deterministic outcomes of quantum measurements. The unitary transformation of the register of size m, which represents the computation performed on 2^m numbers, was decomposed into a sequence of elementary unitary operations; the length of this sequence is proportional to m rather than to 2^m which results in the exponential speed up of the computation.

COMPUTING FUNCTIONS

Let us describe now how quantum computers compute functions. For this we will need two quantum registers of length m and n. Consider a function

$$f : \{0, 1, \dots 2^m - 1\} \longrightarrow \{0, 1, \dots 2^n - 1\}, \qquad (12)$$

where m and n are natural numbers.

A classical computer computes f by evolving each labelled input, $0, 1, \dots 2^m - 1$ into a respective labelled output, $f(0), f(1), \dots f(2^m - 1)$. Quantum computers, due to the unitary (and therefore reversible) nature of their evolution, compute functions in a slightly different way. In order to compute functions which are not one-to-one and to preserve the reversibility of computation, quantum computers have to keep the record of the input. Here is how it is done.

We will use the two quantum registers; the first register to store the input data, the second one for the output data. Each possible input x is represented by $|x\rangle$ - the quantum state of the first register. Analogously, each possible output $y = f(x)$ is represented by $|y\rangle$ - the quantum state of the second register. Vectors $|x\rangle$ belong to the 2^m-dimensional Hilbert space $\mathcal{H}_1$, and vectors $|y\rangle$ belong to the 2^n-dimensional Hilbert space $\mathcal{H}_2$. States corresponding to different inputs and different outputs are orthogonal, $\langle x|x'\rangle = \delta_{xx'}$, $\langle y|y'\rangle = \delta_{yy'}$. The function evaluation is then determined by the evolution of the two registers,

$$|x\rangle |0\rangle \xrightarrow{U_f} |x\rangle |f(x)\rangle. \qquad (13)$$

We can always prepare specific x in the first register and read the value $f(x)$ from the second register. It was shown that as far as the computational complexity is concerned a reversible function evalution, i.e. the one that keeps track of the input, is as good as a regular, irreversible evalution [14]. This means that if a given function can be computed in polynomial time it can also be computed in polynomial time using a reversible computation. The computation we are considering here is not only reversible but also quantum and we can do much more than computing values of $f(x)$ one by one. We can prepare a superposition of all input values as a single state and by running the computation U_f only once, we can compute *all* of the 2^m values $f(0), \ldots, f(2^m - 1)$,

$$\frac{1}{2^{m/2}} \left[\sum_{x=0}^{2^m-1} |x\rangle \right] |0\rangle \xrightarrow{U_f} |0\rangle |f(0)\rangle + |1\rangle |f(1)\rangle + \ldots + |2^m - 1\rangle |f(2^m - 1)\rangle. \tag{14}$$

It looks too good to be true so where is the catch? How much information about f does the state

$$|f\rangle = |0\rangle |f(0)\rangle + |1\rangle |f(1)\rangle + \ldots + |2^m - 1\rangle |f(2^m - 1)\rangle \tag{15}$$

contain?

Unfortunately no quantum measurement can extract all of the 2^m values $f(0), f(1), \ldots, f(2^m - 1)$ from $|f\rangle$. However, there are measurements that provide us with information about joint properties of all the output values $f(x)$ such as, for example, periodicity. We will see in the next section, how a periodicity estimation can lead to fast factorisation.

QUANTUM FACTORISATION

The most spectacular result, due to Peter W. Shor from AT&T Bell Lab, employs quantum computation to perform an efficient factorisation of big composite numbers [9]. It is the problem on which security of many classical public key cryptosystems is based. [2]

We are going to outline briefly how quantum factorisation works. For details see [9] and [16].

Suppose you were given a number N and asked to factorise it into prime factors. In other words you are asked to find whole numbers $\{p\}$ such that any p divides N with the reminder 0. Divisibility by p is a periodic property - every pth number is divided by p. It should not then come as a big surprise that the factorisation problem is related to finding periods of certain functions.

---

[2]RSA - the most popular public key cryptosystem named after the three inventors, Ron Rivest, Adi Shamir, and Leonard Adleman [15] - gets its security from the difficulty of factoring large numbers.

In particular one can show that finding factors of N is equivalent to finding a period of $f_N(x)$, where
$$f_N(x) = a^x \bmod N. \tag{16}$$

The result of this operation is the remainder from the division of a^x by N; a could be any randomly chosen number. The function is periodic and the period r, which depends on a and N, is called the order of x modulo N. For example, the increasing powers of 2 modulo 7 go like $1, 2, 4, 1, 2, 4, 1, ...$ and so on - the order of 2 modulo 7 is 3 (for more information see, for example, [17]).

Quantum computers can compute $f_N(x)$ and find its period much faster than any classical computer. Let us sketch briefly the main steps of the quantum algorithm which for a randomly chosen a finds the smallest r such that
$$f_N(x + r) = f_N(x).$$

As before, our machine has two quantum registers.

- The input register is prepared in the equally weighted superposition of all numbers from 0 to $M - 1$ where $M = 2^m$ is bigger than N (M is of the order of N^2). Then a random a is chosen and the function $f_N(x)$ is computed. The state of the two registers is

$$\frac{1}{\sqrt{M}} \sum_{x=0}^{M-1} |x\rangle |f_N(x)\rangle. \tag{17}$$

- The measurement is performed on the second register yielding a value $|f_N(k)\rangle$ for some k. This k, however, is not known because inverting $|f_N(k)\rangle$ and finding k is a hard problem. What we know is that due to the periodicity of $f_N(x)$ the state of the first register is reduced to the respective relative state labelled by k. It can be written as,

$$\sqrt{\frac{r}{M}} \sum_{x=0}^{M-1} c_x^k |x\rangle, \tag{18}$$

where the coefficients c_x^k, for a fixed k are given by

$$c_x^k = \sum_l \delta_{x, lr+k}, \tag{19}$$

with $l = 0, 1, 2 ... M/r$ (see Fig. 1).

- We sample from the first register with probability $|c_y^k|^2$, where c_y^k is the Fourier transform of c_x^k (see Fig. 1). This operation is a slight modification of the Fourier-Hadamard sampling problem described earlier.

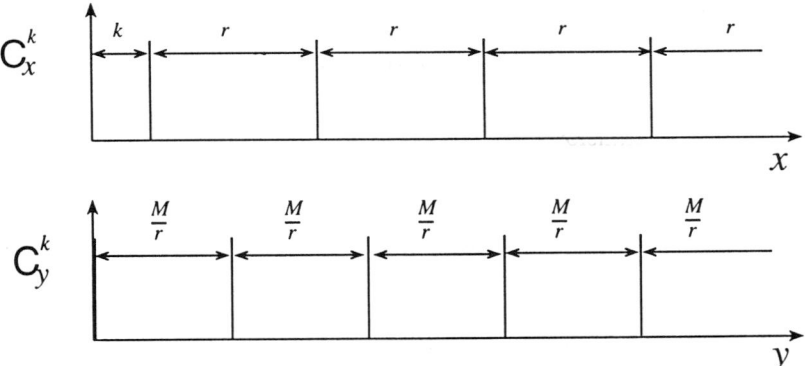

Figure 1: Function c_x^k and its Fourier transform c_y^k.

The rest of the algorithm is purely classical. The result of the sampling is, with a high probability, a number of the form jM/r, where $j = 1, 2, \ldots$. Knowing M and this number we can recover the period r using some mathematical tools (continued-fraction expansion). From r we obtain factors of N completing the computation in not more than $(\log_2 N)^2$ elementary steps! The computation is probabilistic; it may fail to provide the answer from time to time. Still, it is enough to run quantum computer several times to get the proper answer and it is still exponentially faster than any classical factorisation.

LOGIC GATES FOR QUANTUM CIRCUITS

An open question has been whether it would ever be practical to build physical devices to perform such computations, or whether they would forever remain theoretical curiosities. Quantum computers require a coherent, controlled evolution for a period of time which is necessary to complete the computation. Many view this requirement as an insurmountable experimental problem. We believe that progress in the techniques of nano-construction and in manipulation of atomic systems by electromagnetic fields will sooner or later make such devices feasible. Let us then describe a possible route to practical quantum computation.

Like classical computers, quantum computers can be built out of logic circuits. In 1989 Deutsch described quantum circuits composed of elementary logic gates connected together by wires and showed that there exists a universal quantum gate from which any quantum computation can be built [18].

Later Yao proved that such circuits behave properly as far as the computational complexity is concerned [7]. Quantum logic gates perform unitary operations on qubits and in order to implement them it is sufficient, from the experimental point of view, to induce a conditional dynamics of physical bits, i.e. to perform a unitary transformation on one physical subsystem conditioned upon the quantum state of another subsystem,

$$U = |0\rangle\langle 0| \otimes U_0 + |1\rangle\langle 1| \otimes U_1 + \ldots + |k\rangle\langle k| \otimes U_k, \tag{20}$$

where the projectors refer to quantum states of the control subsystem and the unitary operations U_i are performed on the target subsystem. The simplest non-trivial operation of this sort is the quantum controlled-NOT.

The *classical* controlled-NOT gate is a reversible logic gate operating on two bits ϵ_1 and ϵ_2; ϵ_1 is called the controll bit and ϵ_2 the target bit. The value of ϵ_2 is negated if $\epsilon_1 = 1$, otherwise ϵ_2 is left unchanged (in both cases the controll bit ϵ_1 remains unchanged). The quantum controlled-NOT gate $\mathcal{C}_{12}$ is the unitary operation on two qubits *i.e.* states in $\mathcal{H}_2$, which in a chosen orthonormal basis $\{|0\rangle, |1\rangle\}$ reproduces the controlled-NOT operation:

$$|\epsilon_1\rangle|\epsilon_2\rangle \xrightarrow{\mathcal{C}_{12}} |\epsilon_1\rangle|\epsilon_1 \oplus \epsilon_2\rangle, \tag{21}$$

where $\oplus$ denotes addition modulo 2. Here and in the following the first subscript of $\mathcal{C}_{ij}$ always refers to the control bit and the second to the target bit. Thus, for example, $\mathcal{C}_{21}$ denotes the unitary operation defined by

$$|\epsilon_1\rangle|\epsilon_2\rangle \xrightarrow{\mathcal{C}_{21}} |\epsilon_1 \oplus \epsilon_2\rangle|\epsilon_2\rangle. \tag{22}$$

Let us specify some interesting properties of the quantum controlled-NOT gate.

- The quantum controlled-NOT gate transforms superpositions into entanglements,

$$\mathcal{C}_{12} : (a|0\rangle + b|1\rangle)|0\rangle \longleftrightarrow a|0\rangle|0\rangle + b|1\rangle|1\rangle. \tag{23}$$

Thus it acts as "the measurement gate" because if the target bit ϵ_2 is initially in state $|0\rangle$ then this bit is in effect an apparatus that performs a perfectly accurate non-perturbing (quantum nondemolition type [19]) measurement of ϵ_1.

- This transformation of superpositions into entanglements can be reversed by applying the same controlled-NOT operation again. Hence it can be used to implement the Bell measurement [20] on the two bits by disentangling the Bell states. For the four Bell states we get four product states

$$C_{12}\frac{1}{\sqrt{2}}(|0\rangle|0\rangle \pm |1\rangle|1\rangle) = \frac{1}{\sqrt{2}}(|0\rangle \pm |1\rangle)|0\rangle, \qquad (24)$$

$$C_{12}\frac{1}{\sqrt{2}}(|0\rangle|1\rangle \pm |1\rangle|0\rangle) = \frac{1}{\sqrt{2}}(|0\rangle \pm |1\rangle)|1\rangle. \qquad (25)$$

Thus the Bell measurement on the two qubits is reduced to the simple sequence of two independent two dimensional measurements: in the basis $|0\rangle, |1\rangle$ for the control qubit and in the basis $\frac{1}{\sqrt{2}}(|0\rangle \pm |1\rangle)$ for the target qubit. The realisation of the Bell measurement is the main obstacle in the practical implementation of quantum teleportation [21] and the dense quantum coding [22].

- Quantum state swapping can be achieved by cascading three quantum controlled-NOT gates:

$$C_{12}C_{21}C_{12}|\Psi\rangle|\Phi\rangle = |\Phi\rangle|\Psi\rangle, \qquad (26)$$

for arbitrary states $|\Psi\rangle$ and $|\Phi\rangle$.

The quantum controlled-NOT gate is not a universal gate, however, the universal quantum gate can be constructed by a simple extension of the controlled-NOT gate to the controlled-controlled-NOT gate combined with simple unitary operations on a single qubit. Thus once we are able to induce a conditional dynamics in our laboratories we can start thinking about putting gates together into circuits and performing some elementary quantum computations. The task now is to identify physical processes that can lead to a controllable conditional dynamics.

BUILDING QUANTUM LOGIC GATES

In the following we outline two possible experimental realisations of the quantum controlled-NOT gate. The first method is based on the Ramsey atomic interferometry [23–25], the second on the selective driving of optical resonances of two qubits undergoing a dipole-dipole interaction [26].

In the Ramsey atomic interferometry method the target qubit is an atom with selected two circular Rydberg states $|\epsilon_2\rangle$, where $\epsilon_2 = 0, 1$; the control qubit $|\epsilon_1\rangle$ is the quantized electromagnetic field in a high Q cavity C. The field in the cavity contains at most one photon so it can be viewed as a two state system with the vacuum state $|0\rangle$, and the one-photon state $|1\rangle$ as the basis. The cavity C is sandwiched between two auxiliary microwave cavities R_1 and R_2 in which classical microwave fields produce $\pi/2$ rotations of the atomic Bloch vector,

$$|\epsilon_1\rangle_{\text{field}} |\epsilon_2\rangle_{\text{atom}} \longrightarrow |\epsilon_1\rangle_{\text{field}} \frac{1}{\sqrt{2}}(|\epsilon_2\rangle + (-1)^{\epsilon_2} e^{i\alpha} |1 - \epsilon_2\rangle)_{\text{atom}}, \qquad (27)$$

where the phase factor α is different for the two cavities R_1 and R_2. In the central cavity C, a dispersive interaction with the quantized field introduces phase shifts which depend on the state of the atom $|\epsilon_2\rangle$ and on the number of photons in the cavity $|\epsilon_1\rangle$. The interaction does not involve any exchange of excitation, so the number of photons in the cavity remains unchanged.

$$|\epsilon_1\rangle_{\text{field}} |\epsilon_2\rangle_{\text{atom}} \longrightarrow \exp(i(-1)^{1-\epsilon_2}(\epsilon_1 + \epsilon_2)\theta) |\epsilon_1\rangle_{\text{field}} |\epsilon_2\rangle_{\text{atom}}, \qquad (28)$$

where θ, which is the phase shift per photon, can be tuned to be π (θ depends on the atom-cavity crossing time and the atom-field detuning).

The overall process can be viewed as a sequence: half-flopping in R_1, phase shifts in C, and half-flopping in R_2. Depending on the phase shifts the second half-flopping can either put the atom back into its initial state or flop completely into the orthogonal state. The whole interferometer can be adjusted so that when the atom is sent through the cavities R_1, C, and R_2 the two qubits, i.e. the field and the atom, undergo the transformation

$$|\epsilon_1\rangle_{\text{field}} |\epsilon_2\rangle_{\text{atom}} \longrightarrow |\epsilon_1\rangle_{\text{field}} |\epsilon_1 \oplus \epsilon_2\rangle_{\text{atom}}. \qquad (29)$$

The initial and final states of the field in C can be mapped from and to the atomic states, respectively, by a resonant atom-field interaction [24, 27, 28]. This process allows the two qubits to be of the same type i.e. two Rydberg atoms rather than a field and an atom. The practical realisation can be carried out by a modification of the Ramsey atomic interferometry experiments as described in [24, 25].

Our second proposal for the practical implementation of the quantum controlled-NOT gate relies on the dipole-dipole interaction between two qubits. For the purpose of this model the qubits could be either magnetic dipoles, e.g. nuclear spins in external magnetic fields, or electric dipoles, e.g. single-electron quantum dots in static electric fields. Here we describe the model based on interacting quantum dots, however, the two cases are isomorphic from the mathematical point of view and therefore our description is not very restrictive.

Two single-electron quantum dots separated by the distance R are embedded in a semiconductor. The ground state and the first excited state of each

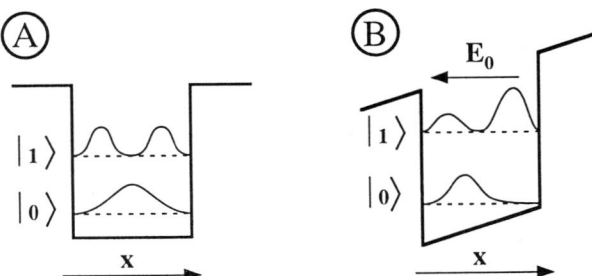

Figure 2: Charge density in the quantum well along the direction in which the field is applied (here x). A permanent dipole is induced when the electric field is turned on (B), whereas the dipole is zero without electric field (A).

dot are labelled as $|0\rangle$ and $|1\rangle$ respectively. The first quantum dot with the resonant frequency ω_1 will act as the control qubit whereas the second one, with the resonant frequency ω_2, as the target qubit. In the presence of an external static electric field, which can be turned on and off adiabatically in order to avoid transitions between the levels, the charge distribution of the ground state, of each dot, is shifted in the direction of the field whilst the charge distribution of the first excited state is shifted in the opposite direction (the *quantum-confined Stark effect*), see Fig. 2. In the simple model in which the state of the qubit is encoded by a single electron per quantum dot, we can choose coordinates in which the dipole moments in states $|0\rangle$ and $|1\rangle$ are $\pm d_i$, where $i = 1, 2$ refers to the control and to the target dot respectively. A more elaborate model would require to take into account the holes in the valence band of the semiconductors. The state of the qubit would be determined by excitons of different energies, and our notations would become more complicated. For simplicity and clarity, we decided to explain the main idea using a slightly simplified model.

The electric field from the electron in the first quantum dot may shift the energy levels in the second one (and vice versa), but to a good approximation it does not cause transitions. That is because the total Hamiltonian

$$\hat{H} = \hat{H}_1 + \hat{H}_2 + \hat{V}_{12} \qquad (30)$$

is dominated by a dipole-dipole interaction term V_{12} that is diagonal in the four-dimensional state space spanned by eigenstates $\{|\epsilon_1\rangle, |\epsilon_2\rangle\}$ of the free Hamiltonian $\hat{H}_1 + \hat{H}_2$, where ϵ_1 and ϵ_2, as before, range over 0 and 1. Specifically,

$$(\hat{H}_1 + \hat{H}_2)|\epsilon_1\rangle|\epsilon_2\rangle = \hbar(\epsilon_1\omega_1 + \epsilon_2\omega_2)|\epsilon_1\rangle|\epsilon_2\rangle, \qquad (31)$$

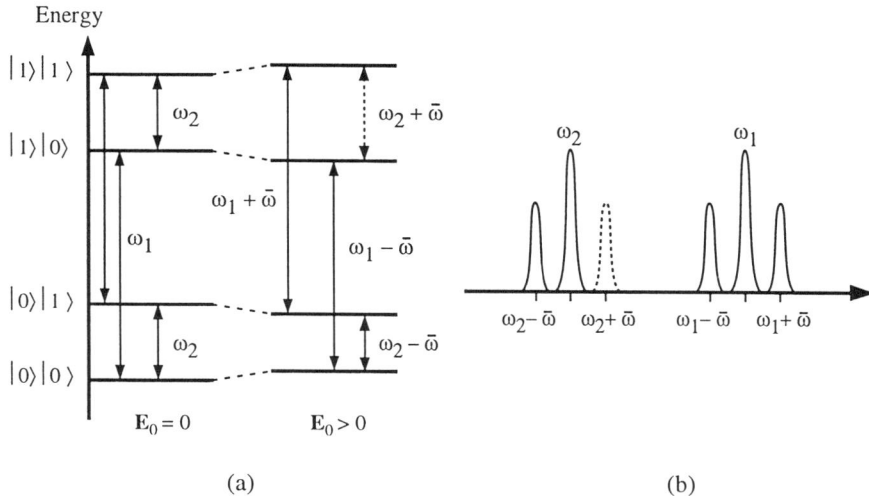

Figure 3: (a) Energy levels of two quantum dots without and with the coupling induced by the presence of a static electric field $\mathbf{E}_0$. (b) Resonance spectrum of the two quantum dots. The dotted line shows the wavelength for which the two dots act as a controlled-NOT gate, with the first dot being the control qubit and the second the target qubit.

and
$$\hat{V}_{12} |\epsilon_1\rangle |\epsilon_2\rangle = (-1)^{\epsilon_1+\epsilon_2} \hbar\bar{\omega} |\epsilon_1\rangle |\epsilon_2\rangle, \tag{32}$$
where
$$\bar{\omega} = -\frac{d_1 d_2}{4\pi \epsilon_0 R^3}. \tag{33}$$

As shown in Fig. 3, it follows that due to the dipole-dipole interaction the resonant frequency for transitions between the states $|0\rangle$ and $|1\rangle$ of one dot *depends on the neighbour's state*. The resonant frequency for the first dot becomes $\omega_1 \pm \bar{\omega}$ according as the second dot is in state $|0\rangle$ or $|1\rangle$ respectively. Similarly the second dot's resonant frequency becomes $\omega_2 \pm \bar{\omega}$, depending on the state of the first dot. Thus a π-pulse at frequency $\omega_2 + \bar{\omega}$ causes the transition $|0\rangle \leftrightarrow |1\rangle$ in the second dot only if the first dot is in $|1\rangle$ state.

Possible physical realisation of the model requires the typical life-time of the excited state $|1\rangle$ ($\approx 10^{-6}$ s in our model) to be much greater than the time-scale of the optical interaction (the length of the π-pulse $\approx 10^{-10}$ ns). For the π-pulse to be monochromatic and selective enough we must also require the length of the pulse to be greater than the inverse of the pulse carrier frequency and the inverse of the dipole-dipole interaction coupling constant ($1/\bar{\omega} \approx 10^{-12}$ s in our model).

FROM GATES TO CIRCUITS

Putting quantum gates into quantum circuits is not easy. The two critical parameters which determine the character of the dynamics are the decoherence time of the computer T_d and the typical time of an elementary coherent computational step — the clock cycle T_c. For a non-trivial quantum computation we require $T_d \gg T_c$, however, even a coherent evolution may create errors e.g. the pulses which are not properly tuned may not be selective enough and may affect different transitions and different quantum dots. Quantum dots can be integrated into circuits by placing a number of them together in an array. The two parameters T_d and T_c can be, to some extent, controlled.

- The minimal clock cycle T_c is of the order of the typical pulse length T_p. The pulses must be approximately monochromatic and selective which requires $T_p \gg 1/\omega$ and $T_p \gg 1/\bar{\omega}$, where ω is the carrier frequency of the pulse. Clearly higher carrier frequencies allow shorter pulses and consequently the shorter clock cycle.

- The decoherence time T_d depends on the interaction with the enviroment. Electrons in quantum dots are coupled to the quantised electromagnetic vacuum which leads to the spontaneous emission. Typical lifetime of the excited state is of the order

$$4\pi\epsilon_0 \frac{3\hbar c^3}{4D^2\omega^3}, \qquad (34)$$

where D is the dipole moment between states $|0\rangle$ and $|1\rangle$, the carrier frequency ω is tuned to the transition frequency between the two states (the off-diagonal dipole moment D is different from the diagonal one d). The interaction between the neighboring quantum dots will also have some off-diagonal terms, which were not included in V_{12}, and which induce the propagation of excitons in the array. This process reduces the lifetime of the excited states of the quantum dots and shortens the decoherence time.

For quantum dots of nanometer size, separated by $\approx 10^{-8}$ m, and the resonant wavelength of the order of $\approx 10^{-6}$ m ($\approx 10^{14}$ Hz) realistic estimates could be $T_c \approx 10^{-6}$ s and $\bar{\omega} \approx 10^{12}$ Hz. Tunable vibronic solid-state laser are available at these wavelengths [29]; for example, the titanium-sapphire lasers with modelocking can produce adequate short pulses $T_p \approx 10^{-13} - 10^{-10}$ s with nanosecond intervals. The applied voltage which activates quantum dots and determines the computational network can be changed on much slower time scale without affecting the speed of computation. Assuming that the clock cycle is of the order 10^{-9} s, we should be able to squeeze about 10^4 coherent steps. Nota bene, one coherent step of the whole device implies many elementary transitions performed in parallel on many activated dots.

Fortunately, the problem of factorising integers, and Shor's quantum algorithm for solving it, seem almost ideally suited for minimising the impact of these limitations while still harnessing quantum parallelism of a tremendous degree. Perhaps in this particular case the experimental realisations may come sooner than we anticipate!

ACKNOWLEDGMENTS

Many thanks to A. Barenco, K Burnett, D. Deutsch, R. Jozsa, C. Macchiavello, M. Palma, K-A. Suominen, and U. Vazirani for discussions, comments, and help during the preparation of this manuscript. This research is supported by The Royal Society, London.

REFERENCES

[1] R. Landauer, IBM J. Res. Develop. **5**, 183 (1961).

[2] D. Deutsch, Proc. R. Soc. London A **400**, 97 (1985).

[3] R. Feynman, Int.J.Theor.Phys. **21**, 467 (1982).

[4] D. Deutsch and R. Jozsa, Proc. R. Soc.Lond. A **439**, 553 (1992).

[5] E. Bernstein and U. Vazirani, *Proc. 25th ACM Symp. on Theory of Computation, 11* (1993).

[6] A. Berthiaume and G. Brassard, *Proc. 7th IEEE Conference on Structure in Complexity Theory* (1992).

[7] A.Yao,*Proc. 34th IEEE Symp. on Foundation of Computer Science* pp.352-360, IEEE Computer Society Press, Los Alamitos, (1993).

[8] D.S. Simon, *On the Power of Quantum Computation*; unpublished manuscript.

[9] P.W. Shor, *Algorithms for quantum computation*, draft of April 18, 1994 manuscript, AT&T Bell Laboratories.

[10] R.D. Silverman, *Math. Comp.* **48**, 329 (1987).

[11] A.K. Lenstra, H.W. Lenstra Jr., M.S. Manasse, and J.M. Pollard, in *Proc. 22nd ACM Symposium on the Theory of Computing*, pp.564-572 (1990).

[12] D. Welsh, *Codes and Cryptography* Clarendon Press, Oxford (1988).

[13] C. H. Papadimitriou, *Computational Complexity* Addison-Wesley Publishing Company (1994).

[14] C. H. Bennett, IBM J. Res. Develop. **17**, 525 (1973).

[15] R. Rivest, A. Shamir, and L. Adleman, *On Digital Signatures and Public-Key Cryptosystems*, MIT Laboratory for Computer Science, Technical Report, MIT/LCS/TR-212 (January 1979).

[16] A. K. Ekert, *Quantum Cryptography and Computation* in *Erice Course in Advances in Quantum Phenomena*, Plenum Press (in print).

[17] M.R. Schroeder, *NumberTheory in Science and Communication*. Springer-Verlag (1984).

[18] D. Deutsch, *Proc. R. Soc. London A* **425**, 73 (1989).

[19] V.B. Braginsky, Yu. I. Vorontsov, and F. Ya. Khalili, Zh. Exksp. Theo. Fiz. **73**, 1340 [Sov. Phys. JETP **46**, 705 (1977)].

[20] S.L. Braunstein, A. Mann, and M. Revzen, Phys. Rev. Lett. **68**, 3259 (1992).

[21] C.H. Bennett, G. Brassard, C. Crépeau, R. Jozsa, A. Peres, and W.K. Wootters, Phys.Rev.Lett. **70**, 1895 (1993).

[22] C.H. Bennett and S.J. Wiesner, Phys.Rev.Lett. **69**, 2881 (1992).

[23] N.F. Ramsey *Molecular Beams*. Oxford University Press (1985).

[24] L. Davidovich, N. Zagury, M. Brune, J.M. Raimond and S. Haroche, Phys. Rev. A, *Teleportation of an atomic state between two cavities using non-local microwave fields*., (to be published).

[25] M. Brune, P. Nussenzveig, F. Schmidt-Kaler, F.Bernardot, A. Maali, J.M. Raimond, and S. Haroche, *Phys. Rev. Lett.* (1994).

[26] K. Obermayer, W.G. Teich, and G. Mahler, Phys.Rev. B **37**, 8096 (1988); W.G. Teich, K. Obermayer, and G. Mahler, *ibid.* p.8111; W.G. Teich and G. Mahler, Phys.Rev. A **45**, 3300 (1992); see also S. Lloyd, Science **261**, 1569 (1993).

[27] A. S. Parkins, P. Marte, P. Zoller, and H. J. Kimble, Phys. Rev. Lett. **71**, 3095 (1993).

[28] J.M. Raimond *et al.*, in *Laser Spectroscopy IX*, edited by M. S. Feld, J. E. Thomas, and A. Mooradian, Academic Press, New York, (1989), p.140.

[29] J. Hecht, *The Laser Guidebook*. McGraw-Hill, Inc. (1992).

Optical Spectroscopy of Individual Molecules Trapped in Solids

W. E. Moerner

IBM Research Division, Almaden Research Center, K13/802D, 650 Harry Road, San Jose, California 95120-6099 USA

Abstract. Optical spectroscopy of single impurity molecules in solids can be used as an exquisitely sensitive probe of the structure and dynamics of the specific local environment around the single molecule (the "nanoenvironment"). Some phenomena that have been studied are the spectral diffusion dynamics of a single molecule, perturbations by external fields, changes in molecular photophysics, shifts in vibrational modes, optical modification of the absorption spectrum, dynamics due to amorphous system physics, and magnetic resonance of a single molecular spin. This demonstrates the vitality of and growing interest in this new field, which may in the future lead to optical storage on the single-molecule level.

1. INTRODUCTION

Over the past five years, the power of optical spectroscopy with high-resolution laser sources has recently been extended into the fascinating domain of individual impurity molecules in solids, where the single molecule acts as a probe of the detailed local environment with unprecedented sensitivity(1)-(6). In ongoing experiments at several laboratories around the world, exactly one molecule hidden deep within a solid sample is probed at a time by tunable laser radiation, which represents detection and spectroscopy of 1.66×10^{-24} moles of material, or 1.66 yoctomoles[1]. Such single-molecule measurements completely remove the normal ensemble averaging that occurs when a large number of molecules are probed at the same time. Thus, the usual assumption that all molecules contributing to the ensemble average are identical can now be directly examined, and strong tests of truly microscopic theories can be completed.

Single-molecule spectroscopy (SMS) in solids is related to, but distinct from, the fascinating and well-established field of spectroscopy of single electrons, atoms, or ions confined in electromagnetic traps or on surfaces(7)-(9). The vacuum environment and confining fields of an electromagnetic trap are quite different from the environment of a single molecule in a solid where the lattice interactions act to constrain the molecule, hindering or preventing molecular rotation. In addition, in the solid the molecule is continuously bathed in the phonon vibrations available at a given temperature. In comparison to another important field, near-

---

1. Since a single molecule is the smallest unit of a molecular substance, a more appropriate unit in this case would be the *guacamole*, which is the quantity of moles exactly equal to the inverse of *avocado*'s number.

field probing of atoms on surfaces with scanning tunneling microscopy or atomic force microscopy(10), where a strong bond between the molecule and the underlying surface as well as tolerance of the perturbing forces from the tunnelling electrons or the tip are required, SMS usually operates noninvasively in the optical far-field with a corresponding loss in spatial resolution, but with no loss of spectral resolution. In recent work, single-molecule imaging has been achieved with near-field *optical* techniques with 100 nm resolution at room temperature(11)(12) and the emission dispersed to study vibrational structure(13), but the spectral resolution available under such conditions was three to four orders of magnitude poorer than the high-resolution SMS described here.

This paper presents an overview of some of the recent advances produced by the spectroscopy of single impurity centers in solids, with emphasis on single, isolated molecular impurities. In the next two sections, the specific requirements for high-resolution SMS will be described and example spectra presented. Section 4 provides a brief historical background with a partial list of the various measurements that have been reported to date. In the remainder of the paper, specific results on imaging, spectral diffusion, optical modification of single molecules (which may lead to single molecule optical storage), correlation properties of the emitted photons, vibrational modes, and magnetic resonance will be described in more detail. The reader may consult one of several recent reviews for more information (4)-(6). The significance of these SMS studies is: (i) new physical effects have been observed in the single-molecule regime, and (ii) as a result of the ability to follow spectral changes of a single molecule in real time, it is now possible in a single nanoenvironment to directly probe the connection between specific microscopic theories(14) of local structure, dynamics, and host-guest interactions and the statistical mechanical averages that are measured in conventional experiments.

2. HOW SINGLE-MOLECULE SPECTROSCOPY IN SOLIDS IS ACCOMPLISHED

One may ask: How is it possible to use optical radiation to isolate a single impurity molecule hidden deep inside a host matrix? This section briefly describes the methods used to ensure that the fluorescence emission[2] from the molecule dominates over all background signals (for full details see Refs.(15) and (16)). First of all, only a small volume of sample on the order of a few hundred μm^3 is probed by using a thin sample and a small (μm-sized) laser spot produced by a lens, the core of an optical fiber or an aperture. In addition, the transparent host material is doped with the impurity at low concentration, in the range 10^{-7} to 10^{-9}

---

2. In fact, the first SMS used a powerful double-modulation modification of laser FM spectroscopy to detect the *absorption* from a single molecule with a signal-to-noise ratio (SNR) of about 5(1)(2). However, since fluorescence excitation(3) provides higher SNR for most systems, we concentrate on fluorescence here.

moles/mole or lower. These two actions alone are insufficient to guarantee that only one impurity molecule in the probed volume is in resonance with the laser at a time. Additional selectivity on the order of a factor of 10^4 or so is provided by carefully selecting the guest and host and using the well-known properties of inhomogeneously broadened absorption lines in solids(17)-(20), summarized below.

For optimal fluorescence excitation, the quantum yield for photon emission per absorption event should clearly be high. Moreover, it has been recognized since the first SMS experiments that in order to efficiently produce absorption events with the incident photons rather than unwanted scattering signals, the peak absorption cross section σ_{pk} should be as large as possible(1)(16)(21)-(23). This is because the probability that a single molecule will absorb a photon from the incident beam is simply σ_{pk}/A, where A is the area of the probing laser beam. Since σ_{pk} depends linearly upon the oscillator strength and inversely upon the optical homogeneous linewidth $\Delta\nu_H$, strong absorptions and narrow lines would be expected to give the largest σ_{pk} values. Strong absorptions are conveniently provided by electric-dipole-allowed singlet-singlet transitions in aromatic molecules where the oscillator strength is often near unity. The narrowest optical linewidths in solids occur for zero-phonon, zero-vibration electronic transitions at low temperatures, because such transitions can only broaden by phonon scattering events. For rigid molecules like pentacene or perylene in a crystalline solid below 10 K, the linewidth of the optical transition near 500 THz is on the order of ten MHz, and σ_{pk} reaches values near 10^{-11} cm^2, approximately 4,000 times the area of a single molecule.

There is an additional crucial phenomenon important for SMS which results from the extreme narrowness of such zero-phonon transitions: inhomogeneous broadening(17)(18) (see Fig. 1, top). This occurs when $\Delta\nu_H$ becomes narrower than the (mostly static) distribution of resonance frequencies in the solid caused by dislocations, point defects, or random internal strain and electric fields and field gradients (inset). Inhomogeneous broadening is a universal feature of high-resolution spectroscopy of defects in solids(19)(20) and of other spectroscopies where zero-phonon lines are probed such as magnetic resonance. (A dynamic inhomogeneous broadening also occurs for Doppler-broadened lines of gases.) Inhomogeneous broadening facilitates selection of a single impurity, because the different guest molecules in the probe volume take on slightly different resonance frequencies according to the local fields at the location of the impurity. One then simply uses the tunability of a narrowband laser to select different single impurity centers. Naturally, this must be done in a region of the optical frequency spectrum where the number of molecules per homogeneous width is on the order of or less than one. This may be accomplished (i) by tuning the laser wavelength out into the wings of the inhomogeneous line (Fig. 1, bottom right), (ii) by using a sample with a very low doping level, or (iii) by producing a large inhomogeneous linewidth so that no two molecules are present at any given frequency. All three of these

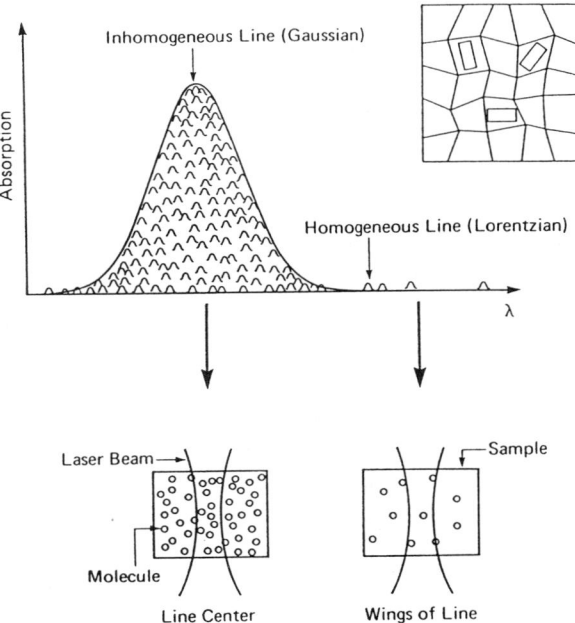

FIGURE 1. Schematic showing an inhomogeneous line at low temperatures and the principle of SMS in solids. The entire line is formed as a superposition of Lorentzian profiles of the individual absorbers, with a distribution of center resonance frequencies caused by random strains and imperfections. In the inset, several dopant molecules are sketched with different nanoenvironments produced by strains, local electric fields, and other imperfections in the host matrix. The lower part of the figure shows how the number of impurity molecules in resonance in the probed volume can be varied by changing the laser wavelength. The laser linewidth is negligible on the scale shown.

methods have been used.

A further requirement on the absorption properties of the probe molecule is the absence of strong bottlenecks in the optical pumping cycle. In organic molecules, intersystem crossing (ISC) from the singlet states into the triplet states represents a bottleneck, because photon emission usually ceases for a relatively long time equal to the triplet lifetime when ISC occurs. This effect results in premature saturation of the emission rate from the molecule and reduction of σ_{pk} compared to the case with no bottleneck(24). Thus, molecules with weak ISC yield and short triplet lifetime are preferable, such as rigid, planar aromatic dyes.

A final requirement for SMS is the selection of a guest-host couple that allows for photostability of the impurity molecule and weak spectral hole-burning, where by spectral hole-burning we include any light-induced change in the resonance frequency of the molecule caused either by frank photochemistry of the molecule or by a photophysical change in the nearby environment(25). For example, most detection schemes with overall photon collection efficiency of 1% to 0.1% require

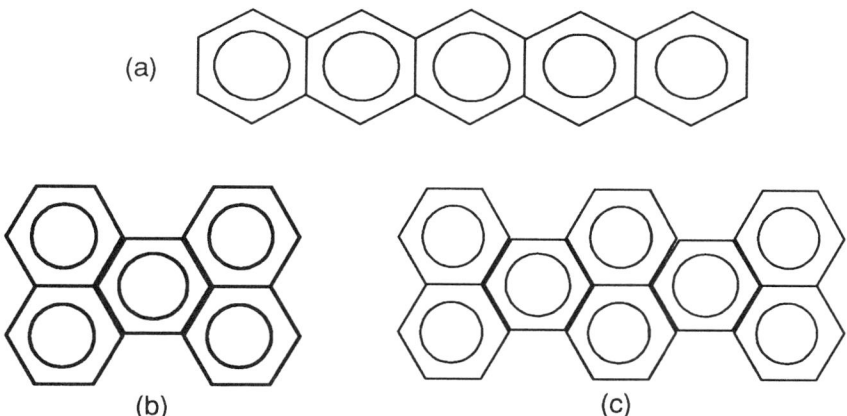

FIGURE 2. Structures of some molecules used thus far for SMS: (a) pentacene, (b) perylene, and (c) terrylene.

that the quantum efficiency for hole-burning, η, be less than 10^{-6} to 10^{-7}. This is necessary to provide sufficient time averaging of the single-molecule lineshape before it changes appreciably or moves to another spectral position.

Using all of these concepts, specific expressions for the signal-to-noise ratio (SNR) for SMS may be derived (see Refs. (4) and (27)), and values on the order of 20 can be expected with 1 s integration time and spot diameters of a few μm for a system like pentacene substitutional impurities in a p-terphenyl crystal.

Most SMS studies to date have concentrated on four host-guest combinations using the three impurity molecules shown in Figure 2: pentacene in p-terphenyl, perylene in poly(ethylene)(26), terrylene in poly(ethylene)(28), and terrylene in the Shpol'skii matrix hexadecane(29)(30). The Shpol'skii matrices in particular should provide a large new class of materials which allow SMS. In very recent unpublished work, several groups have reported new host-guest combinations, so that the total number of systems as of June 1994 is approximately 15.

3. EXAMPLES OF SINGLE-MOLECULE SPECTRA

Figure 3 shows fluorescence excitation spectra for a 10 μm thick sublimed crystal of pentacene in p-terphenyl at 1.5 K (for specific details of the experimental apparatus, see Ref. (15)). In fluorescence excitation, the optical absorption is measured by recording the total fluorescence emission shifted to long wavelengths as the laser frequency is scanned over the absorption line. The 18 GHz spectrum in Fig. 3(a) (obtained by scanning a 3 MHz linewidth dye laser over the entire inhomogeneous line) contains 20,000 points; to show all the fine structure usually requires several meters of linear space. The structures appearing to be spikes are not noise; all features shown are static and repeatable. Near the center of the inhomogeneous line, the "spectral roughness" is a fundamental effect called

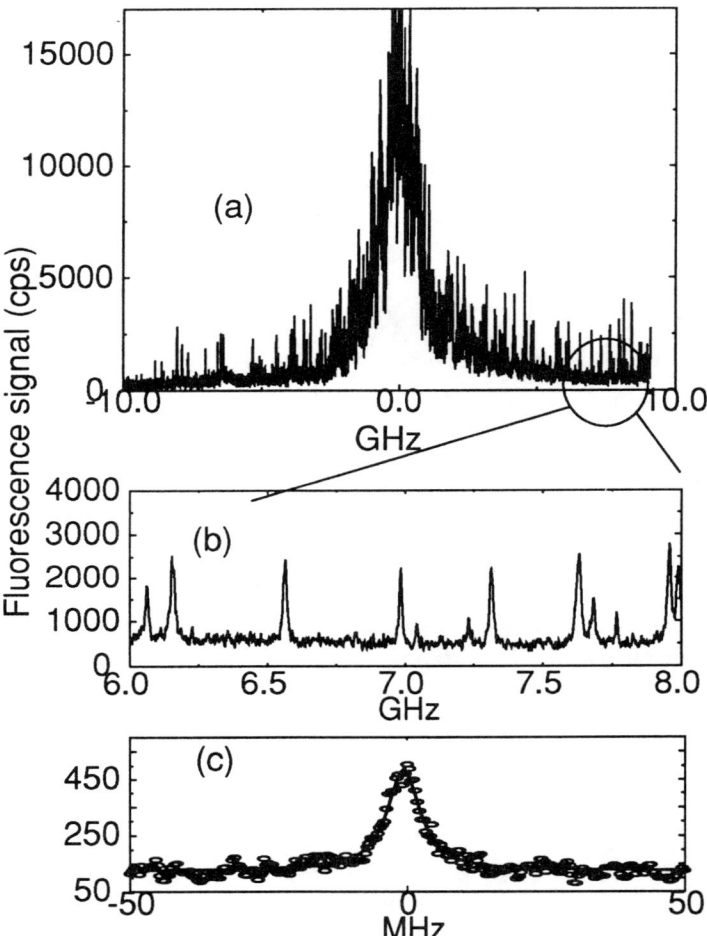

FIGURE 3. Fluorescence excitation spectra for pentacene in p-terphenyl at 1.5 K. (a) Broad scan of the inhomogeneously broadened line, all the spikes are repeatable features. (b) Expansion of 2 GHz showing several single molecules. The laser detuning is referenced to the line center at 592.321 nm. (c) Low-power scan of a single molecule at 592.407 nm showing the lifetime-limited width of 7.8 MHz and a Lorentzian fit.

statistical fine structure (SFS, see Refs. (31)(32)) which arises directly from statistical variations in the spectral density of absorbers with laser frequency. It is immediately obvious that the inhomogeneous line is far from Gaussian in shape and that there are tails extending out many standard deviations from the center both to the red and to the blue. Figure 3(b) shows an expanded region in the wing of the line. Each of the narrow peaks is the absorption profile of a single molecule. The peak heights vary due to the fact that the laser transverse intensity profile is bell-shaped and the molecules are not always located at the center of the laser focal spot. Even though these spectra seem narrow, they are in fact slightly power-broadened by the probing laser.

Upon close examination of an individual single-molecule peak at lower intensity (Fig. 3(c), the quantum-limited linewidth of 7.8 ± 0.2 MHz can be observed(33). Here, the quantum limit applies because the optical linewidth has reached the minimum value allowed by the lifetime of the optical excited state. This value is in excellent agreement with previous photon echo measurements of $\Delta\nu_H$ using large ensembles of pentacene molecules(34)(35). The well-isolated single-molecule spectra in Fig. 3 are wonderful for the spectroscopist: many detailed spectroscopic studies of the local environment can be performed, because such narrow lines are much more sensitive to local perturbations than broad spectral features.

4. HISTORICAL BACKGROUND AND SUMMARY OF RESULTS

The foundations for SMS were laid in 1987 when the spectral roughness called statistical fine structure scaling as the square root of the number of absorbers was first observed for pentacene in p-terphenyl(31)(32). The first single-molecule spectra were also recorded in this material using a sophisticated zero-scattering-background absorption technique, FM spectroscopy, combined with either Stark or ultrasonic modulation of the absorption line(1)(2). The SNR in these FM experiments of about 5 was limited by the shot noise of the laser beam. In 1990, Orrit et al. demonstrated that fluorescence excitation produces superior signals if the emission is collected efficiently and the scattering sources are minimized(3). Subsequent experiments have used fluorescence excitation exclusively.

For example, with SMS methods it has been possible to observe the quantum-limited linewidth of a single molecule(33), temperature-dependent dephasing effects on the single-molecule linewidth(15), optical saturation behavior(15), the effects of applied electric fields in crystals(36) and polymers(28), and shifts due to external hydrostatic pressure(37), all of which represent applications of the tools of the spectroscopist to the narrow spectral absorption line of a single molecule in a solid at low temperatures. With proper time-delayed photon-counting apparatus, the fluorescence emission lifetime of a single molecule can be directly measured(38). More surprisingly, several unexpected effects have also been observed, including spectral diffusion of a single molecule due to dynamics of the nearby host(39), reversible optical modification which could lead to optical storage on a single molecule scale(26)(27), and even magnetic resonance of a single molecular spin(40)-(42). The correlation properties of the emitted photons have been used to good advantage to measure photophysics of single molecules(43), resonance frequency fluctuations due to host degrees of freedom(44), and photon antibunching(45). It has also been possible to obtain fairly detailed information about the nanoenvironment of a single molecule by measuring the vibrational

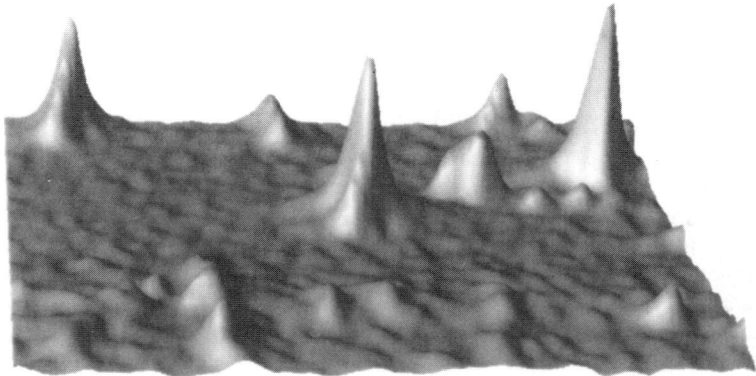

FIGURE 4. Three-dimensional "images" of single molecules. The measured fluorescence signal (z-axis) is shown over a range of 300 MHz in excitation frequency (horizontal axis, center = 592.544 nm) and 40 μm in spatial position (axis into the page).

mode frequencies in the electronic ground state(46)(47)(48).

5. IMAGING SINGLE MOLECULES IN SOLIDS

To survey some of the interesting results which have been obtained with SMS, Figure 4 shows a three-dimensional "image" of single molecules in a solid first obtained for pentacene in p-terphenyl some years ago (15) (rendered using IBM Data Explorer). The z-axis of the image is the usual fluorescence signal, the horizontal axis is the laser frequency detuning (300 MHz range), and the axis going into the page is one transverse spatial dimension (40 μm range). The spatial scan is produced by slowly translating the laser focal spot across the face of the crystal and obtaining spectra at each position. There are three, large, clear single molecule peaks localized in both frequency and position at the center, upper left, and upper right. The resolution of this image in the spatial dimension is clearly limited by the 5 μm diameter laser spot; however, in the frequency dimension the features are fully resolved.

The frequency widths of the peaks in Fig. 4 are slightly larger than the lifetime-limited width (Fig. 3(c)) due to the use of higher probing intensity. The oddly-shaped peak in between the two strong molecules at the center and upper right results from a molecule that is spectrally diffusing due to changes in its local nanoenvironment during the measurement (see the next section).

In recent work at ETH-Zürich, four-dimensional images have been obtained for pentacene in p-terphenyl with two spatial and one frequency dimension (48). This "fluorescence microscopy" experiment utilized a 2-D photocathode/image intensifier/vidicon camera specifically configured as a photon counter in each pixel. The single molecules appear as star-like objects which appear and disappear as the laser frequency is scanned. Future work should consider increasing the spatial resolution by near-field optical techniques(11). At low temperatures, the narrowness of the spectral features compared to room temperature should provide unprecedented detail of the local environment of the impurity molecule.

6. DIRECT OBSERVATIONS OF SPECTRAL DIFFUSION AND HOST DYNAMICS

When a new regime is first opened for study, often new physical effects can be observed. In the course of the early SMS of pentacene in p-terphenyl, an unexpected phenomenon appeared: resonance frequency shifts of individual pentacene molecules in a crystal at 1.5 K(39), called spectral diffusion by analogy to amorphous systems(49). Here, spectral diffusion means changes in the center (resonance) frequency of a defect due to configurational changes in the nearby host which affect the frequency of the electronic transition via guest-host coupling. Experimentally, two distinct classes of single pentacene molecules were identified: class I, which have center frequencies that are stable in time like the three large molecules in Fig. 4, and class II, which showed spontaneous, discontinuous jumps in resonance frequency of 20-60 MHz on a 1-420 s time scale, which are responsible for the distorted single-molecule peak in Fig. 4.

To illustrate the behavior of the fascinating type II molecules, Figure 5(a) shows a sequence of excitation spectra of a single molecule taken as fast as allowed by the available SNR. The laser was scanned once every 2.5 s with 0.25 s between scans, and the hopping of this molecule from one resonance frequency to another is clearly evident. By acquiring hundreds to thousands of such spectra, a trajectory or trend of the resonance frequency can be obtained as shown in Fig. 5(b) (for the same molecule as Fig. 5(a)). For this molecule, the optical transition energy appears to have a preferred set of values and performs spectral jumps between these values that are discontinuous on the 2.5 s time scale of the measurement. The behavior of another molecule is shown in Fig. 5(c) at 1.5 K and in Fig. 5(d) at 4.0 K. This molecule wanders in frequency space with many smaller jumps, and both the rate and range of spectral diffusion increases with temperature suggesting a phonon-driven process. The occurrence of class II defects was quite common in the wings of the inhomogeneous line, but only class I defects were observed close to the center, suggesting the spectral difusion effect is connected to disorder in the crystal structure. In addition, the jumping rate did not depend upon the probing laser power. The spectral diffusion appeared to be a spontaneous process rather than a light-induced spectral hole-burning effect(33).

Since the optical absorption is highly polarized(50) and the peak signal from the molecule does not decrease when the spectral jumps occur, it is unlikely that the molecule is changing orientation in the lattice. Since the resonance frequency of a single molecule in a solid is extremely sensitive to the local strain, the conclusion from these observations is that the spectral jumps are due to internal dynamics of some configurational degrees of freedom in the surrounding lattice. The situation is analogous to that for amorphous systems, which are postulated to contain a

476 Optical Spectroscopy of Individual Molecules

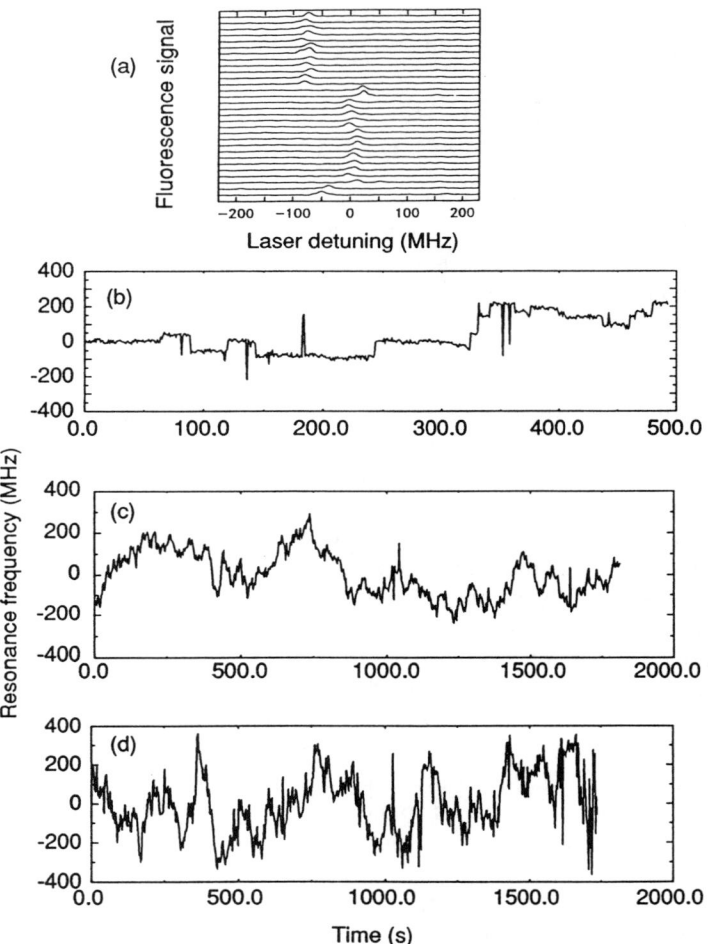

FIGURE 5. Examples of single-molecule spectral diffusion for pentacene in p-terphenyl at 1.5 K. (a) A series of fluorescence excitation spectra each 2.5 s long spaced by 2.75 s showing discontinuous shifts in resonance frequency, with zero detuning=592.546 nm. (b) Trend or trajectory of the resonance frequency over a long time scale for the molecule in (a). (c) Resonance frequency trend for a different molecule at 592.582 nm at 1.5 K and at (d) 4.0 K.

multiplicity of local configurations that can be modeled by a collection of double-well potentials (the two-level system or TLS model(51)). The dynamics results from phonon-assisted tunneling or thermally activated barrier crossing in these potential wells. One possible source for the tunneling states(15) could be discrete torsional librations of the central phenyl ring of the nearby p-terphenyl molecules about the molecular axis. The p-terphenyl molecules in a domain wall between two twins or near lattice defects may have lowered barriers to such central-ring tunneling motions. Further study is necessary to conclusively identify the specific molecular motions responsible for the effect. These direct observations of the dynamics of a nanoenvironment of a single molecule have sparked fascinating new theoretical studies of the underlying microscopic mechanism(14)(52).

Spectral diffusion effects similar to those shown in Fig. 5 also have been observed in SMS studies of perylene in poly(ethylene)(26)(27) and of terrylene in poly(ethylene)(28)(53). As opposed to the crystalline system above, in amorphous materials spectral diffusion is expected(49), and it is observed in a variety of time regimes. On the fast (sub-s) time scale, spectral shifts on the order of 10-100 MHz produce single-molecule lineshapes which fluctuate from measurement to measurement. On longer time scales, jumping behavior can be observed similar to that shown in Fig. 5(b), 5(c), and 5(d). Detailed measurements of terrylene in poly(ethylene)(53) also suggest that at higher laser powers, more spectral diffusion is observed which may be regarded as a type of transient spectral hole-burning (see the next section). Orrit et al.(44)(54) have used photon correlation techniques to study such processes (see Section 7). Finally, for terrylene in hexadecane, a two-state or few-state spectral shifting behavior was observed(30), and this process was shown to be influenced by the intensity of the probing light beam.

These measurements of spectral diffusion in crystals and polymers illustrate the unique power of SMS in detecting changes in the nanoenvironment of a single molecule. The spectral changes can be followed in real time for each individual impurity molecule; no ensemble averages over "equivalent" centers need be made that would obscure the effects.

7. OPTICAL MODIFICATION OF SINGLE MOLECULE SPECTRA AND APPLICATIONS TO OPTICAL STORAGE

A spectral hole usually refers to a dip in the inhomogeneously broadened absorption spectrum of many molecules when light irradiation alters the absorption frequencies of a portion of the resonant absorbers(25). In the single-molecule regime, however, only the isolated absorption line of one center is present. When photochemical or photophysical changes occur as a result of optical excitation, the absorption line simply appears to vanish as the resonance frequency of the single center shifts away from the laser frequency. Since laser-induced resonance frequency shifts form the foundation for the spectral hole-burning process, I will continue to use the term "hole-burning" even for the single-molecule case.

At first glance, the spontaneous spectral diffusion of the last section seems similar to single-molecule hole-burning, and in some cases, it may be hard to distinguish the two(53), because when a single molecule changes resonance frequency, one may not know which mechanism produced the change. Since hole-burning is driven by the light and does not spontaneously occur like the spectral diffusion of the last section, it is best to observe a dependence on the intensity of the laser beam to verify that hole-burning is occurring. The exact microscopic mechanism for light-induced resonance frequency shifts needs further study and may be related to the generation of molecular internal vibrational modes during fluorescence emission or to non-radiative decay of the excited state.

This effect was first convincingly observed in the perylene in poly(ethylene) system(26)(27), where a clear increase in hole-burning rate (i.e., rate of production of light-induced spectral shifts) was observed with increasing laser intensity. This was possible because a large fraction of the single molecules showed reversible hole-burning, where reversibility means the spontaneous return of the resonance frequency to its original value after the spectral shift. Reversibility allows many measurements of the stochastic burning time, yielding the kinetic distribution for the same single molecule(27). Irreversible hole-burning was also observed. In either case, the exact location of the new resonance frequency was unknown, but by analogy with previous nonphotochemical hole-burning studies on large ensembles of molecules(55)(56), the shift may be expected to be up to 100 cm^{-1}.

In principle, single-molecule hole-burning is a controllable process which could allow modification of the transition frequency of any arbitrary chosen molecule in the polymer host. This leads naturally to the possibility of optical storage at the single-molecule level. For highest density, near-field optical excitation with a 50 nm diameter pulled fiber tip could be used, which has demonstrated densities of 45 Gbits/in^2 with magneto-optic recording(57). One can imagine a very thin layer of the material some tens of nm thick with a sufficiently broad inhomogeneous line so that single impurity molecules are isolated and spread over a large range of frequency space. The resonance frequencies constitute the addresses of all the bits to be written in a single focal volume. A binary sequence of "1"'s and "0"'s can be produced by altering or ignoring each single molecule absorption.

Single-molecule optical storage, while highly speculative at the present time, provides several advantages and disadvantages. Among the advantages are (i) the extreme areal density (up to perhaps 10^{13} bits/in^2 for a 50 nm x 50 nm storage volume and doping level of 10^{-3} giving 100 bits per spot) (ii) the increased signal from each molecule resulting from the use of sub-diffraction-limited beams, and (iii) the ease of access of different bits by simply tuning the laser. Some disadvantages are the required low temperatures, stochastically variable burning times, high doping level and variable frequency locations of the individual bits from laser spot to laser spot. The last problem can be overcome in principle by "pre-formatting" the resonance frequencies or even possibly by careful design of the inhomogeneous distribution. In spite of these problems, optical modification of single-molecule spectra not only provides a unique window into the photophysics and low-temperature dynamics of the amorphous state, but it also allows such novel optical storage schemes to be studied and developed.

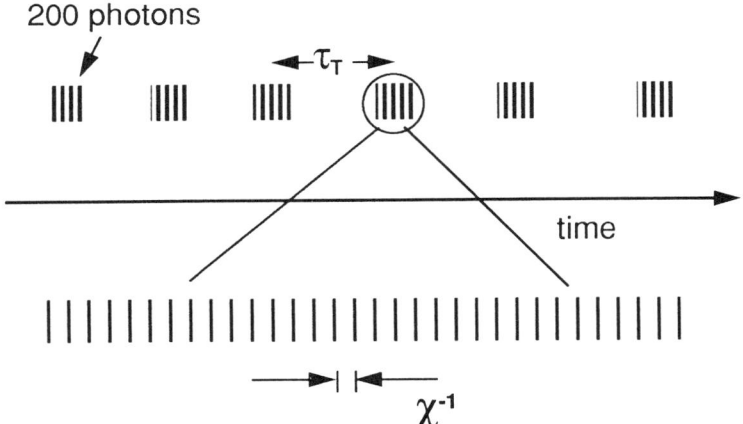

FIGURE 6. Schematic of the photon emission times from a single molecule showing bunching on the scale of the triplet lifetime (upper half) and antibunching on the scale of the inverse of the Rabi frequency (lower half).

8. TIME CORRELATION OF SINGLE MOLECULE EMISSION

The stream of photons emitted by a single molecule contains information about the system encoded in the arrival times of the individual photons. Figure 6 schematically shows the time-domain behavior of the photon stream for a single molecule with a dark triplet state, here taken to be pentacene. While cycling through the singlet states $S_0 \rightarrow S_1 \rightarrow S_0$, photons are emitted until intersystem crossing occurs. Since the triplet yield is 0.5%(58), 200 photons are emitted on average before a dark period which has an average length equal to the triplet lifetime, τ_T. This causes "bunching" of the emitted photons as shown in the upper half of the Figure. The subsequent decay in the autocorrelation of the emitted photons for pentacene in p-terphenyl was first reported by Orrit et al.(3), and this phenomenon has been used to measure the changes in the triplet yield and triplet lifetime from molecule to molecule(43) which occur as a result of distortions of the molecule by the local nanoenvironment. Such correlation measurements can extract information about the single molecule on much shorter time scales (down to the µs range) than the frequency scans described in previous sections; however, the dynamical process must be stationary, that is, the dynamics must not change during the relatively long time (many s) needed to record enough photon arrivals to generate a valid autocorrelation. For the terrylene in poly(ethylene) system which has complex dynamics driven by TLS's in the polymer, the amplitude fluctuations in the single-molecule fluorescence signal resulting from shifts of the resonance frequency sometimes cause a characteristic fall-off in the autocorrelation which yields information about the TLS-phonon coupling(44)(54).

By contrast, in the nanosecond time regime (lower half of Fig. 8), the emitted

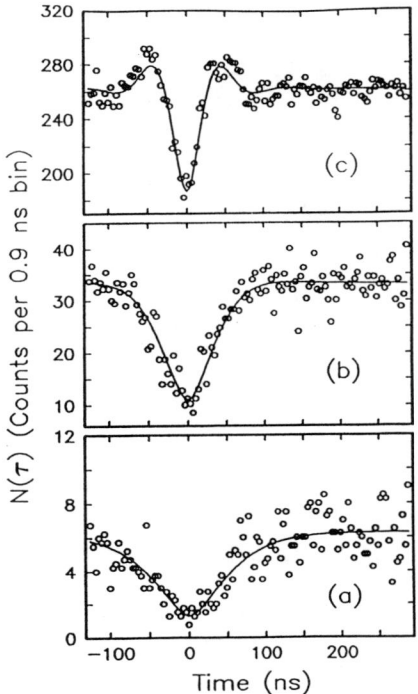

FIGURE 7. Photon antibunching for a single molecule of pentacene in p-terphenyl at 593.4 nm. The intensity correlation $N(\tau)$ is shown for Rabi frequencies χ of (a) 11.2 MHz, (b) 26.2 MHz, and (c) 68.9 MHz. The solid lines are fits as described in Ref. (45).

photons from a single quantum system are expected to show antibunching, which means that the photons "space themselves out in time", that is, the probability for two photons to arrive at the detector at the same time is small. This is a uniquely quantum-mechanical effect, which was first observed for Na atoms in a low-density beam(59). For a single molecule, antibunching is easy to understand as follows. After photon emission, the molecule is definitely in the ground state and cannot emit a second photon immediately. A time on the order of the inverse of the Rabi frequency χ^{-1} must elapse before the probability of emission of a second photon is appreciable. The antibunching in single-molecule emission was first observed at IBM for pentacene in p-terphenyl(45), demonstrating that quantum optics experiments can be performed in solids for the first time (see Fig. 7). Of course, if more than one molecule is emitting, the antibunching effect as well as the bunching effect both quickly disappear since the molecules emit independently. This is further proof that the spectral features are indeed those of single molecules.

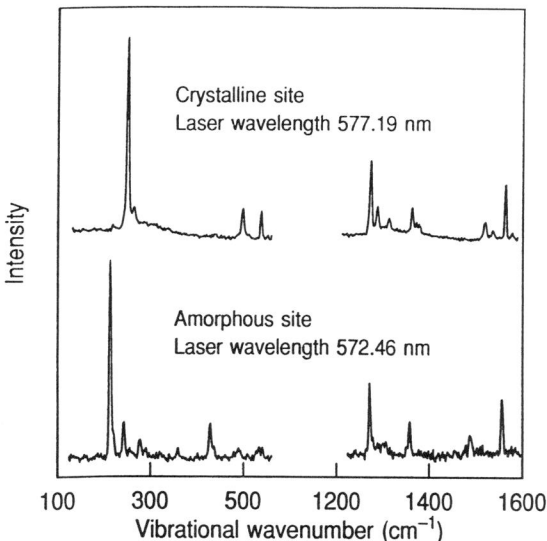

FIGURE 8. Examples of dispersed fluorescence spectra showing even ground-state vibrational mode frequencies and intensities for two molecules of terrylene in poly(ethylene). Upper case: Molecule probably located in a crystalline region; lower case: amorphous region. See Ref. (47).

9. VIBRATIONAL MODES OF INDIVIDUAL MOLECULES

Until recently, all SMS studies utilized the total fluorescence excitation technique where all long-wavelength-shifted photons passing through a long-pass filter are detected. With the addition of a grating spectrograph and a CCD array detector, vibrationally resolved emission spectra from single molecules both in crystals(46) and in polymers(47) have been obtained. Such experiments may also be regarded as resonance Raman studies(48), since the laser is in resonance with the 0-0 electronic transition, and even-parity vibrational modes of the ground state are detected by measuring the shift between the laser wavelength and the wavelength of the emission peak. The ability to examine vibronic or vibrational features of individual absorbers can generate specific details about the identity of the absorber and the nature of the interactions with the nanoenvironment producing shifts or intensity changes in the vibrational spectrum(48).

Figure 8 shows typical dispersed fluorescence spectra for two different single molecules of terrylene in poly(ethylene) at 1.6 K. In addition to small ($\cong$ cm^{-1}) shifts and intensity changes of various modes from molecule to molecule, two rather different classes of spectra were observed(47) as shown in the upper and lower parts of the figure. After considering various possibilities, we concluded that the upper type of spectrum resulted from a terrylene molecule near or inside the crystalline region of the polymer, while the other spectrum resulted from a single terrylene located in an amorphous region. Such results demonstrate the wealth of spectroscopic detail that can be obtained from individual molecules and used to probe truly local aspects of the structure of amorphous solids. For example, the lowest frequency mode in Fig. 8 is a long-axis ring expansion of the molecule; the shift to lower energy in the amorphous site can be understood as resulting from the reduced local density (greater free volume) compared to the crystalline site. Such measurements should stimulate new theoretical calculations of the spectral changes which result from specific local distortions for comparison with the single-molecule spectra.

10. MAGNETIC RESONANCE OF A SINGLE MOLECULAR SPIN

Historically, the standard methods of electron paramagnetic resonance and nuclear magnetic resonance have been limited in sensitivity to about 10^8 electron spins and about 10^{15} nuclear spins, respectively, due chiefly to the weak interaction between the individual spins and the rf electromagnetic fields used to excite the transition. Using a combination of single-molecule optical spectroscopy and optically detected magnetic resonance (ODMR), two groups have independently observed the magnetic resonance transition of a single molecular spin for the first time[40][41] using the pentacene in p-terphenyl model system. In essence, ODMR allows higher sensitivity because the weak spin transition is effectively coupled to a much stronger optical transition with oscillator strength near unity. The method involves selecting a single molecule with the laser as described in Section 2 and monitoring the intensity of optical emission (here, fluorescence from the first excited singlet state) as the frequency of a microwave signal is scanned over the frequency range of the triplet spin sublevels[3]. Since the emission rate is dependent upon the overall lifetime of the triplet (bottleneck) state, the emission rate is affected when the microwave frequency is resonant with transitions among the triplet spin sublevels.

Figure 9 shows examples of the 1480 MHz magnetic resonance transition among the T_x-T_z triplet spin sublevels at 1.5 K for a single molecule of pentacene in p-terphenyl, where the signal plotted is the change in the fluorescence emission rate as a function of the applied microwave frequency[40]. Traces (a) and (b) show the lineshapes when many pentacene molecules are pumped by the laser, where O_1 and O_2 refer to two of the four possible substitutional sites for pentacene in the host crystal. Traces (c)-(g) show the single-molecule lineshapes for four different single molecules. An interesting observation is that the onset of the transition varies from single molecule to single molecule, in a fashion similar to the difference in onsets for the two inhomogeneously broadened site origins. As is the

---

3. Essentially, the triplet state of a molecule is formed by two unpaired electrons with spin parallel to each other leading to a net electronic spin of 1. In zero external magnetic field, the triplet state splits into three spin sublevels due to the anisotropy of the molecular electronic wavefunction and the magnetic structure of the nearby environment.

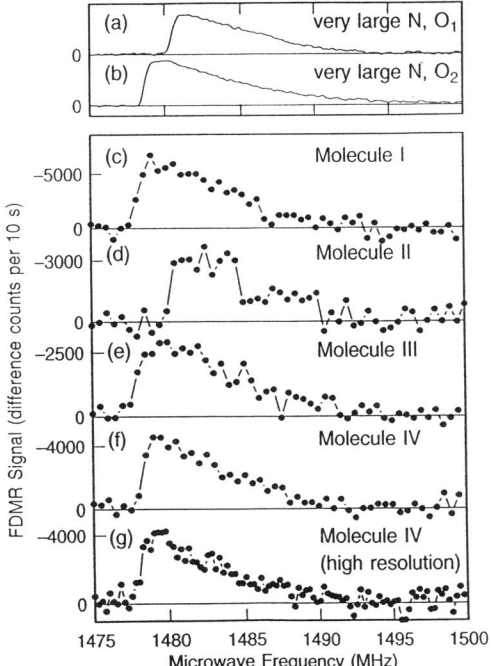

FIGURE 9. Single-spin magnetic resonance for pentacene in p-terphenyl at 1.5 K. The Fluorescence Detected Magnetic Resonance (FDMR) signal is shown as a function of the microwave frequency applied to the sample. (a)-(b) The T_x-T_z transition for a large number of molecules for sites O_1 and O_2. (c)-(g) A selection of single molecule magnetic resonance lineshapes.

case for large N, the lineshape of the microwave transition for a single spin is broadened by hyperfine interactions. This occurs because many different configurations of the proton nuclear spins in the molecule are sampled on the time scale of the measurement of the triplet state transition. In contrast, in the large N experiment, an ensemble average is measured rather than a time average.

These observations open the way for a variety of new studies of magnetic interactions in solids at the level of a single molecular spin such as the use of external magnetic fields or deuteration to reduce proton spin flips and hence the hyperfine broadening. Already, single-spin coherence has been observed(42), because no matter what time the molecule enters the triplet state, the start of the microwave irradiation provides a zero-time reference so that the signals from subsequent entries into the triplet state can add coherently. In future work, the properties of amorphous organic materials may be studied in greater detail, as the selection of a single molecular spin removes all orientational anisotropy as well as all inhomogeneous broadening. Imaging on the spatial scale of a single molecule is possible with a sufficiently large magnetic field gradient. The power of magnetic resonance in general in the study of fine and hyperfine interactions, local structure, and molecular bonding can now be enhanced with these first demonstrations of useful sensitivity in the single-spin regime.

11. OUTLOOK

The attainment of SMS in solids opens up a new frontier of single-absorber experiments in which the measured properties of the absorbing center are not averaged over many "equivalent" absorbers. The significance of such experiments is fourfold. First, the properties of a single absorber are measured without ensemble averaging, which means that tests of specific theoretical models are much stronger. Second, the sensitivity to specific properties of the nanoenvironment such as the local phonon modes and the true local fields is extremely high. This means for example that the identity of the mysterious two-level systems in amorphous materials may finally be determined. Third, it provides a window into the spectral hole-burning process on a molecule by molecule basis. Thus, the exact local coupling through which optical pumping of a single molecule gives rise to changes in the nanoenvironment which shift the resonance frequency may be studied. Fourth, this regime is essentially unexplored, which means that surprises and unexpected physical effects can occur (such as the observation of spectral diffusion in a crystal).

While as a general technique SMS is not applicable to all molecular impurities, it can be applied to the large number of absorbing molecules (and perhaps ions) in solids that have zero-phonon transitions, reasonable absorption strength, and efficient fluorescence. The detectability of the resulting single-center signal, which ultimately depends upon the specific sample and weak or absent spectral hole-burning, must be evaluated in each case. SMS signals should be observable at higher and higher temperatures if the concentration and background are both reduced sufficiently. One method for doing this may be to use near-field excitation to reduce the scattering volume and increase the single-center signal. First results combining near-field excitation with (low-temperature) SMS have recently been reported(60).

Other fascinating future experiments may be contemplated based on the examples presented here. Detailed study of the spectral diffusion process in crystals and polymers will help to eventually identify the actual microscopic nature of the two-level-systems. Nonlinear spectroscopy to measure the AC Stark effect for a single isolated molecule may also be performed. With the proper choice of lifetimes, one would expect quantum jumps and other processes observed for single ions in vacuum electromagnetic traps to become directly observable. The door is also open to true photochemical experiments on single absorbers and even the possibility of optical storage using single molecules. Future efforts to increase the number of probe-host couples which allow single-molecule spectra will lead to an even larger array of novel experiments.

One novel experiment that is now possible would be to use the emission from a single molecule as a light source of sub-nm dimensions for near-field optical microscopy(60). Of course, this would involve the technical difficulty of placement of the single molecule at the end of a pulled fiber tip. Another possibility would be to perform cavity quantum electrodynamics studies with a single molecule in a solid. In all cases, improvements in SNR would be expected to open up a new level of physical detail and possibly new applications. Because this field is relatively new, the possibilities are only limited at present by the imagination and the persistence of the experimenter and the continuing scientific interest in the properties of single quantum systems in solids.

ACKNOWLEDGMENTS

The author thanks Drs. L. Kador, W. P. Ambrose, Th. Basché, P. Tchénio, and Profs. A. B. Myers and U. P. Wild for fruitful collaborations.

REFERENCES

1. W. E. Moerner and L. Kador, *Phys. Rev. Lett.* **62**, 2535 (1989).
2. L. Kador, D. E. Horne, and W. E. Moerner, *J. Phys. Chem.* **94**, 1237 (1990).
3. M. Orrit and J. Bernard, *Phys. Rev. Lett.* **65**, 2716 (1990).
4. W. E. Moerner and Th. Basché, *Angew. Chem.* **105**, 537 (1993); *Angew. Chemie Int. Ed. Engl.* **32**, 457 (1993).
5. M. Orrit, J. Bernard, and R. Personov, *J. Phys. Chem.* **97**, 10256 (1993).
6. W. E. Moerner, *Science* **265**, 46 (1994).
7. W. M. Itano, J. C. Bergquist, and D. J. Wineland, *Science* **237**, 612, (1987) and references therein.
8. F. Diedrich, J. Krause, G. Rempe, M. O. Scully, and H. Walther, *IEEE J. Quant. Elect.* **24**, 1314, (1988) and references therein.
9. H. Dehmelt, W. Paul, and N. F. Ramsey, *Rev. Mod. Phys.* **62**, 525, (1990).
10. G. Binnig, H. Rohrer, *Rev. Mod. Phys.* **59**, 615, (1987).
11. E. Betzig and R. J. Chichester, *Science* **262**, 1422 (1993).
12. W. P. Ambrose, P. M. Goodwin, J. C. Martin, and R. A. Keller, *Phys. Rev. Lett.* **72**, 160 (1994).
13. J. K. Trautman, J. J. Macklin, L. E. Brus, and E. Betzig, *Nature* **369**, 40 (1994).
14. P. D. Reilly and J. L. Skinner, *Phys. Rev. Lett.* **71**, 4257 (1993).
15. W. P. Ambrose, Th. Basché, and W. E. Moerner, *J. Chem. Phys.* **95**, 7150 (1991).
16. W. E. Moerner, *J. Lumin.* **58**, 161 (1994).
17. A. M. Stoneham, *Rev. Mod. Phys.* **41**, 82 (1969).
18. K. K. Rebane, *Impurity Spectra of Solids* (Plenum, New York, 1970), p. 99.
19. D. A. Wiersma, *Adv. Chem. Phys.* **47**, 421 (1981).
20. See *Laser Spectroscopy of Solids*, Springer Topics in Applied Physics **49**, W. M. Yen and P. M. Selzer, Eds. (Springer, Berlin, 1981).
21. W. E. Moerner and L. Kador, *Anal. Chem.* **61**, 1217A (1989).
22. W. E. Moerner, *New J. Chem.* **15**, 199 (1991).
23. K. K. Rebane and I. Rebane, *J. Lumin.* **56**, 39 (1993).
24. H. de Vries and D. A. Wiersma, *J. Chem. Phys.* **72**, 1851 (1980).
25. See *Persistent Spectral Hole-Burning: Science and Applications*, Topics in Current Physics

44, W. E. Moerner, Ed. (Springer, Berlin. Heidelberg, 1988).
26. Th. Basché and W. E. Moerner, *Nature* 355, 335 (1992).
27. Th. Basché, W. P. Ambrose, and W. E. Moerner, *J. Opt. Soc. Am. B* 9, 829 (1992).
28. M. Orrit, J. Bernard, A. Zumbusch, and R. I. Personov, *Chem. Phys. Lett.* 196, 595 (1992).
29. T. Plakhotnik, W. E. Moerner, T. Irngartinger, and U. P. Wild, *Chimia* 48, 31 (1994).
30. W. E. Moerner, T. Plakhotnik, T. Irngartinger, M. Croci, V. Palm, and U. P. Wild, *J. Phys. Chem.* (in press, 1994).
31. W. E. Moerner and T. P. Carter, *Phys. Rev. Lett.* 59, 2705 (1987).
32. T. P. Carter, M. Manavi, and W. E. Moerner, *J. Chem. Phys.* 89, 1768 (1988).
33. W. E. Moerner and W. P. Ambrose, *Phys. Rev. Lett.* 66, 1376 (1991).
34. H. de Vries and D. A. Wiersma, *J. Chem. Phys.* 69, 897 (1978).
35. F. G. Patterson, H. W. H. Lee, W. L. Wilson, and M. D. Fayer, *Chem. Phys.* 84, 51 (1984).
36. U. P. Wild, F. Güttler, M. Pirotta, and A. Renn, *Chem. Phys. Lett.* 193, 451 (1992).
37. M. Croci, H.-J. Müschenborn, F. Güttler, A. Renn, and U. P. Wild, *Chem. Phys. Lett.* 212, 71 (1993).
38. M. Pirotta, F. Güttler, H. Gygax, A. Renn, J. Sepiol, and U. P. Wild, *Chem. Phys. Lett.* 208, 379 (1993).
39. W. P. Ambrose and W. E. Moerner, *Nature* 349, 225 (1991).
40. J. Köhler, J. A. J. M. Disselhorst, M. C. J. M. Donckers, E. J. J. Groenen, J. Schmidt, and W. E. Moerner, *Nature* 363, 242 (1993).
41. J. Wrachtrup, C. von Borczyskowski, J. Bernard, M. Orrit, and R. Brown, *Nature* 363, 244 (1993).
42. J. Wrachtrup, C. von Borczyskowski, J. Bernard, M. Orrit, and R. Brown, *Phys. Rev. Lett.* 71, 3565 (1993).
43. J. Bernard, L. Fleury, H. Talon, and M. Orrit, *J. Chem. Phys.* 98, 850 (1993).
44. A. Zumbusch, L. Fleury, R. Brown, J. Bernard, and M. Orrit, *Phys. Rev. Lett.* 70, 3584 (1993).
45. Th. Basché, W. E. Moerner, M. Orrit, and H. Talon, *Phys. Rev. Lett.* 69, 1516 (1992).
46. P. Tchénio, A. B. Myers, and W. E. Moerner, *J. Phys. Chem.* 97, 2491 (1993).
47. P. Tchénio, A. B. Myers, and W. E. Moerner, *Chem. Phys. Lett.* 213, 325 (1993).
48. A. B. Myers, P. Tchénio, M. Zgierski, and W. E. Moerner, *J. Phys. Chem.* (in press, 1994).
49. F. Güttler, T. Irngartinger, T. Plakhotnik, A. Renn, and U. P. Wild, *Chem. Phys. Lett.* 217, 393 (1994).
50. J. Friedrich and D. Haarer, in *Optical Spectroscopy of Glasses*, I. Zschokke, Ed. (Reidel, Dordrecht, 1986), p. 149.
51. F. Güttler, J. Sepiol, T. Plakhotnik, A. Mitterdorfer, A. Renn, and U. P. Wild, *J. Lumin.* 56, 29 (1993).
52. See *Amorphous Solids: Low-Temperature Properties*, Topics in Current Physics 24, W. A. Phillips, Ed. (Springer, Berlin, 1981).
53. G. Zumofen and J. Klafter, *Chem. Phys. Lett.* 219, 303 (1994).
54. P. Tchénio, A. B. Myers, and W. E. Moerner, *J. Lumin.* 56, 1 (1993).
55. L Fleury, A. Zumbusch, M. Orrit, R. Brown, and J. Bernard, *J. Lumin.* 56, 15 (1993).
56. J. M. Hayes, R. P. Stout, and G. J. Small, *J. Chem. Phys.* 74, 4266 (1981).
57. R. Jankowiak and G. J. Small, *Science* 237, 618-625 (1987).
58. E. Betzig, et al. *Appl. Phys. Lett.*, 61, 142 (1992).
59. H. de Vries and D. A. Wiersma, *J. Chem. Phys.* 70, 5807 (1979).
60. H. J. Kimble, M. Dagenais, and L. Mandel, *Phys. Rev. Lett.* 39, 691 (1977).
61. W. E. Moerner, T. Plakhotnik, T. Irngartinger, U. P. Wild, D. Pohl, and B. Hecht, submitted.
62. K. Lieberman, S. Harush, A. Lewis, and R. Kopelman, *Science* 247, 59 (1990).

Author Index

A

Adams, C., 258
Aspect, A., 193

B

Bardou, F., 193
Becker, St., 30
Beiersdorfer, P., 116
Bettermann, D., 240
Bigelow, N., 193
Blatt, R., 219
Bollen, G., 176
Borneis, S., 30
Bouchaud, J.-P., 193
Bradley, M., 149
Brune, M., 297
Bucksbaum, P. H., 416
Burnett, N. H., 405

C

Carnal, O., 314
Cataliotti, F. S., 81
Christ, M., 240
Chu, S., 258
Cohen-Tannoudji, C., 193
Corkum, P. B., 405

D

Davidovich, L., 297
Davidson, N., 258
De Natale, P., 81
Dietrich, P., 405
DiFilippo, F., 149

E

Ekert, A., 450
Engel, T., 30
Ertmer, W., 240

F

Franzke, B., 176
Fricke, B., 30

G

Georgiades, N., 314
Giusfredi, G., 81
Grieser, M., 30
Grieser, R., 30

H

Habs, D., 30
Hänsch, T. W., 63
Haroche, S., 297
Heinzen, D. J., 369
Hijmans, T. W., 389
Huber, G., 30

I

Inguscio, M., 81
Ivanov, M. Y., 405

J

Jungmann, K. P., 102

K

Kasevich, M., 258
Kimble, H. J., 314
Kirby, K. P., 437
Klaft, I., 30
Kluge, H.-J., 176
Kühl, T., 30

L

Lawall, J., 193
Leduc, M., 193
Lee, H.-J., 258
Lepage, G. P., 18
Luiten, O. J., 389

M

Mabuchi, H., 314
Mandel, L., 279
Marin, F., 81
Marx, D., 30
Merz, P., 30
Moerner, W. E., 467
Müller, J. H., 240

N

Natarajan, V., 149
Neumann, R., 30

P

Palmer, F., 149
Pavone, F. S., 81
Phillips, W. D., 211
Polzik, E. S., 314
Pritchard, D. E., 149

R

Raimond, J. M., 297
Ramsey, N. F., 3
Reynolds, M. W., 389
Rieger, V., 240
Ruschewitz, F., 240

S

Sapirstein, J., 45
Saubamea, B., 193
Schiffer, M., 240
Schmidt-Kaler, F., 297
Scholz, A., 240
Schwalm, D., 30
Seelig, P., 30
Sengstock, K., 240
Setija, I. D., 389
Sterr, U., 240
Stroud, Jr., C. R., 336

T

Tanner, C. E., 130
Thompson, R. J., 314
Turchette, Q. A., 314

V

Verhaar, B. J., 351

W

Walraven, J. T. M., 389
Weitz, M., 258
Werij, H. G. C., 389

Y

Young, B., 258

Z

Zagury, N., 297

AIP Conference Proceedings

		L.C. Number	ISBN
No. 262	Molecular Electronics—Science and Technology (St. Thomas, Virgin Islands, 1991)	92-72210	1-56396-041-9
No. 263	Stress-Induced Phenomena in Metallization: First International Workshop (Ithaca, NY, 1991)	92-72292	1-56396-082-6
No. 264	Particle Acceleration in Cosmic Plasmas (Newark, DE, 1991)	92-73316	0-88318-948-8
No. 265	Gamma-Ray Bursts (Huntsville, AL, 1991)	92-73456	1-56396-018-4
No. 266	Group Theory in Physics (Cocoyoc, Morelos, Mexico, 1991)	92-73457	1-56396-101-6
No. 267	Electromechanical Coupling of the Solar Atmosphere (Capri, Italy, 1991)	92-82717	1-56396-110-5
No. 268	Photovoltaic Advanced Research & Development Project (Denver, CO, 1992)	92-74159	1-56396-056-7
No. 269	CEBAF 1992 Summer Workshop (Newport News, VA, 1992)	92-75403	1-56396-067-2
No. 270	Time Reversal—The Arthur Rich Memorial Symposium (Ann Arbor, MI, 1991)	92-83852	1-56396-105-9
No. 271	Tenth Symposium Space Nuclear Power and Propulsion (Vols. I–III) (Albuquerque, NM, 1993)	92-75162	1-56396-137-7 (set)
No. 272	Proceedings of the XXVI International Conference on High Energy Physics (Vols. I and II) (Dallas, TX, 1992)	93-70412	1-56396-127-X (set)
No. 273	Superconductivity and Its Applications (Buffalo, NY, 1992)	93-70502	1-56396-189-X
No. 274	VIth International Conference on the Physics of Highly Charged Ions (Manhattan, KS, 1992)	93-70577	1-56396-102-4
No. 275	Atomic Physics 13 (Munich, Germany, 1992)	93-70826	1-56396-057-5
No. 276	Very High Energy Cosmic-Ray Interactions: VIIth International Symposium (Ann Arbor, MI, 1992)	93-71342	1-56396-038-9

No. 277	The World at Risk: Natural Hazards and Climate Change (Cambridge, MA, 1992)	93-71333	1-56396-066-4
No. 278	Back to the Galaxy (College Park, MD, 1992)	93-71543	1-56396-227-6
No. 279	Advanced Accelerator Concepts (Port Jefferson, NY, 1992)	93-71773	1-56396-191-1
No. 280	Compton Gamma-Ray Observatory (St. Louis, MO, 1992)	93-71830	1-56396-104-0
No. 281	Accelerator Instrumentation Fourth Annual Workshop (Berkeley, CA, 1992)	93-072110	1-56396-190-3
No. 282	Quantum 1/f Noise & Other Low Frequency Fluctuations in Electronic Devices (St. Louis, MO, 1992)	93-072366	1-56396-252-7
No. 283	Earth and Space Science Information Systems (Pasadena, CA, 1992)	93-072360	1-56396-094-X
No. 284	US-Japan Workshop on Ion Temperature Gradient-Driven Turbulent Transport (Austin, TX, 1993)	93-72460	1-56396-221-7
No. 285	Noise in Physical Systems and 1/f Fluctuations (St. Louis, MO, 1993)	93-72575	1-56396-270-5
No. 286	Ordering Disorder: Prospect and Retrospect in Condensed Matter Physics: Proceedings of the Indo-U.S. Workshop (Hyderabad, India, 1993)	93-072549	1-56396-255-1
No. 287	Production and Neutralization of Negative Ions and Beams: Sixth International Symposium (Upton, NY, 1992)	93-72821	1-56396-103-2
No. 288	Laser Ablation: Mechanismas and Applications-II: Second International Conference (Knoxville, TN, 1993)	93-73040	1-56396-226-8
No. 289	Radio Frequency Power in Plasmas: Tenth Topical Conference (Boston, MA, 1993)	93-72964	1-56396-264-0
No. 290	Laser Spectroscopy: XIth International Conference (Hot Springs, VA, 1993)	93-73050	1-56396-262-4
No. 291	Prairie View Summer Science Academy (Prairie View, TX, 1992)	93-73081	1-56396-133-4
No. 292	Stability of Particle Motion in Storage Rings (Upton, NY, 1992)	93-73534	1-56396-225-X

No. 293	Polarized Ion Sources and Polarized Gas Targets (Madison, WI, 1993)	93-74102	1-56396-220-9
No. 294	High-Energy Solar Phenomena A New Era of Spacecraft Measurements (Waterville Valley, NH, 1993)	93-74147	1-56396-291-8
No. 295	The Physics of Electronic and Atomic Collisions: XVIII International Conference (Aarhus, Denmark, 1993)	93-74103	1-56396-290-X
No. 296	The Chaos Paradigm: Developments an Applications in Engineering and Science (Mystic, CT, 1993)	93-74146	1-56396-254-3
No. 297	Computational Accelerator Physics (Los Alamos, NM, 1993)	93-74205	1-56396-222-5
No. 298	Ultrafast Reaction Dynamics and Solvent Effects (Royaumont, France, 1993)	93-074354	1-56396-280-2
No. 299	Dense Z-Pinches: Third International Conference (London, 1993)	93-074569	1-56396-297-7
No. 300	Discovery of Weak Neutral Currents: The Weak Interaction Before and After (Santa Monica, CA, 1993)	94-70515	1-56396-306-X
No. 301	Eleventh Symposium Space Nuclear Power and Propulsion (3 Vols.) (Albuquerque, NM, 1994)	92-75162	1-56396-305-1 (Set) 156396-301-9 (pbk. set)
No. 302	Lepton and Photon Interactions/ XVI International Symposium (Ithaca, NY, 1993)	94-70079	1-56396-106-7
No. 303	Slow Positron Beam Techniques for Solids and Surfaces Fifth International Workshop (Jackson Hole, WY 1992)	94-71036	1-56396-267-5
No. 304	The Second Compton Symposium (College Park, MD, 1993)	94-70742	1-56396-261-6
No. 305	Stress-Induced Phenomena in Metallization Second International Workshop (Austin, TX, 1993)	94-70650	1-56396-251-9
No. 306	12th NREL Photovoltaic Program Review (Denver, CO, 1993)	94-70748	1-56396-315-9
No. 307	Gamma-Ray Bursts Second Workshop (Huntsville, AL 1993)	94-71317	1-56396-336-1

No. 308	The Evolution of X-Ray Binaries (College Park, MD 1993)	94-76853	1-56396-329-9
No. 309	High-Pressure Science and Technology—1993 (Colorado Springs, CO 1993)	93-72821	1-56396-219-5 (Set)
No. 310	Analysis of Interplanetary Dust (Houston, TX 1993)	94-71292	1-56396-341-8
No. 311	Physics of High Energy Particles in Toroidal Systems (Irvine, CA 1993)	94-72098	1-56396-364-7
No. 312	Molecules and Grains in Space (Mont Sainte-Odile, France 1993)	94-72615	1-56396-355-8
No. 313	The Soft X-Ray Cosmos ROSAT Science Symposium (College Park, MD 1993)	94-72499	1-56396-327-2
No. 314	Advances in Plasma Physics Thomas H. Stix Symposium (Princeton, NJ 1992)	94-72721	1-56396-372-8
No. 315	Orbit Correction and Analysis in Circular Accelerators (Upton, NY 1993)	94-72257	1-56396-373-6
No. 317	Fifth Mexican School of Particles and Fields (Guanajuato, Mexico 1992)	94-72720	1-56396-378-7
No. 318	Laser Interaction and Related Plasma Phenomena 11th International Workshop (Monterey, CA 1993)	94-78097	1-56396-324-8
No. 319	Beam Instrumentation Workshop Santa Fe, NM 1993)	94-78279	1-56396-389-2
No. 320	Basic Space Science (Lagos, Nigeria 1993)	94-79350	1-56396-328-0
No. 321	The First NREL Conference on Thermophotovoltaic Generation of Electricity (Copper Mountain, CO 1994)	94-72792	1-56396-353-1
No. 322	Atomic Processes in Plasmas Ninth APS Topical Conference (San Antonio, TX)	94-72923	1-56396-411-2
No. 323	Atomic Physics 14 Fourteenth International Conference on Atomic Physics (Boulder, CO 1994)	94-73219	1-56396-348-5